TRANSPORT AND OPTICAL PROPERTIES OF NANOMATERIALS

To learn more about AIP Conference Proceedings, including the
Conference Proceedings Series, please visit the webpage
http://proceedings.aip.org/proceedings

TRANSPORT AND OPTICAL PROPERTIES OF NANOMATERIALS

Proceedings of the International Conference

ICTOPON - 2009

Allahabad, India   5 - 8 January 2009

EDITORS

M. R. Singh
Department of Physics and Astronomy
University of Western Ontario, London

R. H. Lipson
Department of Chemistry
University of Western Ontario, London

All papers have been peer-reviewed

SPONSORING ORGANIZATIONS
The University of Western Ontario
The Western Institute for Nanomaterials Science
Centre for Interdisciplinary Studies in Chemical Physics
Allahabad University
Department of Science & Technology, Government of India
Department of Bio-Technology, Government of India
Defense Research Development Organization, Government of India
Indian Space Research Organization
Council of Scientific and Industrial Research
Board of Research in Nuclear Sciences
Indian National Science Academy
Harish Chandra Research Institute, Allahabad, India

Melville, New York, 2009
AIP CONFERENCE PROCEEDINGS ■ VOLUME 1147

Sep/ae
phyp

Editors:

M. R. Singh
Department of Physics and Astronomy
University of Western Ontario
London, ON N6A 3K7
Canada

e-mail: msingh@uwo.ca

R. H. Lipson
Department of Chemistry
University of Western Ontario
London, ON N6A 5B7
Canada

e-mail: rlipson@uwo.ca

L.C. Catalog Card No. 2009905030
ISBN 978-0-7354-0684-1
ISSN 0094-243X
Printed in the United States of America

CONTENTS

PLENARY AND INVITED PAPERS

ELECTRICAL PROPERTIES OF NANOMATERIALS

OTHER TOPICS IN NANOMATERIALS

PREFACE

The first International Conference on the Transport and Optical Properties of Nanomaterials (ICTOPON2009) was held in Allahabad, India on January 5-8, 2009. Allahabad is an historic and holy city whose name is synonymous with the ancient city of "Prayag" described in Hindu mythology and scriptures. It is located at the intersection of three rivers—Ganga, Yamuna, and Saraswati called *Triveni*. One of the greatest fairs in the world—*Kumbh*—is held at *Triveni* every 12 years in January. Allhabad is also the site of a fort built by Emperor Akbar the Great.

The conference brought together more than 200 professional scientists, engineers and graduate students from countries around the world including Belgium, Canada, Egypt, France, Germany, India, Ireland, Israel, Japan, Korea, Netherlands, Norway, Singapore, Saudi Arabia, Spain, Sweden, Turkey, United Kingdom and the USA.

The goal of the conference was to discuss the most recent developments in the areas of growth and characterization of nano-structured materials, the synthesis of novel materials and their incorporation into devices with optical and electronic properties determined by nanoscale features, and the theoretical modeling of electronic, optical, magnetic and thermal properties of such systems. The meeting provided an excellent environment and opportunity to establish networking contacts and partnerships between the conference participants. The high quality of the plenary and invited talks, contributed papers, and poster sessions during the conference provided a strong basis for stimulating practical perspectives and ideas for research and development, as well as a chance to explore possible future collaborations and directions.

Similarly the science presented in this book reflects the current frontiers of innovation in the growth and characterization of nanomaterials, and of their optical transport properties. It provides a strong foundation for stimulating further advances.

Generous financial support for the Conference was provided by the VP Research of the University of Western Ontario, The Western Institute for Nanomaterials Science (WINS), the Faculty of Science at the University of Western Ontario, Western's Centre for Chemical Physics and the University of Allahabad. The support of the Department of Science & Technology, Government of India, the Department of Bio-Technology, Government of India, the Defense Research Development Organization, Government of India, the Indian Space Research Organization, the Council of Scientific and Industrial Research Board of Research in Nuclear Sciences, the Indian National Science Academy and the Harish Chandra Research Institute, Allahabad is also gratefully acknowledged.

The Editors thank the members of the program committee, the session chairs, many of the invited speakers and colleagues from around the world for their assistance in reviewing the manuscripts submitted for publication in this book. They also thank the authors for their diligence in producing the camera-ready copies of their papers and in responding to reviewer comments and suggestions.

Mahi R. Singh
Editor and Conference Chair

R. H. Lipson
Editor and Conference Co-Chair

Committees and Sponsors

M. R. Singh, University of Western Ontario, London, ON, Canada. *(Chair)*

R. H. Lipson, University of Western Ontario, London, ON, Canada. *(Co-Chair)*

H. Prakash, University of Allahabad, India. *(Co-Chair)*

R. Prakash, University of Allahabad, India. *(Conference Secretary)*

L. V. Goncharova, University of Western Ontario, London, ON, Canada. *(Conference Secretary)*

A. Gupta, University of Allahabad, India. *(Conference Treasurer)*

M. Zinke-Allmang, University of Western Ontario, London, ON, Canada. *(Conference Treasurer)*

Patron:

Rajen Harshe, University of Allahabad, India.

International Advisory Committee:

E. Arushanov, (Moldova)
R. N. Bhargava, (U.S.A.)
P. Chaudhary, P. (Singapore)
P. Chu, (U.S.A./Hongkong)
G. De Aquino Farias, (Brazil)
O. De Melo, (Cuba)
E. F. da Silva Jr, (Brazil)
J. C. Inkson, (U.K.)
J. Sajeev (Canada)
S. K. Joshi, (India)
W. Lau, (U.S.A.)
J. Leotin, (France)
R. Nicholas, (U.K.)
P. P. Padilla, (Mexico)
Ploog, Klaus H. (Germany)
M. von Ortenberg, (Germany)
O. N. Srivastava, (India)
K. Takagi, (Japan)
B. Tanatar, (Turkey)
E. Yablonovitch, (U.S.A.)

National Advisory Committee:

B. K. Agrawal, (Allahabad)
S. K. Banerjee, (BARC, Mumbai)
S. K. Brahmchari, (CSIR, New Delhi)
B. N. Jagtap, (Mumbai)
P. Kumar, (Allahabad)
V. Kumar, (NPL)
A. K. Pati, (IOP, Bhubneshwar)
H. Prakash, (Allahabad)
J. Ram, (BHU)
T. V. Ramakrishnan, (BHU)
T. Ramasami, (DST, New Delhi)
A. K. Sood, (IISc., Bangalore)
R. P. Singh, (PTU Jalandhar)
M. D. Tiwari, (IIIT, Allahabad)
M. Vijayan, (IISc., Bangalore)

Local Support Committee:

P. Kumar (Head Physics)
S.G. Prakash (Dean, Science)
R. Gopal (Allahabad)
R. Nakagawa (Canada)
H. Leparskas (Canada).

Sponsors

The University of Western Ontario

Western Institute for Nanomaterials Science

Centre for Interdisciplinary Studies in Chemical Physics

Allahabad University

Department of Science & Technology, Government of India

Department of Bio-Technology, Government of India

Defence Research Development Organization, Government of India

Indian Space Research Organization

Council of Scientific and Industrial Research

Board of Research in Nuclear Sciences

Indian National Science Academy

Harish Chandra Research Institute, Allahabad.

PLENARY AND INVITED PAPERS

Magnetospectroscopy of AlP Quantum Wells

M. Goiran*, J.M. Poumirol,* M.P. Semtsiv¶, W.T. Masselink¶
D. Smirnov°, V. V. Rylkov°°, and J. Léotin*

*Laboratoire National des Champs Magnétiques Intenses, 143 avenue de Rangueil, 31400 Toulouse, France
¶Department of Physics, Humboldt University Berlin, Newtonstrasse 15, D-12489 Berlin, Germany
°National High Magnetic Field Laboratory, Tallahassee, FL 32310, USA
°° Russian Research Center "Kurchatov Institute", Kurchatov sq. 1, Moscow 123182, Russia

Abstract. Cyclotron resonance, quantum Hall effect, and Shubnikov de Haas oscillations measurements of quasi-two-dimensional (2-D) electrons in modulation-doped AlP quantum wells are investigated. This study enables to clarify the electronic structure of quasi-lattice matched AlP/GaP quantum wells that are epitaxially grown on (100)-oriented GaP substrate. The 2-D electrons occupy conduction band minima at X-points of AlP Brillouin zone. When reducing AlP well width in the region between 5 nm and 4 nm, groundstate valley degeneracy swops between $g_v=2$ and $g_v=1$. This indicates a strain and confinement controlled crossover of longitudinal (X_z) and transverse ($X_{x,y}$) valleys centred at the X-point. Cyclotron resonance measurements of AlP wells narrower than 4 nm and wider than 5 nm provide transverse and longitudinal effective masses: $m_t = (0.30 \pm 0.01)m_0$ and $m_l = (0.90 \pm 0.01)m_0$. Knowing the value of AlP-well thickness at valleys crossover, we have calculated the deformation potential for bulk AlP $X_{AlP} = 3.3 eV\pm 0.3 eV$. In addition, g-factor enhancement was measured in a single valley 2D-elctron gas. Finally we address intersubband optical properties of AlP/GaP quantum wells for THz applications.

Keywords: Cyclotron resonance, Quantum Hall effect, quantum wells
PACS: 76.40. +b; 73.43. −f; 73.21.Fg; 78.67.De

INTRODUCTION

Engineering transport and optical properties of semiconductor nanostructures requires a full understanding of their electronic states and dynamical properties. In the following study of AlP quantum wells we demonstrate the importance of high magnetic field measurements of cyclotron resonance (CR) and quantum transport properties to meet this objective [1, 2]. We then address intersubband optical properties of AlP/GaP quantum wells (QWs) for THz applications [3].

CR absorption takes place in a magnetic field B when electrons orbiting in a plane perpendicular to the field have their cyclotron frequency eB/m* equal to the excitation frequency of an incident radiation. In early days, electronic states of bulk silicon and germanium were fully understood from CR experiments after measuring cyclotron effective masses m* in different crystallographic planes. At this time, microwave excitation at gigahertz frequencies and magnetic fields of a few Tesla were used. Silicon electron Fermi surface was established as a set of six ellipsoids elongated along X-axis, centred at the bottom of conduction band valleys that display longitudinal $m_l = 0.98m_0$ and transverse effective masses $m_t = 0.19m_0$ [4]. Similarly, electrons in AlP populate X-valleys, but despite the longstanding importance of AlP

CP1147, Transport and Optical Properties of Nanomaterials—ICTOPON - 2009, edited by M. R. Singh and R. H. Lipson
© 2009 American Institute of Physics 978-0-7354-0684-1/09/$25.00

for optoelectronics applications [5], until recently little of their states were understood since neither valley degeneracy nor effective masses were known [6]. Actually, low mobility electrons in bulk AlP result in short collision times that prevents completion of cyclotron orbits between collisions unless magnetic fields above 100T are used to shorten the cyclotron period.

In the present work, we took another route to tackle CR measurements and quantum transport in AlP. In order to enhance electron mobility, we have grown modulation doped AlP QWs between GaP barriers. We also used a 60 Tesla magnet and THz excitation light. In the past, AlP/GaP Qws have been extensively studied but samples were unintentionally doped for implementing room temperature green light luminescence nanostructures

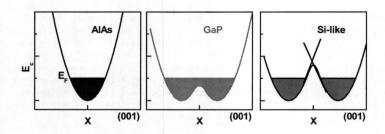

FIGURE 1. Band structure options for AlP multivalley conduction band along 001 direction of the Brillouin zone. Band degeneracy at X-point found in silicon is strongly lifted in III-V materials. Consequently, the absolute conduction band minima falls at X-point in AlAs and very near X-point in GaP, giving a camel's back structure. Valley degeneracy is equal to six in silicon and three in AlAs and GaP.

based on short period superlattices [7]. In this study, although empirical approach was taken in absence of effective masses knowledge, green light emission was finally obtained at low temperatures but room temperature emission was never achieved. To date, better knowledge of AlP conduction band and effective masses enables to engineer optical properties including intersubband optical and transport properties in AlP/GaP nanostructures aimed at THz applications.

In what follows, we first summarize the present knowledge of AlP band structure, then we describe the set of QW samples that have been designed an implemented. After, Quantum Hall Effect (QHE) measurements of 2D-elctron gas are reported in details. The data provide valley degeneracy and furthermore give evidence of valley-symmetry crossover at a critical well thickness. Effective mass measurements are then obtained from CR and Shubnikov de Haas (SdH) experiments. In addition, g-factor enhancement is reported from SdH oscillations measurements under magnetic field tilted at increasing angle from z-axis direction. Intervalley deformation potential is also derived. Finally, single valley intersubband energies are calculated as a function of well thickness and a summary is given.

Band Structure, Valley Symmetry and Degeneracy of AlP Quantum Wells

AlP conduction band minima were first assigned at X-point of the Brillouin zone [8] but exact location of band minima remained uncertain for a long time. Whereas in silicon conduction band minima fall inside the Brillouin zone together with band degeneracy that occur at the X-point, in III-V compounds band degeneracy at the X-point is lifted by the anti-symmetric potential from atoms pair (Fig.1). As a result, absolute conduction band minima may not fall inside the Brillouin zone but either at X-point like in AlAs or nearby like in GaP where a peculiar camel-back structure takes place [9]. As a consequence, valley degeneracy is three for bulk AlAs and GaP while being six for bulk silicon. Actually, both situations may occur for AlP conduction band, therefore valley degeneracy determination is a crucial issue. In the present study, valley degeneracy is directly obtained by QHE measurements of 2D-electron gas confined to AlP QWs having different well thickness.

Let us recall that for 2D-electrons populating X-valleys and confined in the xy plane, the uniaxial confinement potential along z axis lifts the valley degeneracy of bulk crystal and splits the valleys into X_z and X_{xy} valleys. One expects X_z-valley to be the groundstate valley since the effective mass along the confinement direction is larger for Xz valleys. Indeed, QHE shows that 2D-electron gas in silicon MOSFETs populates two degenerate X_z-valleys. Electrons move in xy plane with the isotropic effective mass m_t directly measured by cyclotron resonance absorption.

On the other hand, for 2D-electron gas confined to AlAs QWs, groundstate valley degeneracy depends on well thickness. It was found from QHE experiments that valley degeneracy is one for well thickness below 4nm but two for larger well thickness. Doubling of valley degeneracy at 4nm is caused by strain induced lattice mismatch between AlAs layer and GaAs substrate. Because AlAs lattice parameter is larger by 0.3% an extensive uniaxial strain develops and lifts up X_z valleys with respect to X_x, and X_y valleys by a constant energy, while confinement energy plays in the opposite way and splits the valleys by an energy that increases as well thickness decreases.

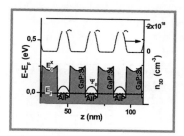 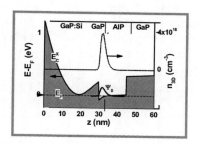

FIGURE 2. Single AlP QW populated by electrons from impurity layer in the top GaP (left picture). Multiple AlP QWs populated by electrons from impurities set in both upper and lower barriers.

Below 4 nm well thickness, strain prevails over confinement and X_z valley becomes ground state. Then the single X_z valley identified by QHE measurements is necessarily centred at the X-point of AlAs Brillouin zone.

A situation similar to AlAs QWs is expected in the case of AlP/GaP Qws. AlP lattice parameter is also larger by 0.3% than in GaP, therefore valley crossover from X_z to $X_{x,y}$ as well thickness is increased to reach a critical value to be determined. Along this line, a wide range of AlP/GaP QWs have been implemented for QHE and CR measurements. The structures are described in the following section.

AlP/GaP Quantum Wells Samples

The samples used in this study were grown using gas source molecular-beam epitaxy (GSMBE) on n-type unintentionally doped GaP(001) substrates ($n = 10^{16}$ cm^3 at 300 K). These substrates are semi-insulating at temperatures below 100 K. Two sets of sample were grown. One set similar to transistor structures has a single AlP well flanked by a top GaP barrier modulation doped with silicon, and undoped bottom barrier. The other set includes multi-quantum wells (MQWs) structures having AlP wells between silicon modulation doped GaP barriers. Parameters of all studied samples are listed in table 1. All samples were characterized at low magnetic fields (0.3T) in the van der Pauw (VdP) geometry. Their parameters are listed in Table 1. The sheet carrier densities are relatively constant within the temperature range between 10 and 77 K, with values in the range 0.5 to 5 10^{12} cm^{-2}/well . Mobilities fall in the range of 10^3 to 2.10^4 cm^2/Vs. For magneto-transport measurements, samples were processed into 2 mm x 6mm Hall-bars.

Quantum Transport and Valley Degeneracy

Magnetotransport measurements up to 20T in superconducting magnet and at helium3 temperatures (280- 800 mK) took place at Tallahassee, NHMFL. Two single QWs samples, A60 and A29, respectively narrow (3nm) and wide (15nm) were measured using a sample holder that could be tilted with respect to the magnetic field direction. Other samples listed in table 1 were measured at Toulouse LNCMP using pulsed magnets reaching up to 60T with a pulse lasting 300 ms. Single QWs were used for magnetotransport and multiple QWs including 50 wells for CR measurements at temperatures above 2K.

TABLE 1. Sample parameters including mobilities and sheet carrier densities per well measured in van der Pauw geometry at 0.3 T magnetic field and a temperature of 10K

Reference	A60	A65	A59	A84	A73	A57	A76	A58	A27	A28	A29	A31
Wells Number	50	1	50	1	1	1	1	50	1	1	1	50
Well width (nm) Barrier width	3 10	3 30	4 10	4 30	5 30	5 50	6.5 30	8 10	15 10	15 30	15 30	10 20
concentration 10^{12} cm^{-2}	1.8	2.0	1.5	1.1	1.4	2.3	1.4	0.7	2.3	2.6	4.5	1.5
Mobility cm-/.V.s	5200	5000	8700	18800	1300	5800	3800	11900	3000	1300	5900	6400

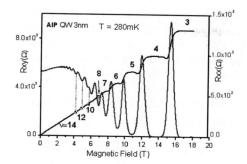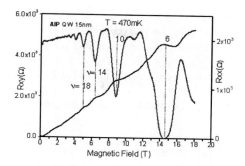

FIGURE 3. Quantum Hall effect and Shubnikov de Haas oscillations for narrow AlP QW (A60, 3nm) on the left picture and a wide QW (A29, 15nm) on the right picture. At low magnetic fields, filling factors written above each plateau jump by two for the narrow QW sample and by four for the wide QW.

a pulse lasting 300 ms. Single QWs were used for magnetotransport and multiple QWs including 50 wells for CR measurements at temperatures above 2K.

For a multivalley 2D-electron gas in a magnetic field, three relevant energies are involved regarding energy levels quantization. First, the cyclotron energy $E_C = heB_\perp/m^*$ separating Landau levels, where m^* is the cyclotron effective mass and $B_\perp$ is the component of the magnetic field perpendicular to 2D- plane.

Second, valley splitting energy caused by the perpendicular component of the magnetic field $B_\perp$[10]. Third, the Zeeman energy $E_Z = g^*\mu_B B$, where B is the total magnetic field that couples to electron spin. In this situation, degeneracy of each Landau Level orbital state is $2.g_v$, therefore the filling factor of a single fully occupied Landau is equal to $2.g_v$. In consequence, since filling factors are directly measured by QHE, this experiment provides the most direct determination of valley degeneracy. Given quantum Hall plateaus resistance R_{xy}, filling factors read $v= (h/e^2).1/ R_{xy}$. Actually, valley degeneracy was evidenced from early QHE measurements in Si-MOSFET 2D-eletron gas [10] and later in AlAs 2D-electron gas [11] and silicon QWs in Si/SiGe structures [12]. Generally, valley and spin degeneracy are not resolved at low magnetic fields therefore filling factor jumps by $2.g_v$ at SdH peak oscillations and R_{xy}. At the SdH peak, the Fermi level stands above a Landau level and below an entirely empty upper level.

Fig. 3 shows transverse and longitudinal resistance (R_{xy}, R_{xx}) at low temperatures for a narrow and a wide QW. samples. The left picture stands for the 3 nm wide AlP QW (A60) and the right picture for the 15nm wide QW sample (A29). Filling factors derived from quantum Hall plateaux resistance R_{xy} are written above each plateau. Remarkably, in the low magnetic field range, filling factors jump by two for narrow QWs, and four for wide QWs. On the other hand, at high magnetic fields filling factors jump by unity. Unity jump start above 8T for the 3nm wide AlP QW and above 20 T for the 15 nm QW as evidenced in pulsed magnetic field measurements [2].

The obvious conclusion to be derived from filling factors jumps is that the narrow QWs has a single ground state valley whereas the wide QW has two degenerate ground state valleys. Obviously, because of uniaxial symmetry along z the single valley in the narrow QW is necessarily elongated along X_z direction   and the twofold valleys are elongated along X_x and X_y directions. Clearly, valley crossover takes place for a critical QW thickness and this thickness was found between 4 and 5 nm after measuring the set of AlP QWs samples listed in table 1.

Effective Mass Measurements from Shubnikov dE Haas Oscillations and Cyclotron Resonance

Let us consider first AlP QWs narrower than 4nm under magnetic field applied along z-axis.   In this system, 2D-electron gas populate X_z-valley and develop circular cyclotron orbits in the xy plane. The cyclotron mass for this motion is the transverse mass m_t. On the other hand, AlP QWs wider than 5nm develop elliptical cyclotron orbits in z,x and z,y planes with cyclotron effective mass $(m_t.m_l)^{1/2}$. It follows that one needs to measure a narrow and a

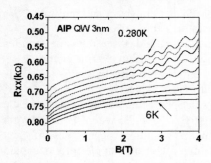 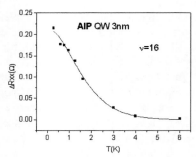

FIGURE 4. Temperature dependence of SdH oscillations amplitude between 0.28K and 6K for a narrow QW, 3nm wide having a single X_z –valley (sample A60). Longitudinal resistance R_{xx} is plotted at the same scale for all curves, but the curves are shifted by a constant value for clarity. The right curve shows a Dingle plot of the temperature dependence of oscillation amplitude at filling factor $\nu = 16$ and B=2.6T. This fit I achieved for an effective mass value $m_t = 0.294 m_0$.

wide QW sample in order to get transverse and longitudinal masses. This was done by measuring cyclotron resonance absorption and temperature dependence of SdH oscillation amplitude at constant magnetic field.

Fig. 4 shows temperature dependence of SdH oscillations amplitude for the X_z–valley, 3nm wide QW (A60). Temperature is set between 0.28K and 6K and magnetic fields are limited below 4T, in a range where spin and valley degeneracy are not resolved. Oscillations peaks amplitude decreases as the temperature is increased up to 6K. Temperature dependence of the amplitude of a given oscillation at field B follows the Dingle expression [13]:

$$\Delta R_{xx}(T,B) \propto \frac{T}{\sinh(\frac{2\pi^2 k_B T m^*}{eB\hbar})}$$

The best fit of the data is an average of effective masses obtained from Dingle plots taken at filling factors ranging from 16 to 22. One then gets the value $m_t = (0.294 \pm 0.005) m_0$ for the transverse effective mass. Figure 4 shows an example of Dingle plot taken at filling factor $\nu = 16$. In the same way, a 15nn wide QW (A29) having elliptical cyclotron orbit was similarly measured and analysed to obtain a cyclotron effective mass $m^* = (m_t m_l)^{1/2} = (0.503 \pm 0.005) m_0$. Given the above transverse mass, a longitudinal mass m_l is deduced $m_l = (0.90 \pm 0.005) m_0$

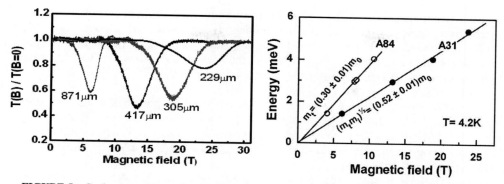

FIGURE 5a. Cyclotron resonance lines derived from transmission intensity across the 2D-elecron gas populating Xx and Xy valleys of a wide AlP MQW sample, 10 nm wide (sample A31). Temperature is 4K. The plotted signal has been corrected by subtraction of a monotonous background signal.
FIGURE 5b. Magnetic field dependence of cyclotron energy for wide QW sample A31 and narrow MQWs sample A84.

On the other hand, cyclotron resonance was measured at 4K for multiquantum wells having well thickness of 10 nm and 4 nm. We used excitation radiation in the wavelength range 200-900 μm provided by molecular gas lasers optically pumped by a CO_2 laser. The transmitted radiation through the sample placed at the centre of the pulsed magnet is measured by InSb hot electron bolometer. Fig 5. shows cyclotron resonance lines measured at 4K during pulsed magnet shots by recording transmission intensity of excitation radiation at wavelength 871, 417, 305, 229μm. The signal is plotted for a MQW sample A31comprising QWs, 10nm wide. This signal has been corrected by subtraction of a monotonous background signal [1]. Similarly, a narrow QW sample A84, 5 nm wide, was measured. Figure 5b on the right shows a linear plot of cyclotron energy versus magnetic field. The narrow QW sample gives the transverse cyclotron effective mass $m_t = (0.30 \pm 0.01)m_0$ and the wide MQWs sample the ellipsoidal orbit mass $m^* = (m_t\, m_l)^{1/2} = (0.52 \pm 0.01)m_0$.

A fairly good agreement is obtained between SdH and CR measurements of effective masses. The linear plot of cyclotron energy versus magnetic field in figure 5b indicates that the band behaves as a parabolic band. However since the energy range probed by SdH and CR measurements is only near Fermi energy of a few mev, band parabolicity is only demonstrated in this limited energy range.

Deformation Potential Measurement

Because AlP lattice parameter is larger by 0.3% than the GaP one, a compressive biaxial strain in the xy plane gives rise to a uniaxial crystal potential that shift the X_z valley upwards in energy. The strain-induced splitting of X_z and $X_{x,y}$ valleys is $\Delta X = \Xi^{AlP}(e_l - e_t)$ [14], where Ξ^{AlP} is the deformation potential, e_l and e_t are strain values perpendicular and parallel to the epitaxial layers. Using the lattice constants of AlP and GaP, respectively 0.54672 and 0.54505 nm [1], we calculate $e_l = 2.9.10^{-3}$ and $e_t = -3.1.10^{-3}$. At valley crossover point, strain induced splitting and confinement cancel each other. Thus, by calculating the confinement splitting for the crossover thickness, strain induced splitting is obtained and consequently the deformation potential energy. We found that level splitting ΔX due to quantum confinement ranges between 17 meV and 23 meV when AlP well width is varied from 5 nm to 4 nm. We finally estimated deformation potential $\Xi^{AlP} = \Delta X /(e_l - e_t) = 3.3\text{eV} \pm 0.3\text{eV}$.

We should emphasize that given the deformation potential value and the ability to tune the valley splitting energy by applying in-plane strain along crystallographic axis, one could achieve a controlled tuning of valley population in the same way as successfully demonstrated in the case of AlAs QWs by Shayegan et al. [15] Recently, the rich physics involving both spin and valley degree of freedom in AlAs 2D-electron was explored in magnetotransport experiments in which the occupancy of X_x an X_y valleys was continuously adjusted in wide QWs. [16].

g-factor Enhancement

Figure 3 clearly evidences spin splitting of the Landau levels at magnetic fields above 6T in a single valley narrow QW sample (A60). The appropriate method to measure the g-factor, that means Zeeman energy splitting $E_Z = g^*\mu_B B$ at given magnetic field, is the well known coincidence method. This method is illustrated in reference [16] and depicted in fig 6a, 6b and 6c. As the magnetic field gets tilted by an angle θ against the normal to the sample, spin up and spin down LLs come closer to each other and finally coincide at a critical angle θ_c. At this

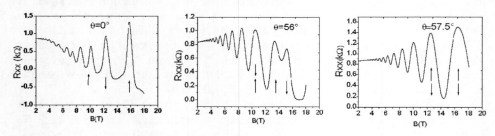

FIGURE 6. Shubnikov de Haas oscillations for a 2D-electron gas in a single valley AlP QW, 3 nm wide, at 280 mK. Magnetic field direction is along the growth axis(left) and then tilted by 50°26 and 57°5 (right). Landau level spin state at Fermi level is indicated by arrows. Spin up and down levels are entirely brought into coincidence at an angle of 57°5.

angle, cyclotron energy $\hbar eB_\perp/m^*$ which has been reduced and multiplied by cosθ becomes equal to the Zeeman energy $g^*\mu_B B$. In this situation, spin up and spin down levels from neighbouring LLs come into coincidence, therefore critical angle detection is directly obtained by peaks coincidence. One can then derive the g-factor from the simple relationship $g = 2m_0\cos \theta_c /m^*$.

Figure 6a, 6b and 6c show Shubnikov de Haas oscillations on the longitudinal resistance R_{xx} measured at temperature 280mK with a sample having sheet carrier concentration $n_s = 2.10^{12}\text{cm}^{-2}$. Magnetic field is respectively set along the z-direction, tilted by 50°26 and by 57°5. Above 6T, oscillations in figure 6a get split into two components related to LLs filled respectively by spin down and spin up electrons that cross the Fermi level. Arrows on the figures pointing up and down indicate LLs spin states. Fig 6b and 6c demonstrate the approach and the coincidence of initially double peaks. Given the effective mass $m^* = 0.30m_0$, and the critical angle $\theta_c = 57°5$ the deduced g-value is $g = 2m_0\cos\theta_c/m^* = 3.57 +/- 0.05$.

A theoretical estimate of the g-factor is given by the relationship [18]: $g = 2 - 2E_p$. $\Delta /3E_g(E_g + \Delta) = 1.87$ where E_g is the bandgap energy, Δ □is the spin-orbit splitting energy, and E_p is the energy equivalent of the principal interband momentum matrix element. One finds the measured value $g = 3.57$ significantly enhanced with respect to predicted value $g = 1.87$. Similarl enhancement was already found in other multivalley systems like Si and AlAs 2D-electron gas [18]. g-factor enhancement caused by electron- electron interaction is still a matter of continuing studies. Since the interaction strength parameter r_s given by the ratio of Coulomb energy to Fermi energy increases with effective mass and average separation between electrons, AlP 2D electron system is another good candidate like AlAs to investigate spin and valley susceptibility in a 2D-electron gas multivalley system [19].

.

Intersubband THz applications

Although optoelectronic devices like lasers, detectors and modulators have been demonstrated to cover a wide spectral range 0.4 - 300 μm, they suffer from a spectral gap in the range between 25 and 50 μm. This is due a to phonon absorption band in widely used III-V materials GaAs, InP, AlAs, InAs, GaSb as shown in fig.7a . Because AlP and GaP have a higher energy phonon band (in 20-25 μm spectral range), AlP/GaP heterostructures are good candidates for implementing THz devices based on intersubband transitions. Recently, first studies of intersubband optical transitions in AlP/GaP quantum wells were made using QW structure that have been doped by silicon inside the AlP well [3]. Samples were multiquantum wells having a narrow QW width (2.6nm) and therefore an X_z groundstate valley. They were polished into a 45° multiple-reflection prism and optical absorption was measured using polarised radiation. Strong p-polarized absorption peaks were identified as transition from the first

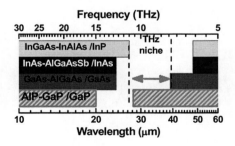
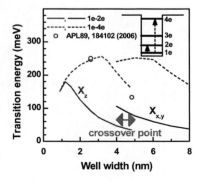

FIGURE 7a. LO-TO phonon absorption bands for heterostructures based on GaAs, AlAs, AlP, InAs and GaP. A THz niche above 30 μm wavelength is left for AlP/GaP quantum wells.
FIGURE 7b. Intersubband transition energies for X_z and $X_{x,y}$ valleys in AlP/GaP quantum well as a function of well thickness. Valley crossover takes place at thickness between 4 and 5nm. Full line corresponds to fundamental transition and dashed line for transition to the 4th subband.

to the fourth subband of the X_z valley. To calculate intersubband energies, it was assumed that the longitudinal effective mass was of the order of free electrons mass. Best fitting of the data gave longitudinal effective mass m_l=1.1m_0 and intersubband offset between AlP and GaP equal to 280meV. Indeed, because of the valley crossover, intersubband energies became totally wrong for the case of wide QWs.

In the present study, we give a full picture of intersubband transitions as a function of well thickness based on longitudinal and transverse effective masses and well thickness at valley crossver. Fig. 7b shows fundamental transition energies for X_z and X_{xy} valleys on both sides of crossover point. Transitions from the first to the fourth subbands are also given in dashed lines. Intersubband calculations reported on fig.7b indicate that a well thickness larger than 5nm is best appropriate to cover the wavelength range above 25μm. One should noice the possibility to implement a two-colour system at 14 and 30 μm wavelegnth by making a device with crossover thickness at crossover thickness. Another intersting feature of AlP/GaP QWs is the ability to tune the valleys order by gluing the sample on a piezo transucer actuator. This was demonstrated in the case of AlAs QWs by appplying in-plane uniaxial stress along x- or y-directions [15].

CONCLUSIONS

This paper reports on a study of the multivalley conduction band structure of AlP quantum wells based on magnetospectroscopy measurements of AlP/QWs in high magnetic fields. Quantum Hall Effect measurements of 2D-electron gas allowed to elucidate symmetry and degeneracy of the QW groundstate valley and to evidence valley-symmetry crossover as quantum well thickness is increased above 5nm. Cyclotron Resonance and Shubnikov de Haas experiments gave transverse and longitudinal effective masses, respectively equal to m_t=(0.30 ± 0.01)m_0 and m_l = (0.90 ±0.01)m_0. In addition, intervalley deformation potential is also derived and the value obtained is Ξ^{AlP}=3.3eV±0.3eV. Measurements of g-factor based on Subnikov de Haas oscillations taken under tilted magnetic fields direction show enhancement by nearly a factor of two, emphasizing a similar behaviour as AlAs. Finally single valley intersubband energies are calculated as a function of well thickness in view of modelling THz intersubband devices.

ACKNOWLEDGMENTS

The authors acknowledge the European program EuroMagNet-FP6 (Contract RII3-CT2004-506239).

REFERENCES

1. M. P. Semtsiv, S. Dressier, W. T. Masselink, V. V. Rylkov, J. Galibert, M. Goiran and J. Léotin, *Phys. Rev. B* **74**, 0413031 (2006).
2. M. P. Semtsiv, O. Bierwagen, W. T. Masselink, M. Goiran, J. Galibert and J. Léotin, *Phys. Rev. B* **77**, 165327 (2008).
3. M. P. Semtsiv, U. Müller, and W. T. Masselink, N. Georgiev, T. Dekorsy and M. Helm, *Appl. Phys. Lett.* **89**, 184102 (2006).
4. G. Dresselhaus, A. F. Kip and C. Kittel, *Phys. Rev.* **98**, 368 (1955).
5. Hilsum, *Physics of III.V Compounds*, vol.1, (1966).
6. I. Vurgaftman, J. R. Meyer, and L. R. Ram-Mohan, *J. Appl. Phys.* **89**, 5815 (2001).
7. F. Issiki, S. Fukatsu and Y. Shiraki, *Appl. Phys. Lett.* **67**, 1048 (1995).
8. F. Bassani and M. Yoshimine, *Phys. Rev.* **130**, 20 (1963).
9. P. J. Dean and D. C. Herbert, *J. Lumin.* **14**, 55 (1976).
10. T. Ando, A.B. Fowler and F. Stern, *Rev. Mod. Phys.* **54**, 437 (1982).
11. F. W. van de Stadt, P. M. Koenraad, J. A. A. J. Perenboom and J. H. Wolter, *Surf. Sci.* **361/362**, 521(1996).
12. D.-H. Shin, C.E. Becker, J.J. Harris, J.M. Fernandez, N.J. Woods, T.J. Thornton, D.K. Maude and J.-C. Portal, *Semicond. Sci. Technol.* **13**, 1106 (1998).
13. R. Dingle, *Proc. R. Soc. A* **211**, 517 (1952).
14. C. G. Van de Walle, *Phys. Rev. B* **39**, 1871 (1989).
15. M. Shayegan, E.P. De Poortere, O. Gunawan, Y.P. Shkolnikov, E. Tutuc and K. Vakili, *Phys. Stat. Sol. (b)* **243**, 3629-3642 (2006) (review article)
16. T. Goknmen, Medini Padmanabhan and M. Shayegan, *Phys. Rev. Lett.* **101**, 146405 (2008).
17. R. J. Nicholas, R. J. Haug, K. v. Klitzing and G. Weimann, *Phys. Rev.* **37**, 1294 (1988).
18. L. M. Roth, B. Lax and S. Zwerdling, *Phys. Rev.* **114**, 90 (1959).
19. T. Gokmen, M. Padmanabhan, E. Tutuc, M. Shayegan, S. De Palo, S. Moroni and G. Senatore, *Phys. Rev. B* **76**, 233301 (2007).

The Affinity Between High Magnetic Fields And Nano-Structured Systems

Michael von Ortenberg

Institute of Physics, Humboldt University at Berlin, Newtonstrasse 15, 12489 Berlin, Germany

Abstract. High magnetic fields B define for particles bearing one elementary charge e an externally tunable length scale by the magnetic length $\lambda = (\hbar /eB)^{\frac{1}{2}}$. For B = 100 T, corresponding to 1 Megagauss, $\lambda(100 \text{ T}) = 2.56$ nm. So high magnetic fields provide an ideal tool to probe nano-structured systems. We compare several semiconductor systems in 3D, 2D, 1 D, and 0D as a function of the external magnetic fields between 0 and 100 T. Phenomena such as magnetic freeze-out, magneto condensation, and magnetically induced 2D $\rightarrow$ 3D transition will be reviewed in several examples, such as for GaAs by magneto luminescence investigations and HgSe:Fe by Shubnikov-de Haas effect and Megagauss IR-magneto transmission.

Keywords: Megagauss magnetic fields, nano semiconductor systems
PACS: 75.75.+a

INTRODUCTION

Modern digital electronics and information technology depends crucially on nano-structured semiconductor systems. Only the successive reduction in size of semiconductor structures made it possible to obtain the packing densities in artificial memories we have today. Actually the storage volume contains now only some lattice cells. The electrons are confined inside this storage volume so that their wave function is essentially determined by this boundary problem. To learn more about this confinement we have to look for an alternative physical method to confine electrons in an arbitrary way with external control. External magnetic fields force electrons in their motion perpendicular to the field on a circular orbit whose circumference has to be an integer multiple of the *de Broglie*-wavelength. Thus the radius of the smallest orbit for a given external magnetic field B is essentially determined by the magnetic length:

$$\lambda(B) = (\hbar/eB)^{\frac{1}{2}} \tag{1}$$

Here $\hbar$ is the reduced Planck's constant and e the elementary charge. In Fig. 1 we have plotted the magnetic length $\lambda(B)$ as a function of the magnetic field B. It should be noted, that the magnetic length depends only on the external magnetic field and natural constants. This means the magnetic length is independent of the actual mass parameter of the particle. Thus a proton and an electron have for a given magnetic field exactly the same shape of the wave function. The energy of the particle, however, depends of course on the mass parameter m:

$$E = (\hbar/\lambda)^2 *(N+\tfrac{1}{2}) /2m = \hbar eB*(N+\tfrac{1}{2})/2m \qquad N = 0, 1, 2, \ldots . \tag{2}$$

CP1147, *Transport and Optical Properties of Nanomaterials—ICTOPON - 2009*, edited by M. R. Singh and R. H. Lipson
© 2009 American Institute of Physics 978-0-7354-0684-1/09/$25.00

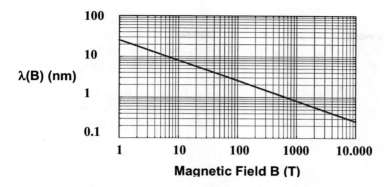

FIGURE 1. The magnetic length λ(B) as a function of the external magnetic field B.

For a magnetic field of B = 100 T = 1 Megagauss the magnetic length λ(B=100T) = 2.56 nm. As shown in Fig. 1 the magnetic length λ(B) provides an externally tunable length scale in the regime of nm. For small magnetic fields an electron confined into a nano-structured volume is determined essentially by the potential steps experienced at the boundaries. If at high magnetic fields the magnetic length gets comparable to the confinement dimensions, the magnetic field quantization becomes competitional and dominates finally for very strong magnetic fields. So the change of the quantization mechanism as a function of the external magnetic field is a sensitive criterion that magnetic length and confinement dimensions are equal. Many different experimental setups allow the direct observation of the take-over of the magnetic field quantization from the confinement mechanism.

It should be noted that only for magnetic field intensities close to 10,000 T the magnetic length gets comparable to the natural lattice constants and hence only for such high fields the phenomena described by *Hofstadter's Butterfly* can be expected [1].

GENERATION OF HIGH MAGNETIC FIELDS

At present magnetic fields up to 1,000 T can be generated and applied in experiments for solid state physics [2]. Generally we distinguish stationary and pulsed fields, among the latter non-destructive and destructive methods. Stationary fields are produced by solenoids, either by monolithic coils wound as classical coils or in an arrangement of different coils in concentric configuration and each one having an individual power supply. In *hybrid systems* the inner coil is resistive, whereas the outer coil is produced by use of superconducting wires. Using *hybrid systems* at present the highest stationary fields can be generated touching now the 50 T limit. It should be noted that for resistive magnets power supplies of some ten mega-W are necessary, so that the cooling of these magnets is a technological challenge.

The most serious problem in high magnetic field generation is caused, however, by the Maxwell-stress: the Lorentz-force produced by the field and the field creating current is outward directed with respect to the coil axis and tries to explode the coil [3]. For B = 100 T a total pressure difference between inside and outside of the coil of 40 ton/cm^2 tries to blow up the coil. At present despite of long-during and intense efforts no solution of this problem has been found. The highest non-destructive field has been generated in the National High Magnetic Field Laboratory, Los Alamos, USA with a peak value of about 90 T [4]. Increasing the field to 100 T will also increase the pressure inside the coil by more than 20%!

Using alternatively *destructive coils* much higher fields can be generated: in these experiments the coil is allowed to blow-up by the Maxwell-stress [5]. The trick of the experiment is, however, to perform the measurement very quickly *before* the coil will be blown up. This time window is in the order of only some microseconds. The experimental techniques related to this kind of field generation are the *single-turn coil* up to about 300 T [6] and the explosive flux-compression, which has been applied to experiments up to 1.000 T [2], however field generation without experimental investigation has been successful up to 2,800 T [7]. In the following discussion we will demonstrate all different kinds of field generation in experiments with nano-structured systems.

NANO-STRUCURED SYSTEMS IN HIGH MAGNETIC FIELDS

At present semiconductor structures can be produced to represent systems of different dimensionality: depending on the suppression of the motion in one or more Cartesian-coordinate direction we have Q3D (**Q**uasi **3-D**imensional), Q2D, Q1D, and Q0D systems as represented by bulk material, quantum well, quantum wire, and quantum dot, respectively.

High Magnetic Fields in Q3D

A Q3D-system is represented by a bulk sample, where the dimensions in all Cartesian-coordinate directions do not manifest in a special confinement quantization. The magnetic field determines therefore completely the motion in the plane perpendicular to the field direction $B/|B|$. As a result for a parabolic system the energy of an electron in the presence of B is given by the sum of the energy of the motion perpendicular and parallel to B: :
$$E = \hbar eB/m^*(N+\tfrac{1}{2}) + p_B^2/2m, \quad N = 0, 1, 2,\ldots \quad (3)$$
The resultant density of states features the discrete spectrum of the *Landau*-levels

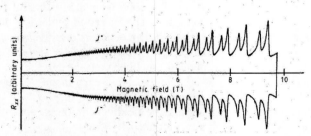

FIGURE 2. The experimental data of 3D analogue of the Quantum Hall Effect in HgSe:Fe [8].

16

indicates by the *Landau*-quantum number N with the superposition of a one-dimensional density of states for the motion parallel to the magnetic field. It should be noticed that this level structure can under special conditions even in the measurements of the Hall-effect in Q3D be observed featuring the *three-dimensional analogue of the Quantum Hall Effect* as shown in Fig, 2 for HgSe:Fe [8].

High Magnetic Fields in Q2D

If you reduce the spatial extension in one direction of a bulk sample in such a way that the sample thickness gets comparable to the *de Broglie*-wavelength of the electrons for the motion perpendicular to the layer, this motion is limited to discrete modes related to corresponding subband energies. Considering an otherwise free electron in one subband level the particle can move without restriction only in two dimensions, so that we have a Q2D system. In Fig. 3 we have simulated the gradual manifestation of the Q2D in the measurement of the *three-dimensional analogue of the Quantum Hall Effect* by subsequent reduction of the layer thickness of the HgSe:Fe sample [8].

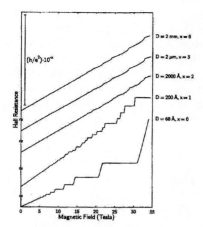

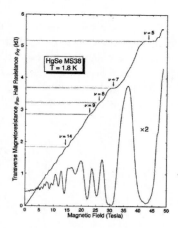

FIGURE 3. The simulation of the 3D → 2D transition of the Quantum Hall Effect in HgSe:Fe [8].

FIGURE 4. Data of the 2D Quantum Hall Effect in HgSe:Fe [9].

The simulation reflects the gradual transition of the density of states from Q3D to Q2D. In Fig. 4 we reproduce the direct measurement of the Hall effect for a HgSe:Fe layer of 20 nm thickness. The well-known plateaus of the Quantum Hall Effect are clearly present. Please notice that in simulation as well as in the experiment the external magnetic field is oriented perpendicular to the layer, so that only the Q2D is affected by the magnetic field resulting in a totally discrete density of states.

High Magnetic Fields in Q1D

Applying *Molecular Beam Epitaxy* on prestructured substrates also for HgSe:Fe quantum grooves can be grown. Here the different growing speed of the material on lattice faces for different orientation is used. In Fig. 5 the schematic of such as structure (Fig. 5a) as well as the result of a scanning microscope (Fig. 5b) is reproduced [10]. At the ridge of the roof-like structure the quantum wire has been grown.

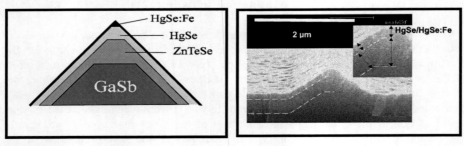

FIGURE 5a   **FIGURE 5b**

The schematic (a) and experimental realization (b) of the HgSe:Fe quantum wire [10].

We have investigated the transverse magneto resistance of the quantum wires and compared with the corresponding results taken on an unstructured layer of the same material as shown in Fig. 6. At high magnetic fields both experiments show pronounced *Shubnikov-de Haas* oscillations indicating that in both experiments the cyclotron motion is well established due to the only small scattering of the electrons [11]. Evidently for this magnetic field regime the cyclotron motion is not blocked by the

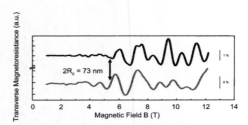

FIGURE 6. The SdH-oscillations in the upper part for the quantum wire have on onset beyond 5.5 T corresponding to a cyclotron diameter of 73 nm in contrast to the oscillation in a corresponding pure 2D system of a layer [11].

walls of the quantum groove and we have to assume that the cyclotron diameter is smaller than the groove width. Reducing the magnetic field, however, below 20 T the manifestation of the *Shubnikov-de Haas* oscillations is suppressed for the quantum-groove sample whereas not so for the unstructured layer sample. Evidently for the groove sample, the cyclotron motion is blocked by scattering of the electron at the groove walls. So we have a sensitive method to find out how large the cyclotron diameter has to be to fit just into the groove. We may call the fitting-in of the cyclotron orbit into any given potential structure by use of the external magnetic field as a *magneto condensation*. From the observed transition field of B_{trans} we derive an effective groove width of 73 nm, which agrees in an excellent way with the value obtained from the scanning microscope of 85 nm [11]. Please note, that at high magnetic fields the quantization by the groove walls becomes irrelevant because the extension of the electron wave function is much smaller than any wall separation so that the electrons cannot be distinguished from a Q2D system: the application of the magnetic field has driven the system from Q1D → Q2D effectively.

High Magnetic Fields in Q0D

The Hydrogen-Like Impurity in High Magnetic Fields

Q0D systems are confined in each of the three Cartesian directions, so that the electron is totally determined by the *Schroedinger* equation with the corresponding potentials. The most ancient Q0D system in semiconductors is the hydrogen-like shallow impurity having an effective binding energy of:

$$E_{Bind} = \frac{1}{2} * e^4 m / (4\pi\varepsilon\hbar)^2 \tag{4}$$

with an effective Bohr radius:

$$R_{Bohr} = 4\pi\varepsilon\hbar^2 / (me^2) \tag{5}$$

In contrast to the chemical hydrogen problem in the vacuum, in the semiconductor we have to use an effective mass and an effective dielectric constant. So for indiumantimonide $m = 0.0135 * m_0$ and $\varepsilon = 16.6 * \varepsilon_0$ resulting in $E_{Bind}^{InSb} = 0.664$ meV and $a_0^{InSb} = 65.2$ nm. For a magnetic field of 10 T is $\lambda(10 \text{ T}) = 8.1$ nm already much smaller then the effective Bohr radius. Since for the hydrogen-like impurity there are no strict confinement walls but a gradual change of the electrostatic potential, with increasing magnetic field the electron drops deeper into the *Coulomb* well and the effective binding energy is increased. This effect is the origin of the well known *magnetic freeze out* [12]. We will follow this effect now in relation to *man made impurities* as are quantum dots and define the corresponding effect as *magneto condensation*.

Quantum Dots of InAs on GaAs in High Magnetic Fields

A quantum dot is the man-made equivalent in a solid-state matrix to the chemical atoms in vacuum. Sophisticated methods have been developed to grow quantum dots in connection with most of the known semiconductors. One of the mostly investigated material combinations is InAs/GaAs. Quantum dots of InAs on GaAs were investigated by magneto-luminescence [13]. The special feature of these dots is that the form a flat pyramid with a basis length of $2b = 12$ nm and a height of 6 nm., so that the projected cross areas parallel and perpendicular of the symmetry axis of the dot are quite different. This means that with respect to *magneto-condensation* we expect different behavior for the magnetic field orientation parallel and perpendicular to the

19

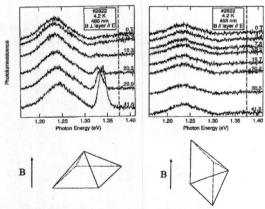

FIGURE 7. Magneto-condensation into InAs quantum dots in high magnetic fields [13].

pyramid axis. In Fig. 7 we have reproduced the corresponding results. We have measured the magneto-luminescence in pulsed fields up to about 40 T. Whereas for the magnetic field parallel to the pyramid axis we observe with increasing field different behavior for the magnetic field orientation parallel and perpendicular to the pyramid axis. In Fig.7 we have reproduced the corresponding results. We have measured the magneto-luminescence in pulsed fields up to about 40 T. Whereas for the magnetic field parallel to the pyramid axis we observe with increasing field starting at about 20 T the manifestation a new bound state related to the quantum dot, there is no indication of this state for the field orientation perpendicular to the pyramid axis: the cyclotron orbit does not yet fit into the decreased cross area of the rotated quantum dot. This experiment is a classical demonstration of *magneto condensation*.

Quantum Dots of HgSe:Fe in High Magnetic Fields

Using sophisticated techniques also quantum dots of HgSe:Fe can be grown on ZnTe-substrate as shown by the AFM-picture in Fig. 8 [14]. The very thin wetting layer of this material prevents its population by electrons, so that all electrons are confined into the dots. It should be noted that actually 30 to 300 electrons – depending on the actual dot size – are simultaneously present within one dot and interact. We have measured the cyclotron resonance of the dot electrons in Faraday configuration and compared with the corresponding measurements of an unstructured layer system as shown in Fig. 9. To obtain clear results and good transmission we had to apply 10.6 µm-wavelength radiation beyond the plasma edge of the high-carrier system with $n \approx 5*10^{18}$ cm^{-3} volume concentration [15]. To observe the cyclotron resonance using 13.1 meV-radiation we need magnetic fields in the Megagauss regime with the expected cyclotron mass of $m_c = 0.065*m_0$. We applied a *single-turn-coil* generator to obtain fields up to 100 T [3]. The two resonance curves in the transmission in Fig. 9 result in a cyclotron mass of $0.065*m_0$ for the lower resonance in a HgSe:Fe layer and in a

20

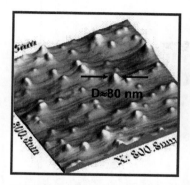

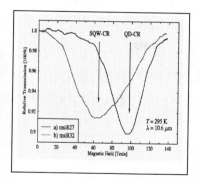

FIGURE 8. HgSe:Fe *Quantum Dots* [14].

FIGURE 9. Cyclotron resonance (*CR*) in *Q*uantum Dot (*QD*) and *S*ingle *Q*uantum *W*ell (*SQW*) [15].

dramatically increased value of $m_c = 0.1 \, {}^{*}m_0$ for the electrons in the quantum dot. This mass increase of about 50% has tentatively been explained by the different dimensions of the lattice constants for the unstressed material of the HgSe:Fe layer and the stressed material in the quantum dot [16]. For the resonance fields the values of the magnetic length are $\lambda(60 \text{ T}) = 3.3$ nm and $\lambda(100 \text{ T}) = 2.56$ nm. In both cases many orbits fit into the quantum dot without being scattered at the dot walls.

REFERENCES

1. D. R. Hofstadter, *Phys. Rev.* B **14** 2339 (1976).
2. M. von Ortenberg, N. Puhlmann, I. Stolpe, A. Kirste, H.-U. Mueller, S. Hansel, O. M. Tatsenko, I. M. Markevtsev, N. A. Moiseenko, V. V. Platonov, A. I. Bykov, V. D. Selemir, *Proc. 26ᵗʰ Int. Conf. Phys. Semicond.*, Edinburgh (2002) F1.2
3. O. Portugall, N. Puhlmann, H.-U. Mueller, M. Barczewski, I. Stolpe, M. von Ortenberg, *J. Phys. D: Appl. Phys.* **32**, 2354-2366 (1999)
4. S. A. Crooker, N. Samarth, *Appl. Phys. Lett.* **90**, 102109 (2007)
5. F. Herlach and M. von Ortenberg, *IEEE Trans. Magn.* **32**, 2438 (1996)
6. M. von Ortenberg, *J. Physics: Conf. Ser.* **51**, 371 (2006)
7. A. Boyko, A. I. Bykov, M. I. Dolotenko, N. P. Kolokolchikov, I. M. Markevtsev, O. M. Tatsenko, A. M. Shuvalov, *Proc. VIII Int. Conf. on Megagauss Magnetic Field Generation and Related Topics,* Tallahassee 1998, ed. H.-J. Schneider-Muntau, World Scientific (2004) p. 61
8. M. von Ortenberg, O. Portugall, W. Dobrowolski, A. Mycielski, R. R. Galazka, F. Herlach, *J. Phys. C: Solid State Phys.* **21**, 5393-5401 (1988)
9. G. Machel, *PhD thesis*, Humboldt University at Berlin, Germany (1997)
10. H. Wissmann, T. Tran-Anh, S. Rogaschewski, M. von Ortenberg, *J. Crystal Growth* **201/202**, 619-622 (1999)
11. H. Wissmann, T. Tran-Anh, A. Kirste, O. Portugall, S. Rogaschewski, M. von Ortenberg, *Proc. 24th Int. Conf. on the Physics of Semicond.*, edited by D. Gershoni, World Scientidfic (1999) #7
12. M. von Ortenberg, *J. Phys. Chem. Solids* **34**, 397-411 (1973)
13. M. von Ortenberg, K. Uchida, N. Miura, F. Heinrichsdorff, D. Bimberg, *Physica* **B246/247**, 88-92 (1998)
14. M. von Ortenberg, *J. Alloys and Compounds* **371**, 42-47 (2004)
15. T. Tran-Anh, S. Hansel. A. Kirste, H.-U. Mueller, M. von Ortenberg, *Physica* **E20**, 448 (2004)
16. M. von Ortenberg, C. Puhle, S. Hansel, *Proc. 27th Int. Conf. on the Physics of Semicond.*, Flagstaff, USA, 357 (2004)

Time-Stepping and Convergence Characteristics of the Discontinuous Galerkin Time-Domain Approach for the Maxwell Equations

Jens Niegemann and Kurt Busch

Institut für Theoretische Festkörperphysik, Universität Karlsruhe (TH), D-76128 Karlsruhe, Germany
DFG-Center for Functional Nanostructures (CFN), Universität Karlsruhe (TH), D-76128 Karlsruhe, Germany

Abstract. We provide details on the performance characteristics of the Discontinuous Galerkin Time-Domain (DGTD) method when applied to solving the linear Maxwell equations. It is shown that even for very simple, grid-aligned structures, the DGTD easily outperforms the well-established Finite-Difference Time-Domain (FDTD) method.

Keywords: Computational Photonics
PACS: 42.25.-p,0 2.70.-c, 02.70.Dh

INTRODUCTION

Over the past years, the research field of photonics has received an ever increasing amount of interest and tremendous progress has been accomplished. Similar advances are expected for the years to come. This origin for this is twofold: (i) Advanced micro- and nano-fabrication techniques along with the synthesis of novel optical materials facilitate the realization of highly complex structures of unprecedented quality. Sophisticated spectroscopy methods allow for detailed investigations of such systems and their exploitation for advanced functional elements and devices. (ii) The theoretical description of the linear, nonlinear and quantum optical properties of photonic elements has been considerably refined and together with the ever increasing power of modern computers, quantitative predictions up to the level of blueprints for devices have become possible.

Broadly speaking, the computational methods for solving the Maxwell equations may be divided into frequency-domain and time-domain methods. Most frequency-domain methods such as the Boundary-Element (BE) and Finite-Element (FE) Methods [1] lend themselves to highly accurate representations of complex geometries and, therefore, provide excellent results for stationary problems. However, certain applications such as transient wave propagation phenomena are more naturally treated within a time-domain framework. Surprisingly and in contrast to the numerous frequency-domain methods, the choice of time-domain methods for the treatment of photonic problems is rather limited, the Finite-Difference Time-Domain (FDTD) method undoubtedly being the most popular one [2]. While being easy to implement and ideally suited for modern computer architectures, FDTD suffers from the fact that is only second-order accurate in space and time and – due to the use of a regularly spaced grid – exhibits notorious

CP1147, *Transport and Optical Properties of Nanomaterials—ICTOPON - 2009*, edited by M. R. Singh and R. H. Lipson
© 2009 American Institute of Physics 978-0-7354-0684-1/09/$25.00

problems when complex geometries need to be handled with high accuracy.

The obvious way out is to combine the respective advantage of FE and FDTD methods. But how to do this is far from obvious. For instance, a traditional FE time-domain (FETD) approach inevitably leads to an implicit time-stepping scheme [1], where, at each time-step, a system of equations has to be solved. Since typical applications quickly lead to 10^6-10^7 or even more unknowns, the performance characteristics of traditional FETD becomes unacceptably large for realistic systems. Fortunately, very recently Hesthaven and Warburton have shown how to realize a FE-like approach that remain explicit in time which is based on discontinuous Galerkin techniques [3], the resulting method is generally referred to as the Discontinuous-Galerkin Time-Domain (DGTD) approach.

DGTD AND THE CHOICE OF A TIME-STEPPING SCHEME

Within a DGTD formalism [3, 4], the computational domain is tesselated into finite elements, typically triangles in 2D and tetrahedra in 3D. On each element, the various components of the electromagnetic field are expanded into nodal-based interpolating polynomials of adjustable order. In contrast to ordinary finite element techniques, within DGTD one allows that field values at the edges of the element are discontinuous. Instead, the coupling between neighboring elements is facilitated through the introduction of a so-called numerical flux [3] which enforces the boundary conditions between adjacent elements in a weak sense. In essence, this leads to a strictly local coupling of elements that, in contrast to FETD, allows to derive *explicit update-equations*. As a result, the combination of flexible FE-like spatial discretization and explicit time-stepping capabilities makes DGTD an excellent candidate for efficient and accurate computations of complex photonic nano-structures.

We have described details of our implementation of the DGTD spatial discretization in Ref. [4] and, below, we will refer to Eqs. (5) and (6), of that paper as the semi-discrete system (SDS) which has to be supplied with a suitable time-stepping algorithm. In this proceeding, we present the details of the time-stepping scheme details of our DGTD-implementation [4] and discuss its performance characteristics for a few, rather simple systems. The simplest possible time-stepping scheme would be to employ a finite-difference discretization just as in the FDTD method. Unfortunately, this leads to a rather poor performance, since this scheme does not exploit the advantage of high-order spatial discretization that is offered by the DGTD method. In turn, this would dramatically limit the efficiency of our solver. Instead, we have adopted a more sophisticated integration scheme known as the low-storage Runge-Kutta (LSRK) method. In contrast to classical Runge-Kutta schemes, which are known for their versatility and excellent performance, the LSRK methods are tuned to require less memory. This becomes important, when we attempt to simulate very large systems where the number of unknowns reaches the order of about 10^7 or more. Remarkably, as we will demonstrate below, LSRK methods are also advantageous for smaller systems where memory consumption is not issue.

Combining the SDS (Eqs. (5) and (6) of Ref. [4]) for all elements into a single system

$$\frac{\partial}{\partial t} \mathbf{y}(t) = f(\mathbf{y}, t), \tag{1}$$

we can integrate in time by using the 2M-LSRK due to Williamson [5] as

$$\mathbf{y}_0 = \mathbf{y}(t)$$
$$\left.\begin{array}{l} \mathbf{k}_i = \tilde{a}_i\mathbf{k}_{i-1} + \Delta t f(\mathbf{y}_{i-1}, t + \tilde{c}_i\Delta t) \\ \mathbf{y}_i = \mathbf{y}_{i-1} + \tilde{b}_i\mathbf{k}_i \end{array}\right\} \quad \forall i = 1\ldots\sigma$$
$$\mathbf{y}(t+\Delta t) = \mathbf{y}_\sigma.$$

Here, σ denotes the number of stages while $\tilde{a}_i$, $\tilde{b}_i$ and $\tilde{c}_i$ are the coefficients which define a particular scheme. Furthermore, we have introduced the supervector $\mathbf{y}$ that denotes the combination of all $\vec{E}^k$ and $\vec{H}^k$, for all elements k, and the function f represents the combination of the right-hand sides of the SDS (Eqs. (5) and (6) of Ref. [4]) with the modification that the first and second equation of the SDS have been, for all elements k, divided by, respectively, ε^k and μ^k.

In Ref. [6], Carpenter and Kennedy propose a set of coefficients for a fourth-order LSRK scheme which we use in our implementation. Unfortunately, and in contrast to the classical Runge-Kutta methods, it is not possible to create such a fourth-order scheme with only four stages. Instead, one requires at least $\sigma = 5$. As an explicit scheme, the LSRK scheme is only conditionally stable. Similar to the case of FDTD, we, therefore, need to fulfil a CFS-criterion which limits the allowed time step. However, in contrast to FDTD, we can not express this criterion explicitly. Instead, we can find the appropriate condition by considering the ordinary differential equation

$$\frac{\partial}{\partial t}y = \lambda y. \tag{2}$$

Inserting this equation into the LSRK-scheme described above yields

$$y(t+\Delta t) = \Delta t \underbrace{\left(1 + \alpha_1\lambda + \alpha_2\lambda^2 + \cdots + \alpha_s\lambda^s\right)}_{=:G(\lambda)} y, \tag{3}$$

where α_n are numerical coefficients depending on the parameters $\tilde{a}_i$ and $\tilde{b}_i$ of the particular scheme. Obviously, in order to avoid that our solution shows unphysical growth, we must fulfil the condition

$$\Delta t\, G(\lambda) \leq 1. \tag{4}$$

A contour of the function $\Delta t G(\lambda) = 1$ is plotted in Fig. 1 for a classical fourth-order Runge-Kutta scheme and for our LSRK integrator. Clearly, the LSRK has a significantly larger region of stability. Leaving aside the lower memory requirements, this distinct advantage of the LSRK is the primary reason for employing a LSRK method instead of a classical Runge-Kutta integrator. Returning to our full problem (1), we note that λ plays the role of to the eigenvalues of the discrete operator on the r.h.s. of the SDS (Eqs. (5) and (6) of Ref. [4]). Thus, we are required to pick Δt sufficiently small so as to ensure that the largest eigenvalue still fits into the region of stability.

Unfortunately, we can not afford to calculate the spectrum of each operator before we conduct a time-domain calculation. Therefore, we require a more accessible way

24

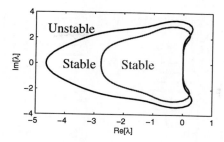

FIGURE 1. Stability region of the classical Runge-Kutta (red) and a LSRK scheme (blue).

to obtain a sufficiently small Δt without wasting computational resources by picking it too small. Empirically, we know that the largest eigenvalue is connected to the shortest distance between two gridpoints in our system. This scale can be separated into two parts. First, we have the nodes on the standard element (recall that the nodes on each element are determined from the nodes of the standard element through an appropriate mapping). As a rather primitive measure for the minimal distance between them, we consider the LGL-nodes along the edges. There, we easily calculate the distance Δu_p of the first two nodes. This length will depend on the expansion order p as $\Delta u_p \propto p^{-2}$. Second, in order to turn this dimension measure Δu_p into a physical length, we require a measure for the size of the smallest element in our computations. Following the suggestions of Hesthaven and Warburton [7], we pick the incircle or insphere of the triangles or tetrahedra, respectively. This length is denoted by r^k for each element k. As a result, we determine the timestep as

$$\Delta t = s \Delta u_p \min_{\Omega} \left(r^k \right), \tag{5}$$

where s is a number of order 1, which we determine from a set of numerical experiments.

EMPTY METALLIC CAVITIES

In contrast to the FDTD method, the performance of the DGTD scheme is affected by a number of external parameters. Specifically, the shape and quality of the mesh can significantly influence the results. For this section, we employ regular meshes with perfect magnetic conductor (PMC) boundary conditions. This test system of an empty metallic cavity provides an analytic solution since all the eigenmodes are easily determined. This allows us to complete the determination of the optimal time step Δt and to conduct various tests that are required before proceeding to realistic systems.

According to the above discussion, we still need to empirically determine an appropriate value for the factor s in Eq. (5). To do so, we simulate take as an initial condition the fundamental eigenmode of the cavity and let it evolve for 50 periods T_0 for different meshes, different orders p, and different values of s. In Fig. 2, we display the resulting maximal error after $T = 50 T_0$ as a function of s.

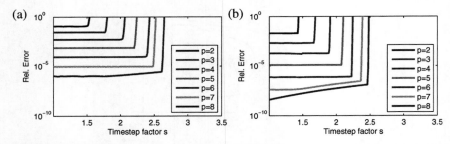

FIGURE 2. Error for empty cavity calculations as a function of the time step. The data in (a) was obtained with 4 triangles per wavelength, while (b) contains data for 8 triangles per wavelength.

From these results, one can directly extract the point of instability. Further, we find that once the scheme is stable, most of the results do not change with different time step. Thus, we conclude that the error is mostly dominated by the spatial discretization. Only for relative errors below 10^{-5} do we observe a further reduction of the error with decreasing time steps. As a consequence, we should attempt to make the timestep as large as possible without crossing the instability limit. Only for highly accurate calculations, do we have to reduce s slightly. From the data in Fig. 2, we find that the maximal value s_{max} still depends on both, the spatial order and the size of the mesh. In Fig. 3 we plot the maximal values of s_{max} for different orders and meshes.

We observe how the maximally allowed time step factor increases with the order and decreases with refinement of the mesh. The variations, however, are not too large, indicating that the scaling discussed in the previous section works reasonably well. In order to improve the performance, we fit a quadratic polynomial to the maximally allowed timestep for each order of the smallest mesh. To be on the safe side, we then subtract a certain margin of error and obtain the expression

$$s_{max}(p) = 0.8 + 0.27p - 0.011p^2. \tag{6}$$

in Fig. 3, the resulting values are indicated as blue circles. Those values will used for all subsequent computations.

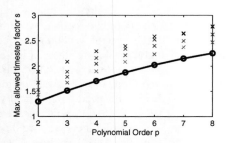

FIGURE 3. The crosses indicate maximally allowed timestep factors s_{max} for different meshes and orders. The solid line corresponds to values for s_{max} as given by Eq. (6).

26

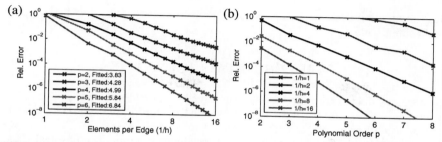

FIGURE 4. (a) Relative error as a function of mesh refinement. Straight lines in the semi-logarithmic plot indicate algebraic convergence. The fitted slopes are given in the legend. (b) Relative error as a function of increasing order of the polynomial expansion p. The straight lines in this double logarithmic plot demonstrate exponential convergence.

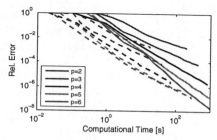

FIGURE 5. The relative error as a function of the computational time. The dashed lines correspond to calculations with a time step factor of $s = s_{max}$, while the data for the solid lines was obtained with $s = 0.1 s_{max}$.

However, before we proceed to more sophisticated setups, we should also analyse the errors as a function of the mesh refinement and the spatial order. To completely eliminate the influence of the time integration, we will work with a time step factor of $s = 0.1 s_{max}$. In Fig. 4, we display the error as a function of the mesh size (Fig. 4(a)) and of the polynomial order (Fig. 4(b). In particular, from the straight lines in the semi-logarithmic plot 4(a) we confirm the algebraic h-convergence. Upon conducting a linear fit, we find a slope roughly equivalent to $p + 1$. Furthermore, from the double logarithmic plot in Fig. 4(b) we observe exponential p-convergence.

The qualitatively different convergence behaviour naturally leads to the question as to which refinement should be preferred. To investigate this, we plot the relative error as a function of the computational time that has been spent to obtain the results in Fig. 5. From these results, we observe that for this test problem the p-refinement does indeed result in faster calculations for any given error. But it should be noted, that this can not directly be transfered to calculations of realistic systems. In particular, if curved interfaces are involved, the error can be strongly dominated by the approximation due to straight-sided elements. Such an approximation of rounded surfaces through polygons will be particularly detrimental for plasmonic systems. In this case, only h-refinement will lead to an improvement. Similar considerations are also valid, if we have to deal with field singularities.

HALF-FILLED METALLIC CAVITIES

As a next step we demonstrate the superior convergence properties of the DGTD method, when interfaces are involved. It is well-known that a fourth-order FDTD scheme is reduced to second order of accuracy if material interfaces are present [2]. Here, we demonstrate that the DG method does not exhibit this restriction. The physical setup is a metallic cavity where one half of the cavity is filled with a material with dielectric constant $\varepsilon = 2$. As in the previous section, we launch the cavity's fundamental mode and employ time step of $s = 0.1 s_{max}$ to avoid error contributions from the time integration. The important feature is that both, h- and p-convergence remain entirely unaffected by the presence of an interface (cf. Fig. 6). This is in stark contrast to the FDTD results and highlights the versatility of the DGTD method. The important feature is that both, h- and p-convergence remain entirely unaffected by the presence of an interface (cf. Fig. 6). This is in stark contrast to the FDTD results and highlights the versatility of the DGTD method.

Finally, we conduct a direct comparison between DGTD and FDTD method. To this end, we plot in Fig. 7 the DGTD results obtained with the maximal time-step factor as a function of CPU time. In the same plot, we also depict the computational time required for the corresponding FDTD calculations. From this plot, we deduce that the DGTD method is clearly superior for this particular problem. If we require a relative error below 1%, then DGTD delivers the solution about two orders of magnitude faster than FDTD. However, this advantage largely stems from the higher-order nature of our method. As discussed previously, for more complicated structures containing curved interfaces, we might not be able to fully exploit the p-convergence. For such cases, the advantage of the DGTD method is less spectacular then described above but from our extensive experience [4, 8, 9] we maintain the conclusion that a rather distinct advantage

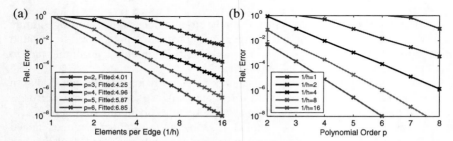

FIGURE 6. (a) Relative error of the DGTD method for a half-filled metallic cavity as a function of mesh refinement. Straight lines in the semi-logarithmic plot indicate algebraic convergence. The values of the fitted slopes are provided in the legend. (b) Relative error as a function of increasing order of the polynomial expansion. The straight lines in this double logarithmic plot demonstrate exponential convergence.

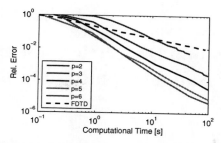

FIGURE 7. The relative error as a function of the computational time for a half-filled metallic cavity. The solid lines correspond to DGTD computation with a time-step factor of $s = s_{max}$, while the dashed line represents results from an FDTD computations.

of DGTD over FDTD remains.

In conclusion, we have determined detailed performance characteristics of the DGTD method with fourth-order LSRK time-stepping scheme for simple test systems and have compared the results with those from corresponding FDTD computations. For these systems, we can fully exploit the p-convergence associated with the finite-element like discretization of the DGTD method. In turn, this leads to significant advantages of the DGTD method over FDTD method in terms of efficiency which can be as large two orders of magnitude. For more realistic systems with complex scatterer geometries this advantage becomes somewhat less spectacular but is still quite significant. In addition, since within the DGTD method, the time-stepping scheme is essentially disentangled from the spatial discretization (as opposed to the situation in FDTD), more sophisticated methods such as Krylov-subspace based operator exponential techniques [10] could provide an even more efficient solver than our present version that is based on fourth-order LSRK [6].

We acknowledge support by the Deutsche Forschungsgemeinschaft (DFG) and the State of Baden-Württemberg through the DFG-Center for Functional Nanostructures (CFN) within subproject A1.2.

REFERENCES

1. J. Jin, *The Finite Element Method in Electromagnetics*, Wiley-IEEE Press, Second Edition, 2002.
2. A. Taflove, and S. C. Hagness, *Computational Electrodynamics: The Finite-Difference Time-Domain Method*, Artech House Publishers, Third Edition, 2005.
3. J. Hesthaven, and T. Warburton, *J. Comput. Phys.* **181**, 186 (2002).
4. J. Niegemann, M. König, K. Stannigel, and K. Busch, *Photon. Nanostruct.: Fundam. Appl.* **7**, 2 (2009).
5. J. H. Williamson, *J. Comput. Phys.* **35**, 48 (1980).
6. M. H. Carpenter, and C. A. Kennedy, *Tech. Report NASA-TM-109112* (1994).
7. J. Hesthaven, and T. Warburton, *Nodal Discontinuous Galerkin Methods*, Springer, 2007.
8. J. Niegemann, W. Pernice, and K. Busch, *submitted* (2009).
9. K. Stannigel, J. Niegemann, M. König, and K. Busch, *submitted* (2009).
10. J. Niegemann, L. Tkeshelashvili, and K. Busch, *J. Comput. Theor. Nanosci.*, **4**, 627 (2007).

The Study of a Dielectric Catastrophe in Nonlinear Nanophotonic Quantum Wires

Mahi R. Singh

Department of Physics and Astronomy, The University of Western Ontario, London, Canada N6G 3K7.

Abstract. We study a dielectric catastrophe in a new class of materials called nanophotonic quantum wires. They are made from two types of photonic crystals where one is embedded in the other. It is considered that the embedded crystal is made from nonlinear dielectric spheres. The wire is doped with an ensemble of nanoparticles which are interacting with each other via the dipole-dipole interaction. The absorption coefficient of the wire has been investigated in the presence a weak probe laser and a strong pump laser. The probe field measures the absorption spectrum and the pump field changes the refractive index of nonlinear photonic crystals due to the Kerr effect. The density matrix method is used to calculate the absorption coefficient. Numerical simulations have been performed for the dielectric constant and the photon bound states. It is found that the dielectric constant of the system has a singularity when the resonance energy of a nanoparticle lies near the bound photon states of the wire. This is known as the dielectric catastrophe. This phenomenon can be used to make all-optical switching devices.

Keywords: Nanophotonics, photonic crystals, quantum dots, absorption, nanoparticles, wires, dipole-dipole interaction, dielectric catastrophe, all-optical switching, optical devices.
PACS: 42.70Qs, 42.55Tv, 78.20.Bh

INTRODUCTION

We study the dielectric catastrophe in nanophotonic quantum wires which are made from two photonic crystals. The nonlinear photonic crystal is embedded into the linear photonic crystal (see fig.1). It is known that photonic crystals have an energy gap in their photonic energy spectrum. Recently, there has been a considerable interest to study photonic crystals because they can control the properties of electromagnetic light in the same way as semiconductor materials control the propagation of electrons [1-5]. Nanophotonic wires are a promising class of structures which have very interesting optical properties and can be used to make new types of optical devices [5-10]. The study of these systems will also enhance our knowledge of nanoscale science and optoelectronic technology.

The nanophotonic wires are doped heavily by nanoparticles such as quantum dots (see fig.1). The nanoparticles are interacting with each other via the dipole-dipole interaction (DDI) [11]. The dielectric catastrophe found in this paper is due to the DDI effect. The DDI has been investigated in quantum optics, photonic crystals and nanostructures [11-24]. For example, the DDI effect has been studied in the quantum coherence and interference phenomenon which is one of the most important mechanisms for controlling spontaneous emissions in atomic gases and solids. The spontaneous emission is the main source of noise in quantum computation, teleportation, and quantum information processing [12, 13].

CP1147, *Transport and Optical Properties of Nanomaterials—ICTOPON - 2009,* edited by M. R. Singh and R. H. Lipson
© 2009 American Institute of Physics 978-0-7354-0684-1/09/$25.00

The study of the DDI in quantum optics has led to many fascinating effects such as ultrafast optical switching and the enhancement of inversionless gain and index of refraction [14, 15]. The effect of the DDI has also been investigated in quantum entanglement [17, 18], quantum jumps [19] and resonance fluorescence [20] for multi-level atoms.

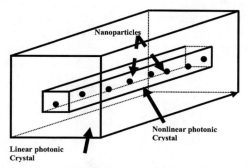

FIGURE 1. A schematic diagram is presented for a nanophotonic wire. The nonlinear photonic crystal is embedded within the linear photonic crystal. The nanoparticles are doped within the nonlinear photonic crystals.

The effect of the DDI has also been investigated in photonic and polaritonic band gap materials. For example, John and Quang [23] have studied the self induced transparency due to the DDI in photonic crystals doped with two-level atoms. Singh [24] has studied the effect on the DDI on the enhancement of the refractive index in a photonic band gap material where the material was doped with five-level nano-particles. The inhibition of the two-photon absorption due to the DDI has been found in doped photonic crystals [11].

We have applied a strong pump laser field in the wire. According to the Kerr effect the refractive index of the nonlinear material is modulated due to the strong electric field produced by the pump laser field [25]. Therefore, due to the modulation of the refractive index, the bound photon energies are thereby dynamically shifted.

Recently, the Kerr effect has been investigated in nonlinear photonic crystals. For example, Scalora *et al.* [26] have developed a theory of optical switching in one-dimensional nonlinear photonic crystals using the nonlinear Kerr effect. On the other hand, Tran [27] studied a switching mechanism that utilizes a bound photon state within the gap of a photonic crystal. Recently Singh and Lipson [28] have developed a possible switching mechanism for nonlinear photonic crystals doped with an ensemble of non-interacting three-level nanoparticles. The intense pump laser field is used to change the refractive index of the nonlinear photonic crystal. It is found that in the absence of the strong laser field the system transmits the probe laser field. However, upon application of an intense pump laser field the system becomes absorbing due to a nonlinear Kerr effect.

We consider the nonlinear photonic crystal wire is made from nonlinear dielectric spheres and the background material is taken as a linear dielectric material (such as air). The refractive index of spheres is modulated by the pump laser field whereas the refractive index of the background material is not affected. An ensemble of three-level nanoparticles is doped in the nonlinear photonic crystal (see fig. 2). The concentration of the nanoparticles is such that they interact with one another via the DDI.

We have chosen the conduction band edge of the nonlinear photonic crystal (NPC) in the wire to lie below that of the linear photonic crystal (LPC). Due to this band engineering, the embedded crystal acts as a photonic nanowire. Photons with energies lying between the two conduction bands propagate in the wire and are reflected from the LPC. In other words, photons are confined within the wire and have quantized bound photon modes [9].

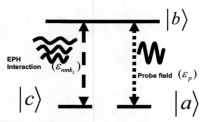

FIGURE 2. A schematic diagram is plotted for of a three-level nanoparticle. The three levels are denoted by $|a\rangle$, $|b\rangle$ and $|c\rangle$. A probe laser field is applied to monitor the transition $|a\rangle \leftrightarrow |b\rangle$. The transition $|b\rangle \leftrightarrow |c\rangle$ couples to the bound photon modes of the wire.

We applied a probe laser field to monitor the absorption coefficient of the doped nanophotonic wire. Nanoparticles have two transitions a↔b and a↔c as shown in fig. 2. The first transition is responsible for the absorption process which is monitored by the probe field. The second transition is resonantly coupled with one of the bound photon modes via the electron-bound photon (EBP) interaction. The density matrix method has been applied to calculate the absorption coefficient.

Numerical simulations for the real susceptibility and the imaginary susceptibility have been performed in the presence of the DDI. It is found the dielectric constant of the system has a singularity when the resonance energy of a nanoparticle lies near the bound photon states of the wire. This is known as a dielectric catastrophe. The dielectric catastrophe can be switched on and off by the pump laser. This phenomenon can be used to make smaller and faster optical switching devices.

Dipole-Dipole Interaction in Nanophotonic Wires

Both photonic crystals in the wire are fabricated from dielectric spheres and a background material. Dielectric spheres of the NPC are made from a nonlinear material whose refractive index is affected by the application of the pump laser field as [28]

$$n_{nL} = n_n^s + n_3 I \tag{1}$$

Here, n_3 is the Kerr constant, I is the intensity of the pump laser and n_n^s is the refractive index of a sphere. Using the isotropic band structure model of photonic crystals [28] the block wave vectors k_n and k_l for the NPC and the LPC, respectively are written as

$$k_n = F_n(\varepsilon_k) \qquad k_l = F_l(\varepsilon_k) \tag{2}$$

$$F_n(\varepsilon_k) = \frac{1}{L_n} \arccos\left(\sum_\pm \pm \frac{(n_n^s + n_3 I \pm 1)^2}{4(n_n^s + n_3 I)} \cos\left(\frac{2\varepsilon_k \left[(n_n^s + n_3 I) a_n \pm b_n \right]}{\hbar c} \right) \right) \tag{3}$$

$$F_l(\varepsilon_k) = \frac{1}{L_l} \arccos\left(\sum_\pm \pm \frac{(n_l^s \pm n_l^b)^2}{4 n_l^s n_l^b} \cos\left(\frac{2\varepsilon_k \left(n_l^s a_l \pm n_l^b b_l \right)}{\hbar c} \right) \right) \tag{4}$$

Here, n_i^s (i=n, l) are the refractive indices for a sphere. The background materials for both crystals are taken as air. a_i and b_i are the radius of a sphere and thickness of the background material, respectively. L_i is the lattice constant and defined as $L_n = 2 a_i + 2 b_i$. Using the transfer matrix method we obtained the bound photon states in the wire as [9]

$$k_{nx} \tan\left(\frac{k_{nx} d}{2} - \frac{m\pi}{2} \right) - \sqrt{F_n^2(\varepsilon) - F_l^2(\varepsilon) - (k_{nx})^2} = 0$$
$$k_{py} \tan\left(\frac{k_{py} d}{2} - \frac{p\pi}{2} \right) - \sqrt{F_n^2(\varepsilon) - F_l^2(\varepsilon) - (k_{py})^2} = 0 \tag{5}$$

Note that the wave vectors k_{mx} and k_{py} are quantized. They are components of the block wave vector k_n. The cross sectional area of the NPC is taken as a square and the length of one side is denoted as d. The bound photon energy ε_{mpk_z} are calculated from eqn. (2) as

$$F_n\left(\varepsilon_{mpk_z} \right) = \sqrt{k_{nx}^2 + k_{py}^2 + k_z^2} \tag{6}$$

Using eqns. (5) and (6) we have evaluated the bound photon energies ε_{mpk_z}.

    An ensemble of nanoparticles has been doped into the NPC. Nanoparticles have three energy levels which are denoted as $|a\rangle$, $|b\rangle$ and $|c\rangle$ (see fig. 2). The transition energies for the transitions $|a\rangle \leftrightarrow |b\rangle$ and $|b\rangle \leftrightarrow |c\rangle$ are labelled as ε_{ba} and ε_{bc}, respectively. To monitor the absorption due to the transition $|a\rangle \leftrightarrow |b\rangle$ a probe laser field of energy ε_p and amplitude E_p is applied. The absorption coefficient for the transition $|a\rangle \rightarrow |b\rangle$ due to the probe laser field is related to the imaginary part of the susceptibility. It is obtained as [13]:

$$\chi = \frac{\mu_{ab}\rho_{ab}}{\varepsilon_0 E_p} \tag{7}$$

Here, ρ_{ab} and μ_{ab} are the density matrix and the electric moment for the transition $|a\rangle \leftrightarrow |b\rangle$, respectively. It is considered that induced dipoles in nanoparticles are interacting with each other through the DDI. Nanoparticles are also interacting with bound photon states of the wire via the EBP interaction in the dipole approximation. The total Hamiltonian of the system is written as

$$H_0 = \sum_{i=a,c}\varepsilon_{bi}\sigma_{bi}^z + \sum_{mpk_z}\varepsilon_{mpk_z}\left(p_{mpk_z}^+ p_{mpk_z}\right) - \Omega_p\sigma_{ab}^+ e^{-i\delta_{ab}t/\hbar}$$

$$-\sum_{mp}\sum_{k_z}\left(\frac{\hbar^2\varepsilon_{bc}\mu_{bc}}{\sqrt{2\varepsilon_0\varepsilon_{mpk_z}V}}\right)\left[p_{mpk_z}\sigma_{bc}^+ e^{i\delta_{mpk_z}t/\hbar}\right] + \frac{N\mu_{ab}^2\rho_{ab}}{3\varepsilon_0}\sigma_{ab}^+ e^{i(\delta_{ab})t/\hbar} + h.c. \tag{8}$$

Here, the first term corresponds to the energy levels of a nanoparticle, the second term is the energy of the bound photons modes in the nanowire, the third term is the EBP interaction and the last term is the DDI effect. The terms $\delta_{ab} = \varepsilon_p - \varepsilon_{ab}$ and $\delta_{mp} = \varepsilon_{mpk_z} - \varepsilon_{bc}$ are called the probe detuning parameter and the EBP detuning parameter, respectively. ε_0 is the average dielectric constant of the nanowire. The Rabi energy of the probe laser is defined as $\Omega_p = \mu_{ab}E_p/2$. σ_{ij}^z and σ_{ij}^+ are the Pauli spin operators, while p_{mpk_z} and $p_{mpk_z}^+$ are the photon annihilation and the creation operators, respectively. The concentration of nanoparticles is N. The volume of the nanowire is V and h.c. stands for the hermitian conjugate.

With the help of eqns. (7) and (8) and using the density matrix method one can obtain the following expression of the susceptibility. The susceptibility is a complex quantity. The real part χ_d' gives the information about the dielectric constant of the system and the imaginary part χ_d'' gives the absorption coefficient. They are written as

$$\chi_d' = \chi_0\left(\frac{F'(1-C_dF')-C_d(F'')^2}{(1-C_dF')^2+(C_dF'')^2}\right)$$

$$\chi_d'' = \chi_0\left(\frac{F''}{(1-C_dF')^2+(C_dF'')^2}\right) \tag{9}$$

$$F' = \left(\frac{\gamma_b(\delta_{ab}+\Delta_{bc})}{(\delta_{ab}+\Delta_{bc})^2+(\Gamma_{bc}-\gamma_b)^2}\right), \quad F'' = \left(\frac{\gamma_b(\Gamma_{bc}-\gamma_b)}{(\delta_{ab}+\Delta_{bc})^2+(\Gamma_{bc}-\gamma_b)^2}\right) \tag{10}$$

Here χ_0 is a constant and defined in reference [9]. C_d is called the DDI parameter and is responsible for the DDI effect [11]. Other parameters are obtained as

$$\Delta_{bc} = \sum_{mp} \frac{g_{mp}\sin(\theta/2)}{\left[\left(\delta_{ab}-\delta_{mp}\right)^2+\left(\gamma_c\right)^2\right]^{1/4}}$$

$$\Gamma_{bc} = \sum_{mp} \frac{g_{mp}\cos(\theta/2)}{\left[\left(\delta_{ab}-\delta_{mp}\right)^2+\left(\gamma_c\right)^2\right]^{1/4}}, \theta = \arctan\left(\frac{\gamma_c}{\delta_{ab}-\delta_{mp}}\right)$$

(11)

Here, g_{mp} is the EBP coupling parameter and is defined in reference [9]. γ_b and γ_c are the decay linewidths for levels $|b\rangle$ and $|c\rangle$, respectively, in the absence of the EBP interaction.

Numerical Simulations

In the last section we have studied the real and imaginary susceptibilities for nanophotonic wires. We have considered that both crystals are made from dielectric spheres. Background materials for both crystals are taken as air. Refractive indices for spheres in the NPL and the LPC are assigned as n_n^s =1.5 and n_l^s =1.65, respectively. The lattice parameters are chosen as L_n = 250 nm, a_n= 0.43 L_n, L_l = 250 nm and a_l = 0.34 L_n. Other parameters are taken as γ_b=1.0 meV, γ_c=0.001 γ_b [28] and g_{mp} =γ_b.

We have calculated the conduction band edges for both crystals. They are found as ε_n^c =1.075 eV and ε_l^c =1.281 eV for the NPC and the LPC, respectively. Note that the conduction band edge of the NPC lies below that of the LPC. This band structure engineering makes photons confine in the wire.

We have taken the length of the square wire as d =1000 nm. The Kerr parameter is taken from reference [28] as n_{3s} =+0.01 μcm^2/W and the intensity of the pump laser field is assigned a value of I = 0.2 W/μcm^2. Using these parameters we have calculated the bound photons states. One bound photon mode is found within the wire and is located at ε_{00} = 1.202 eV. However, when we switch off the pump laser field (i.e. I=0) the bound photon state moves to a new location at ε_{00} = 1.203 eV.

The imaginary susceptibility has been calculated as a function of the probe detuning parameter δ_{ab}. The resonance energy ε_{bc} of the nanoparticles is taken at the bound photon state ε_{00} = 1.202 eV. This gives the EBP detuning parameter a value of $\delta_{mp} = \varepsilon_{mpk_z} - \varepsilon_{bc}$ = 0. It is found that the imaginary susceptibility becomes zero at zero probe detuning. This is a transparent state which occurs due to the strong EBP coupling. This coupling is strong due to the large density of states of photons at the bound states. Similar results are also found in reference [9].

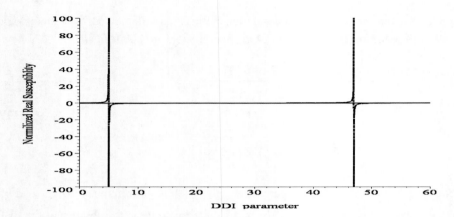

FIGURE 3. The normalized real susceptibility is plotted as a function of the DDI parameter. The left peak and the right peek represent when the pump laser is on ($I = 0.2$ W/μcm^2) and when it is off ($I=0$), respectively.

The normalized real susceptibility (χ'_d / χ_0) is calculated as a function of the DDI parameter when the absorption coefficient is zero. The results are shown in fig. 3. The left and the right peaks correspond to $I = 0.2$ W/μcm^2 and $I=0$, respectively. It is interesting to note that the real susceptibility has a singularity at $C=47\gamma_b$ and $C=5\gamma_b$. This is called the dielectric catastrophe phenomena. It is known in solid state physics that when the system goes from the ferroelectric phase to the paraelectric phase, the dielectric (i.e. susceptibility) has a singularity. It implies that the dielectric catastrophe phenomenon in the present system occurs due to the DDI effect. The dielectric catastrophe can be understood as follows. The real susceptibility is evaluated where the absorption coefficient is zero. This gives F''=0. Putting this condition into eqn. (9) we find

$$\chi'_d = \chi_0 \left(\frac{|F'|}{\left(1 - C_d |F'|\right)} \right) \tag{12}$$

Note that eqn. (12) has a singularity for a certain value of the DDI parameter.

Note that when the pump field is switched on, the dielectric catastrophe occurs at $C=47\gamma_b$. However, when the pump field is switched off the dielectric catastrophe disappears at $C=47\gamma_b$ and reappears at $C=5\gamma_b$. This means that the dielectric catastrophe can be switched ON and OFF by switching the pump field.

CONCLUSIONS

The dielectric catastrophe has been studied in nanophotonic quantum wires. They are made from embedding one photonic crystal into another photonic crystal. It is considered that the embedded crystal is made from nonlinear dielectric spheres. The wire is doped with an ensemble of nanoparticles which are interacting with each other via the dipole-dipole interaction. The absorption coefficient of the wire has been investigated in the presence a weak probe laser and a strong pump laser. Numerical simulations have been performed for the dielectric constant and the photon bound states. It is found that the dielectric constant of the system has a singularity when the resonance energy of a nanoparticle lies near the bound photon states of the wire. This is known as the dielectric catastrophe. This phenomenon can be used to make all-optical switching devices.

ACKNOWLEDGMENTS

This work was supported by the NSERC of Canada.

REFERENCES

1. S. John, Phys. *Phys. Rev. Lett.* **58**, 2486 (1989).
2. E. Yablonovitch, *Phys. Rev. Lett.* **58**, 2059 (1989).
3. J. D. Joannopoulos and J. N. Winn, *Photonic Crystals: Molding the Flow of Light* (Princeton University Press, Princeton, 1995).
4. V. I. Rupasov and Mahi R. Singh, *Phys. Rev. Lett.* **77**, 338 (1996).
5. P. Lambropoulos et al., *Rep. Prog. Phys.* **63**, 455 (2000) *and references within.*
6. P. N. Prasad, *Nanophotonics* (Willey Interscience, New Jersey, 2004).
7. Yuri Nakayama et al., *Nature* **447**, 1096 (2007).
8. B. G. Lee et al., *IEEE Photonic. Tech. L.* **20**, 398 (2008).
9. Mahi R. Singh, *Appl. Phys. B* **93**, 91 (2008).
10. Mahi R. Singh, *Microstructure Journal* (2008) *in press.*
11. Mahi R. Singh, *Phys. Lett. A* **372**, 5083 (2008).
12. D. Suter, *The Physics of Laser-Atom Interactions* (Cambridge University Press, 1997).
13. M. O. Scully and M.S. Zubairy, *Quantum Optics* (Cambridge University Press, Cambridge, 1997).
14. J. P. Dowling and C. M. Bowden, *Phys. Rev. Lett.* **70**, 1421 (1993).
15. A. S. Manka et. al., *Quantum Opt.* **6**, 371 (1994).
16. O. G. Calderon et. al., *Eur. Phys. J. D* **25**, 77 (2003).
17. M.D. Lukin and P.R. Hemmer, *Phys. Rev. Lett.* **84**, 2818 (2000).
18. O. Cakir et. al., *Phys. Rev. A* **71**, 034303 (2005).
19. C. Skornia et. al., *Phys. Rev. A* **64**, 053803 (2005).
20. J. Evers et. al., *Phys. Rev. A* **73**, 023804 (2006).
21. S. N. Wuister et. al., *J. Chem. Phys.* **121**, 4310 (2004); T. Unold et. al., *Phys. Rev. Lett.* **94**, 137404 (2005).
22. G. Ya Slepyan et. al., *Phys. Rev. B***70**, 45320 (2004); E. Paspalakis et. al., *Phys. Rev. B***73**, 73305 (2006).
23. S. John and T. Quang, *Phys. Rev. Lett.* **76**, 2484 (1996).
24. Mahi. R. Singh, *Phys. Rev. A***75**, 043809 (2007).
25. B. A. Saleh and M. C. Teich, *Fundamental of Photonic* (John Wiley and Sons, Inc, New York, 1991) chapter 19.
26. M. Scalora et. al., *Phys. Rev. Lett.* **73**, 1368 (1994).
27. P. Tran, *J. Opt. Soc. B* **14**, 2589 (1997).
28. Mahi R. Singh and R. H. Lipson, *J. Phys. B: At. Mol. Opt. Phys.* **41**, 015401 (2008) *and references within.*

Optical Surface Vortices and Their Use in Nanoscale Manipulation

M. Babiker[a], V. E. Lembessis[b] and D. L. Andrews[c]

[a] Department of Physics, University of York, Heslington, York YO10 5DD, England, UK
[b] New York College of Athens, Athens, GR 105 58, Greece
[c] School of Chemical Sciences, University of East Anglia, Norwich NR 4 7TJ, England, UK

Abstract. Following a brief overview of the physics underlying the interaction of twisted light with atoms at near-resonance frequencies, the essential ingredients of the interaction of atoms with surface optical vortices are described. It is shown that surface optical vortices can offer an unprecedented potential for the nanoscale manipulation of absorbed atoms congregating at regions of extremum light intensity on the surface.

Keywords: Surface Optical Vortices; Nanoscale; Optical Manipulation; Atoms at Surfaces.
PACS: 42.50Tx; 42.50.Wk; 78.68+m; 37.10.De; 37.10.Vz

INTRODUCTION

With a combination of simple optical elements and a suitable laser, it is now possible to produce beams of light possessing unusual twisting properties [1-4]. In particular, the family of optical modes known as Laguerre-Gaussian beams, any desired member of which can now be readily created in the laboratory, is distinguished by an optical torque that is determined by the beam structure and phase properties. The discovery that these so-called 'twisted beams', also known as 'optical vortices', can produce turning effects on individual atoms, ions and molecules, and even larger particles, is arguably one of the most significant advances in optics in recent years. The twisting property is directly related to the orbital angular momentum carried by the light beams, which is in addition to any spin angular momentum associated with their wave polarisation. Since the discovery [1], this previously scarcely explored feature of laser light has generated much interest, not just from a fundamental point of view, but also with a view to its possible applications, especially the micro- and nano-manipulation of matter. The best known such application of twisted light to date has earned it the soubriquet 'optical spanner' [4], signifying a device capable of rotating small particles. Optical spanners are based on similar principles to optical tweezers [5] in which conventional laser light is capable of influencing the translation of small particles using the radiation pressure force. The rotations generated in optical spanners are due to the angular momentum property of the light beam. The essential physics underlying atomic manipulation by twisted light is based on the principles of cooling and trapping of neutral atoms and ions by ordinary laser beams. Theoretical work [6] has already established the existence of a light-induced torque acting on the

CP1147, *Transport and Optical Properties of Nanomaterials—ICTOPON - 2009*, edited by M. R. Singh and R. H. Lipson
© 2009 American Institute of Physics 978-0-7354-0684-1/09/$25.00

centre of mass of an atom possessing a transition at near-resonance with the frequency of a Laguerre-Gaussian light beam carrying an orbital angular momentum $\hbar l$. The saturation form of this torque is $T \sim \hbar \Gamma l$, where Γ is the width of the upper atomic state. The effects of this twisting effect on the motion of atoms and ions have been explored in a number of studies [7-10] leading to predictions of a localization of atoms in high intensity or dark regions of the light beam, suggesting a variety of possibilities for atom manipulation and trapping at the micro and nano-scales.

Recent experimental work on the optical interference between different Laguerre-Gaussian light beams has revealed a rich variety of intensity distributions. Two co-propagating beams of equal and opposite values of l, but with slightly different frequencies produce a so-called 'optical Ferris wheel', in which the pattern rotates at frequency that depends on l and the frequency separation of the two beams [11].

SURFACE VORTICES

The already established properties and behavior of optical vortices suggest that interesting effects might arise in an interference situation that involves twisted light suffering internal reflection at the surface between an optically dense medium and a vacuum. To our knowledge, this simple and easily realizable experimental scenario has not been investigated, nor has there been an exploration of its possible implications for surface atom manipulation using Laguerre-Gaussian light. Evanescent light due to conventional beams has, however, been successfully used in atomic mirrors [12] its use for manipulation at surfaces being considered for small objects [13-15], rather than atoms and molecules at near-resonance . For Laguerre-Gaussian beams, we show here that the evanescent light emerging in the vacuum region decays exponentially with distance normal to the surfaces. Significantly, its in-plane field distribution possesses the angular momentum properties of the incident beam. The emanating fields also include a plane surface wave with a wave-vector equal to the in-plane component of the axial wave-vector of the incident beam. The overall field distributions form a surface optical vortex, endowed with an orbital angular momentum. A system of well defined incident beams, both of which are counter-propagating or both are co-propagating, can, in principle, be made to generate a corresponding set of evanescent surface optical vortices. There is, therefore, considerable scope for a variety of interference effects, possibilities for surface atom manipulation, and new surface optical tools akin to optical spanners and tweezers.

To begin, we specify the electric field of a Laguerre-Gaussian beam characterized by the Laguerre polynomial integer parameters l and p; the beam is assumed to be traveling along the z-axis in a medium of a constant refractive index n (the frequency dependence of the latter left implicit). Let ω be the frequency of the light and $k = n\omega/c$ the axial wavevector. For light plane-polarized along the y axis the field vector can be written in cylindrical coordinates as;

$$\boldsymbol{E}_{klp}^{l}(r,\phi,z) = \hat{\boldsymbol{y}}\, F_{klp}(r,z)e^{i(kz-\omega t)}e^{il\phi} \tag{1}$$

where F_{klp} is an envelope function of r and z . For the amplitude we have;

$$F_{klp}(r,z) = \frac{\varepsilon_{k00}}{(1+z^2/z_R^2)^{1/2}} \left(\frac{r\sqrt{2}}{w(z)}\right)^{|l|} L_p^{|l|}\left(\frac{2r^2}{w^2(z)}\right) e^{-r^2/w^2(z)} e^{i\Theta_{GR}(z)} \qquad (2)$$

Here ε_{k00} is the amplitude for the corresponding plane wave of wave-vector k; and $w(z)$ is the beam waist at axial coordinate z such that $w^2(z) = 2(z^2 + z_R^2)/kz_R$ where z_R is the Rayleigh range. The last phase factor in Eq. 2 exhibits the Guoy phase as well as the change in beam curvature with axial position. We also have;

$$\Theta_{GR}(z) = \frac{kr^2 z}{2(z^2 + z_R^2)} + (2p + |l| + 1)\tan^{-1}\left(\frac{z}{z_R}\right). \qquad (3)$$

Note that the $z = 0$ plane corresponds to the minimum beam waist $w(z) = w_0$, and that on this plane the Guoy and curvature term in Eq. 3 both vanish.

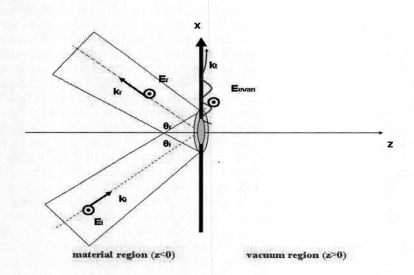

FIGURE 1: Laguerre-Gaussian light internally reflected at an angle ϑ greater than the critical angle (schematic). The incident beam is arranged such that at $\vartheta = 0$ the beam waist coincides with the surface at $z = 0$. The evanescent light possesses angular momentum properties, but is confined near the surface, exponentially decaying in the direction normal to the surface.

Such a light field can be arranged, as shown in **FIGURE 1**, to strike the internal planar surface of a medium in which it is propagating, the surface assumed to be in contact with a vacuum. If the interface with the vacuum occupies the plane $z = 0$ and the angle of incidence, ϑ, exceeds the total internal reflection angle, an evanescent mode is created in the vacuum. The main requirement for the formation of this wave is the applicability of the standard phase-matching boundary condition, along with a requirement for the electric field vector component tangential to the surface to be continuous across the boundary. To be able to define the evanescent electric field we must first obtain expressions that are appropriate for a beam incident at an angle ϑ. This is achievable by a simple rotational transformation of the expressions given in Eqs. (1) to (3).

The fields of a Laguerre-Gaussian beam propagating in a general direction can be constructed from the formalism outlined above by the application of two coordinate transformations, the first transformation rotating the beam as a rigid body about the y-axis by angle α and the second one rotates the resultant beam in the z plane about the x axis by an angle β. The compound transformation amounts to the following relations;

$$x \rightarrow x' = x\cos\alpha + z\sin\alpha$$
$$y \rightarrow y' = -x\sin\alpha\sin\beta + y\cos\beta + z\cos\alpha\sin\beta$$
$$z \rightarrow z' = -x\sin\alpha\cos\beta - y\sin\beta + z\cos\alpha\cos\beta \qquad (4)$$

By a suitable choice of the angles α and β we can determine the field distributions of the incident ($\alpha = \vartheta; \beta = 0$) and the internally reflected light ($\alpha = \pi - \vartheta; \beta = 0$), as exhibited in **FIGURE 1**. The continuity of the electric field vector tangential to the surface, along with the exponential decay with the coordinate z together determines the form of the evanescent filed in the vacuum region. We have;

$$\boldsymbol{E}_{klp}^{evan}(x,y,z) = 2\boldsymbol{E}_{klp}^{I}(x \rightarrow x\cos\vartheta; y; z \rightarrow -x\sin\vartheta)e^{-zk_0\sqrt{n^2\sin^2\vartheta-1}}e^{-ik_0nx\sin\vartheta} \qquad (5)$$

The explicit form of the evanescent electric field that displays the angular momentum properties, as well as the mode characteristics is as follows;

$$\boldsymbol{E}_{klp}^{evan}(x,y,z) = \hat{\boldsymbol{y}}\frac{\varepsilon_{k00}}{(1+x^2\sin^2\vartheta/z_R^2)^{1/2}}\left(\frac{\sqrt{2(x^2\cos^2\vartheta+y^2)}}{w_0^2(1+x^2\sin^2\vartheta/z_R^2)^{1/2}}\right)^{|l|}$$
$$\times \exp\left(\frac{-(x^2\cos^2\vartheta+y^2)}{w_0^2(1+x^2\sin^2\vartheta/z_R^2)^{1/2}}\right)L_p^{|l|}\left(\frac{x^2\cos^2\vartheta+y^2}{w_0^2(1+x^2\sin^2\vartheta/z_R^2)^{1/2}}\right) \qquad (6)$$
$$\times \exp\left(-zk_0\sqrt{n^2\sin^2\vartheta-1}-ik_0nx\sin\vartheta\right)\exp[il\tan^{-1}(y/x\cos\vartheta)]$$

Note that the above expression for the evanescent light bears a vorticity nature in that it is characterized by the azimuthal phase dependence $\exp[il\tan^{-1}(y/x\cos\vartheta)]$ of the incident and internally reflected light. The important point to bear in mind in this context is that the field distribution associated with this vortex is concentrated on the surface, rather than axially, as in the normal Laguerre-Gaussian light in an unbounded

41

space.

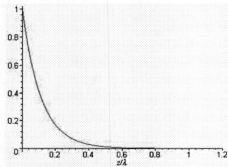

FIGURE 2: Intensity variations of the evanescent surface optical vortex (arbitrary units) with distance normal to the surface (in units of wavelength λ). This variation is independent of the values of l and p.

The typical exponential decay of the evanescent light is shown in **FIGURE 2**. It is seen that the length scale of the exponential decay along z spans a small fraction of the wavelength. In consequence, the evanescent light plays an unimportant role in the trapping normal to the surface – since adsorbed atoms are subject to the much stronger attractive van der Waal's potential.

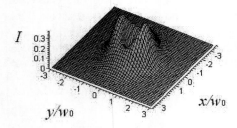

FIGURE 3: Intensity distribution (arbitrary units) on the surface $z = 0$ for the surface vortex arising from internally reflecting a Laguerre-Gaussian light beam for which $l = 1$ and $p = 0$. Distances are measured in units of the beam waist w_0. The light has a wavelength $\lambda = 589.16$ nm and beam waist is taken as $w_0 = 35\,\lambda$, corresponding to a long Rayleigh range $z_R \approx 2.27 \times 10^{-3}$ m. The refractive index is assumed to be $n = \sqrt{2}$ and the angle of internal reflection is taken as $\vartheta = \pi/3$, which is greater than the critical angle.

The intensity distribution corresponding to the evanescent light created by an incident Laguerre-Gaussian beam for which $l = 1$; $p = 0$ is shown in **FIGURE 3**, plotted in the x-y plane at $z = 0$. It is seen that the evanescent light possesses well-defined intensity maxima and minima that can be used to trap adsorbed atoms that have transition frequencies appropriately detuned from the frequency ω of the light. Note that the profile of the intensity distribution is no longer circular, but elliptical, because the light strikes the surface at the angle of incidence ϑ, with an ellipticity increasing with increasing ϑ.

42

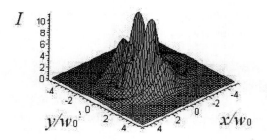

FIGURE 4: As in Figure 3, but here the Laguerre-Gaussian beam is such that $l = 3$ and $p = 2$.

The case $l=3;\ p=2$, with the same parameters as in earlier figures, is shown in **FIGURE 4**. This case presents a more complex field distribution for the corresponding surface optical vortex. Clearly the field distributions for any values of the parameters l and p can be obtained in a similar manner, suggesting that it is in principle possible to create a surface vortex of any order, confirming that the surface optical vortex phenomenon is quite general.

Consider now the case of co-propagating incident beams of opposite helicity creating an interference of two surface vortices, in a manner similar to that discussed recently for beams in an unbounded space, leading to the phenomenon of an optical Ferris wheel [11]. It is easy to see that the total electric field generated in the vacuum region contains the following factor;

$$\cos[l \tan^{-1}(y/x\cos\vartheta)] \tag{7}$$

The appearance of the cosine term in Eq. 7 indicates an interference of the two evanescent light beams in the azimuthal direction.

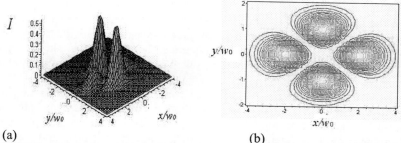

(a)

(b)

FIGURE 5: (a) Intensity distribution (arbitrary units) on the surface at $z = 0$ arising from two co-propagating internally reflected Laguerre-Gaussian light beams for which $l_1 = 2; p_1 = 0$ and $l_2 = -2; p_2 = 0$ (b) the corresponding contour plot. The variables and parameters are as in FIGURES

3 and 4. If the beams are slightly different in frequencies such that $\omega_1 - \omega_2 = \delta$, the intensity distribution pattern is predicted to rotate at a frequency $\delta/(2\,|\,l_1 - l_2\,|)$.

FIGURE 5 shows the field distributions for the case $l_1 = 2; p_1 = 0$ and $l_2 = -2; p_2 = 0$. If the frequencies of the beams differ slightly, the pattern will rotate at a frequency $\delta/(2\,|\,l_1 - l_2\,|)$ with δ the frequency difference, in a similar manner to that illustrated in the case of optical Ferris wheel in unbounded pace [11].

CONCLUSIONS

In conclusion, we have outlined the essential physics underlying the new phenomenon of surface optical vortices, associated particularly with Laguerre-Gaussian light. We have shown that the intensity distribution for a typical surface optical vortex has a limited spatial extent in the direction normal to the surface in the vacuum region, extending significantly over a distance of the order of a fraction of a wavelength. However, the in-plane distributions mimic those of the incident internally reflected light and carry the signature of orbital angular momentum of the incident beam. Since the optical vortices of a Laguerre-Gaussian kind form a family of vortices of any desired order defined by the integers l and p, the phenomenon is quite general.

In addition to the intrinsic importance of surface optical vortices as physical entities in their own right – especially the fact that they carry the property of orbital angular momentum – there are some applications that can be envisaged at this stage. We have briefly mentioned the manipulation of adsorbed atoms and molecules held on the surface by the strong van der Waal's force, but these are commonly able to migrate along the surface. There has recently been strong interest in the manipulation of such atoms on surfaces, but none of the methods suggested so far include near-resonance optical manipulation along the surface. Optical vortices have the potential for the manipulation of adsorbed atoms, forcing them to congregate in extremum regions of intensity on the surface. The same principle can be employed to create patterned surfaces by employing carefully designed sets of incident beams creating a network of evanescent light wells. Moreover, the same system can be used to manipulate larger objects on the surface as optical tweezers and optical spanners, by moving the spot at which the incident beam strikes the interface. Finally, it can be observed that surface optical vortices of other kinds can be generated from other optical beams carrying orbital angular momentum, and that enhancement effects of the evanescent light can be realized by introducing a metallic film, leading to an evanescent surface plasmon mode [16] in which the filed intensities can be at least an order of magnitude larger. Work on surface plasmon optical vortices and guided modes is now in progress and the results will be reported in due course.

ACKNOWLEDGMENTS

VEL wishes to thank the ESF for financial support under the programme QUDEDIS Exchange Grant 1750, while this work was carried out.

REFERENCES

1. L. Allen, M. W. Beijesbergen, R. J. C. Spreeuw, and J. P. Woerdman, *Phys. Rev. A* **45**, 8185 (1992).
2. L. Allen, M. J. Padgett, and M. Babiker, *Prog. Optics*, **vol. XXXIX**, pp291-372 (1999).
3. L. Allen, S. M. Barnett, and M. J. Padgett, `**Optical Angular Momentum**` (Institute of Physics Publishing, 2003).
4. D. L. Andrews, Ed. `**Structured Light and its Applications, an Introduction to Phase-structured Beams and Nanoscale Optical Forces** ` (Academic, Burlington MA, 2008).
5. Special journal issue on optical tweezers, *J. Mod. Opt.* **50**, 1501ff (2003).
6. M. Babiker, W. L. Power and L. Allen, *Phys. Rev. Lett.* **73**, 1239 (1994).
7. M. Babiker, C. R. Bennett, D. L. Andrews, and L. C. D'avila Romero, *Phys. Rev. Lett.* **89**, 143601 (2002).
8. D.L. Andrews, A.C. Carter, M. Babiker, and M. Al-Amri, *Proc. SPIE* **6131**, 101-110 (2006).
9. A.R. Carter, M. Babiker, M. Al-Amri, and D. L. Andrews, *Phys. Rev. A* **72**, 043407 (2005).
10. A. R. Carter, M. Babiker, M. Al-Amri, and D. L. Andrews, *Phys. Rev. A* **73**, 021401 (2006).
11. S. Franke-Arnold, J. Leach, M. J. Padgett, V. E. Lembessis, D. Ellinas, A. J. Wright, J. M. Girkin, P. Ohberg and A. S. Arnold, *Opt. Express* **15**, 8619 (2007).
12. See C. R. Bennett, J. B. Kirk and M. Babiker, *Phys. Rev. A* **63**, 033405 (2001), and references therein
13. V. Garce's-Chavez, K. Dholakia and G. C. Spalding, *Appl. Phys. Lett.* **86**, 031106 (2005).
14. P. J. Reece, V. Garce's-Chavez and K. Dholakia, *Appl. Phys. Lett.* **88**, 221116 (2006).
15. L. C. Thomson, G. Whyte, M. Mazilu and J. Courtial, *J. Opt. Soc. Am. B* **25**, 849 (2008).
16. W. L. Barnes, A. Dereux and T. W. Ebbesen, *Nature* **424**, 824 (2003).

Optical Spectra of the Jaynes-Cummings Ladder

Fabrice P. Laussy* and Elena del Valle[†]

*School of Physics and Astronomy, University of Southampton, United Kingdom
[†]Departamento de Fisica Teorica de la Materia Condensada (C–V), Universidad Autonoma de Madrid, Spain

Abstract. We explore how the Jaynes-Cummings ladder transpires in the emitted spectra of a two-level system in strong coupling with a single mode of light. We focus on the case of very strong coupling, that would be achieved with systems of exceedingly good quality (very long lifetimes for both the emitter and the cavity). We focus on the incoherent regime of excitation, that is realized with semiconductors quantum dots in microcavities, and discuss how reasonable is the understanding of the systems in terms of transitions between dressed states of the Jaynes-Cummings Hamiltonian.

Keywords: Jaynes-Cummings, Strong-Coupling, Microcavity, Quantum Dot, cavity QED, Spectroscopy
PACS: 42.50.Ct, 78.67.Hc, 32.70.Jz, 42.50.Pq

INTRODUCTION

The crowning achievement of cavity Quantum Electrodynamics is the so-called *Strong Coupling* (SC), a regime dominated by quantum interactions between light (photons) and matter (an atom, or its semiconductor realization, a quantum dot (QD), sometimes called an artificial atom) [1]. In this regime, both the atom and the photon loose their individual identity, and vanish to give rise to new particles—sometimes called *polaritons*—with new properties. In this text, we shall be concerned with their optical spectral properties.

Strong light-matter coupling originated with atomic cavities [2, 3], and was later reproduced (among other systems) with semiconductor cavities [4], more promising for future technological applications (see for instance [5]). It is only recently, however (c. 2004), that this regime was reached for the zero-dimensional semiconductor case [6, 7, 8] and the number of reports has not been overwhelming ever since [9, 10, 11, 5]. In these systems, the QD excitation—the exciton—strongly couples to the single mode of a microcavity (realized as a pillar, a photonic crystal or other variants). As far as cavity QED is concerned, these systems are in principle superior to their two-dimensional counterpart (where SC is routinely achieved), because only a few excitations enter the problem in an environment with much reduced degrees of freedom, as opposed to planar polaritons whose most adequate description is in terms of continuous fields.[1]

The landmark of SC is the Rabi doublet, where two modes (light and matter) at

---

[1] Here we must outline that we mean "quantum" in the sense of quantization of the fields, and thus breaking the classical picture. 2D polaritons have demonstrated stunning properties rooted in quantum physics, such as Bose-Einstein condensation or superfluid motion [12, 13]. However, those are manifestations of macroscopic coherence where large numbers of microscopic particles exhibit the behaviour of a continuous field (classical or not).

CP1147, *Transport and Optical Properties of Nanomaterials—ICTOPON - 2009*, edited by M. R. Singh and R. H. Lipson
© 2009 American Institute of Physics 978-0-7354-0684-1/09/$25.00

resonance, do not superimpose but split, each line corresponding to one of the polaritons that have overtaken the bare modes. It was early understood that this splitting is by itself, however, not a proof of quantization [14]. Parenthetically, 2D polaritons display most eloquently this splitting (see footnote 1).

Antibunching in the optical emission of a strongly coupled QD-microcavity system has been demonstrated [9, 10], further supporting quantization of the fields, but this is not completely conclusive, as although it proves that the dynamics involves a single quantum of excitation between two isolated modes (by itself already a considerable achievement), it does not instruct on the modes themselves (consider the vacuum Rabi problem of two harmonic oscillators, that gives the same result). After all, dimming classical light until single photons remain, would exhibit antibunching, but this says nothing about the emitter itself (which is ultimately quantum anyway; if it's coming from the sun, say, it originates from the spontaneous emission of an atom, or to much lower probability, from stimulated emission).

A genuine, or quantum, SC [15], should culminate with a direct, explicit demonstration of quantization, with one quantum more or less changing the behaviour of the system. The most fundamental model to describe light (bosons)/matter (fermions) interactions is the celebrated *Jaynes-Cummings* (JC) Hamiltonian [16], where such a quantum sensitivity is strongly manifest, and has been observed more or less directly in various systems [17, 18]. Recently, direct spectroscopic evidence has been reported for atoms and superconducting circuits, in elaborate experiments [19, 20] that remind the heroic efforts of Lamb to reveal the splitting of the orbitals of hydrogen. With semiconductor QDs, there has been so far, to the best of our knowledge, no explicit demonstration of JC nonlinearities. In a previous work [21], we analyzed the peculiarities of these systems both with respect to their particular physics (involving a steady state under incoherent excitation rather than spontaneous emission or coherent excitation) and their parameters. We concluded that even present-day structures could evidence anharmonicity of the quantized levels with particular pumping schemes and a careful analysis of the data. With much better structures but still conceivably realizable in the near future, strong qualitative signatures emerge and there is no need to go further than a simple observation of anharmonically spaced multiplets. Here we take the further step to go towards unrealistically good (as of today!) structures, with negligible exciton decay and quality factors orders of magnitude higher than state of the art system. In this regime of very strong coupling, one expects a priori Lorentzian emission of the dressed states [22]. We show with an exact quantum-optical computation of the spectra the value and limitations of this approximation. Our results below can be seen as the ideal quantum limit of the Jaynes-Cummings model of light and matter, where the quantization of the fields appears in its full bloom.

THE JAYNES-CUMMINGS PHYSICS

The Jaynes-Cummings Hamiltonian is a textbook model, that admits essentially analytical solutions. It reads:

$$H = \omega_a a^\dagger a + \omega_\sigma \sigma^\dagger \sigma + g(a^\dagger \sigma + a \sigma^\dagger) . \tag{1}$$

Here, $\omega_{a,\sigma}$ are the free energies for the modes a (boson) and σ (fermion) and g is their coupling strength. To describe realistically an experiment, one needs at least to include dissipation. Excitation is then typically assumed as an initial state, or with a coherent (Hamiltonian) pumping. To address the semiconductor case, we considered incoherent excitation [21], i.e., a rate P_σ of QD excitation.[2] The equation of motion for the density matrix ρ then reads $\partial_t \rho = i[H, \rho] + \sum_{c=a,\sigma} \gamma_c (2c\rho c^\dagger - c^\dagger c\rho - \rho c^\dagger c) + P_\sigma (2\sigma^\dagger \rho \sigma - \sigma\sigma^\dagger \rho - \rho\sigma\sigma^\dagger)$. So far we have not been able to provide an analytical solution for the steady state density matrix of this equation, but in the tradition of the JC, many exact results can nevertheless be extracted, for instance the eigenvalues of H in presence of dissipation, in Fig. 1(a). These give the energies of the states that are renormalized by the light-matter interaction, Eq. (1). For a given number n of excitations, there are two new eigenstates $|n+\rangle$ and $|n-\rangle$ that take over the bare modes $|n \text{ photon(s)}, 0 \text{ exciton}\rangle$ and $|n-1 \text{ photon(s)}, 1 \text{ exciton}\rangle$. The polaritons are, in their canonical sense, the states $|1, \pm\rangle = (|10\rangle \pm |01\rangle)/\sqrt{2}$. For each n, the two new states are splitted by $2\sqrt{n}g$. This is the main manifestation of quantization in the JC Hamiltonian: the difference from $\sqrt{n}$ to $\sqrt{n+1}$, when n is small, can be detected in a careful experiment. This structure of renormalized states is called the *Jaynes-Cummings ladder*. The transitions between its stairs account for the lines that are observed in an optical luminescence spectrum. This generates the structure shown on Fig. 1(b).[3]

The transitions between the same kind of dressed states ("same-states transitions") of two adjacent steps (or *manifolds*, mathematically speaking), e.g., from $|n+\rangle$ to $|n-1, +\rangle$, emit at the energy $\sqrt{n+1} + \sqrt{n}$ (in units of g), while the transitions between different kind of dressed states ("different-states transitions")e.g., from $|n+\rangle$ to $|n-1, -\rangle$ emit at $\sqrt{n+1} - \sqrt{n}$. Different-states transitions emit beyond the Rabi doublet while the same-states ones pack-up in between.

To obtain the exact emission spectra of this system, one needs to compute two times correlators $G_a^{(1)}(t, \tau) = \langle a(t+\tau)a(\tau)\rangle$ for the cavity emission and $\langle \sigma(t+\tau)\sigma(\tau)\rangle$ for the direct exciton emission (in the steady state, the limit $t \to \infty$ is taken). These two cases correspond to the geometry of detection in an ideal spectroscopic measurement: the cavity spectrum $S_a(\omega)$ (the τ-Fourier Transform of $G_a^{(1)}$) corresponds to detection of the cavity mode—this would be along the cavity axis of a pillar microcavity, for instance—while $S_\sigma(\omega)$ (FT of $G_\sigma^{(1)}$) corresponds to direct emission of the exciton, like the top emission of a photonic crystal or the side of a pillar. We present both cases in the results below.

---

[2] We also considered a rate P_a of incoherent *photon* excitation, due to other dots or any other source populating the cavity. For more dissipative cases, this term bears a crucial importance on the spectral shapes. However in the very strong coupling, its role is of a less striking character and we therefore ignore it in the present discussion.

[3] This is only the imaginary part. The real part, that corresponds to broadening of the transitions, is given in Ref. [21].

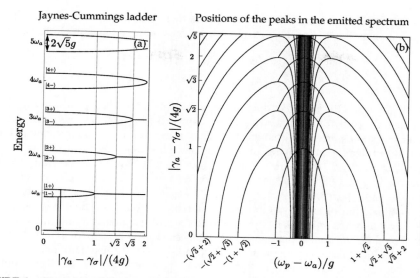

FIGURE 1. The Jaynes-Cummings ladder (a) and the positions of peaks in its emitted spectrum (b), obtained from the difference in energy of transitions between any two branches of two consecutives "stairs" (or manifolds) of the ladder. Without dissipation, the splitting of the nth stair is $2\sqrt{n}g$ and transitions produce an infinite sequence of peaks at $\pm\sqrt{n}\pm\sqrt{n+1}$, an infinitely countable number of them piling up towards 0 from above and below. With increasing $|\gamma_a - \gamma_\sigma|$, the steps go one after the other into weak-coupling, producing a complex diagram of branch-coupling in the resonances of the emitted spectrum (b).

RESULTS

We consider very good cavities with $g = 1$ (defining the unit), γ_a of the order of 10^{-3}, 10^{-2} and $\gamma_\sigma = 0$ (results are not significantly modified qualitatively for nonzero values of γ_σ of the order of γ_a). We consider small electronic pumping P_σ of the same order of γ_a. At higher pumpings, the system goes into the classical regime, with lasing, very high populations, and continuous fields replacing discretization [21]. We therefore focus on small pumping rates, as we are interested in manifestations of quantization. Those are neatly displayed, as seen on Fig 2, where very sharp (owing to the small decay rates) lines reconstruct the transitions of the JC ladder. In the plot of the cavity emission, Fig 2(a), we have marked each peak with its corresponding transition between two quantized, dressed states of the JC hamiltonian. These peaks correspond one-to-one with those of the exciton emission, that are, however, weighted differently [21]. A simple argument explains the different weights in the intensity of the lines in the cavity emission I_a as compared to their counterpart in the exciton emission I_σ:

$$I_a^{(\pm\to\mp)} \propto |\langle n-1, \mp|a|n,\pm\rangle|^2 = |\sqrt{n} - \sqrt{n-1}|^2/4, \tag{2a}$$

$$I_a^{(\pm\to\pm)} \propto |\langle n-1, \pm|a|n,\pm\rangle|^2 = |\sqrt{n} + \sqrt{n-1}|^2/4, \tag{2b}$$

i.e., intensity is enhanced for same-peaks but smothered for different-peaks transitions, while on the other hand, the QD emission is level regardless of the configuration:

$$I_\sigma^{(\pm \to \mp)} \propto |\langle n-1, \mp| \sigma |n, \pm\rangle|^2 = 1/4, \tag{3a}$$

$$I_\sigma^{(\pm \to \pm)} \propto |\langle n-1, \pm| \sigma |n, \pm\rangle|^2 = 1/4. \tag{3b}$$

In both cases, the Rabi doublet dominates strongly over the other peaks. In Fig. 2(a), for instance, the peaks at ± 1 extend for about 9 times higher than is shown, and already the outer transitions are barely noticeable. This is because the pumping is small and so also the probability of having more than one photon in the cavity (it is in this configuration of about 10% to have 2 photons, see Fig. 5). One could spectrally resolve the window $[-g/2, g/2]$ over a long integration time and obtain the multiplet structure of nonlinear inner peaks, with spacings $\{\sqrt{n+1} - \sqrt{n}, n > 1\}$ (in units of g), observing direct manifestation of single photons renormalizing the quantum field. Or one could increase pumping (as we do later) or use a cavity with smaller lifetime. In this case, less peaks of the JC transitions are observable because of broadenings mixing them together, dephasing and, again, reduced probabilities for the excited states, but the balance between them is better. In Fig. 3, where γ_a is now $g/100$, the Rabi doublet (marked R) is dominated by the nonlinear inner peaks in the cavity emission, and a large sequence of peaks is resolved in the exciton emission.

Going back to the case of Fig. 2, but increasing pumping, we observe the effect of climbing higher the Jaynes-Cummings ladder. Results are shown in Fig. 4 in logarithmic scale, so that small features are magnified. First row is Fig. 2 again but in log-scale, so that the effect of this mathematical magnifying glass can be appreciated. Also, we plot over the wider range $[-15g, 10g]$. Note how the fourth outer peak, that was not visible on the linear scale, is now comfortably revealed with another three peaks at still higher energies. As pumping is increased, we observe that the strong linear Rabi doublet is receding behind nonlinear features, with more manifolds indeed being probed, with their corresponding transitions clearly observed (one can track up to the 19th manifold in the last row). This demonstrates obvious quantization in a system with a large number of photons. The distribution of photons in these three cases is given in Fig. 5, going from a thermal-like, mostly dominated by vacuum, distribution, to coherent-like, peaked distribution stabilizing a large number of particles in the system. At the same time, note the cumulative effect of all the side peaks from the higher manifolds excitations, absorbing all quantum transitions into a background that is building up shoulders, with the overall structure of a triplet. This is the mechanism through which the system bridges from a quantum to a classical system. The new spectra are reminiscent of the Mollow triplet of resonance fluorescence [23]. These are obtained both in the cavity and the exciton emission, but much more so in the latter. With more realistic parameters, it is indeed only visible in the exciton emission.

In the very strong coupling limit, one could content with a basic understanding of the transitions mechanisms (given by Eqs. (2-3)) and the knowledge of the distribution of particles (cf. Fig. 5), to provide a fairly good account of the final result [22]. This is true for the most basic understanding of the transitions only, essentially valuable to identify peaks in terms of transitions in the JC ladder. Some features of the problem, like interferences, are missing completely. Observe for instance the sharp dip that is retained

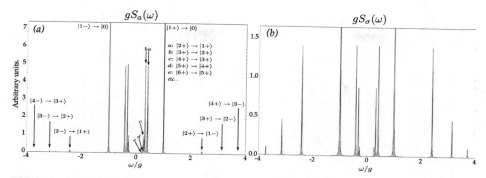

FIGURE 2. Fine structure of the "light-matter molecule": emission spectra in the cavity (a) and direct exciton emission (b) of the strongly-coupled system with $(\gamma_a, \gamma_\sigma)/g = (10^{-3}, 0)$ at $P_\sigma/g = 10^{-3}$.

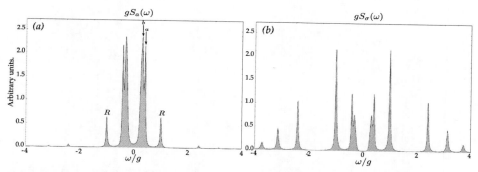

FIGURE 3. Same as Fig. 2 but now with $\gamma_a/g = 10^{-2}$. Less peaks are resolved because of broadening but nonlinear peaks (a, b) are neatly observable. In fact, now inner nonlinear peaks dominate in the cavity emission (the linear Rabi peaks are denoted R). In the exciton direct emission, the Rabi doublet remains the strongest.

in all the exciton spectra. This is a quantum optical result that can be recovered only with an exact treatment of the emission. The qualitative result remains strong enough to support the underlying physics with little needs of these refinements. In the case of less ideal cavities, however, the evidence is not so strong for field-quantization, and both a proper description as well as an understanding of its specificities is then required to support one's conclusions.

CONCLUSION

We have overviewed the problematic of nonlinearities in optical spectra as it is posed by cavity QED, where both fields (atomic and photonic) are quantized. We explored the very strong coupling regime in a system with very small decay rates. We confirm with an exact quantum optical treatment that a qualitative picture in terms of transitions in the

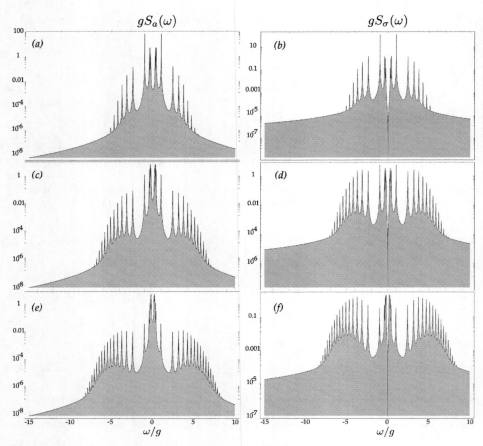

FIGURE 4. Spectra of emission in log-scales as a function of pumping P_σ/g, for 10^{-3} (upper row), 5×10^{-3} (middle) and 10^{-2} (lower row).

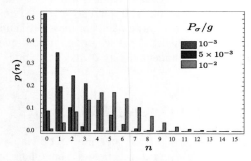

FIGURE 5. Probability $p(n)$ of having n photon(s) in the cavity, for the three cases shown on Fig. 4. Quite independently of the distribution of photon numbers in the cavity, field-quantization is obvious.

Jaynes-Cummings ladder reasonably accounts for the observations. The successful observations of full quantization of the light-matter field merely require high experimental sensitivity and very good samples. In less ideal conditions, a finer analysis is required, as was discussed elsewhere [21].

ACKNOWLEDGMENTS

This work has been initiated in UAM Madrid in the group of Prof. Carlos Tejedor. Results have been computed with the Iridis cluster facility of the University of Southampton.

REFERENCES

1. A. Kavokin, J. J. Baumberg, G. Malpuech, and F. P. Laussy, *Microcavities*, Oxford University Press, 2007.
2. M. G. Raizen, R. J. Thompson, R. J. Brecha, H. J. Kimble, and H. J. Carmichael, *Phys. Rev. Lett.* **63**, 240 (1989).
3. H. Mabuchi, and A. C. Doherty, *Science* **298**, 1372 (2002).
4. C. Weisbuch, M. Nishioka, A. Ishikawa, and Y. Arakawa, *Phys. Rev. Lett.* **69**, 3314 (1992).
5. A. Laucht, F. Hofbauer, N. Hauke, J. Angele, S. Stobbe, M. Kaniber, G. Böhm, P. Lodahl, M. C. Amann, and J. J. Finley, *arXiv:0810.3010v2* (2008).
6. J. P. Reithmaier, G. Sek, A. Löffler, C. Hofmann, S. Kuhn, S. Reitzenstein, L. V. Keldysh, V. D. Kulakovskii, T. L. Reinecke, and A. Forchel, *Nature* **432**, 197 (2004).
7. T. Yoshie, A. Scherer, J. Heindrickson, G. Khitrova, H. M. Gibbs, G. Rupper, C. Ell, O. B. Shchekin, and D. G. Deppe, *Nature* **432**, 200 (2004).
8. E. Peter, P. Senellart, D. Martrou, A. Lemaître, J. Hours, J. M. Gérard, and J. Bloch, *Phys. Rev. Lett.* **95**, 067401 (2005).
9. K. Hennessy, A. Badolato, M. Winger, D. Gerace, M. Atature, S. Gulde, S. Fält, E. L. Hu, and A. Ĭmamoğlu, *Nature* **445**, 896 (2007).
10. D. Press, S. Götzinger, S. Reitzenstein, C. Hofmann, A. Löffler, M. Kamp, A. Forchel, and Y. Yamamoto, *Phys. Rev. Lett.* **98**, 117402 (2007).
11. M. Nomura, Y. Ota, N. Kumagai, S. Iwamoto, and Y. Arakawa, *Appl. Phys. Express* **1**, 072102 (2008).
12. J. Kasprzak, M. Richard, S. Kundermann, A. Baas, P. Jeambrun, J. M. J. Keeling, F. M. Marchetti, M. H. Szymanska, R. André, J. L. Staehli, V. Savona, P. B. Littlewood, B. Deveaud, and Le Si Dang, *Nature* **443**, 409 (2006).
13. A. Amo, D. Sanvitto, F. P. Laussy, D. Ballarini, E. del Valle, M. D. Martin, A. Lemaître, J. Bloch, D. N. Krizhanovskii, M. S. Skolnick, C. Tejedor, and L. Viña, *Nature* **457**, 291 (2009).
14. Y. Zhu, D. J. Gauthier, S. E. Morin, Q. Wu, H. J. Carmichael, and T. W. Mossberg, *Phys. Rev. Lett.* **64**, 2499 (1990).
15. G. Khitrova, H. M. Gibbs, M. Kira, S. W. Koch, and A. Scherer, *Nature Phys.* **2**, 81 (2006).
16. B. W. Shore, and P. L. Knight, *J. Mod. Opt.* **40**, 1195 (1993).
17. M. Brune, F. Schmidt-Kaler, A. Maali, J. Dreyer, E. Hagley, J. M. Raimond, and S. Haroche, *Phys. Rev. Lett.* **76**, 1800 (1996).
18. D. M. Meekhof, C. Monroe, B. E. King, W. M. Itano, and D. J. Wineland, *Phys. Rev. Lett.* **76**, 1796 (1996).
19. I. Schuster, A. Kubanek, A. Fuhrmanek, T. Puppe, P. W. H. Pinkse, K. Murr, and G. Rempe, *Nature Phys.* **3**, 382 (2008).
20. J. M. Fink, M. Göppl, M. Baur, R. Bianchetti, P. J. Leek, A. Blais, and A. Wallraff, *Nature* **454**, 315 (2008).
21. E. del Valle, F. P. Laussy, and C. Tejedor, *arXiv:0812.2694* (2008).
22. F. P. Laussy, A. Kavokin, and G. Malpuech, *Solid State Commun.* **135**, 659 (2005).
23. B. R. Mollow, *Phys. Rev.* **188**, 1969 (1969).

Nonlinear Electrodynamic Properties Of Graphene

S. A. Mikhailov

Institute of Physics, University of Augsburg, D-86135 Augsburg, Germany

Abstract. A brief overview of theoretical and experimental results on the linear and nonlinear electrodynamic properties of graphene is given. Due to the massless energy dispersion of quasi-particles graphene manifests strongly nonlinear electromagnetic response. It is shown that graphene can be used in electronics as a frequency multiplier for microwave and terahertz generation.

Keywords: graphene, electrodynamic properties, frequency multiplication, nonlinear cyclotron resonance, nonlinear pendulum
PACS: 81.05.Uw, 78.67.-n, 73.50.Fq

INTRODUCTION

Graphene is a recently discovered [1, 2] two-dimensional material with unique physical properties [3, 4]. This is a monolayer of carbon atoms packed in a dense honey-comb lattice [5]. The spectrum of graphene quasi-particles, electrons and holes, is described by an effective Dirac equation with a vanishing effective mass [3]. The massless energy dispersion of graphene quasi-particles leads to many unusual and interesting physical properties of this material, such as the minimal electrical conductivity [1, 2, 6–11], unconventional quantum Hall effect [1, 2, 7, 12–15] and other. Optical properties of graphene have been also studied experimentally [16–22] and theoretically [7, 14, 15, 23–37].

It has been also shown that the linear energy dispersion of quasi-particles should lead to a strongly nonlinear electromagnetic response of graphene [38–41]. As a consequence, the higher harmonics generation can be observed in this material, which may be used in microwave and terahertz electronics. Another interesting nonlinear phenomenon has been recently predicted in graphene in finite magnetic fields [42, 43]. Being irradiated by a monochromatic electromagnetic wave at the cyclotron frequency, a graphene electron scatters the radiation in a broad frequency range around the incident-wave frequency which should lead to a broad cyclotron resonance line even in the absence of any scattering in the system. In this paper we give a brief overview of these and some other linear and nonlinear electrodynamic phenomena in graphene.

DYNAMICAL CONDUCTIVITY OF GRAPHENE

The Dirac Hamiltonian of graphene has the form $\hat{H}_0 = V \sigma_\alpha \hat{p}_\alpha$, where $V \approx 10^8$ cm/s is the Fermi velocity, $\hat{\mathbf{p}}$ is the momentum operator and σ_α are Pauli matrixes. It determines

CP1147, *Transport and Optical Properties of Nanomaterials—ICTOPON - 2009*, edited by M. R. Singh and R. H. Lipson
© 2009 American Institute of Physics 978-0-7354-0684-1/09/$25.00

the spectrum of graphene quasi-particles

$$E_{\mathbf{k}l} = (-1)^l \hbar V k = (-1)^l \hbar V \sqrt{k_x^2 + k_y^2} \tag{1}$$

and the corresponding spinor wave-functions

$$|\mathbf{k}l\rangle = \frac{1}{\sqrt{2S}} \begin{pmatrix} e^{-i\vartheta_{\mathbf{k}}/2} \\ (-1)^l e^{i\vartheta_{\mathbf{k}}/2} \end{pmatrix} e^{i\mathbf{k}\cdot\mathbf{r}}. \tag{2}$$

Here $l = 1, 2$ is the band index, $\mathbf{k} = (k_x, k_y) = k(\cos\vartheta_{\mathbf{k}}, \sin\vartheta_{\mathbf{k}})$ is the quasi-momentum and S is the sample area. Response of graphene to an external time-dependent electric potential $\phi(\mathbf{r}, t) = \phi_{\mathbf{q}\omega} e^{i\mathbf{q}\cdot\mathbf{r} - i(\omega + i0)t}$ is described, within the linear response theory, by the single particle Liouville equation $i\hbar\partial\hat{\rho}/\partial t = [\hat{H}, \hat{\rho}]$, where $\hat{\rho}$ is the density matrix and $\hat{H} = \hat{H}_0 - e\phi(\mathbf{r}, t)$. After standard manipulations one gets the frequency- and wavevector-dependent current

$$\mathbf{j}_{\mathbf{q}\omega} = \frac{e^2}{2S} \phi_{\mathbf{q}\omega} \sum_{\mathbf{k}\mathbf{k}'ll'} \langle \mathbf{k}l | \hat{\mathbf{v}} e^{-i\mathbf{q}\cdot\mathbf{r}} + e^{-i\mathbf{q}\cdot\mathbf{r}}\hat{\mathbf{v}} | \mathbf{k}'l'\rangle \frac{\mathscr{F}_0(E_{\mathbf{k}'l'}) - \mathscr{F}_0(E_{\mathbf{k}l})}{E_{\mathbf{k}'l'} - E_{\mathbf{k}l} - \hbar(\omega + i0)} \langle \mathbf{k}'l' | e^{i\mathbf{q}\cdot\mathbf{r}} | \mathbf{k}l\rangle \tag{3}$$

which can be used for calculations of response functions [7, 14, 23, 24, 32] and of the spectra of collective excitations (plasmons) [28, 29]; here $\mathscr{F}_0$ is the Fermi distribution function and $\hat{v}_\alpha = V\sigma_\alpha$ is the velocity operator.

The response of graphene to a uniform external electric field ($\mathbf{q} = \mathbf{0}$) is described by the high-frequency conductivity $\sigma_{\alpha\beta}(\omega) = \sigma(\omega)\delta_{\alpha\beta}(\omega)$ which consists of the intraband

$$\sigma_{\alpha\beta}^{intra}(\mathbf{0}, \omega) = \frac{-ie^2}{\hbar^2(\omega + i\gamma)S} \sum_{\mathbf{k}l} \frac{\partial E_{\mathbf{k}l}}{\partial k_\alpha} \frac{\partial \mathscr{F}_0(E_{\mathbf{k}l})}{\partial E} \frac{\partial E_{\mathbf{k}l}}{\partial k_\beta}, \tag{4}$$

and the interband conductivity

$$\sigma_{\alpha\beta}^{inter}(\mathbf{0}, \omega) = \frac{ie^2\hbar}{S} \sum_{\mathbf{k}l \neq l'} \langle \mathbf{k}l | \hat{v}_\alpha | \mathbf{k}l'\rangle \frac{\mathscr{F}_0(E_{\mathbf{k}l'}) - \mathscr{F}_0(E_{\mathbf{k}l})}{E_{\mathbf{k}l'} - E_{\mathbf{k}l} - \hbar(\omega + i0)} \frac{1}{E_{\mathbf{k}l'} - E_{\mathbf{k}l}} \langle \mathbf{k}l' | \hat{v}_\beta | \mathbf{k}l\rangle. \tag{5}$$

In (4) we have replaced $i0$ by $i\gamma$ in order to phenomenologically take into account the scattering rate of electrons. At low temperatures $kT \ll \mu$ the intraband conductivity has a Drude form

$$\sigma_{intra}(\omega) = \frac{in_e e^2 V}{(\omega + i\gamma)p_F} = \frac{e^2}{\pi\hbar} \frac{i\mu}{\hbar(\omega + i\gamma)}, \tag{6}$$

similarly to the systems of electrons with the parabolic dispersion (μ is the Fermi energy). The interband conductivity reads

$$\sigma_{inter}(\omega) = \frac{e^2}{4\hbar} \left[\frac{\sinh\frac{\hbar|\omega|}{2kT}}{\cosh\frac{\mu}{kT} + \cosh\frac{\hbar|\omega|}{2kT}} + i\mathscr{P} \int_0^\infty dx \frac{\cosh\frac{\mu}{kT} + e^{-x}}{\cosh\frac{\mu}{kT} + \cosh x} \frac{\frac{\hbar\omega}{\pi kT}}{x^2 - (\frac{\hbar\omega}{2kT})^2} \right], \tag{7}$$

where $\mathscr{P}$ means the Cauchy principal value.

At low temperatures the real part of the inter-band conductivity $\sigma'_{inter}(\omega) = (e^2/4\hbar)\Theta(\hbar|\omega| - 2\mu)$ has a step at $\hbar|\omega| = 2\mu$ related to the absorption edge at the Fermi level. At high frequencies, $\hbar\omega \gg \max\{kT, \mu\}$, the dynamical conductivity is universal, $\sigma_\infty(\omega) \approx e^2/4\hbar$. It can be directly measured in the transmission experiment of visible light [18, 22, 44]; the transmission coefficient $T = (1 + \pi\alpha/2)^{-2}$ is then determined by only the fine structure constant $\alpha = e^2/\hbar c$. At low frequencies, the inter-band conductivity tends to zero while the intra-band contribution gives

$$\sigma(0) = (e^2/\pi\hbar)(\mu/\hbar\gamma). \tag{8}$$

While the high-frequency limit $\sigma_\infty(\omega) \approx e^2/4\hbar$ and the overall frequency dependence of the conductivity are in good agreement with experiments [18, 22, 44, 20], the theoretical prediction (8) for $\sigma(0)$ is in strong contradiction with experimental observations. In the limit $\mu \to 0$ the experiments show, in contrast to (8), a finite value $\sigma_{min} \approx 2e^2/\pi\hbar$ which does not depend on temperature and on the mobility of samples [1, 2]. The nature of the minimal conductivity of graphene was a subject of intensive theoretical studies [6–11] but the obtained results are controversial and still disagree with experiments.

NONLINEAR ELECTRODYNAMIC EFFECTS IN GRAPHENE

In this Section we assume that the Fermi level lies in the electron band, $\mu > 0$, the temperature is low, $kT \ll \mu$, and the radiation frequency is smaller than the Fermi-energy absorption edge, $\hbar\omega \ll 2\mu$. Under these conditions one can neglect the inter-band transitions and consider the intra-band response quasi-classically [38, 39]. A quantum-mechanical treatment of the problem [41] leads to the same results.

The physics of the nonlinear electromagnetic response of graphene is very simple [38]. Consider a particle with the spectrum (1) in the external electric field $\mathbf{E}(t) = \mathbf{E}_0 \cos\omega t$. Its momentum, according to the Newton's law, is $\mathbf{p}(t) = -(e\mathbf{E}_0/\omega)\sin\omega t$. The velocity of a *massive* particle is proportional to the momentum $\mathbf{p}$, therefore it responds at the same frequency ω. In graphene, however, the velocity of quasi-particles $\mathbf{v}$ is not proportional to the momentum, $\mathbf{v} = \nabla_\mathbf{p} E_\mathbf{pl} = V\mathbf{p}/p$, and

$$v(t) \approx V \text{sgn}(\sin\omega t) = V\frac{4}{\pi}\left(\sin\omega t + \frac{1}{3}\sin 3\omega t + \frac{1}{5}\sin 5\omega t + \ldots\right). \tag{9}$$

The irradiation of a graphene sheet by an external electromagnetic wave with the frequency ω should thus lead to the higher harmonics generation. The above consideration is valid for any frequency ω satisfying the condition $\hbar\omega \lesssim 2\mu$. For typical experimental parameters this leads to the restriction $\omega/2\pi \lesssim 10 - 30$ THz. Thus, graphene could work as a generator (multiplier) of electromagnetic waves in the technologically very important area of sub-terahertz and terahertz radiation. Below we describe the non-linear electromagnetic response of graphene, both in zero and in finite magnetic fields, in some more detail.

Frequency Multiplication in Strong Electric and Zero Magnetic Field

In the quasi-classical approximation the graphene response is described by the kinetic Boltzmann equation for the electron distribution function $f_{\mathbf{p},2}(t) = f_{\mathbf{p}}(t)$,

$$\frac{\partial f_{\mathbf{p}}(t)}{\partial t} - \frac{\partial f_{\mathbf{p}}(t)}{\partial \mathbf{p}} e\mathbf{E}(t) = 0. \tag{10}$$

In this equation we ignore the scattering of electrons by impurities and phonons. The field $\mathbf{E}(t)$ in (10) is the total (self-consistent) electric field $\mathbf{E}(t) = \mathbf{E}^{ext}(t) + \mathbf{E}^{ind}(t)$ which consists of the sum of the external field of the incident electromagnetic wave $\mathbf{E}^{ext}(t)$ and of the induced field $\mathbf{E}^{ind}(t) = -2\pi\mathbf{j}(t)/c$ produced by the electric current $\mathbf{j}(t) = -(4e/S)\sum_{\mathbf{p}} \mathbf{v}f_{\mathbf{p}}(t)$. Eq. (10) has the solution $f_{\mathbf{p}}(t) = \mathscr{F}_0 \ \mathbf{p} + e\int_{-\infty}^{t} \mathbf{E}(t')dt'$ and the self-consistent ac electric current is

$$\mathbf{j}(t) = -\frac{eV}{(\pi\hbar)^2} \int \frac{\mathbf{p}d\mathbf{p}}{p} \mathscr{F}_0 \left(\mathbf{p} + e\int_{-\infty}^{t} \mathbf{E}(t')dt' \right). \tag{11}$$

The above equations describe the nonlinear response of graphene to the external electromagnetic wave. The electric field of the wave is included non-perturbatively and and the radiative damping effects are taken into account.

The results of such a theory depend on two dimensionless parameters, $\mathscr{E} = eE_0V/\mu\omega$ and $\Gamma/\omega = (e^2/\hbar c)(2\mu/\hbar\omega)$. The first parameter characterizes the strength of the external electric field; it equals the energy, which electrons get from the external radiation during one oscillation period (eE_0V/ω), divided by their average energy (μ). If this parameter is small, the system demonstrates the linear response (see Fig. 1 below and Refs. [38, 39]). The calculated ac current contains only the fundamental frequency harmonic ω. If $\mathscr{E} \gg 1$, the higher harmonics generation becomes efficient and the current is $j(t) \approx en_sV\,\mathrm{sgn}(\sin\omega t)$, Eq. (9). The second parameter is the ratio of the radiative decay rate Γ to the frequency of radiation ω and characterizes the suppression of the higher harmonics due to the radiative friction. The system efficiently generates higher harmonics if $\mathscr{E} \gg 1$ and $\mathscr{E} \gg \Gamma/\omega$, which under typical experimental conditions corresponds to the fields $E_0 \gtrsim 100 - 1000$ V/cm.

Frequency Multiplication in Strong Electric and Finite Magnetic Fields

The Boltzmann equation cannot be analytically solved in the presence of a finite external magnetic field $\mathbf{B} = (0,0,B)$. To solve the nonlinear response problem in the magnetic field we use the molecular dynamics approach numerically solving the system of classical equations of motion of N quasiparticles with the linear spectrum (1),

$$\frac{d\mathbf{p}_i}{dt} = -e\mathbf{E}^{ext}(t) - \frac{e}{c}V\frac{\mathbf{p}_i\times\mathbf{B}}{p_i} + \frac{2\pi e}{c}\mathbf{j} - \gamma\mathbf{p}_i, \quad i = 1,\dots,N; \quad \frac{\mathbf{j}}{en_sV} = \frac{1}{N}\sum_{i=1}^{N}\frac{\mathbf{p}_i}{p_i}. \tag{12}$$

In these equations we have included the radiative damping effect (the term proportional to $\mathbf{j}$) and the scattering due to impurities and phonons (the term $-\gamma\mathbf{p}_i$). The solution

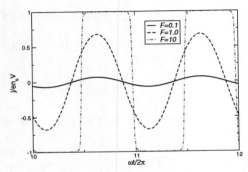

FIGURE 1. Time dependence of the current j_x at $\mathcal{B} = \gamma/\omega = 0$, $\Gamma/\omega = 1$ and different values of $\mathcal{E}$ (the electric field is linearly polarized in the x-direction). The response is linear at $\mathcal{E} \lesssim 1$ and strongly nonlinear at $\mathcal{E} \gg 1$. Calculations have been done for $N = 316$ particles.

of (12) depends on four dimensionless parameters, $\mathcal{E} = eE_0V/\mu\omega$, $\mathcal{B} = eBV^2/\omega\mu c$, Γ/ω and γ/ω. Figure 1 shows results of solving these equations at zero magnetic field $\mathcal{B} = 0$, $\Gamma/\omega = 1$ and $\gamma/\omega = 0$, for the cases of weak, moderate and strong electric fields, $\mathcal{E} = 0.1$, 1, and 10 (the results obtained by the molecular dynamics and the Boltzmann methods coincide with each other). If $B = 0$, the graphene response is linear in weak and nonlinear in strong electric fields. In finite magnetic fields the situation is less trivial, Figure 2. If the electric field is strong (left panel, $\mathcal{E} = 10$), the diagonal current $j_x(t)$ is similar to that at $B = 0$, and the Hall current $j_y(t)$, which also contains higher Fourier harmonics, appears in addition to the diagonal one. In lower electric fields (right panel, $\mathcal{E} = 2$) additional low-frequency (smooth variations of the oscillation amplitudes) and high-frequency harmonics appear in the time dependence of the current. Further reduction of the electric field amplitude leads to even more complicated, chaotic dependencies of the electric current with many higher harmonics [42]. In order to understand the physics of such a chaotic behavior of the graphene response at weak electric and finite magnetic fields, we turn to the problem of the cyclotron resonance of a single massless quasi-particle in graphene.

Cyclotron Resonance of a Massless Quasi-Particle in Graphene

Here we consider the motion of a single massless quasi-particle under the action of the circularly polarized weak external ac electric field $\mathbf{E}_0(t) = E_0(t)(\cos\omega t, \sin\omega t)$, $\mathcal{E} \ll 1$. The equations of motion and the initial conditions read

$$d\mathbf{p}/dt = -(e/c)\mathbf{v} \times \mathbf{B} - e\mathbf{E}(t), \quad \mathbf{p}|_{t=0} = \mathbf{p}_0, \tag{13}$$

the frequency ω coincides with the cyclotron frequency of the particle at the initial instant of time, $\omega_c = eBV/cp_0$, and the scattering is ignored. Introducing the variables $p(t)$ and $\phi(t)$ according to the formulas $[p_x(t), p_y(t)] = p(t)[-\sin(\omega t + \phi(t)), \cos(\omega t +$

58

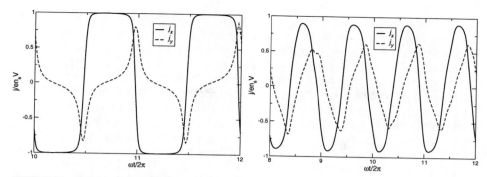

FIGURE 2. Time dependence of the current at $\mathscr{B}=1$, $\gamma/\omega=0$, $\Gamma/\omega=1$ and $\mathscr{E}=10$ (left panel), $\mathscr{E}=2$ (right panel). Calculations have been done for $N=316$ particles.

$\phi(t))]$, we rewrite (13) as

$$\dot{p}(t)=eE_0\sin\phi(t), \quad \dot{\phi}(t)=-\omega+\frac{eVB}{cp(t)}+\frac{eE_0}{p(t)}\cos\phi(t); \quad p|_{t=0}=p_0, \quad \phi|_{t=0}=\phi_0. \quad (14)$$

If $\mathscr{E}\ll 1$, the absolute value of the momentum does not substantially deviate from its initial value p_0, $p(t)=p_0[1+q(t)]$, where $|q(t)|\ll 1$. Then we can write

$$\dot{q}(t)=\omega\mathscr{E}\sin\phi, \quad \dot{\phi}(t)=-\omega q+\omega\mathscr{E}\cos\phi. \quad (15)$$

If the second term in the last equation here is small as compared to the first one, equations (15) are reduced to the well known problem of the nonlinear pendulum,

$$\ddot{\phi}(t)=-\omega^2\mathscr{E}\sin\phi. \quad (16)$$

Dependent on the initial conditions for ϕ_0 Eq. (16) may have different solutions. For example, if $\phi_0\lesssim 1$ we get

$$\phi(t)=\phi_0\cos(\omega\sqrt{\mathscr{E}}t), \quad q(t)=\phi_0\sqrt{\mathscr{E}}\sin(\omega\sqrt{\mathscr{E}}t), \quad \phi_0\lesssim 1 \quad (17)$$

(now one sees that the second term in (15) could be neglected if $\sqrt{\mathscr{E}}\ll\phi_0\lesssim 1$). The phase $\phi(t)$, as well as the momentum $q(t)$ oscillate in time with the low frequency $\omega\sqrt{\mathscr{E}}\ll\omega$ proportional to the square root of the electric field. Important is that the amplitude of the phase oscillations, in contrast to the amplitude of the momentum $q(t)$, does not depend on $\mathscr{E}$. This leads to that non-trivial result that the velocity of quasi-particles

$$\mathbf{v}(t)=V\mathbf{p}(t)/p(t)=V[-\sin(\omega t+\phi(t)),\cos(\omega t+\phi(t))] \quad (18)$$

remains a non-monochromatic function of time even in the limit of the vanishing external electric field $\mathscr{E}\to 0$. Substituting the phase $\phi(t)$ from (17) to (18) and expanding the velocity in the Fourier series we get (for instance for the v_y component)

$$\tilde{v}_y(\Omega)=\frac{V}{2}\sum_{k=-\infty}^{\infty}J_k(\phi_0)\left[i^k\delta(\Omega-\omega-k\omega\sqrt{\mathscr{E}})+(-i)^k\delta(\Omega+\omega-k\omega\sqrt{\mathscr{E}})\right]. \quad (19)$$

Apart from the main Fourier harmonics $\Omega = \omega$ with the amplitude $\sim J_0(\phi_0)$ there always exist the satellite components with the frequencies $\Omega_k = \omega \pm k\omega\sqrt{\mathscr{E}}$. The amplitudes of the satellite harmonics $\sim J_k(\phi_0)$ do not depend on the electric field. Such a non-analytic behavior of the response is a mathematical consequence of the singularity at $p \to 0$ of the Lorentz-force terms in the equations of motion (13), (12) and means that the linear response theory is not applicable to graphene in finite magnetic fields [43]. Experimentally this means that the measured linewidth of the cyclotron resonance may be very broad even in perfectly pure graphene. The linewidth of the cyclotron resonance in graphene and similar systems is determined not only by the scattering rate of electrons or by the radiative damping but also by the nonlinear frequency transformation effects appearing in the system with a strongly non-parabolic (linear in graphene) dispersion. It is worth noting that in the experiments [16, 17] the measured cyclotron resonance line was found to be very broad.

From these results one has to draw an important conclusion concerning the methods of measurement of the electromagnetic response of graphene. A system of massive quasi-particles responds to the monochromatic electromagnetic excitation at the same frequency, i.e. from $S_{inc}(\Omega) \propto \delta(\Omega - \omega)$ follows $S_{scat}(\Omega) \propto \delta(\Omega - \omega)$, where $S_{inc}(\Omega)$ and $S_{scat}(\Omega)$ are the spectra of the incident and the scattered (reflected, transmitted) electromagnetic waves. Experimentally measuring the cyclotron resonance in such a system it is sufficient to use a narrow-band source and a broad-band detector of radiation or a broad-band source and a narrow-band detector. In materials with a strongly non-parabolic energy dispersion (graphene, graphite, etc) the irradiation of the system by a monochromatic wave, $S_{inc}(\Omega) \propto \delta(\Omega - \omega)$, leads to the scattered radiation at many different frequencies around $\Omega \approx \omega$, so that $S_{scat}(\Omega) \neq \delta(\Omega - \omega)$, Ref. [43]. In order to get full information on the electrodynamic response of graphene one needs to make an experiment using a narrow-band source *and* a narrow-band detector.

SUMMARY

We have discussed some linear and nonlinear electrodynamic properties of graphene. The results outlined here are definitely not complete and further experimental and theoretical studies are needed. It is clear however that graphene is a very interesting nonlinear material with unique physical properties which may bring new important results both for our fundamental knowledge and for useful electronics applications.

ACKNOWLEDGMENTS

The work was supported by the Swedish Research Council, INTAS and the Deutsche Forschungsgemeinschaft.

REFERENCES

1. K. S. Novoselov, A. K. Geim, S. V. Morozov, D. Jiang, M. I. Katsnelson, I. V. Grigorieva, S. V. Dubonos and A. A. Firsov, *Nature* **438**, 197–200 (2005).

2. Y. Zhang, Y.-W. Tan, H. L. Stormer and P. Kim, *Nature* **438**, 201–204 (2005).
3. M. I. Katsnelson, *Materials Today* **10**, 20–27 (2007).
4. A. K. Geim and K. S. Novoselov, *Nature Materials* **6**, 183–191 (2007).
5. P. R. Wallace, *Phys. Rev.* **71**, 622–634 (1947).
6. M. I. Katsnelson, *Europ. Phys. J. B* **51**, 157–160 (2006).
7. N. M. R. Peres, F. Guinea and A. H. Castro Neto, *Phys. Rev. B* **73**, 125411 (2006).
8. P. M. Ostrovsky, I. V. Gornyi and A. D. Mirlin, *Phys. Rev. B* **74**, 235443 (2006).
9. K. Ziegler, *Phys. Rev. Lett.* **97**, 266802 (2006).
10. S. Ryu, C. Mudry, A. Furusaki and A. W. W. Ludwig, *Phys. Rev. B* **75**, 205344 (2007).
11. K. Nomura and A. H. MacDonald, *Phys. Rev. Lett.* **98**, 076602 (2007).
12. K. S. Novoselov, Z. Jiang, Y. Zhang, S. V. Morozov, H. L. Stormer, U. Zeitler, J. C. Maan, G. S. Boebinger, P. Kim and A. K. Geim, *Science* **315**, 1379–1379 (2007).
13. V. P. Gusynin and S. G. Sharapov, *Phys. Rev. Lett.* **95**, 146801 (2005).
14. V. P. Gusynin and S. G. Sharapov, *Phys. Rev. B* **73**, 245411 (2006).
15. V. P. Gusynin, S. G. Sharapov and J. P. Carbotte, *J. Phys. Condens. Matter* **19**, 026222 (2007).
16. R. S. Deacon, K.-C. Chuang, R. J. Nicholas, K. S. Novoselov and A. K. Geim, *Phys. Rev. B* **76**, 081406 (2007).
17. Z. Jiang, E. A. Henriksen, L. C. Tung, Y.-J. Wang, M. E. Schwartz, M. Y. Han, P. Kim and H. L. Stormer, *Phys. Rev. Lett.* **98**, 197403 (2007).
18. R. R. Nair, P. Blake, A. N. Grigorenko, K. S. Novoselov, T. J. Booth, T. Stauber, N. M. R. Peres and A. K. Geim, *Science* **320**, 1308–1308 (2008).
19. E. A. Henriksen, Z. Jiang, L. C. Tung, M. E. Schwartz, M. Takita, Y.-J. Wang, P. Kim and H. L. Stormer, *Phys. Rev. Lett.* **100**, 087403 (2008).
20. Z. Q. Li, E. A. Henriksen, Z. Jiang, Z. Hao, M. C. Martin, P. Kim, H. L. Stormer and D. N. Basov, *Nature Physics* **4**, 532–535 (2008).
21. L. M. Zhang, Z. Q. Li, D. N. Basov, M. M. Fogler, Z. Hao and M. C. Martin, *Phys. Rev. B* **78**, 235408 (2008).
22. K. F. Mak, M. Y. Sfeir, Y. Wu, C. H. Lui, J. A. Misewich and T. F. Heinz, *Phys. Rev. Lett.* **101**, 196405 (2008).
23. L. A. Falkovsky and A. A. Varlamov, *Europ. Phys. J. B* **56**, 281–284 (2007).
24. L. A. Falkovsky and S. S. Pershoguba, *Phys. Rev. B* **76**, 153410 (2007).
25. J. Nilsson, A. H. Castro Neto, F. Guinea and N. M. R. Peres, *Phys. Rev. Lett.* **97**, 266801 (2006).
26. V. P. Gusynin, S. G. Sharapov and J. P. Carbotte, *Phys. Rev. Lett.* **96**, 256802 (2006).
27. D. S. L. Abergel and V. I. Fal'ko, *Phys. Rev. B* **75**, 155430 (2007).
28. B. Wunsch, T. Stauber, F. Sols and F. Guinea, *New J. Phys.* **8**, 318 (2006).
29. E. H. Hwang and S. Das Sarma, *Phys. Rev. B* **75**, 205418 (2007).
30. V. Ryzhii, *Jpn. J. Appl. Phys.* **45**, L923–L925 (2006).
31. V. Ryzhii, A. Satou and T. Otsuji, *J. Appl. Phys.* **101**, 024509 (2007).
32. S. A. Mikhailov and K. Ziegler, *Phys. Rev. Lett.* **99**, 016803 (2007).
33. T. Stauber, N. M. R. Peres and A. K. Geim, *Phys. Rev. B* **78**, 085432 (2008).
34. T. Stauber, N. M. R. Peres and A. H. Castro Neto, *Phys. Rev. B* **78**, 085418 (2008).
35. A. Satou, F. T. Vasko and V. Ryzhii, *Phys. Rev. B* **78**, 115431 (2008).
36. F. T. Vasko and V. Ryzhii, *Phys. Rev. B* **77**, 195433 (2008).
37. T. Morimoto, Y. Hatsugai and H. Aoki, *Phys. Rev. B* **78**, 073406 (2008).
38. S. A. Mikhailov, *Europhys. Lett.* **79**, 27002 (2007).
39. S. A. Mikhailov and K. Ziegler, *J. Phys. Condens. Matter* **20**, 384204 (2008).
40. S. A. Mikhailov, *Physica E* **40**, 2626–2629 (2008).
41. F. J. López-Rodríguez and G. G. Naumis, *Phys. Rev. B* **78**, 201406(R) (2008).
42. S. A. Mikhailov, *Microelectron. J.* **40**, 712–715 (2009).
43. S. A. Mikhailov, Non-linear effects in the cyclotron resonance of a massless quasi-particle in graphene, arXiv:0812.0855v1, 2008.
44. A. B. Kuzmenko, E. van Heumen, F. Carbone and D. van der Marel, *Phys. Rev. Lett.* **100**, 117401 (2008).

Colloidal Synthesis and Optical Properties of Lead-Iodide-Based Nanocrystals

Marek Osiński

*Center for High Technology Materials, University of New Mexico, 1313 Goddard SE
Albuquerque, New Mexico 87106-4343, USA*

Abstract. Nanosize devices have attracted tremendous interest over the last few years for a wide range of biomedical, biochemical sensing, and optoelectronic applications. In this paper, we will review recent progress in understanding the behavior of lead-iodide-based nanocrystals and discuss their optical properties. Due to a much better overlap of the electron and hole wavefunctions, the optical transitions in nanocrystals have a much higher probability than in bulk materials, leading to optical emission from materials that in their bulk form do not even emit light at room temperature. Possible applications of lead- halide-based nanocrystals are also discussed.

Keywords: Lead-halide-based nanocrystals; Synthesis of nanocrystals; Optical properties of nanocrystals
PACS numbers: 61.46.Df, 78.67.Bf, 81.07.Bc, 81.16.Be

1. INTRODUCTION

Semiconductor nanocrystals (NCs) or quantum dots (QDs) have been extensively investigated over the last decade for a variety of biomedical, biochemical sensing, and optoelectronic applications. An area that has received relatively little attention so far is the use of NCs as gamma or X-ray detectors in applications such as positron emission tomography (PET), digital radiography, dosimetry, and nuclear medicine. In a typical radiation detection system, conversion of the incident energy of ionizing radiation is accomplished by using scintillating materials that emit photons in the visible/UV spectral range, subsequently collected by a photosensitive element.

Compared to currently used bulk crystal scintillators, NCs offer the prospect of significantly improved performance. Due to their small size, they are expected to have better solubility in organic polymer or inorganic sol-gel host materials and to cause much less scattering, which should result in higher efficiency of the scintillator. Due to three-dimensional confinement and much better overlap of electron and hole wavefunctions, the optical transitions are expected to have higher efficiency and faster rate than in bulk scintillators, which should eliminate the major problem of relatively slow response of scintillator detectors.

Heavy-meal halides, and in particular lead-based compound NCs, are of interest as potential novel scintillation materials due to their high density and the high atomic

CP1147, *Transport and Optical Properties of Nanomaterials—ICTOPON - 2009,* edited by M. R. Singh and R. H. Lipson
© 2009 American Institute of Physics 978-0-7354-0684-1/09/$25.00

number of Pb. In addition, all naturally occurring isotopes of lead are non-radioactive, thus there is no intrinsic radiation background associated with the use of lead in a scintillator. While the bulk materials may have poor efficiency of light emission at room temperature, the effects of quantum confinement are expected to greatly enhance the probability of radiative transitions, as well as reduce the radiative recombination lifetime. Hence, NCs offer a possibility of utilizing the materials that may not even emit light at room temperature.

In this paper, results of our recent investigations of lead-halide-based NCs are presented. In particular, we describe the synthesis, structural and optical characterization of PbI_2, $PbClOH$, $PbIOH$, $Pb_3O_2Cl_2$, and $Pb_3O_2I_2$. The Ncs were characterized using transmission electron microscopy (TEM), scanning electron microscopy (SEM), energy dispersive spectroscopy (EDS), steady state UV-VIS optical absorption and photoluminescence (PL) spectroscopy, and PL lifetime and quantum efficiency measurements. Subsequently, we describe the results of radiation hardness testing of the synthesized material, and show evidence of room-temperature scintillation under gamma irradiation.

SYNTHESIS AND CHARACTERIZATION OF PbI_2 NANOCRYSTALS

Synthesis of PbI_2 nanocrystals

The procedure for synthesizing PbI_2 NCs has been adopted from Finlayson and Sazio [1]. It involves dissolution of bulk lead iodide in a coordinating solvent tetrahydrofuran (THF), subsequent re-crystallization with the addition of anhydrous methanol, and addition of dodecylamine (DDA) to obtain solvent-stabilized PbI_2 NCs. THF, anhydrous methanol, and DDA were purchased from Sigma Aldrich and used directly.

In a typical procedure, 100 mg of high purity (99.999%) lead (II) iodide powder was initially dissolved in 15 mL of THF under continuous stirring at room temperature and under atmospheric pressure. The above conditions are important, since solubility is a strong function of temperature and pressure. Subsequently, the solution was sonicated in centrifuge tubes in order to obtain a saturated solution. Then, to remove any insoluble suspension still present, the saturated solution was centrifuged and the clear deep yellow supernatant was decanted out into a flask. Finally, while stirring this solution continuously under nitrogen atmosphere, 10 mL of anhydrous methanol was gradually added to the flask.

A change in color from deep yellow to a colorless solution was noticed upon addition of methanol. Since PbI_2 is only slightly soluble in methanol, this change was interpreted as indication of the formation of nascent NCs due to re-precipitation in the solution. For this reason, the volumetric ratio of THF to methanol was very important in determining the growth kinetics and nature of the resulting NCs. The re-precipitation process was allowed to continue for 24 hours under constant stirring in nitrogen atmosphere. After that, the process was quenched by addition of DDA at a ratio of 1 mg per 1 mL of the resulting nanoparticulate colloidal solution, and the solution was stored in a vial at room temperature. Aliquots were taken at pre-determined time intervals to study the time evolution of the synthesis process.

Transmission electron microscopy and corresponding energy dispersive spectroscopy analysis

For structural characterization, TEM samples were prepared 9 days after synthesis by placing a drop of the colloidal solution in a 200-mesh carbon coated copper grid and the solvent was allowed to dry, fixing the NCs on the grid. High-resolution transmission electron microscope, JEOL-2010 operating at 200 kV, was used with the OXFORD Link ISIS EDS apparatus.

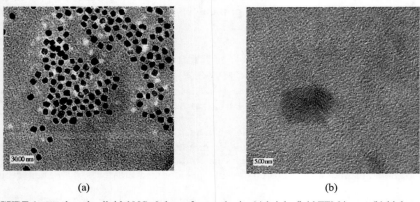

(a)   (b)

FIGURE 1. PbI$_2$-based colloidal NCs 9 days after synthesis: (a) bright-field TEM image (b) high-resolution TEM image.

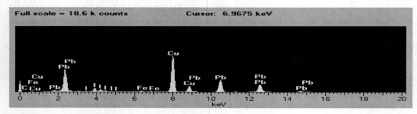

FIGURE 2. EDS spectrum of the PbI$_2$-based NCs obtained 9 days after synthesis, used in conjunction with the transmission electron microscope, shows multiple lead and iodine lines.

Bright field TEM images (Figure 1a) show relatively monodisperse nanoparticles of about 7 to 15 nm in size. The high-resolution TEM images (Figure 1b) indicate particles appearing to have a hexagonal crystalline structure. While the TEM images confirm presence of nanoparticles and their crystalline nature, the EDS analysis performed at the TEM facility confirms presence of both lead and iodine in the NCs (Figure 2).

Optical characterization of PbI$_2$ nanocrystals

The absorption measurements were conducted using a dual-beam CARY 400 UV-VIS spectrophotometer. The sample was prepared by adding three drops of the NC solution to a 3:2 ratio mixture of THF and methanol. Another sample was prepared using only the THF/methanol/DDA mixture and was placed in the reference compartment of the spectrophotometer. The PL spectra were collected using a Horiba Jobin Yvon Fluorolog-3 spectrofluorometer. PL was measured for the colloidal NC solution as well as for the THF/methanol/DDA mixture of solvents.

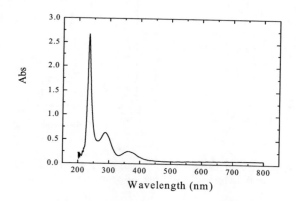

FIGURE 3. Absorption spectrum of the lead-iodide-based nanocrystals measured 2 days after synthesis.

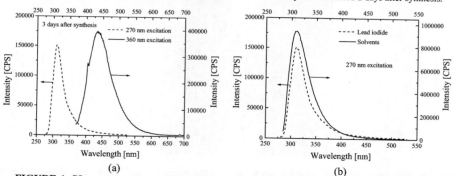

(a)   (b)

FIGURE 4. PL emission spectra for the (a) colloidal nanocrystal solution (b) solvent mixture: THF, anhydrous methanol, and DDA.

The absorption spectrum measured 2 days after synthesis (Figure 3) clearly shows three discrete ultraviolet absorption peaks. The longest-wavelength peak at ~360 nm corresponds to direct band-to-band transitions in the material (bandgap energy of 2.55 eV was reported for bulk PbI_2 [2]). When used for the excitation of the sample, it produces blue PL with the peak centered at 437 nm (Figure 4a). The middle peak in the absorption spectrum was identified as originating from the THF/methanol/DDA mixture of solvents, wiyth the corresponding PL spectra shown in Figure 6b. The origin of the shortest wavelength peak in the absorption spectrum remains unclear at present.

Time evolution of PbI₂ nanocrystals

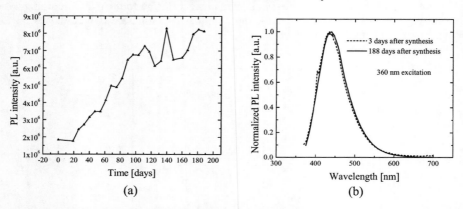

(a)   (b)

FIGURE 5. (a) Observed increase in PL intensity from PbI_2-based NCs over 188 days after synthesis; (b) Spectral change in PL peak over that period of time.

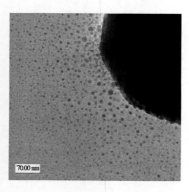

FIGURE 6. Bright-field TEM image of PbI_2-based NC sample 62 days after synthesis (magnification 40,000×). The electron beam current was 108 μA.

Observation of PL spectra over a prolonged period of time led to discovery of steady increase in the light intensity from the samples. Figure 5a shows about fourfold increase in PL intensity over a period of 188 days after synthesis. No shift in spectral position of the peak was observed during that time (Figure 5b). In order to better understand possible origin of that phenomenon, we have performed another TEM study, which revealed that the increase in PL intensity correlated with formation of much larger crystals, illustrated in Figure 6..Their elemental analysis using EDS setup on TEM was not possible, as they were too thick to provide data in the transmission mode. We have therefore used SEM for further analysis. Figures 7a,b show three-dimensional morphology of the PbI_2-based material as observed by SEM 113 days after synthesis. The SEM images revealed crystals of ~2 μm in size. An ditrigonal pyramidal class structure was inferred from these images. The SEM EDS analysis (Figures 7c,d) provides evidence on the presence of lead and iodine in the crystals. Furthermore, the elemental analysis of the sample revealed that there was an equal percentage of oxygen along with lead and iodine.

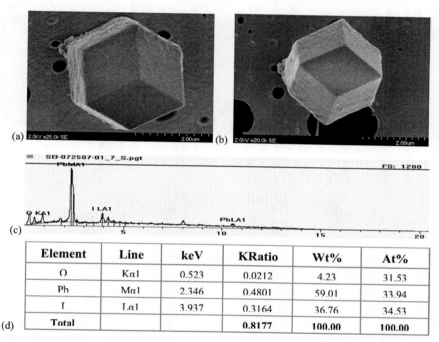

Element	Line	keV	KRatio	Wt%	At%
O	Kα1	0.523	0.0212	4.23	31.53
Pb	Mα1	2.346	0.4801	59.01	33.94
I	Lα1	3.937	0.3164	36.76	34.53
Total			0.8177	100.00	100.00

FIGURE 7. (a) and (b) SEM images of micro-scale PbI_2-based crystals showing ditrigonal pyramidal class structure, (c) EDS spectrum showing presence of lead, iodine, and oxygen, (d) elemental analysis table showing percentage composition of elements.

According to the procedure established by Horiba Jobin Yvon [Porres 2006] and based on the method of deMello *et al.* [3], quantum efficiency of the PbI$_2$-based material was measured in a dilute solution of the sample using the integrating sphere capability on the Horiba Jobin Yvon Fluorolog-3 spectrofluorometer. As distinct from comparative methods of measuring quantum efficiency, integrating sphere approach allows for absolute measurement of quantum efficiency over a wide spectral range.

Quantum efficiency for the blue PL of the PbI$_2$-based material was measured at two different times after synthesis. Quantum efficiencies of 6.7% and 15.6% were recorded after 115 and 197 days after synthesis, respectively, which is consistent with the PL intensity increasing over time (Figure 6a).

The PL lifetime of NCs is expected to be shorter than that of bulk material. PL lifetime measurements for the PbI$_2$-based material, illustrated in Figure 8, were taken on the Horiba Jobin Yvon Fluorolog-3 spectrofluorometer in a configuration allowing for time-correlated single photon counting. Very short PL lifetimes of ~4 ns and ~4.2 ns were obtained from the measurements taken, respectively, 148 and 190 days after synthesis. This is again consistent with the observed increase in PL intensity over time, and can be explained by a diminishing with time contribution of nonradiative recombination into carrier lifetime.

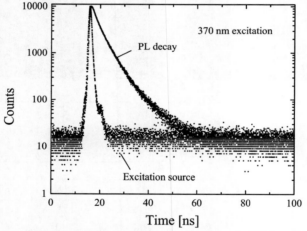

FIGURE 8. Results of PL lifetime measurements for PbI$_2$-based NCs.

DISCUSSION AND CONCLUSIONS

The original procedure of Finlayson and Sazio [1] was intended for synthesis of PbI$_2$ NCs. Addition of DDA was supposed to stop growth by capping the NCs. We found this quenching procedure to be inefficient, as synthesized NCs left in THF/methanol/DDA solvent kept growing over time, reaching a micrometer size in about 3-month period.

PbI_2 is known to have hexagonal unit cell, on which, in principle, ditrigonal pyramidal class of crystals can be formed. Although over 40 polytypes of PbI_2 have been reported in the literature [4], we were unable to identify the microcrystals as belonging to any polytype of PbI_2. Another possibility, strongly suggested by the SEM analysis of micro-scale crystals, is PbIOH (iodolaurionite). Its composition is consistent with the results of our SEM EDS elemental analysis (hydrogen does not show up on EDS spectra), and it belongs to orthorhombic crystalline system. We are also considering $Pb_3O_2I_2$ (iodomendipite) of the lead oxyhalide family as a candidate material for the spontaneously formed optically active microcrystals that evolved from the original PbI_2 synthesis. Nanocrystals made of these materials are being synthesized now in order to verify these hypotheses.

ACKNOWLEDGMENTS

This paper summarizes the work performed by many members of the author's research group, and in particular Brian A. Akins, Antonio C. Rivera, Tosifa Memon, Krishnaprasad Sankar, Nathan J. Withers, and Dr. Gennady A. Smolyakov. Thanks are due to Dr. Timothy J. Boyle of the Advanced Materials Laboratory, Sandia National Laboratories, for allowing access to X-ray diffractometer and spectrophotometer. Funding from the National Science Foundation (Grants No. IIS-0610201 and DGE-054950) and the Defense Threat Reduction Agency (Grant No. HDTRA-1-08-1-0021) is gratefully acknowledged.

REFERENCES

1. C. E. Finlayson and P. J. A. Sazio, *J. Phys. D: Appl. Phys.* **39**, 1477-1480 (2006).
2. M. V. Artemyev, Y. P. Rakovich, and G. P. Yablonski, *J. Cryst. Growth* **171**, 447-452 (1997).
3. J. C. deMello, H. F. Wittmann, and R. H. Friend, *Adv. Mater.* **9**, 230 (1997).
4. M. Chand and G. C. Trigunayat, *Acta Cryst.* **B31**, 1222-1223 (1975).

Nonlinear Optics in Confined Structures for a Better Understanding of Vacuum Field Fluctuations

Jean Desforges, Tahar Ben-Messaoud, Martin Leblanc and Serge Gauvin

Physics and Astronomy Department, University of Moncton, NB, Canada E1A 3E9

Abstract. The confinement of light between two highly reflective mirrors, the so-called optical microcavities, can lead to very interesting novel effects. It is well known that the resonant recirculation of light inside a microcavity can be used to study the strong coupling of light waves with nonlinear materials. Under these confinement conditions, nonlinear optical effects like second harmonic generation, parametric fluorescence, and so on, become more easily observable. Microcavities can also be used to investigate vacuum quantum fluctuation, i.e. the temporary appearance of particle-antiparticle pairs out of nothing allowed by the Heisenberg's uncertainty principle. In such confined media, the fluctuating quantum states can be manipulated and intensified at specific wavelength by properly selecting the size of the microcavity resonator. In this work, we give an overview of the research done by our team using microcavities. We also discuss how these microcavities are fabricated using Bragg mirrors obtained from direct current magnetron sputtering.

Keywords: Bragg mirrors, microcavity, nonlinear optics, vacuum fluctuations.
PACS: 42.65.-k, 78.67.Pt, 81.10.-h

INTRODUCTION

Optical microcavities are micrometer size devices that can trap incoming light waves between two highly reflective mirrors. They are used in a wide range of applications [1-5], from realization of narrow spot-size laser to devices related to long-distance transmission over optical fibers [6]. In general, microcavities are useful for any kinds of light-emitting devices. Two interesting subjects can be investigated with the help of microcavities. The first one is the strong coupling of light waves with nonlinear materials. Nonlinear materials are materials which can be polarised easily so that their polarization vector responds non-linearly with the electric field of incident light waves. Under confinement conditions, nonlinear optical effects become more easily observable. We have studied second harmonic generation (SHG) using femtosecond laser pulses interacting with a micrometer size layer of 2-methyl-4-nitroaniline (MNA), a highly nonlinear organic material. Some results are discussed.

The second important area where microcavities can be useful is in the investigation of vacuum quantum fluctuation. Following Heisenberg's uncertainty principle, vacuum can be seen as full of fluctuating electromagnetic waves for which excitation levels of the quantized field are associated with virtual elementary particles. In

CP1147, *Transport and Optical Properties of Nanomaterials—ICTOPON - 2009,* edited by M. R. Singh and R. H. Lipson
© 2009 American Institute of Physics 978-0-7354-0684-1/09/$25.00

confined media, the fluctuating quantum states can be manipulated and intensified at specific wavelength by properly selecting the size of the microcavity resonator. Electromagnetic waves whose wavelengths fit with the cavity resonator contribute to the vacuum energy whereas electromagnetic waves that do not fit with the size of the cavity are not counted in the vacuum energy calculation. There are therefore more states outside the cavity then inside. The existence of the quantum vacuum is demonstrated by the well known Casimir force that pulls together the plates of the cavity [7-8].

The aim of this work is to give an overview of the research done by our team using microcavities. We discuss how these microcavities are fabricated using Bragg mirrors obtained from direct current magnetron sputtering. Some interesting results obtained recently are presented. The paper ends with a discussion of vacuum quantum fluctuations and how this highly interesting field can be investigated.

BRAGG MIRRORS FROM DC MAGNETRON SPUTTERING

Bragg mirrors are made of alternating dielectric layers with high and low refractive index n and they work like one-dimensional photonic crystals. The thickness of each layer is chosen such that the optical path represents a quarter wavelength of the incident wave to be reflected. Considering the change of sign of the reflection coefficients at each interface, all reflected components interfere constructively. A larger difference in refractive index provides a larger optical band gap. The degree of reflectivity achieved depends on the number of layers used to make the mirror. However, the number of layers also affects the total absorption and interface diffusion. Figure 1 shows an example of optical band gap resulting from a 5-5 stacks multilayer of ZrO_2 and SiO_2. The position of the band gap is seen to be centered near 1100 nm but due to the small number of layers, the transmittance in the gap cannot reach very low values.

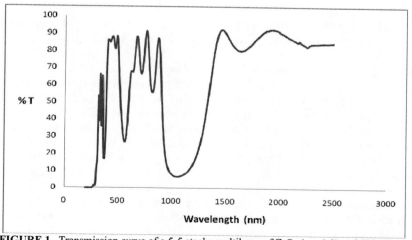

FIGURE 1. Transmission curve of a 5-5 stacks multilayer of ZrO_2 ($n = 1.9$) and SiO_2 ($n = 1.5$) obtained from DC magnetron sputtering.

Bragg mirrors can be obtained from different deposition methods and are easily available from industry. The most commonly used deposition techniques are Electron-Beam Evaporation [9-10], Chemical Vapor Deposition [11-12], RF and DC magnetron sputtering [13-15]. We recently start to synthesize our own Bragg mirrors using DC magnetron sputtering with the aim to study and improve their performance. Sputtering deposition occurs as follow. Low pressure plasma is created from Argon injection into the process chamber from which energetic ions are made to accelerate toward the target material. Sputtered atoms are then ejected from the material and migrate randomly towards the substrate. We talk about reactive sputtering when another gas (typically oxygen) is also injected in the process chamber in the path of the sputtered atoms in their way to the substrate. So far, we have made Bragg mirrors from multilayer of zirconium oxide (ZrO_2) and silicon oxide (SiO_2) but we plan to explore mirrors made from other oxides as well. Good mirrors (high reflectivity, low absorption, reduced interface roughness, good volume homogeneity inside each layer, etc.) are needed in order to perform high precision nonlinear optic experiments.

Some Advantages of the Sputtering Method

Each deposition method has its own advantages. Among the advantages of the sputtering method are the following. The deposition process is relatively fast and do not require high temperature. High melting point materials are easily sputtered. Films have a composition close to that of the source material. Sputtered films have a good adhesion on the substrate. Sputtering usually results in lest contamination than evaporation methods. Sputtering can easily be performed top-down. A very important issue to mention here is that optical and morphological properties of the films not only depend on the deposition method used but are also strongly influenced by the deposition conditions (partial gas pressure, process chamber temperature, deposition rate, etc.). This is why a lot of work can still be done to find the best fabrication conditions to get better mirrors.

Effect of Substrate Temperature on the Optical and Morphological Properties of Thin Zirconium Oxide Films

This is an example of studies we have performed on simple depositions of zirconium oxide (250-300 nm thick) to investigate homogeneity and surface roughness of films obtained from DC reactive magnetron sputtering. It is well known [16] that heating the films after the deposition process enhance the oxygen diffusivity and help to complete oxidation, resulting in more uniformity in the films. Furthermore, the heating process helps to fill out possible voids inside the films and favor the formation of more closely packed thin films [17]. As a result, the refractive index of the film increases while the film thickness decreases. We have also observed these results from ellipsometry measurements on our samples. One of the drawback of such post-deposition heating process is that it tends to increase the surface roughness which in turn can have a dramatic consequence on the surface diffusion. That is why we have investigated the effect of heating the substrate during the deposition. The refractive index and the film thickness have the same behavior as before (the variations are even

72

more apparent). However, the main point is the evolution of the surface roughness in regard of substrate temperature. As shown in Figure 2, from Atomic Force Microscopy measurements, the best conditions for deposition would be achieved around 200°C in order to meet optimum optical properties.

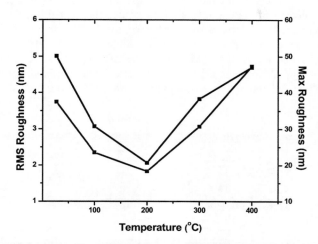

FIGURE 2. (a) RMS roughness (upper curve, left scale) and (b) maximum roughness (lower curve, right scale) as a function of *in situ* deposition temperature of thin films of zirconium oxide obtained from DC magnetron sputtering. The thickness of the each sample was taken close to 300 nm.

NONLINEAR OPTICAL EFFECTS

Nonlinear optics concerns the interaction of intense laser light with highly nonlinear materials. This interaction is possible because the nonlinear susceptibility of the material generate a source term in the wave equation associated with the nonlinear optical media [18]. Nonlinear optical effects (second harmonic generation, parametric fluorescence, and other similar frequency mixing processes) benefit from field enhancement in a resonating cavity. As a first step toward extension to the study of more general parametric processes in confined environments, we have performed SHG experiments using femtosecond laser pulses interacting with a micrometre size layer of MNA. To obtain good homogenous crystals from nonlinear organic materials like MNA is, however, a challenging matter. Let say few words on the delicate subject of organic crystal growth.

Organic Crystal Growth

The MNA crystalline thin films are grown from the melt between two Bragg mirrors in a modified Bridgman technique [19-21]. We use an edge capillary filling technique to ensure two-dimensional growth. A small mirror is placed on top of a larger one and a very small amount of MNA (100 µg) is deposited as powder on the large mirror. When the system is heated to melt the powder, the melt flows between the mirrors by capillary action. MNA is then cooled down and allowed to crystallize. However, under spontaneous crystallization, organic materials like MNA, grow preferentially in one direction and tend to produce needles! An « Inverted bell » technique is then used for recrystallization and it acts as follow (see Figure 3) [22]. The sample is placed inside a small oven which can generate a temperature gradient that keeps the centre of the sample slightly colder than its immediate surrounding while the temperature is elevated by computer control. The size of the cold zone is then slowly reduced so that selection of the highest quality monocrystal is made possible. Slow expansion of the cold zone is the last step allowing a good control of the crystal growth. Centimeter size samples have been obtained with this method which is more than sufficient to perform high quality SHG experiments.

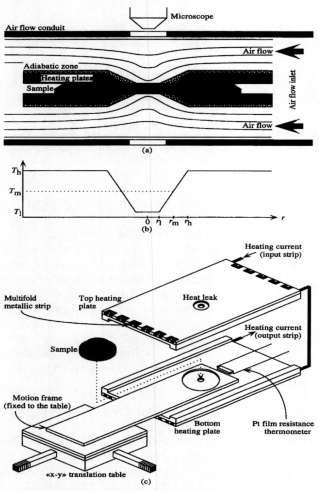

FIGURE 3. « Inverted bell » technique apparatus. (a) Design. (b) Idealized thermal profile. (c) Schematic representation of the furnace core

74

SHG in MNA Thin Crystal from Short Laser Pulses

Our research group has access to a femtosecond pulsed laser whose signal beam is generated from an optical parametric amplification (OPA). This femtosecond laser allows for sophisticated experiments but ultra short pulses may complicate theoretical calculations. Indeed, short pulses are frequently used to carry out second-harmonic generation (SHG) experiments. In that case, a broad distribution of wavelengths can be present in the crystal allowing frequency mixing. In a recent work, we have studied SHG in MNA crystalline thin films grown between two glass plates. To get a useful physical picture of the involved mechanisms, we made use of a simple formalism based on monochromatic plane wave (MPW). Assuming that the Gaussian shape of the fundamental radiation source can be treated as a sum of coherent monochromatic beams, the SH intensity was calculated from the sum of a series of non-mixing monochromatic components However, this simple model need validation and we have published [23] experimental evidence that even for highly nonlinear material as MNA, wave mixing has negligible effect on the SHG using 250 fs fundamental pulses. This pulse duration is indeed not in the ultra short pulses regime and we found that our simple model agree well with experimental data for a pump pulse with a number of cycles ≥ 6. Comparison of predictions from our simple model with experimental results is shown in Figure 4.

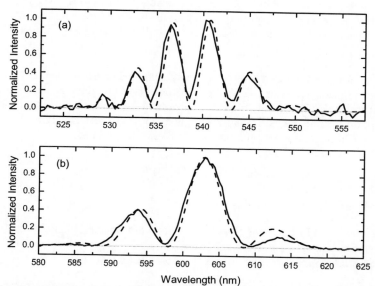

FIGURE 4. Comparison between simple model calculations (dotted line) and experimental data (solid line) for normalized SH spectra obtained from 24 µm thick MNA crystal. The pump pulse is centered at (a) 1077 nm. (b) 1203 nm.

VACUUM QUANTUM FLUCTUATIONS

As mentioned in the introduction, confinement considerably affect the cavity modes profile inside a microcavity and electromagnetic wave components are enhanced or inhibited depending on whether or not their wavelength fit with the size of the cavity resonator. The role of optical confinement on nonlinear processes follows the same path. For instance, one can expect to observe enhancement or inhibition of SHG from optical confinement depending upon the wavelength of the laser pump. In addition to the role played by the confinement, any noise present in the external wave will manifests in the intensity of the internal SH field generated.

Our approach in looking for experimental evidences that non classical effects are present is the following. Enhancement of typical nonlinear processes is investigated (here, SHG in MNA). We have made theoretical predictions of SHG calculated from classical matrix formalism [24-25]. Any deviation in the classical prediction might be attributable to non classical effects typical to high confinement. Indeed, SHG measurements showed an energy conversion factor that is 650 fold the value calculated from the classical formalism!

CONCLUSIONS

We are studying nonlinear optical processes inside microcavities. Bragg mirrors are obtained from DC magnetron sputtering and some studies of their performances have been done. Nonlinear organic materials are crystallized using the « Inverted bell » technique.

Optical confinement is used to investigate vacuum field fluctuations. Enhancement of typical nonlinear processes is investigated. SHG measurements showed an energy conversion factor that is 650 fold the value calculated from classical matrix formalism. The deviation might be attributable to non classical effects typical to high confinement. Other parametric processes experiments are planned.

ACKNOWLEDGMENTS

This work was supported in part by:

Natural Sciences and Engineering Research Council of Canada (NSERC)
New Brunswick Innovation Foundation (NBIF)
Faculté des Études Supérieures et de la Recherche (FESR)

REFERENCES

1. H. Yokoyama, *Science* **256**, 66-70 (1992).
2. R.W. Martin *et al.*, *Mater. Sci. Eng.* B **93**, 98-101 (2002).
3. C. J. Hood, H. J. Kimble and J. Ye, *Phys. Rev. A* **64**, 033804-033810 (2001).
4. G. Li *et al.*, *Appl. Optics* **45**, 7628-7631 (2006).

5. C. Ye and J. Zhang, *Phys. Rev. A* **73**, 023818-023822 (2006).
6. K. J. Vahala, *Nature* **424**, 839-846 (2003).
7. H. B. G. Casimir, *Proc. Kon. Nederl. Akad. Wetensh. B* **51**, 793 (1948).
8. B. Jancovici and L. Samaj, *Europhys. Lett.* **72** (1), 35-41 (2005).
9. M. Jerman, Z. Qiao and D. Mergel, *Appl. Optics* **44**, 3006-3012 (2005).
10. S. Shao *et al.*, *Thin Solid Films* **445**, 59-62 (2003).
11. W.-H. Nam and S.-W. Rhee, *Chem. Vapor Depos.* **10**, 201-205 (2004).
12. O. Bernard *et al.*, *Appl. Sur. Sci.* **253**, 4626-4640 (2007).
13. P. Gao *et al.*, *Vacuum* **56**, 143-148 (2000).
14. C.Y. Ma *et al.*, *Appl. Sur. Sci.* **253**, 8718-8724 (2007).
15. O. Buiu *et al.*, *Thin Solid Films* **515**, 623-626 (2006).
16. M. H. Suhail, G. Mohan Rao and S. Mohan, *J. Vac. Sci. Technol., A* **9** 2675-2677 (1991).
17. S. Venkataraj *et al.*, *J. Appl. Phys.*, **92**, 3599-3607 (2002).
18. R. W. Boyd, *Nonlinear Optics*, 2nd edition, Academic Press, San Diego, 2003
19. S. Gauvin and J. Zyss, *J. Cryst. Growth* **166**, 507-527 (1996)
20. R.A. Laudise, *The Growth of Single Crystals*, Prentice-Hall Ed., Englewood Cliffs, 1970
21. *Crystal Growth: An Introduction*, P. Hartman Ed., North-Holland, Amsterdam, 1973
22. Reprinted from *J. Cryst. Growth* **166**, S. Gauvin and J. Zyss, *Growth of organic crystalline thin films, their optical characterization and application to nonlinear optics*, p.514, Copyright (1996), with permission from Elsevier.
23. T. Ben-Messaoud, J. Desforges and S. Gauvin, *J. Nonlinear Opt. Phys.* **15**, 491-499 (2006).
24. D. S. Bethune, *J. Opt. Soc. Am. B* **6**, 910-916 (1989).
25. N. Hashizune, M. Ohashi, T. Kondo and R. Ito, *J. Opt. Soc. Am. B* **12**, 1894-1904 (1995).

Exciton- and Light-induced Current in Molecular Nanojunctions

B.D. Fainberg[a,b], P. Hanggi[c], S. Kohler[c] and A. Nitzan[b]

[a]*Faculty of Sciences, Holon Institute of Technology, 52 Golomb St., Holon 58102, Israel*
[b]*School of Chemistry, Tel-AvivUniversity, Tel-Aviv 69978, Israel*
[c] *Institute for Physics, University of Augsburg, Augsburg, D-86135, Germany*

Abstract. We consider exciton- and light-induced current in molecular nanojunctions. Using a model comprising a two two-level sites bridge connecting free electron reservoirs we show that the exciton coupling between the sites of the molecular bridge can markedly effect the source-drain current through a molecular junction. In some cases when excited and unexcited states of the sites are coupled differently to the leads, the contribution from electron-hole excitations can exceed the Landauer elastic current and dominate the observed conduction. We have proposed an optical control method using chirped pulses for enhancing charge transfer in unbiased junctions where the bridging molecule is characterized by a strong charge-transfer transition.

Keywords: Molecular nanojunctions, excitons, light-induced current.
PACS: 71.35.Aa, 73.63.Rt, 73.23.Hk, 85.65.+h

INTRODUCTION

Electron transport through molecular wires has been under intense study in the last few years [1-3]. Theoretical modeling of electron transport [2,3] starts from the wire Hamiltonian as a tight-binding model composed of N sites that contain electron transfer (tunneling) interactions between nearest sites. For a molecular wire, this constitutes the so-called Huckel description where each site corresponds to one atom. Necessary conditions for finite current in this model are, first, the existence of such interactions between (quasi-) resonant states of nearest sites, and second, a biased junction. In this presentation we consider additional interactions, which enable us to remove one of these conditions for current to occur.

COHERENT CHARGE TRANSPORT THROUGH MOLECULAR WIRES: CURRENT FROM ELECTRONIC EXCITATION IN THE WIRE

For a typical distance of 5 Å between two neighboring sites, which can be either atoms or molecules in molecular assemblies, energy-transfer interactions – excitation (deexcitation) of a site accompanied by deexcitation (excitation) of its nearest neighbor - are well-known in the exciton theory [4]. Electron transfer is a tunneling process that depends exponentially on the site-site distance, while energy transfer is associated with dipolar coupling

CP1147, *Transport and Optical Properties of Nanomaterials—ICTOPON - 2009*, edited by M. R. Singh and R. H. Lipson
© 2009 American Institute of Physics 978-0-7354-0684-1/09/$25.00

that scales like the inverse cube of this distance, and can therefore dominate at larger distances. To the best of our knowledge, there were no previous treatments of transport in molecular wires that take into account simultaneous effects of both electron and energy transfer. Here we address this problem by using the Liouville-von Neumann equation (LNE) for the total density operator to derive an expression for the conduction of a molecular wire model that contains both electron and energy-transfer interactions, then analyze several examples with reasonable parameters. We show that the effect of exciton type interactions on electron transport through molecular wires can be significant, sometimes even dominant, in a number of situations. In particular, the current occurs even when the above mentioned first condition is not fulfilled.

Model

We consider a molecular wire that comprises a dimer represented by its highest occupied molecular orbitals (HOMO), $|g\rangle$, and lowest unoccupied molecular orbitals (LUMO), $|e\rangle$, positioned between two leads represented by free electron reservoirs L and R (Fig.1). The electron reservoirs (leads) are characterized by their electronic chemical potentials μ_L and μ_R, where the difference $\mu_L - \mu_R = e\Phi$ is the imposed voltage bias. The corresponding Hamiltonian is

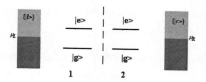

FIGURE 1. A model for energy-transfer induced effects in molecular conduction. The right ($R=|\{r\}\rangle$) and left ($L=|\{l\}\rangle$) manifolds represent the two metal leads characterized by electrochemical potentials μ_R and μ_L respectively. A molecular dimer is represented by its HOMOs, $|1g\rangle$ and $|2g\rangle$, and LUMOs, $|1e\rangle$ and $|2e\rangle$.

$$\hat{H} = \hat{H}_{wire} + \hat{H}_{leads} + \hat{H}_{contacts} \qquad (1)$$

where the different terms correspond to the wire, the leads ($\hat{H}_{leads} = \sum_{k\in\{L,R\}} \varepsilon_k \hat{c}_k^+ \hat{c}_k$), and the wire-lead couplings ($\hat{H}_{contacts} = \hat{V}_M + \hat{V}_N$), respectively.

$$\hat{V}_M = \sum_{nkf} V_{nf,k}^{(MK_n)} \hat{c}_k^+ \hat{c}_{nf} + H.c.$$ describes electron transfer between the molecular bridge and the leads that gives rise to net current in the biased junction while

$$\hat{V}_N = \sum_{n,\, k \neq k'} V_{kk'}^{(NK_n)} \hat{c}_k^+ \hat{c}_{k'} b_n^+ + H.c.$$ describes energy transfer between the bridge and

electron-hole excitations in the leads. Here $H.c.$ denotes Hermitian conjugate, L and R denote the left and right leads, respectively, and $K_1 = L$, $K_2 = R$.

$$\hat{H}_{wire} = \sum_{\substack{m=1,2 \\ f=g,e}} \varepsilon_{mf} \hat{c}_{mf}^+ \hat{c}_{mf} - \sum_{f=g,e} \Delta_f (\hat{c}_{2f}^+ \hat{c}_{1f} + \hat{c}_{1f}^+ \hat{c}_{2f}) + \hbar J(b_1^+ b_2 + b_2^+ b_1) + \sum_{m=1,2} U_m N_m (N_m - 1)$$

(2)

The operators $\hat{c}_{mf}^+$ ($\hat{c}_{mf}$) create (annihilate) an electron in the orbital $|mf\rangle$, and ε_{mf} denotes the respective on-site energy, $\hat{n}_{mf} = \hat{c}_{mf}^+ \hat{c}_{mf}$. The second and the third terms on the RHS of Eq. 2 describe electron and energy transfer between the sites, respectively. Since we aim at exploring blocking effects, the last term on the RHS of Eq. 2 takes account of the Coulomb repulsion on a site in the limit of large interaction strengths U_m where $N_m = \sum_{f=g,e} \hat{n}_{mf}$ is the operator counting the excess electrons on the sites. The excitonic operators are equal to $b_m^+ = \hat{c}_{me}^+ \hat{c}_{mg}$. The effect of the corresponding interaction in the bridge ($= \hbar J b_1^+ b_2 + H.c.$) on the charge transport properties is the subject of our discussion.

Master Equation in the Eigenbasis of Many-electron Wire Hamiltonian

The central idea of using LNE for the computation of stationary currents is to consider $\hat{H}_{contacts}$ as a perturbation. For the total density operator $\wr$ one can obtain by standard techniques the approximate equation of motion [2,3]. The information of interest is limited only to the wire part of the density operator $\sigma(t)$, which can be obtained by defining a projection operator P that projects the complete system onto the relevant (molecule) part and by tracing out the reservoir degrees of freedom ($K = L, R$): $P\rho(t) = \rho_K Tr_K \rho(t) = \rho_K \otimes \hat{\sigma}(t)$ with reservoir density matrix ρ_K. As to ρ_K, we employ the grand-canonical ensemble of non-interacting electrons in the leads at temperature T, characterized by electrochemical potentials μ_K. Therefore, the lead electrons are described by the equilibrium Fermi function $f_K(\varepsilon_k) = [\exp((\varepsilon_k - \mu_K)/k_B T) + 1]^{-1}$. From this follows that all expectation values of the lead operators can be traced back to the expression $\langle \hat{c}_k^+ \hat{c}_{k'} \rangle = f_K(\varepsilon_k) \delta_{kk'}$ where $\delta_{kk'}$ is the Kronecker delta. As a matter of fact, we get

$$\frac{d\sigma(t)}{dt} = -\frac{i}{\hbar}[\hat{H}_{wire}, \sigma(t)] - \frac{1}{\hbar^2} \sum_{S=M,N} Tr_K \int_0^\infty dx [\hat{V}_S, [\hat{V}_S^{int}(-x), \rho(t)]]$$ (3)

where $\hat{V}_S^{int}(-x) = \exp[-\frac{i}{\hbar}(\hat{H}_{wire} + \hat{H}_{leads})x]\hat{V}_S \exp[\frac{i}{\hbar}(\hat{H}_{wire} + \hat{H}_{leads})x]$, and we used the noncrossing approximation [5]. For the evaluation of Eq. 3 it is essential to use an exact expression for the zero-order time evolution operator $\exp[-\frac{i}{\hbar}(\hat{H}_{wire} + \hat{H}_{leads})x]$. The use of any approximation bears the danger of generating artifacts, which, for instance, may lead to a violation of fundamental equilibrium properties [6]. To do so, we first define new operators $b_f^+ = \hat{c}_{2f}^+\hat{c}_{1f}$ and $b_f = \hat{c}_{1f}^+\hat{c}_{2f}$ ($f=e,g$) describing charge transfer $1 \to 2$ and $2 \to 1$, respectively, in the donor-acceptor (DA) two-level system. Then the non-diagonal part of $\hat{H}_{wire}$, Eq. 2, can be rewritten in terms of operators b_f only

$$\hat{H}_{wire}^{(nondiag)} = -\sum_{f=g,e} \Delta_f (b_f^+ + b_f) - \hbar J (b_e^+ b_g + b_g^+ b_e) \qquad (4)$$

By expanding σ in the many-electron eigenstates of the uncoupled sites, one obtains a $2^4 \times 2^4 = 256$ density matrix $\sigma_{\{n_{mf}\},\{n'_{mf}\}}$. Fortunately, the following consideration can be essentially simplified by using the pseudospin description based on the symmetry properties of Lie group SU(2). A two states DA system can be described by the pseudospin vector, using Pauli matrices $\hat{\sigma}_{1,2,3}$ and the unit matrix I [7]. The components of the Bloch vector in the second quantization picture are given by $r_1^f = b_f^+ + b_f, r_2^f = i(b_f - b_f^+), r_3^f = \hat{n}_{2f} - \hat{n}_{1f}$. Owing to the commutation of operators $\lambda_f = \hat{n}_{2f} + \hat{n}_{1f}$ (the electron number operator for the f-th DA system) and r_i^f, λ_f is conserved under unitary transformations related to the diagonalization of $\hat{H}_{wire}$. Therefore, a total $2^4 \times 2^4$ space can be partitioned into nine smaller subspaces according to the values of $\lambda_f = 0,1,2$: four one-dimensional subspaces for $\lambda_f = 0,2$ (type I); four two-dimensional subspaces for $\lambda_f = 1$ and $\lambda_{f'} = 0$, 2 where $f \neq f'$ (type II) ; and one four-dimensional subspace for $\lambda_e = \lambda_g = 1$ (type III). One can show that

$$(r_1^f)^2 = (r_2^f)^2 = (r_3^f)^2 = \lambda_f - 2\hat{n}_{2f}\hat{n}_{1f} = \begin{cases} 0 \text{ for } \lambda_f = 0,2 \\ 1 \text{ for } \lambda_f = 1 \end{cases}, \qquad (5)$$

Using Eqs. 2,4 and 5, we can write $\hat{H}_{wire}$ in terms of the Bloch vector components as follows

$$\hat{H}_{wire} = \frac{1}{2}\lambda_e(\varepsilon_{1e} + \varepsilon_{2e}) + \sum_{m=1,2} U_m N_m (N_m - 1) + \begin{cases} 0 \text{ for subspaces (I),} \\ \frac{1}{2}r_3^f(\varepsilon_{2f} - \varepsilon_{1f}) - \Delta_f r_1^f \text{ for subspaces (II),} \\ \frac{1}{2}\sum_{f=g,e} r_3^f(\varepsilon_{2f} - \varepsilon_{1f}) - \sum_{f=g,e}\Delta_f r_1^f - \\ -\frac{\hbar J}{2}(r_1^e r_1^g + r_2^e r_2^g) \text{ for subspace (III)} \end{cases},$$

(6)

where without loss of generality we put $(\varepsilon_{1g} + \varepsilon_{2g})/2 = 0$, and the energy of the wire depends in the main on λ_e. In the following, we specify the master equation, Eq. 3, for studying two limiting cases. The first limit $U_m = 0$ describes non-interacting electrons at each sites. The second limit is the one of strong Coulomb repulsion at each site in which U_m is much larger than any other energy scale of the problem. Then, only the states with at most one excess electron on the site are relevant. In both cases, a diagonal representation of the first term on the RHS of Eq. 3 is achieved by a decomposition into the eigenbasis (α, β) of the many-electron wire Hamiltonian. In this basis, the fermionic interaction picture operators read

$$\langle \lambda_e, \lambda_g \mid \hat{c}_{nf}^{int}(-x) \mid \lambda_f + 1 \rangle_{\alpha\beta} = [\hat{Y}^+(\lambda_e, \lambda_g) \tilde{\chi}^+(\lambda_e, \lambda_g) \hat{c}_{nf} \tilde{\chi}(\lambda_f + 1) \hat{Y}(\lambda_f + 1)]_{\alpha\beta}$$

$$\times \exp[\frac{i}{\hbar}(E_\beta(\lambda_f + 1) - E_\alpha(\lambda_e, \lambda_g))x]$$

(7)

where $\hat{Y}(\lambda_e, \lambda_g)$ are unitary transformations related to the diagonalization of $\hat{H}_{wire}$; $\hat{\chi}(\lambda_e, \lambda_g) = (\{\mid n_{1g}, n_{2g}, n_{1e}, n_{2e}\}\})$ is the column matrix of the many-electron eigenstates of the uncoupled sites for $n_{1e} + n_{2e} = \lambda_e$ and $n_{1g} + n_{2g} = \lambda_g$; $(\lambda_f + 1) \equiv (\lambda_e + 1, \lambda_g)$ if $f = e$ and $(\lambda_f + 1) \equiv (\lambda_e, \lambda_g + 1)$ if $f = g$; ⊿ denotes the transpose matrix $\hat{\chi}$. The unitary transformations $\hat{Y}(\lambda_e, \lambda_g) = I$ for subspaces (I). As to subspaces (II), Hamiltonian corresponding to the second line of the RHS of Eq. 6 where $\lambda_f = 1 \neq \lambda_{f'}$ can be diagonalized, using the unitary transformation

$$\begin{pmatrix} R_1^f \\ R_2^f \\ R_3^f \end{pmatrix} = \hat{T}^f \begin{pmatrix} r_1^f \\ r_2^f \\ r_3^f \end{pmatrix} \equiv \begin{pmatrix} \cos 2\vartheta_f & 0 & -\sin 2\vartheta_f \\ 0 & 1 & 0 \\ \sin 2\vartheta_f & 0 & \cos 2\vartheta_f \end{pmatrix} \begin{pmatrix} r_1^f \\ r_2^f \\ r_3^f \end{pmatrix}$$

(8)

where $\tan 2\vartheta_f = -2\Delta_f /(\varepsilon_{2f} - \varepsilon_{1f})$. The matrix elements of $\hat{T}^f$ are connected with the unitary transformations $\hat{Y}(\lambda_e, \lambda_g)$ for subspaces (II) by formula $T_{nj}^f = (1/2)Tr(\hat{\sigma}_n \hat{Y}^+ \hat{\sigma}_j \hat{Y})$ where $\hat{\sigma}_n$ and $\hat{\sigma}_j$ are Pauli matrices. The calculation of $\hat{Y}^+(1,1)$ for subspace (III) is more involved. Employing the master equation Eq. 3 and keeping for brevity only terms with $S = M$, we obtain in the wide-band limit for the steady-state condition

$$\frac{i}{\hbar}(E_\alpha - E_\beta)\sigma_{\alpha\beta} = \frac{1}{2}\sum_{nf\alpha'\beta'}\Gamma_{M,nf}\{\hat{c}_{nf,\alpha\alpha'}\sigma_{\alpha'\beta'}\hat{c}_{nf,\beta'\beta}^+[2 - f_{K_n}(E_{\beta'} - E_\beta) - f_{K_n}(E_{\alpha'} - E_\alpha)]$$

$$+\hat{c}_{nf,\alpha\alpha'}^+\sigma_{\alpha'\beta'}\hat{c}_{nf,\beta'\beta}[f_{K_n}(E_\beta - E_{\beta'}) + f_{K_n}(E_\alpha - E_{\alpha'})] - \{\hat{c}_{nf,\alpha\alpha'}\hat{c}_{nf,\alpha'\beta'}^+ f_{K_n}(E_{\alpha'} - E_{\beta'})$$

$$+\hat{c}_{nf,\alpha\alpha'}^+\hat{c}_{nf,\alpha'\beta'}[1 - f_{K_n}(E_{\beta'} - E_{\alpha'})]\}\sigma_{\beta'\beta}$$

$$-\sigma_{\alpha\alpha'}\{\hat{c}_{nf,\alpha'\beta'}\hat{c}_{nf,\beta'\beta}^+ f_{K_n}(E_{\beta'} - E_{\alpha'}) + \hat{c}_{nf,\alpha'\beta'}^+\hat{c}_{nf,\beta'\beta}[1 - f_{K_n}(E_{\alpha'} - E_{\beta'})]\}\}$$

(9)

where $\Gamma_{M,nf} = \frac{2\pi}{\hbar} \sum_{k \in K_n} |V_{nf,k}^{(MK_n)}|^2 \delta(\varepsilon_k - \varepsilon_{nf})$.

Calculation of Current

The current through the dashed line (see Fig.1) is given by $\hat{I} = e\frac{d}{dt}\hat{N} = \frac{ie}{\hbar}[\hat{H},\hat{N}]$ where $\hat{N} = \sum_{k \in L} \hat{c}_k^+ \hat{c}_k + \hat{n}_{1g} + \hat{n}_{1e}$ is the operator of the electron number on the left from the dashed line. Calculating the commutator on the RHS of the last equation, we get $\hat{I} = \frac{ie}{\hbar}\sum_{f=g,e}\Delta_f(b_f - b_f^+) = \frac{e}{\hbar}\sum_{f=g,e}\Delta_f r_2^f$. Using Eq. 6, we obtain

$$\hat{I} = \frac{e}{\hbar}\sum_{f=g,e}\Delta_f r_2^f(\lambda_f = 1) \qquad (10)$$

Obviously $\lambda_f = 1$ in Eq. 10 is another way of saying that the current in channel f exists only for the case of one of states $\{f\}$ is occupied and another one of $\{f\}$ is unoccupied.

Strong Coulomb Repulsion at Sites

In the limit of strong Coulomb repulsion, U_m is assumed to be so large that at most one excess electron resides on each site. Thus, the available Hilbert space for uncoupled sites is reduced to three states $\hat{\chi}(0,0)$, $\hat{\chi}(2,0)$ and $\hat{\chi}(0,2)$ for subspaces I; two states $\hat{\chi}(1,0)$ and $\hat{\chi}(0,1)$ for subspaces II; and the state $\hat{\chi}(1,1) = \begin{pmatrix} |1_{1g}, 0_{2g}, 0_{1e}, 1_{2e}\rangle \\ |0_{1g}, 1_{2g}, 1_{1e}, 0_{2e}\rangle \end{pmatrix}$ for subspace III, which in this case becomes two-dimensional one. Consider subspaces (II). The matrix $\hat{r}^f$, Eq. 8, with matrix elements $T_{nj}^f = (1/2)Tr[\hat{\sigma}_n \hat{Y}^+(\lambda_f = 1; \lambda_{f'} = 0)\hat{\sigma}_j \hat{Y}(\lambda_f = 1; \lambda_{f'} = 0)]$ describes a rotation by mixing angle $2\vartheta_f$ around axis $"y"$. $\hat{Y}(\lambda_f = 1; \lambda_{f'} = 0)$ is an unitary operator defined by

$$\hat{Y}^+(\lambda_f = 1; \lambda_{f'} = 0) = \begin{pmatrix} \cos\vartheta_f & \sin\vartheta_f \\ -\sin\vartheta_f & \cos\vartheta_f \end{pmatrix} \qquad (11)$$

which enables us to obtain eigenstates

$$\begin{pmatrix} \Phi_+(\lambda_f = 1; \lambda_{f'} = 0) \\ \Phi_-(\lambda_f = 1; \lambda_{f'} = 0) \end{pmatrix} = \hat{Y}^+(\lambda_f = 1; \lambda_{f'} = 0)\hat{\chi}(\lambda_f = 1; \lambda_{f'} = 0) \qquad (12)$$

and eigenvalues

83

$$E_{\pm}(\lambda_f = 1; \lambda_{f'} = 0) = \frac{1}{2}[\lambda_e(\varepsilon_{1e} + \varepsilon_{2e}) + (\varepsilon_{2f} - \varepsilon_{1f}) \pm \sqrt{(\varepsilon_{2f} - \varepsilon_{1f})^2 + 4\Delta_f^2}])$$

(13)

for subspaces (II). As to subspace III, operator $\hat{Y}^+(1,1)$ for $\varepsilon_{ng} = 0$, $\varepsilon_{ne} = \varepsilon_e$ and $\Delta_g = 0$ is reduced to

$$\hat{Y}^+(1,1) = \frac{1}{\sqrt{2}}\begin{pmatrix} 1 & 1 \\ -1 & 1 \end{pmatrix}$$

(14)

It enables us to obtain the corresponding eigenstates $\hat{\Phi}(1,1) = Y^+(1,1)\hat{\chi}(1,1)$ and eigenvalues $E_{1,2} = \varepsilon_e \mp J\hbar$. Using Eqs. 10 and 14 and taking the expectation value of current, we get for $\Delta_g = 0$

$$\langle I \rangle = \frac{2e}{\hbar}\Delta_e \operatorname{Im}\sigma_{-+}(1,0)$$

(15)

Current from Electronic Excitations in the Wire

Recently Galperin, Nitzan and Ratner [8] predicted the existence of non-Landauer current induced by the electron-hole excitations in the leads. Here we show that the non-Landauer current is induced also by the exciton type excitations in the wire itself. Consider a strong bias limit in the Coulomb blockade case, where $\mu_L > \varepsilon_e$ and $\mu_R < \varepsilon_g$, and the states ε_f are positioned rather far ($\Box\, k_B T, | J |, | \Delta_e |$) from the Fermi levels of both leads so that $f_L(\varepsilon) = 1$ and $f_R(\varepsilon) = 0$ on the RHS of Eq. 9. The Landauer current in the case under consideration ($\Delta_g = 0$) occurs in channel "e" when it is isolated from channel "g" that is realized for $\Gamma_{M,1g} = \Gamma_{M,2g} = 0$ and $\lambda_g = 0$. The latter equality enables us to avoid blocking the current in channel "e" due to strong Coulomb repulsion at sites. Indeed, the Landauer current does not exist for $\Gamma_{M,1e} = \Gamma_{M,2e} = 0$ even when $\Gamma_{M,1g}, \Gamma_{M,2g} \neq 0$, since $\Delta_g = 0$. In contrast, the solution of Eq. 9 in the rotating-wave approximation (RWA) [3] gives for this case

$$\sigma_{-+}(1,0) = \frac{i\hbar}{-8\Delta_e}\{\Gamma_{M,1g}Tr\sigma(1,0) + \Gamma_{M,2g}Tr\sigma(1,1)\}$$

(16)

and $Tr\sigma(1,0) = (\Gamma_{M,2g}/\Gamma_{M,1g})Tr\sigma(1,1)$. Then using the normalization condition $Tr\sigma(1,0) + Tr\sigma(1,1) = 1$ and Eq. 15, we obtain

$$\langle I \rangle_{RWA} = -\frac{e}{2} \frac{\Gamma_{M,2g}\Gamma_{M,1g}}{\Gamma_{M,2g}+\Gamma_{M,1g}} \tag{17}$$

The last equation describes a non-Landauer current caused by the electron transfer occurring in different channels: the intersite transfer in the bridge takes place in channel "e", and the bridge-metals charge transfer occurs in channel "g". The inter-channel mixing is induced by the energy-transfer term ~ J (see Fig.2). Consider a cycle corresponding to the charge transfer of e. Initially electrons populate states $|1g^{<}$ (due to coupling to the left lead, $\Gamma_{M,1g} \neq 0$) and $|2e^{<}$ ($\Gamma_{M,2e} = 0$) of the sites (Fig.2a). The energy transfer induces the following transitions: $|2e\rangle \rightarrow |2g\rangle$ and $|1g\rangle \rightarrow |1e\rangle$ (vertical arrows). The system arrives to the state shown in Fig.2b. Due to the coupling to the right lead ($\Gamma_{M,2g} \neq 0$), electron from state $|2g\rangle$ moves into the right lead (the horizontal arrow), and after this the system is described by Fig.2c. Due to releasing the right site and the hopping matrix element Δ_e, electron from state $|1e\rangle$ passes into state $|2e\rangle$ (the upper horizontal arrow), releasing the left site. Then an electron from the left lead moves into state $|1g\rangle$ ($\Gamma_{M,1g} \neq 0$, the lower horizontal arrow), and the system returns into the initial state described by Fig.2a.

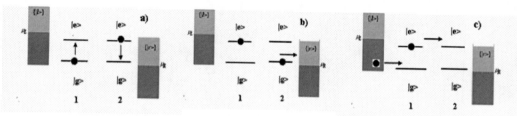

FIGURE 2. Different stages of the energy-transfer induced current. a) energy transfer, $\sigma(1,1)\neq0$. b) the charge transfer to the right lead. c) the intersite charge transfer, $\sigma(1,0)\neq0$; the charge transfer from the left lead.

OPTICAL CONTROL OF CURRENT WITH CHIRPED PULSES

In the second part of the paper we describe a theory for light-induced current by strong optical pulses in unbiased molecular tunneling junctions as a special case where the second condition for finite current (see Introduction) - a biased junction - is not fulfilled. We consider a class of molecules characterized by strong charge-transfer transitions into their first excited state [9]. We have proposed a novel control mechanism by which the charge flow is enhanced by chirped pulses. For linear chirp and when the energy transfer between the molecule and electron-hole excitations in the leads is absent, this control model can be reduced to the Landau-Zener transition to a decaying level. The details can be found in Ref. [5].

ACKNOWLEDGMENTS

This work was supported by the German-Israeli Fund (P.H., S.K. and A.N.), ISF (B.F. and A.N.), the Deutsche Forschungsgemeinschaft through SPP 1243 and the German Excellence Initiative via the "Nanosystems Initiative Munich (NIM)" (P.H., S.K., B.F.).

REFERENCES

1. *Proc. Natl. Acad. Sci. U.S.A.* **102**, 8800 (2005), special issue on molecular electronics, edited by C. Joachim and M. A. Ratner.
2. S. Kohler, J. Lehmann, and P. Hanggi, *Phys. Reports* **406**, 379-443 (2005).
3. F. J. Kaiser, M. Strass, S. Kohler and P. Hanggi, *Chem. Phys.* **322**, 193-199 (2006).
4. A. S. Davydov, "Theory of Molecular Excitons", New York, Plenum, 1971.
5. B. D. Fainberg, M. Jouravlev, and A. Nitzan, *Phys. Rev. B* **76**, 245329 (2007).
6. T. Novotny, *Europhys. Lett.* **59**, 648-654 (2002).
7. F. T. Hioe and J. H. Eberly, *Phys. Rev. Lett.* **47**, 838-841 (1981).
8. M. Galperin, A. Nitzan and M. A. Ratner, *Phys. Rev. Lett.* **96**, 166803 (2006).
9. M. Galperin and A. Nitzan, *Phys. Rev. Lett.* **95**, 206802 (2005).

Coulomb drag effect in parallel quantum dots

B. Tanatar* and V. Moldoveanu†

*Department of Physics, Bilkent University, Bilkent, 06800 Ankara, Turkey
†National Institute of Materials Physics, P.O. Box MG-7, 077125 Bucharest-Magurele, Romania

Abstract. We study theoretically the electronic transport in parallel few-level quantum dots in the presence of both intradot and interdot long-range Coulomb interaction. Each dot is connected to two leads and the steady-state currents are calculated within the Keldysh formalism using the random-phase approximation for the interacting Green functions. Due to the momentum transfer mechanism between the two systems it is possible to get a nonvanishing current through an unbiased Coulomb-blockaded dot if the other dot is set in the nonlinear transport regime. The transitions between the levels of the passive dot reduce the drag current and lead to negative differential conductance.

Keywords: Coulomb drag effect, Quantum dots, Nanoelectronic devices
PACS: 73.23.-b, 73.63.-b, 73.21.La, 85.35.-p

INTRODUCTION

More than a decade ago Gramila *et al.* [1] measured for the first time a finite current through an unbiased two-dimensional semiconductor layer when a current passes a nearby identical layer. Nowadays it is generally accepted that the mechanism behind the observed drag current is the inelastic Coulomb scattering. [2] The electrons exchange momentum during these scattering processes and this induces a voltage drop in the passive layer, henceforth a current appears. In the standard Coulomb drag the induced current flows in the same direction as the one in the driven subsystem. Later on other experiments were performed with one-dimensional quantum wires (see [3, 4] and references therein) and even a negative Coulomb drag was reported. [5]

On the other hand a different class of experiments with Coulomb-coupled mesoscopic interferometers show that the resonant tunneling through a quantum dot can be recorded by a so-called charge detector and, conversely, the latter scrambles the coherence properties via measurement-induced dephasing. [6] Other experiments considered speculate the charge sensing effect in quantum point contacts [7] in order to extract time-resolved tunneling in quantum dots. [8] More recently, McClure *et al.* [9] reported on the current cross-correlation in Coulomb-coupled quantum dots, when each dot is individually biased.

Starting from these findings one can naturally wonder if an analog Coulomb drag could appear in quantum dot systems. More precisely, the question is whether an unbiased dot coupled to two leads and set in the Coulomb blockade regime could be set to transmit when a (possibly high) bias is applied on a second nearby dot coupled to other two leads.

On the theoretical side, the Coulomb drag effect is mostly described via the linear response theory with respect to the bias applied on the drive subsystem, and the effective layer interaction is usually computed within the random-phase approximation (see e.g.

CP1147, *Transport and Optical Properties of Nanomaterials—ICTOPON - 2009*, edited by M. R. Singh and R. H. Lipson
© 2009 American Institute of Physics 978-0-7354-0684-1/09/$25.00

the review paper [2] and the recent paper by Asgari *et al.* [10]). As for the current correlations one could rely on scattering formalism for interacting systems [11] or on the quantum Master equation up to the second order in the tunneling Hamiltonian. [12] In the first approach one computes the first-order contribution of the Coulomb interaction to the two-particle S-matrix, which would therefore describe the charge sensing effect but misses the inelastic scattering processes responsible for the Coulomb drag. Also, the intradot interaction is neglected.

In this paper we present transport calculations for few-level Coulomb-coupled quantum dots obtained from a recently developed random-phase approximation method for the non-equilibrium Green functions. [13] The method can deal with dephasing effects and when applied to a setup similar to the one proposed by McClure *et al.* [9] it clearly supports the existence of a Coulomb drag effect in a double dot system. We also identify some specific features like negative differential conductance regime for the unbiased system due to the intradot transitions or temperature effects.

The rest of the paper is organized as follows. Section II presents the model and gives the main equations. Section III contains the numerical results and their discussion. We conclude in Section IV with a brief summary.

THE MODEL

We shall use a tight-binding Hamiltonian to describe the two subsystems. The Hamiltonians of the non-interacting quantum dots are denoted by H_i ($i = 1, 2$) and the leads are modelled as one-dimensional chains, their Hamiltonian being denoted by H_L. The lead-dot coupling H_T and the Coulomb interaction H_I are included adiabatically through a smooth switching function $\chi(t)$. The on-site energies in each quantum dot are given by $\varepsilon_m^{(i)}$ and the creation/annihilation operators by $c_{m_i}^\dagger/c_{m_i}$. The index i indicates that the site m belongs to QD_i. The states in the leads are denoted by q_l where $l = L1, R1, L2, R2$. We use creation/annihilation operators $d_{q_l}^\dagger$ and d_{q_l}. Then, the Hamiltonian of the coupled system reads therefore as follows:

$$H(t) = H_1 + H_2 + H_L + \chi(t)(H_I + H_T) \tag{1}$$

where

$$H_i = \sum_{i=1}^{2} \sum_{m,n \in QD_i} (\varepsilon_m^{(i)} \delta_{mn} + t_{mn}^{(i)}) c_{m_i}^\dagger c_{n_i}, \tag{2}$$

$$H_T = \sum_{i,n} \sum_{l,q} (V_{n_i q_l} c_{n_i}^\dagger d_{q_l} + h.c), \tag{3}$$

$$H_I = \sum_{i,j} \sum_{m,n} W_{0,n_i m_j} c_{n_i}^\dagger c_{n_i} c_{m_j}^\dagger c_{m_j}, \tag{4}$$

We introduced the hopping integrals $V_{n_i q_l}$ between the leads and the sample. $W_{0,n_i m_j} = U/|r_n^{(i)} - r_m^{(j)}|$ is the bare Coulomb potential depending on the constant U and on the distance between two sites from the double-dot system. By taking a nearest-neighbor coupling between the leads and the dots and $V_{n_iq_l}$ has four non-vanishing elements $V_{L1(R1)}$ and $V_{L2(R2)}$. Also, $t_{mn}^i = t_D$ if m, n are nearest neighbors and zero otherwise. Each lead has its own chemical potential included in the corresponding Fermi functions. Within the Green-Keldysh formalism the steady-state current entering QD_1 from the left lead (t_L is the hopping energy on leads) is given by (see [13])

$$J_1 = \frac{e}{h} \int_{-2t_L}^{2t_L} dE \, \text{Tr}\{\Gamma_{L1} G^R \Gamma_{R1} G^A (f_{L1} - f_{R1}) - \Gamma_{L1} G^R \text{Im}(\Sigma_I^< + 2f_{L1}\Sigma_I^R)G^A\}. \quad (5)$$

The Green functions in Eq. (5) should be computed from the Dyson equation in the Keldysh space

$$G = G_0 + G_0(\Sigma_L + \Sigma_I)G, \quad (6)$$

written in terms of the Green function of the disconnected non-interacting system G_0. $\Sigma_{I,L}$ are the interaction self-energy and the leads' self-energy, respectively. $\Gamma_{L1} = \Gamma_{R1}$ are coupling matrices. The random-phase approximation scheme is implemented as follows. We define the contour-ordered polarization operator Π_{mn} which is used to construct the screened interaction W, with the retarded, lesser and greater components given by the Dyson and Keldysh equations:

$$W^R(E) = W_0 + W_0 \Pi^R(E) W^R(E) \quad (7)$$
$$W^{<,>}(E) = W^R(E) \Pi^{<,>}(E) W^A(E). \quad (8)$$

The numerical simulation presented in Section III were mostly performed within the non-selfconsistent scheme which takes however into account the creation of electron-hole pairs in the double dot system. The self-consistent scheme requires more computational effort because of the convergence condition. We compare the two approaches at the end of Section III and show that they lead to qualitatively similar results. The expansion of the effective interaction W with respect to W_0 leads easily to the standard form of the interaction self-energy:

$$\Sigma_I = \Sigma_H + \Sigma_X + \Sigma_C, \quad (9)$$

where the Hartree, exchange and correlation contributions have been separated. [15] Note that (i) Σ_H does not contribute to the second term in the current since it has no imaginary part; (ii) in the G_0W approximation Σ_X cannot couple sites from different dots because the noninteracting Green function has a block-diagonal form. The effect of the interdot interaction is then expected to come from higher-order diagrams containing the electron-hole bubble.

The lesser Green function can be used to compute the occupation number N_i of each dot or the density of states ρ_i (DOS), according to the definitions:

$$N_i = \frac{1}{2\pi} \sum_{m \in QD_i} \int_{-2t_L}^{2t_L} dE \, \text{Im} G_{mm}^< = \int_{-2t_L}^{2t_L} dE \rho_i(E) \quad (10)$$

In this work we consider only the long-range Coulomb interaction and disregard the on-site Hubbard interaction as well as the spin degree of freedom.

RESULTS

The numerical calculations we shall present below were performed for 3-site quantum dots that are coupled to one-dimensional leads. The hopping constant on the leads is denoted by t_L and is taken as the energy unit for the bias and the interaction strength. The current is given in units of et_L/h. The first quantum dot QD_1 is coupled to unbiased leads, while the second dot is subjected to a bias $V_2 = \mu_{L2} - \mu_{R2}$. We will study the transport properties of QD_1 when the second dot is driven out of equilibrium by varying V_2 such that the bias window covers the quantum dot levels one by one. This setup corresponds more to the experiment of McClure $et\ al.$ [9]

The passive dot is set in the Coulomb blockade regime, by taking the chemical potentials of the leads $\mu_{L1} = \mu_{R1} = \mu_0 = 1.0$; in this situation the highest level of the dot is well above μ_0 and the other two below it. The transport through the active dot is now controlled by the applied bias. The latter is varied by gradually decreasing μ_{R2} from 3 to -3 while the chemical potential of the left lead is fixed (i.e. $\mu_{L2} = 3$). Through the variation of μ_{R2} the levels of the active dot become available for transport and as a consequence we expect a Coulomb-induced response of the passive dot. Figs. 1(a) and (b) confirm that a current passing through the active dot leads to a drag current in the passive dot. Indeed, the onset of a current in QD_2 at $V_2 \sim 1.25$ corresponds to the appearance of a current in QD_1. We note that the drag current has the same sign as the driving current J_2. This suggests that the mechanism behind the Coulomb drag in parallel double quantum dots is the same as in the bilayer systems.

One can also observe that this setup leads to a sawtooth like behavior of the dragged current rather than to a series of peaks. This is because in this situation the levels of the active dots enter the bias window one by one and remain there. Note that the dragged current however saturates when the lowest level of the active dot passes above μ_{R2}. The sawtooth like behavior is due to the fact that in some regions $dJ_1/dV_2 < 0$, that is, the differential conductance of the passive dot is negative. In order to understand this aspect analysed the occupation number N_1 as a function of V_2. Fig. 2(a) shows that N_1 decreases when one level of QD_2 enters the bias window, but also that between two steps of J_2 the occupation number of QD_1 $increases$ slightly as well, especially in the case $U = 0.1$. This means that in these regions some charge actually accumulates in the dot. This happens when the energy transferred in the inelastic scattering processes increases and matches the gaps $E_3 - E_2$, $E_2 - E_1$, or $E_3 - E_1$, $E_{1,..3}$ being the levels of the passive dot. We have actually checked this fact by looking at the density of states in QD_1 (not shown). Also, if we consider only single site dots and repeat the numerical calculations of the dragged current we get a single step in J_1 and no negative differential conductance; this confirms our explanations about the competing role of the intradot transitions (no transitions are possible in the single-level case).

We remark that the drag current saturates when the lowest level of the active dot passes above μ_{R2}. The occupation number N_2 given in Fig. 2(b) confirms this scenario: N_1 decreases in step and reaches a constant value 1.5 which corresponds to half-filling of the active dot in the steady-state regime.

Fig. 3(a) shows the effect of the lead-dot coupling that characterizes the passive dot. In configuration A the relaxation processes from the level located above the chemical potential of the leads are enhanced, because the tunneling rates in and out of the dot

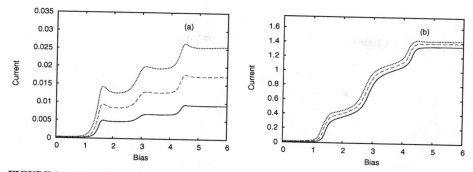

FIGURE 1. (Color online) (a) The currents through the passive dot as a function of the bias V_2 applied on the active dot for different values of the interaction strength. Full line - $U = 0.05$, dashed line - $U = 0.075$, dotted line - $U = 0.1$. (b) The current through the active dot as a function of V_2. Other parameters: $V_{L1} = V_{R1} = 0.25$, $V_{L2} = V_{R2} = 0.5$, $kT = 0.001$.

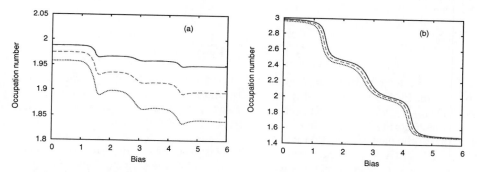

FIGURE 2. (Color online) a) The occupation number of the passive dot as a function of V_2. (b) The occupation number N_2 as a function of the applied potential. Other parameters: $V_{L1} = V_{R1} = 0.25$, $V_{L2} = V_{R2} = 0.5$, $kT = 0.001$.

increase as the lead-dot coupling increases. Also, the electronic number within the dots is not quantized anymore and the level broadening increases. The drag effect can still be observed and at higher coupling one gets a step-like structure of the current, the regions with negative differential conductance being again washed out. Note that since the active dot is not affected when the couplings $V_{L1,R1}$ are varied the onset of the drag current appears roughly at the same value of V_2.

As we have already mentioned, the present scheme can be improved by computing self-consistently the Green functions. We find good convergence for interaction strengths up to $U = 0.2$ and rather moderate coupling to the leads. The self-consistency condition requires a larger number of steps in the energy integrals and it is therefore time and memory consuming. We compare in Fig. 3(b) the self-consistent and non-selfconsistent currents for $U = 0.1$. We find that the self-consistent current curves are shifted with respect to the ones presented here and that the drag current tends to be larger on the first peak. Nevertheless, the results are qualitatively similar.

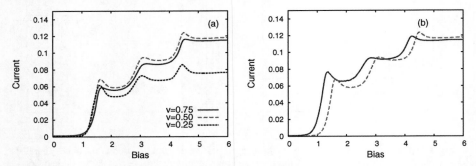

FIGURE 3. (Color online) (a) The drag current J_1 at different couplings between the passive dot and the leads. (b) Comparison between self-consistent and non-selfconsistent schemes. Full line - self-consistent result for the current through the passive dot; long-dashed line - non-selfconsistent result. Other parameters: $U = 0.1$, $kT = 0.001$, $V_{L1} = V_{R1} = 0.5$, $V_{L2} = V_{R2} = 0.5$

CONCLUSION

In this work we have investigated the Coulomb drag effect in a parallel double-quantum dot system. Each dot is coupled to two leads, one of them being set in the Coulomb blockade regime while the other is conducting due to an applied bias on the leads. Using the non-equilibrium Keldysh Green's function formalism and the random-phase approximation for the electron-electron interaction we show that a current passes through the blockaded (passive) dot when increasing the bias on the second (active) dot. The dragged current increases rather abruptly when one more level of the active dot enters the bias window. Between these jumps the passive dot also exhibits negative differential conductance (i.e. $dJ_1/dV_2 < 0$). We explain that this feature is due to the transitions between the levels of the passive dot which become possible at higher values of the driving bias and of the interaction strength. The numerical calculations given here were performed for 3-site quantum dots, but we find similar results for other configurations. We have also analysed the dependence of the Coulomb drag on the interaction strength, lead-dot coupling and temperature. In particular we have compared the selfconsistent and non-selfconsistent RPA.

ACKNOWLEDGMENTS

This work is supported by TUBITAK (No. 108T743) and TUBA. V.M. thanks the hospitality of Department of Physics, Bilkent University and acknowledges support from TUBITAK-BIDEP and Romanian Ministry of Education and Research through the PNCDI2 program.

REFERENCES

1. T. J. Gramila, J. P. -Eisenstein, A. H. MacDonald, L. N. Pfeiffer, and K. W. West, Phys. Rev. Lett. **66**, 1216 (1991).
2. A.G. Rojo, J. Phys.: Condens. Matter **11**, R31 (1999).
3. P. Debray, V. Gurevich, R. Klesse, and R.S. Newrock, Semicond. Sci. Technol. **17**, R21 (2002).
4. D. Laroche, E. S. Bielejec, J. L. Reno, G. Gervais, and M. P. Lilly, Physica E **40**, 1569 (2008).
5. M. Yamamoto, M. Stopa, Y. Tokura, Y. Hirayama, and S. Tarucha, Science **313**, 204 (2006).
6. E. Buks, R. Schuster, M. Heiblum, D. Mahalu, and V. Umansky, Nature (London) **391**, 871 (1998).
7. A.C. Johnson, C.M. Marcus, M.P. Hanson, and A.C. Gossard, Phys. Rev. Lett. **93**, 106803 (2004).
8. S. Gustavsson, R. Leturcq, B. Simovic, R. Schleser, T. Ihn, P. Studerus, K. Ensslin, D. C. Driscoll and A. C. Gossard, Phys. Rev. Lett. **96**, 076605 (2006).
9. D.T. McClure, L. DiCarlo, Y. Zhang, H.-A. Engel, C.M. Marcus, M.P. Hanson, and A.C. Gossard, Phys. Rev. Lett. **98**, 056801 (2007).
10. R. Asgari, B. Tanatar, and B. Davoudi, Phys. Rev. B **77**, 115301 (2008).
11. M.C. Goorden and M. Büttiker, Phys. Rev. B **77**, 205323 (2008); Phys. Rev. Lett. **99**, 146801 (2007).
12. S. Haupt, J. Aghassi, M.H. Hettler, and G. Schön, arXiv:0802.3579.
13. V. Moldoveanu and B. Tanatar, Phys. Rev. B **77**, 195302 (2008).
14. A.-P. Jauho and H. Smith, Phys. Rev. B **47**, 4420 (1993).
15. The explicit expression can be traced from Ref. [13]: Σ_C is given in Eqs. (21)-(24), while the exchange and Hartree contributions are given by Eq. (25) and (26).
16. V. S. Khrapai, S. Ludwig, J. P. Kotthaus, H. P. Tranitz and W. Wegscheider, Phys. Rev. Lett. **99**, 096803 (2007), Phys. Rev. Lett. **97**, 176803 (2006).

Transport and thermodynamic properties of 2DEG in the presence of tilted magnetic field

G.A. Farias[a], A.C.A. Ramos[b], T.F.A. Alves[a], R.N. Costa Filho[c], and N.S. Almeida[a]

[a]*Departamento de Física, Universidade Federal do Ceará, Caixa Postal 6030, Campus do Pici, 60455-760, Fortaleza- Ceará.*
[b]*Universidade Federal do Ceará, Campus do Cariri, 63030-200 Juazeiro do Norte,Ceará, Brazil*
[c]*Department of Physics and Astronomy, University of Western Ontario, London, Ontario, Canada, N6A3K7*

Abstract. We study the effects of a tilted magnetic field acting on a non-interacting two-dimensional electron gas (2DEG). We show that the existence of an angle between the external magnetic field and the normal direction to the 2DEG plane introduces severe changes on the electronic properties of the electron gas, with consequences on the thermodynamic and transport properties. Here, we consider that the electrons of the gas feel the presence of a tilted magnetic field and are confined by a non-symmetric potential. Then, taking into account the Zeeman and Rashba spin-orbit interactions, we obtain the Fermi energy, magnetization and the Hall conductivity for magnetic fields with different intensities and tilt angles. The influence of a weak periodic potential on these quantities is also analyzed. Our results show that some specific physical properties can be given to the 2DEG (in a reversible way) with a convenient choice of the intensity and tilt angle.

Keywords: 2DEG, Tilted Magnetic Field, Thermodynamic Properties.
PACS: 71.10.Ca, 71.70.-d, 71.70.Di, 72.80.-r

INTRODUCTION

New and sophisticated growing techniques have allowed the production of heterostructures with electrons confined in a very tiny region. For most purposes, these confined carriers, here named 2DEG, can be seen as a two dimensional system in the real three dimensional world[1]. They can be observed, for example, in heterostructures as GaAs/AlGaAs grown by molecular bean epitaxy (MBE) or metal-organic chemical vapor deposition (MOCVD). In these systems, the relative position of the GaAs and AlGaAs conduction band provides a trap to the electrons and it is responsible for their confinement in a region perpendicular to the growing direction[2,3]. Nowadays the thickness of the confining region can easily be smaller than one hundred angstroms, making it much smaller than the size of the lateral dimensions.

In this work we define the z axis parallel to the growing direction with the carriers confined between the planes $z = -\delta/2$ and $\delta/2$. These systems exhibit unique physical properties around the liquid helium temperature and would be overqualified for

CP1147, *Transport and Optical Properties of Nanomaterials—ICTOPON - 2009*, edited by M. R. Singh and R. H. Lipson
© 2009 American Institute of Physics 978-0-7354-0684-1/09/$25.00

technological application if these characteristics could be observed and used at higher temperature. However, to achieve this goal, it is necessary a quite precise description of the system, with special attention to the relative importance of the interactions felt by the particles. In this paper we present an effort to understand the effect of each interaction and its relative importance on the physical characteristic of these systems. Special attention will be given to the effects produced by a tilted magnetic field since the component perpendicular to the growing direction produces important modifications on the degeneracy of the states, with consequences on almost all physical observables of these systems. This fact gives a spin dependence to some physics observables of the system and it is due to the Zeemann interaction which, in a reversible way, modifies the electronic spin states. A second point to be considered is the form of the confining potential. The band offset of the semiconductor heterojuntions that host the electron gas can produce a field responsible for changes in the degeneracy of the states, even in the absence of external dc magnetic field. This is known as *Rashba effect*[2] and it can be enforced by an external electric field applied in the z direction. In other words, this interaction can also be reversible and externally controlled. In order to achieve a more complete description of these systems we must consider the electrons in presence of a periodic potential. The periodic structures of the materials that compose the heterostructure already introduce this kind of interaction. However, it can also be externally built (or reinforced) by some photolitographic techniques[3].

In this paper we consider a quasi two dimensional electron gas subjected to a tilted dc magnetic field, under the influence of the Rashba interaction and a weak periodic potential. Here, our main purpose is to find out how each one of these interactions affects the transport and thermodynamic properties of the gas. To do that, we characterize the system by a set of parameters (electron density, strength of the Rashba interaction and periodic potential, as well as the intensity and direction of the magnetic field) to obtain the energy spectra. With these results we calculate the Fermi energy, density of states, free energy, magnetization, susceptibility and Hall conductivity at temperature near to 0K for each set of parameters.

THE CONSEQUENCES OF THE RASHBA INTERACTION

For moderate or low electronic densities (around 10^{12} electrons per cm^2) the 2DEG can be described as a non interacting particles system and can be studied through the one particle approximation. In this case, the eigenstates should be obtained from the one electron Hamiltonian that can be written as:

$$H = \frac{1}{2m*}(\vec{p} + e\vec{A})^2 - \frac{e\hbar}{2m*}\vec{\sigma}.\vec{B} + \frac{\alpha_R}{\hbar}[\vec{\sigma} \times (\vec{p} + e\vec{A})]_z \qquad (1)$$

where the first term is the electron kinetic energy, with m* and e denoting the effective mass and charge of the electron, respectively, $\vec{P}$ is the linear moment and $\vec{A}$ the vector potential that, for a magnetic field in the plane x-z [$\vec{B} = B(\sin\theta\,\hat{x} + \cos\theta\,\hat{z})$], is given by $\vec{A} = -B(y\cos\theta\,\hat{x} - z\sin\theta\,\hat{y})$. Here we have named θ the angle between the field and the growing direction. The second term gives the Zeeman interaction, with $\vec{\sigma}$ denoting the Pauli matrices. The Rashba interaction is described by the third term where we have used the parameter α_R to indicate its strength. The Hamiltonian given by Eq. 1 can be rewritten to read:

$$H = \frac{1}{2m*}\left[(p_x^2 + p_y^2 - 2ep_x y\cos\theta + e^2 B^2 y^2 \cos^2\theta) + (p_z^2 + e^2 B^2 z^2 \sin^2\theta - 2ep_y z\sin\theta)\right]$$

$$+\frac{\alpha_R}{\hbar}\left[\sigma_x(p_y - eBz\sin\theta) - \sigma_y(p_x - eBy\cos\theta)\right] - \frac{e\hbar B}{2m*}(\sigma_x\sin\theta + \sigma_z\cos\theta).$$

$$(2)$$

The dependence of the Hamiltonian on z and p_z suggests that the z dependence of the wave function has a Gaussian form. We use this fact to define the effective Hamiltonian for the electrons in the 2DEG as:

$$H_{eff} = \frac{1}{2\pi\delta^2}\int_{-\infty}^{\infty} dz e^{-z^2/\delta^2} H$$

$$= \frac{p_y^2}{2m*} + \frac{1}{2}m*\omega_c\cos\theta(y - y_c)^2 + \frac{\alpha_R}{\hbar}\left[\sigma_x p_y + m*\omega_c\cos\theta\sigma_y(y - y_c)\right]$$

$$- \frac{\hbar\omega_c}{2}(\sigma_x\sin\theta + \sigma_z\cos\theta) + \frac{1}{2}m*\omega_c^2\delta^2\sin^2\theta.$$

$$(3)$$

where $\omega_c = eB/m*$, and $y_c = p_x/m*\omega_c\cos\theta$. Next we consider the width of the Gaussian equal to the thickness of the region where the electrons are confined. This approach allows us to work with the carriers as a truly two dimensional system, but takes into account the finite value of the thickness.

An electron constrained to move in a two dimensional region and in the presence of a dc magnetic field perpendicular to the confining region, has analytical and well known solutions. They are equivalent to one dimensional harmonic oscillator around a position that depends on the component of the linear moment perpendicular to it. We named these eigenvectors as $|n,\sigma>$ [$H_0 | n,\sigma >= (n + 1/2 + \sigma/2)\hbar\omega_c$, where H_0 is H_{eff} for θ and α_R equal to zero], to write the eigenvectors for the Hamiltonian given by Eq. 3 as:

$$|\Psi(p_x)> = \frac{e^{ik_x x}}{\sqrt{L_x}}\sum_{n,\sigma} C_{n,\sigma} | n,\sigma >$$

$$(4)$$

A bit of algebra allows us to reduce our problem to the search for eigenvalues and eigenvectors for the tridiagonal matrix given by:

$$\begin{bmatrix} h_{11} - E & h_{1,2} & 0 & \cdots \\ h_{2,1} & h_{2,2} - E & h_{2,3} & \cdots \\ 0 & h_{3,2} & h_{3,3} - E & \cdots \\ & & h_{4,3} & \\ \cdot & \cdot & \cdot & \\ \cdot & \cdot & \cdot & \\ \cdot & \cdot & \cdot & \end{bmatrix}\begin{pmatrix} v_1 \\ v_2 \\ v_3 \\ \cdot \\ \cdot \\ \cdot \end{pmatrix} = 0$$

$$(5)$$

where the elements of the matrix are given by:

$$h_{n,m} = \begin{cases} \left[(\frac{n+1}{2})\,\hbar\omega_c\cos\theta + \frac{1}{2}m*\omega_c^2\delta^2\sin^2\theta\right]\delta_{n,m} - \left[\frac{\hbar\omega_c}{2}\sin\theta\right]\delta_{m,n+1} & \text{for n odd} \\ \left[(\frac{n}{2}-1)\hbar\omega_c\cos\theta + \frac{1}{2}m*\omega_c^2\delta^2\sin^2\theta\right]\delta_{n,m} - i\left[\alpha_R\sqrt{neB\cos\theta/\hbar}\right]\delta_{m,n+1} & \text{for n even} \end{cases}$$

(6)

This kind of secular equation was solved some time ago by Swalen and Pierce[2]. We use the procedure described in the Ref. 3 to write the ith eigenvalue as:

$$E_i = h_{i,i} - \cfrac{h_{i,i-1}h_{i-1,i}}{h_{i-1,i-1}-E_i - \cfrac{h_{i-1,i-2}h_{i-2,i-1}}{h_{i-2,i-2}-E_i - \cfrac{h_{i-2,i-3}h_{i-3,i-2}}{h_{i-3,i-3}-E_i - ....}}}$$

(7)

By following similar procedure the eigenvectors are obtained from:

$$\frac{v_{k\pm1}^{(i)}}{v_k^{(i)}} = \cfrac{h_{k\pm1,k}}{h_{k\pm1,k\pm1}-E_i - \cfrac{h_{k\pm2,k+1}h_{k\pm1,k+2}}{h_{k\pm2,k\pm2}-E_i - ....}}$$

(8)

The iterative method was used to obtain numerical values for the eigenvalues with error smaller than 10^{-4} percent. The density of states is given by $D(E) = \frac{L_xL_y}{l_c^2 2\pi}\sum_i\delta(E-E_i)$, where $l_c^2 = \hbar/m*\omega_c\cos\theta$ and L_i (i=x,y) the lateral lengths. However, in order to take into account the finite value of width of the energy levels, we write[5,6] it as $D(E) = \frac{L_xL_y}{l_c^2 2\pi\Gamma}\sum_i e^{-(E-E_i)^2/2\Gamma^2}$. We notice that the effect of this interaction is to produce modulations in D(E) which are a consequence of the fact that its action depends on the dynamical observables of each state.

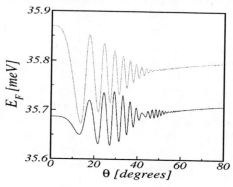

FIGURE 1. Fermi energy for B=1T, as a function of θ for a system with n_s=10^{12} electrons per cm^2, Γ=0.5meV; α_R=0 (dotted line) and α_R= 10^{-11}eV/m (full line).

Similar signature can be observed for the Fermi energy. In Fig. 1 we show the behavior of this quantity with the tilt angle for a magnetic field of moderate intensity (one Tesla), and density $n_s=10^{12}$ electrons per cm^2. In this picture one can see not only the influence of the Rashba interaction, but also the effect produced by the change of the direction of the magnetic field.

The Hall conductivity σ_{xy} is given by[5,6]:

$$\sigma_{xy} = \frac{ie^2\hbar}{l_c^2 2\pi} \sum_{i,i'} \frac{F(E_i) - F(E_{i'})}{(E_i - E_{i'})^2} v_x^{i',i} v_y^{i,i'}$$

(9)

where $F(E)$ is the Fermi-Dirac distribution, and $v_k^{i,j}$ is the matrix element of the velocity operator $v_k = \dfrac{\partial H}{\partial p_k}$ between the states i and j.

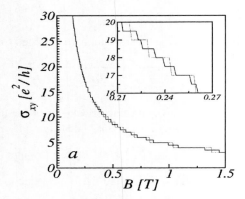

FIGURE 2. Hall conductivity as a function of the intensity of the magnetic field (for θ=0) in a system with $n_s=2.10^{11}$ electrons per cm^2, $\Gamma=0.5$meV; $\alpha_R=0$ (dotted line) and $\alpha_R=10^{-11}$eV/m (full line).

It can be observed in the Fig 2 that the main effect of the Rashba interaction is to introduce an intermediary step between any two consecutive values obtained in the absence of this interaction ($\alpha_R=0$). This same behavior is obtained when the tilt angle has a finite value. In the Fig. 3 we show the behavior of σ_{xy} as a function of the tilt angle and the same general form is also obtained, but one can observe a different structure introduced by the direction of the magnetic field.

The previous results allow us to obtain the free energy. From that, we calculated the magnetization of the system[7] . In Fig 4 one can appreciate the behavior of the magnetization as a function of the intensity of the magnetic field and the effect introduced by the Rashba interaction. As can be seen, its main effect is a modulation of the magnetization as that one observed in the Fermi energy.

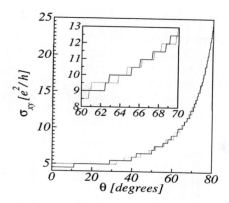

FIGURE 3. Hall conductivity as a function of θ for B=1T in a system with $n_s=2.10^{11}$ electrons per cm², Γ=0.5meV; $α_R=0$ (dotted line) and $α_R= 10^{-11}$eV/m (full line).

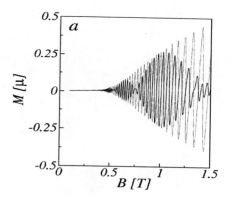

FIGURE 4. Magnetization (in units of eh/m*c) as a function of B, for θ=30° in a system with $n_s=10^{12}$ electrons per cm², Γ=0.5meV; $α_R=0$ (dotted line) and $α_R= 10^{-11}$eV/m (full line).

THE EFFECTS OF A PERIODIC POTENTIAL

We assume that the periodicity of the potential is in the y direction and can be described by:

$$V_p = V_0 \cos(Ky) \qquad (10)$$

where the amplitude V_0 is supposedly much smaller than the separation of any two energy levels, and $2π/K$ is the periodicity of the potential. We use the first order perturbation theory to estimate the effects of this potential on the physical properties of the system. We calculate the correction on the energy levels (eigenvalues) and eigenstates to recalculate the physical observables studied in the previous Section. In all numerical calculations we considered $V_0 = 0.03\,\mathrm{meV}$ and $K = 2\pi/35\,\mathrm{cm}^{-1}$. The results are shown in Figs 5 to 7. The oscillatory behavior of the Fermi energy when the field is increased and/or tilted is preserved, but the abrupt jump in the Hall conductivity disappears being replaced by a transition region, as can be seen in the Fig. 7. We notice that the size of this region depends on the periodicity of the potential.

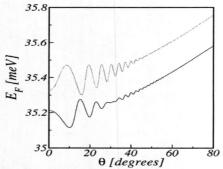

FIGURE 5. Fermi energy for B=1T, as a function of θ, for a system with $n_s=10^{12}$ electrons per cm², Γ=0.5 meV; $\alpha_R=0$ (dotted line) and $\alpha_R=10^{-11}$eV/m (full line).

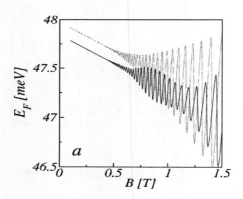

FIGURE 6. Fermi energy as a function of B for θ=0° in a system with $n_s=10^{12}$ electrons per cm², B=1T, Γ=0.5 meV; $\alpha_R=0$ (dotted line) and $\alpha_R=10^{-11}$eV/m (full line).

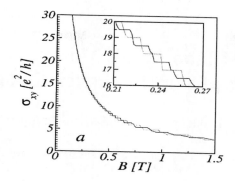

FIGURE 7. Hall conductivity as a function of the intensity of the magnetic field (for θ=0) in a system with n_s=2.10^{11} electrons per cm^2, Γ=0.5 meV; α_R=0 (dotted line) and α_R= 10^{-11}eV/m (full line).

CONCLUSIONS

The results discussed here show that these systems are quite sensitive to interactions present or externally introduced in the system. This sensibility, together with the fact that they have a very high mobility, would make them a good candidate for many technological applications. However, the behavior of the physical quantities discussed in this paper almost disappears when the temperature approaches to 10K. At the moment, the biggest challenge is to find out the conditions (parameters) to these systems have at higher temperatures the characteristics observed around 0K.

ACKNOWLEDGMENTS

The authors thank the Brazilian National Research Councils (CNPq) and the Fundação Cearense de Apoio a Pesquisa (FUNCAP) for the financial support.

REFERENCES

1. T. Ando, A.B. Fowler and F. Stern, *Rev. Mod. Phys.* **54**, 437 (1982).
2. Y.A. Bychkov, and E.I. Rashba, *J. Phys. C: Solid State Phys.* **17**, 6039 (1984).
3. R.M. Winkler, J.P. Kotthaus, and K. Ploog, *Phys. Rev. Lett.* **62**, 85302 (2003).
4. J.D. Swalen and L.J. Pierce, *J. of Math. Phys.* **2**, 736 (1961).
5. X. F. Wang and P. Vasilopoulos, *Phys. Rev. B* **67**, 085313 (2003).
6. X. F. Wang, P. Vasilopoulos and F. M. Peeters, *Phys. Rev. B* **71**, 125301 (2005).
7. see for example, L.D. Landau and E.M. Lifshitz "*Statistical Physics*" translated by E. Peierls and R.F. Peierls, Addison-Wesley Publishing Co., Inc., Reading, Mass. (1958).

Multi Million Atoms Molecular Statics Simulations of Semiconductor Nanostructures

V. Haxha, R. Garg, and M. A. Migliorato

School of Electrical and Electronic Engineering, University of Manchester,
Sackville Street, Manchester, M60 1QD

Abstract. Atomistic Molecular Dynamics and Molecular Statics simulations are nowadays capable of accessing highly sophisticated high performance computer architectures. Therefore it is becoming natural to push these methods towards realistic experimental sizes in order to design simulations with predictive power. In order to simulate elastic strain in semiconductor nanostructures and their immediate environment one needs to typically include at least 1 million atoms. Such sizes require ad hoc methods that in spite of their empirical nature can still prove accurate compared to ab initio methods, at least in the subset of the physical properties that one is simulating. We will show how the Abell-Tersoff empirical potential can be convincingly used to simulate the elastic behavior of even epitaxial quantum dots comprising 3 million atoms and made of complex quaternary semiconductor alloys.

Keywords: Atomistic simulations, semiconductors
PACS: 61.43.Bn, 61.72.uj

INTRODUCTION

Atomistic empirical potential methods for Molecular Dynamics (MD) and Molecular Statics (MS) simulations of low dimensional III-V semiconductor materials have attracted a substantial amount of interest in recent years.[1,2,3,4,5,6,7,8,9,10] The motivation for this spurs from the yet limited system sizes that are possible to simulate through *ab initio* methods. Furthermore interest also lies on the need for nanostructures to accurately determine the changes in bandstructure due to elastic deformation,[11] the fact that e.g. epitaxial quantum dots show stochastic variations of the local stoichiometry on the nanometer scale and the fact that atomistic symmetry groups need to be correctly reproduced.[12] Compared to other conventionally used methods such as Valence Force Field[13] and Continuum Elasticity Theory, properly parameterised empirical potentials the Abell-Tersoff potentials (ATPs),[14,15,16,17] can accurately capture the essential elastic properties of a material, with the same accuracy as an *ab initio* calculation but with a substantially lower computational demand. Previously we have presented a comprehensive collection of highly optimised parameter sets for zinc blende III-V semiconductor materials obtained by fitting to the cohesive energy, lattice parameter, the 3 elastic constants, the internal sublattice displacement and non linear effects.[18]

CP1147, *Transport and Optical Properties of Nanomaterials—ICTOPON - 2009*, edited by M. R. Singh and R. H. Lipson
© 2009 American Institute of Physics 978-0-7354-0684-1/09/$25.00

QUANTUM DOTS FOR TELECOMMUNICATION

Of epitaxial self assembled quantum dot (QD) islands, the most interesting from the modelling point of view, are the so called long wavelength QDs. In fact typically InAs/GaAs QDs grown by means of Molecular Beam Epitaxy (MBE) at roughly the first stages of 2D to 3D transition, emit at around 900nm. However by overgrowing larger islands can be produced that can extend their room temperature lasing to 1300nm.[19] Such MBE growth is not straightforward and it was found that particular care needs to be taken during the capping process. Strain Relieving Layers (SRLs) have been used successfully to improve the crystal quality and consequently lasing efficiency.[20,21]

Recently, cappings of 6nm GaAsSb with different Sb content has also been tried. When GaAs is used as a capping material strong intermixing[22] is generally expected but for the case of InGaAsSb this does not happen as expected.[23] Experimental data (previously reported[23]) of the chemical analysis of the In and Sb profiles in a region on the side of the QDs shown a distinct reduction of intermixing and a tendency for incorporation as ternary alloys.

ATOMISTIC SIMULATIONS

In order to better understand the influence of the Sb atomic species on the elastic behavior structure, we implemented a series of Molecular Statics simulations of the complete QD island, covered by 6 nm of capping layer and embedded in a GaAs matrix. The experimental structures we wished to compare where grown by MBE with the following sequence: a GaAs [001] substrate was covered by a GaAs buffer, followed by a 2.8 ML InAs layer, 6 nm GaSbAs and were finally capped with 50 nm GaAs. Three samples were grown with Sb composition of 12, 15 and 20%.

The atomistic models are built as follows: we initially design a GaAs (001) substrate section of dimensions 40x40x10nm. This is followed by a thin InAs 2D layer (wetting layer) of thickness of about 2 monolayers. The QD island sits on top of the 2D layer and is made of pure InAs. The island dimensions are roughly 28nm in diameter and 10nm in height. The dimensions are taken from the data acquired during STM analysis.[23] The shape of the island is taken form that proposed by Costantini et al^{24} where a detailed analysis of the facets observable in STM microscopy was presented. The region around the island and up to the height of the QD (SRL) was formed by pure GaAs, InGaAs, GaSbAs or InGaSbAs. The entire structure was then capped with a 10nm layer of pure GaAs.

In the [100] and [010] crystallographic directions we applied periodic boundary conditions. The size of the simulation box in these directions is sufficient to avoid the influence of the periodic images of the QD islands. In the [001] directions we applied a frozen approach for the bottom layer (only for the atoms in the bottom most two monolayers) and left the topmost monolayers free to move (floating layer). This is necessary because we cannot be sure a priori of the lattice parameter of the SRL region if Sb is present in the InGaAs alloy. Therefore unbounding the surface allows the structure to freely relax to its energy minimum without artificial constraints.

In order to minimize the computation time we pre-strain the regions that are definitely going to show strain, like e.g. the InAs island. This step proves to produce greatly accelerated relaxation.[25]

The structures comprise usually about 2.8 million atoms. To our knowledge these are the largest atomistic simulations of QDs ever attempted. The structures have been relaxed using Molecular Statics within the IMD™ software.[26] We implemented a parallel configuration and were normally able to conclude each simulation within a few hours on a 64 processors machine. The strain is directly extracted from the atomic bonds by comparing the strained lattice to an ideal unstrained one and taking the local composition into account.

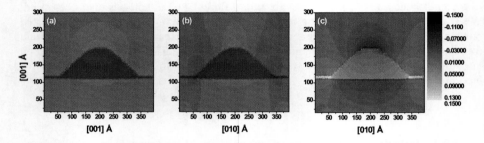

FIGURE 1. Strain maps for a GaAs/InAs/GaAs Quantum Dot island. (a) , (b) and (c) are the components of the strain ε_{xx} , ε_{yy} and ε_{zz}.

RESULTS

In this section we present the strain maps obtained for several of our atomistic models. We concentrate our attention to the diagonal components that compared to the off diagonal components are more significant in the description of tetragonal distortion i.e. the most important distortion in pseudomorphic layers.

In Fig. 1 we show such maps for the simple case of a GaAs/InAs/GaAs Quantum Dot island. The difference in the parallel strain components is due to having used an asymmetric shape in accordance to the shapes reported in the literature.[24] The propagation of the strain inside the GaAs barrier is very pronounced as expected.

The fact that the addition of an SRL during growth is a very effective way of reducing the strain in the GaAs matrix is confirmed by our simulations. In Fig. 2 we show the strain maps for the case of an SRL with 10% In and 20% Sb, 50% In and no Sb, 30% Sb and no In. In all cases the reduction of the propagating strain is evident, as well as a modification of the strain inside the islands.

However in the case of the SRL containing antimony is an idealized situation. We have previously reported that when trying to cover the islands with a GaAsSb SRL strong segregation effects are present and the SRL ends up with a mixed composition of InGaSbAs. This happens because the In in the wetting layer tends to diffuse and in part slows down the incorporation of Sb. The SRL therefore happens to be In rich at the bottom and Sb rich at the top, with the profiles being roughly described by segregation models.[23]

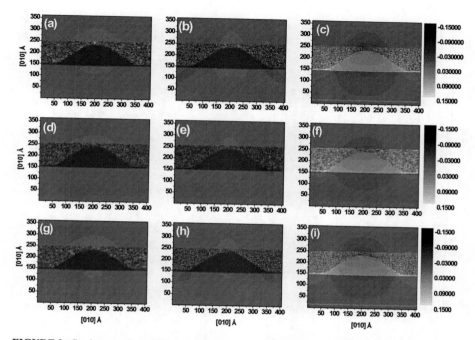

FIGURE 2. Strain maps for a InGaSbAs/InAs/GaAs Quantum Dot island. The SRL is composed by InGaSbAs with 10% In and 20% Sb ((a),(b) and (c)), InGaAs with 50% In ((d),(e) and (f)) or GaSbAs with 30% Sb ((g),(h) and (i)). The first, second and third columns are the components of the strain ε_{xx}, ε_{yy} and ε_{zz}, respectively.

The question we need to answer is whether such a segregated and intermixed profile has a different effect on the strain in the GaAs matrix and/or the island.

In order to answer this question we simulated the intermixing of a GaSbAs SRL with nominally 12, 15 and 20% Sb. All other parameters are kept the same. The compositions are taken by carefully comparing experimental data from STM measurements and theoretical models of segregation, discussed elsewhere.[23] The experimental cross sectional STM topographies also clearly indicate that the material in the SRL does not form a uniform 2D layer but rather wraps around the QD island. This effect as well was reproduced in our input models.

The results of the simulations indicate that the strain distribution is strongly altered by the presence of an intermixed SRL compared with having a uniform composition. Of the three compositions tested we found that the highest the Sb content, the lower the strain in the matrix, although at the expense of higher strain in the island.

If we then compare the highest Sb composition in both the uniform and non uniform case, it is obvious that the SRL with high intermixing is more effective in reducing the strain in the GaAs regions both above and below the QD island.

The lower strain in the GaAs matrix with a high Sb content can be exploited in the growth of high quality multilayered QDs. In fact strain often acts as a "seed" [27] and favors growth of islands on top of existing islands. It is normally accepted that for laser devices uncorrelated structures have better performances.

Therefore the reduced strain would allow islands to be more closely packed, ultimately producing lasers with a higher volume of the active region compared to the surrounding material.

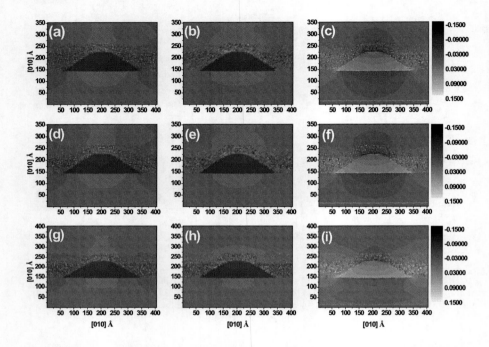

FIGURE 3. Strain maps for a InGaSbAs/InAs/GaAs Quantum Dot island. The SRL is intermixed and composed nominally by GaSbAs with 12% Sb ((a),(b) and (c)), 15% Sb ((d),(e) and (f)) and 20% Sb ((g),(h) and (i)). The first, second and third columns are the components of the strain ε_{xx}, ε_{yy} and ε_{zz}, respectively.

CONCLUSIONS

We reported a series of atomistic simulation of InAs QDs and show how different strain reducing layers influence the strain status in the island and the GaAs matrix. We used realistic models of the islands and the strain reducing layer. In particular we used models that accurately reproduce the experimentally determined compositional gradients. We conclude that because of its lower strain in the GaAs matrix a high Sb content is better suited to allow the growth of high quality multilayered QDs.

ACKNOWLEDGMENTS

We acknowledge the support of the Engineering and Physical Sciences Council (EPSRC) and of the Royal Academy of Engineering (RAEng). We would also like to thank Prof Koenraad and Ilias Drouzas of the Technical University of Eindhoven for valuable discussions.

REFERENCES

1. R. Smith 1992 *Nucl. Instr. Meth. Phys. Res. B* **67**, 335 (1992).
2. M. Sayed, J.H. Jefferson, A.B. Walker and A.G. Cullis, *Nucl. Instr. Meth. Phys. Res. B* **102**, 218 (1995).
3. P. A. Ashu, J. H. Jefferson, A. G. Cullis, W. E. Hagston, and C. R. Whitehouse, *J. Crystal Growth* **150**, 176 (1995).
4. M. Nakamura, H. Fujioka, K. Ono, M. Takeuchi, T. Mitsui and M. Oshima, *J. Crystal Growth* **209**, 232 (2000).
5. F. Benkabou, H. Aourag, P. J. Becker, and M. Certier, *Mol. Sim.* **23**, 327 (2000).
6. M. A. Migliorato, A.G. Cullis, M. Fearn and J. H. Jefferson, *Phys. Rev. B* **65**, 115316 (2002).
7. W. H. Moon and H. J. Hwang, *Phys. Lett. A* **315**, 319 (2003).
8. S. Goumri-Said, M. B. Kanoun, A. E. Merad, G. Merad and H. Aourag, *Chem. Phys.* **302**, 135 (2004).
9. S. R. Billeter, A. Curioni, D. Fischer, W. Andreoni, *Phys. Rev. B* **73**, 155329 (2006).
10. R. Drautz , D.A. Murdick, D. Nguyen-Manh, X.W. Zhou, H.N.G. Wadley, D.G. Pettifor, *Phys. Rev. B* **72**, 144105 (2005).
11. F. H. Pollak, *Semiconduct. Semimet.* **32**, 17 (1990).
12. G. Bester and A. Zunger, *Phys. Rev. B* **71**, 045318 (2005).
13. P. N. Keating, *Phys. Rev.* **145**, 637 (1966); *Phys. Rev.* **149**, 674 (1966).
14. G. C. Abell, *Phys. Rev. B* **31**, 6184 (1985).
15. J. Tersoff, *Phys. Rev. Lett.* **56**, 632 (1986).
16. J. Tersoff, *Phys. Rev. B* **37**, 6991 (1988).
17. J. Tersoff, *Phys. Rev. B* **39**, 5566 (1989).
18. D. Powell, M. A. Migliorato, and A. G. Cullis, *Phys. Rev. B* **75**, 115202 (2007).
19. J. M. Ulloa, I.W. Drouzas, P.M. Koenraad, D.J. Mowbray, M.J. Steer, H.Y. Liu, M. Hopkinson, *Appl. Phys. Lett.* **90**, 213105 (2007).
20. H. Y. Liu, Y. Qiu, C.Y. Jin, T. Walther, A.G. Cullis, *Appl. Phys. Lett* **92**,111906 (2008).
21. H. Y. Liu, M. J. Steer, T. J. Badcock, D. J. Mowbray, M. S. Skolnick, F. Suarez, J. S. Ng, M. Hopkinson, and J. P. R. David, *J. Appl. Phys.* **99**, 046104 (2006).
22. Heyn Ch. and W. Hanse, *J. Cryst. Growth.* **251**, 140 (2003).
23. V. Haxha, R. Garg, M. A. Migliorato, I. W. Drouzas, J. M. Ulloa, P. M. Koenraad, M. J. Steer, H. Y. Liu, M. J. Hopkinson, and D. J. Mowbray, *Microelectron. J.* **40**, 533 (2009).
24. G. Costantini, C. Manzano, R. Songmuang, O. K. Schmidt, and K. Kern, *Appl. Phys. Lett.* **82**, 3194 (2003).
25. M. A. Migliorato, A. G. Cullis, M. Fearn, J. H. Jefferson, *Phys. Rev. B* **65**, 115316 (2002).
26. J. Stadler, R. Mikulla, and H.-R. Trebin, *Int. J. Mod. Phys. C* **8**, 1131 (1997).
27. M. A. Migliorato, L. R. Wilson, D. J. Mowbray, M. S. Skolnick, M. Al-Khafaji, A. G. Cullis, and M. Hopkinson, *J. Appl. Phys.* **90**, 6374 (2001).

Diffusion In Nano-Scale Metal-Oxide/Si And Oxide/SiGe/Si Structures

Lyudmila V. Goncharova,[a] Nathan Yundt[b] and Eric G. Barbagiovanni[a]

[a]Department of Physics and Astronomy; University of Western Ontario, London, N6A 3K7 Ontario, Canada
[b]Department of Chemistry, University of Western Ontario, London, N6A 5B7 Ontario, Canada

Abstract. MOSFETs incorporating SiGe and $In_xGa_{1-x}As$ channels are an attractive option to further increase complementary MOS logic performance since these alloys offer electron mobilities significantly higher than Si. The poor electrical quality and low chemical stability of native oxides, leading to Fermi level pinning and high defect densities, often prevents the fabrication of competitive devices. Oxygen transport in model 2-10 nm thick systems, including Si, SiGe ultra-thin alloys, and hafnium oxide films grown by atomic layer deposition was studied by high-resolution ion scattering (MEIS) and conventional Rutherford Backscattering Spectroscopy (RBS) in combination with isotope tracing. We found much lower activation barrier value in the case of HfO_2/Si which is attributed to distinctly different oxygen incorporation mechanisms. The growth rate is two orders of magnitude faster for SiGe ultra-thin alloys and it cannot be described by the Deal-Grove model and its modifications, nor the reactive layer model. Potential outcomes of these reactions at the interfaces and associated band alignments with the introduction of metal gate materials will be discussed.

Keywords: alternative gate dielectrics, ion beam analysis, oxygen diffusion.
PACS: 61.72.-y; 77.55.+f; 79.20.Rf..

INTRODUCTION

As the existing complementary metal-oxide semiconductor (CMOS) technology continues to scale down, and device characteristics have reached the nanometer scale, tremendous efforts have been devoted to maintain the device performance gain by using innovations in the materials and integration schemes[1]. Interface passivation and defect minimization remain central to the viability of the integration of new materials. Transition metal (Hf, Zr, La) oxides and silicates with a dielectric constant higher than that of SiO_2 are currently being implemented and investigated as the gate dielectric materials in complementary metal-oxide-semiconductor (CMOS) devices[2]. The integration layer thickness and composition, and the device properties, are strongly effected by surface preparation [3], growth chemistry, thermal treatment and, often by the nature of the metal gate[4].

Many fundamental properties of O diffusion in ultra-thin (~10-30Å) layers are under debate, in particular the nature of the diffusing oxygen species, the role of oxygen vacancies[5], interstitial oxygen and other defects. Oxygen transport across the oxide

CP1147, *Transport and Optical Properties of Nanomaterials—ICTOPON - 2009*, edited by M. R. Singh and R. H. Lipson
© 2009 American Institute of Physics 978-0-7354-0684-1/09/$25.00

layer usually involves surmounting large potential barriers (~ 1-3eV), making the oxidation process kinetically limited. At high processing temperatures (~ 1000°C), thermal energy is large enough to provide appreciable atomic transport thereby allowing oxide growth. At lower temperatures (<500°C), thermal energy may be insufficient for a migrating species to overcome the transport potential barrier. Therefore, another driving force (besides temperature) for oxidation is usually involved, such as the defect structure of an oxide film and of its interface (effective electrical field or strain).

Thermal oxidation and diffusion processes become more complex when SiGe ultra-thin layer is introduced into device structure. Oxidation leads to a relaxation of the strained SiGe layer and to a segregation of Ge[6]. According to reports on oxidation at low temperatures (<1000°C) [7-9], the oxidation characteristics complexly depend on various conditions, *i.e.*, temperature, time, Ge fraction[6], presence of H_2O in the oxidizing environment[8,10].

We have examined a series of different dielectric/Si(001) (dielectric: SiO_2, HfO_2) and SiGe/Si(001) stack structures before and after annealing, and specifically looked at diffusion of oxygen in these 2-3nm thick films using an isotopic tracing approach. Our work follows the interfacial layer composition change in HfO_2/Si and SiO_2/SiGe/Si(001) with the emphasis on examining different oxygen and germanium diffusion processes and the behaviour of individual layers. We can also correlate high-resolution ion scattering profiles describing these diffusion processes with the variation of electronic band structures and band gaps measured by Bersch *et al*[11].

EXPERIMENTAL

2-3nm thick Hf oxide films were deposited on a 1 nm SiO_2/Si(001) film using atomic layer deposition (ALD) at 325 °C with O_3 as an oxidation agent[12]. SiGe /Si(001) alloys were prepared by 25keV Ge^+ ion implantation (1×10^{15} ions/cm^2) with a sample 60° tilt. This Ge implantation creates ~ 20 nm thick $Si_{1-x}Ge_x$ layer with the maximum Ge content at a depth of ~12-13nm. Reoxidation in $^{18}O_2$ (98% isotopically enriched) was performed *in-situ* in two different UHV chambers (p_{base} ~10^{-9} Torr) by stabilizing the sample at a temperature in the ~ 490 – 950 °C range (measured by an optical pyrometer and or K-type thermocouple), followed by $^{18}O_2$ or $^{16}O_2$ gas introduction at a pressure of 0.01Torr (5-30 min).

Rutherford backscattering spectroscopy and medium energy ion scattering (MEIS) were used to measure the ^{18}O and ^{16}O profiles to quantitatively determine the depth distribution of both oxygen species through the dielectric film. Oxidation rates were compared, and factors affecting oxygen exchange[13] such as film composition, and phase separation were examined.

MEIS was used to determine the depth profile of all elements in the dielectric (HfO_2 or SiO_2) ultra-thin layer. We used an H^+ beam with an energy of 130 keV. This energy was chosen to separate completely ^{18}O and ^{16}O peaks[14]. Depth profiles of the elements were obtained by computer simulations of the backscattered ion energy distributions. The depth resolution was ~ 3Å in the near surface region and ~ 8 Å at a depth of 30 Å[15]. RBS was applied in case of the thicker SiO_2 layer growth.

RESULTS

Figure 1 shows the part of the MEIS backscattered ion spectrum for the as-deposited HfO_2 film, where the oxygen yield has contributions from both HfO_2 and the SiO_2 interfacial layer. The proton energies corresponding to the high energy edges of Hf, O and C peaks are in excellent agreement with binary collision model calculations, which means that all these elements are located on the surface. The Si peak is shifted lower in energy to where it should be, assuming that the HfO_2 layer is continuously covering $SiO_2/Si(001)$. Scaling of Hf and O peaks shows that O is contained mostly within HfO_2 layer, with close to the oxygen-rich $HfO_{2.10\pm0.05}$ stoichiometry. Small hydrocarbon contamination of the top surface is apparent from a small C surface peak. (Note, that prior to re-oxidation in $^{18}O_2$ samples were typically annealed at 250-300°C to remove surface carbon.)

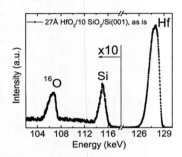

FIGURE 1. Backscattering MEIS spectrum for an as-deposited $HfO_2/SiO_y/Si$ (001) film in double channeling alignment for H^+ of 130keV, and scattering angle = 125.3°.

Figure 2a shows the part of the backscattered H^+ spectrum corresponding to both the isotopic oxygen peak positions for the as-deposited and $^{18}O_2$ re-oxidized HfO_2 films. Based on a full analysis of the Hf, Si and O peak shapes and energies, we conclude that the as-deposited HfO_2 film has an excess of oxygen and a 6-7Å interfacial SiO_2 layer. SiO_2 is also clearly visible in transmission infra-red spectra (not shown). A pronounced ^{18}O peak is observed after 10min of $^{18}O_2$ exposure (p_{18O2}=0.01Torr, 763K). Concurrent with the development of the ^{18}O peak, the intensity of the ^{16}O peak goes down. This observation shows that the ^{18}O peak is not due to the $^{18}O_2$ molecular dissolution in hafnium oxide, but rather due to an exchange reaction, i.e., ^{16}O leaves the surface and ^{18}O goes into the high-κ film[12]. After longer (40 min) $^{18}O_2$ exposure there is a larger increase in the ^{18}O aerial density and decrease in the ^{16}O aerial density, however, the total oxygen content (sum of ^{16}O and ^{18}O), as calculated from the oxygen peak area, remains the same. Note that for the same ^{18}O and ^{16}O content and distribution, the ^{18}O peak should have a $(^{18}/_{16})^2 \sim 1.27$ higher intensity than the ^{16}O peak, because of the different scattering cross sections. Kinetics of the exchange process differs between Hf oxide and silicate films of comparable thicknesses. The equilibrium $^{16}O/^{18}O$ exchange fraction is achieved in 5-10 mins of $^{18}O_2$ exposure for the oxide films, where it is much slower (more than 30min) for the silicate films (Figure 2b).

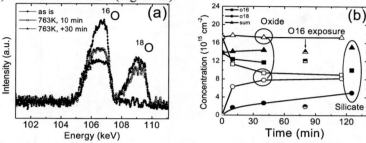

FIGURE 2. Variation of ^{18}O and ^{16}O peaks for (a) the $HfO_2/SiO_2/Si$ (001) as a function of re-oxidation time. (b) Oxygen exchange kinetics curves in hafnium oxide (open symbols) and hafnium silicate (dark symbols) films.

As the annealing temperature increases more significant exchange and interfacial growth is observed. The evolution of Si and O ion scattering peaks at the different temperatures (same $p_{18O2}=0.01$Torr, anneal time 30 min) is presented in Figure 3a. The rise of the Si peak area can be directly associated with SiO_x growth, and we will analyze it separately[16]. The ratio of Si and sum of oxygen peaks, and the depth profiles shown in Fig. 3b immediately suggest that the composition of the SiO_2/Si system does not have an atomically sharp transition at the interface but changes through $\sim$10-15Å thick region[17]. The detailed shape of the oxygen distribution is not certain due to straggling effects for buried layers. Silicon sub oxide formation, or interfacial layer roughness can both contribute to this effect[18].

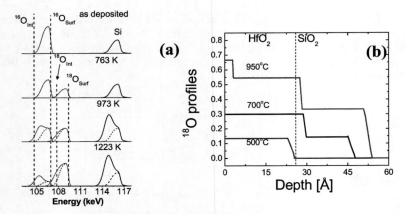

Figure 3. (a) Schematic representation of MEIS spectra for Si, ^{18}O and ^{16}O energy range of isotopic exchange and incorporation in hafnium oxide film corresponding to (a) as-deposited film, re-oxidized at (b) 763 K, (c) 973 K and (d)1223 K; (b) calculated ^{18}O depth profiles.

A much higher oxidation rate was observed for Ge-implanted samples. Since SiO_2 layer thicknesses were in the 100-2000Å range conventional RBS measurements were performed instead of MEIS. RBS spectra before and after high temperature exposure to oxygen (Figure 4) show a wide oxygen peak and stepped ion yield drop at the surface Si region (< 860keV). Additionally there is evidence of Ge redistribution in the growing SiO_2 layer. A single Ge peak at ~ 1180keV is noticeable in the as-implanted $Si_{1-x}Ge_x$/Si(001) sample data. Energy positions and peak yields are consistent with a SRIM[19] simulated profile. After oxygen exposure, the Ge surface peak disappears, whereas another lower energy (~ 1100keV) and broader peak appears, corresponding to Ge pile up at the growing SiO_2/Si interface.

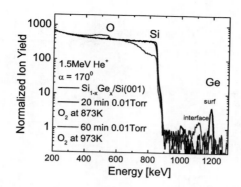

FIGURE 4. He⁺ RBS spectra of SiGe/Si(001) samples before and after high temperature oxygen exposure.

DISCUSSION

The kinetic information of the interfacial reactions can be obtained by studying the time, temperature, and pressure dependences of the interface $^{18}O/^{16}O$ concentration. We will focus on the temperature dependence here. Figure 5 shows an Arrhenius plot of the interface $^{18}O + ^{16}O$ for samples with 45Å of starting SiO_2, 27Å HfO_2, and 20Å/200Å $Si_{1-x}Ge_x$ after similar oxidizing conditions. Since we are concerned with ultrathin films where oxidation is presumably "reaction limited", the Deal-Grove model predicts the interface reaction to be independent of the starting oxide thickness and the increase of ^{18}O at the interface should depend on both time and pressure linearly.

One can see that the rate of oxide growth near the interface is almost one order of magnitude lower for the oxynitride films, and almost two orders of magnitude higher for 20Å/200Å $Si_{1-x}Ge_x$ films (compared to pure SiO_2 on Si(001) interface). The Arrhenius plots show straight lines with apparent activation energies of 2.7±0.1eV (SiO_xN_y/Si), 3.0±0.1eV (SiO_2/Si), 1.7±0.1eV ($Si_{1-x}Ge_x$ /Si), and 0.5±0.1eV (HfO_2/Si). Note the much lower activation barrier value in the case of HfO_2/Si which is attributed to distinctly different oxygen incorporation mechanisms[12].

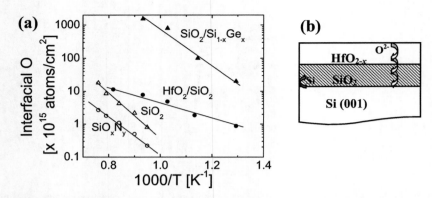

FIGURE 5. (a) Semi-logarithmic dependence of the amount of oxygen atoms incorporated near SiON interface of HfO_2/SiO_y (0.01Torr, 30min), SiO_xN_y and SiO_2 (7Torr, 60min), $SiO_2/Si_{1-x}Ge_x$ (0.02Torr, 60min) films on the inverse temperature after re-oxidation in oxygen. SiO_xN_y and SiO_2 data are reproduced with the permission from Ref.[20]; (b) Diagram of diffusion at $HfO_2/SiO_2/Si$ interface.

In the case of $Si_{1-x}Ge_x$ /Si, the activation barrier is also lowered. It was proposed that Ge acts as a catalyst during oxidation of Si, i.e., it enhances the reaction rate while remaining at the interface almost without any loss. Observed enhancement of the oxidation rate is consistent with earlier studies[9,10]. It was shown that the thin Ge layer at the interface between Si and its oxide remains unchanged with oxide thickness, and photoemission data indicated no evidence of GeO_x present. An alternative enhancement mechanism involves a catalytic action in which Ge promotes decomposition of O_2; any GeO formed, being less stable than SiO_2 is quickly reduced by Si to elemental Ge. Additional photoemission measurement will be conducted to resolve atomistic details of oxidation.

Our earlier results indicated that atomic oxygen diffusion via an oxygen lattice exchange mechanism is the predominant diffusion mechanism in hafnia[12], consistent with theoretical calculations[21]. Previous experiments with ZrO_2 films[14] and ultra-fine grained ZrO_2[22], and density functional calculations of oxygen incorporation and diffusion energies in monoclinic hafnia (HfO_2)[21] have suggested that oxygen incorporates and diffuses in atomic (ionic, non-molecular) form. Furthermore, O^{2-} becomes a more thermodynamically stable interstitial by accepting two electrons. Calculations[21] show that diffusion via oxygen lattice exchange should be the favored mechanism, however the barriers for interstitial oxygen diffusion in HfO_2 are small, and defects could be mobile under high temperature processing conditions. In contrast, molecular oxygen incorporation is preferred for the less dense SiO_2 structure, with diffusion proceeding through interstitial sites[23].

Presuming that the transported species are individual oxygen atoms (or ions), the availability of atomic oxygen at the surface is one of the factors affecting the extent of the exchange in the oxide. O_2 is expected to be adsorbed molecularly on perfect surfaces of HfO_2 and to dissociate primarily at O-vacancy defect sites. The amount of available atomic O depends on the rate of O_2 dissociation at the surface, and is therefore related to the number of oxygen vacancies at the top surface. This brings an interesting possibility of blocking oxygen dissociation by adsorption of 1ML of SiO_2. The covalent bonding of this layer inhibits oxygen dissociation, and therefore inhibits its further diffusion and interfacial SiO_2 growth. Additionally, as relaxation of atoms along the diffusion path is important for a lattice exchange mechanism, as-deposited disordered HfO_2 films might be expected to display a lower diffusion barrier than crystalline films.

CONCLUSIONS

Understanding fundamental processes of diffusion growth during fabrication of high-κ gate stacks is vital to establishing an atomic level control of interfacial layers and minimizing defects. The results of isotopic oxidation were examined in this work. We find that neither the Deal-Grove model and its modifications, nor the reactive layer model, offer an accurate description of very thin (<5nm) films. The complex oxidation behavior is likely to be a combination of interfacial, near-interfacial, and surface reactions. New features that cannot be explained in the traditional oxidation models are observed: (a) the isotopic exchange throughout whole Hf oxide film, and (b) the enhancement of interfacial SiO_2 layer growth compared to pure Si, both occurring in the case of Hf oxide films and, especially in the presence of Ge in the top Si layer. We also observe oxygen loss from the surface, suggesting SiO desorption or an O_2 surface exchange reaction. Contrary to the traditional viewpoint, this oxygen loss takes place during oxide film growth, on the "oxidation part" of the pressure-temperature phase diagram.

ACKNOWLEDGMENTS

Partial support of this project from the Natural Sciences and Engineering Research Council (NSERC) of Canada is gratefully acknowledged. The authors are grateful for the technical assistance of Mr. Jack Hendriks at Tandetron Accelerator facility, UWO.

REFERENCES

1. D. G. Schlom, S. Guha, and S. Datta, *MRS Bulletin* **33**, 1017 (2008).
2. J. J. Peterson, C. D. Young, J. Barnett *et al.*, *Electrochem. Solid State Lett.* **7**, G164 (2004).
3. Y. Wang, M.-T. Ho, L. V. Goncharova *et al.*, *Chem. Mater.* **19**, 3127 (2007).
4. Z. Chen, V. Misra, R. Haggerty *et al.*, *Phys. Stat. Sol. B* **241**, 2253 (2004); L. V. Goncharova, M. Dalponte, T. Gustafsson *et al.*, *J. Vacuum Sci. Technol. A* **25**, 261 (2007); D. Lim, R. Haight, M. Copel *et al.*, *Appl. Phys. Lett.* **87**, 72902 (2005).

5. H. S. Baik, M. Kim, G.-S. Park *et al.*, *Appl. Phys. Lett.* **85**, 672 (2004).
6. M. Tanaka, T. Ohka, T. Sadoh *et al.*, *J. Appl. Phys.* **103**, 054909 (2008).
7. O.W. Holland, C.W. White and D. Fathy, *Appl. Phys. Lett.* **51**, 520 (1987); J. Rappich, I. Sieber, and R. Knippelmeyer, *Electrochem. Solid St.* **4**, B11 (2001).
8. F. K. LeGoues, R. Rosenberg, and B.S. Meyerson, *Appl. Phys. Lett.* **54**, 644 (1989).
9. F. K. LeGoues, R. Rosenberg, T. Nguyen *et al.*, *J. Appl. Phys.* **65**, 1724 (1989).
10. D. K. Nayak, K. Kamjoo, J. S. Park *et al.*, *Appl. Phys. Lett.* **57**, 369 (1990).
11. E. Bersch, S. Rangan, R.A. Bartynski *et al.*, *Phys. Rev. B* **78**, 085114 (2008).
12. L.V. Goncharova, M. Dalponte, T. Gustafsson *et al.*, *Appl. Phys. Lett.* **89**, 044108 (2006).
13. We define oxygen exchange as substitution of the oxygen atoms within the framework of the thin film by oxygen coming from the gas phase. The number of atoms gained from the gas phase equals to the number of oxygen atoms lost, so total oxygen content of the dielectric remains constant during exchange. This exchange process has to be differentiated from interfacial incorporation and growth, where oxygen reacts with the film or the substrate, increasing the total oxygen content.
14. B. W. Busch, W. H. Schulte, E. Garfunkel *et al.*, *Phys. Rev. B* **62**, R13 290 (2002).
15. W.H. Schulte, B.W. Busch, E. Garfunkel *et al.*, *Nucl. Instrum. Meth. B* **183**, 16 (2001).
16. L. V. Goncharova, M. Dalponte, T. Gustafsson et al., (to be published).
17. An estimate of the error of the width of this region is ~ 5Å; taking into account an additional shift of 2Å from the asymmetry of energy straggling, we have a ~13+-5Å thick transition region, where the oxide structure deviate from the amorphous SiO_2.
18. E.P. Gusev, H.C. Lu, T. Gustafsson *et al.*, *Phys. Rev. B* **52**, 1759 (1995).
19. J.F. Ziegler, J.P. Biersack, and M.D. Ziegler, *SRIM - The Stopping and Range of Ions in Matter* (http://www.srim.org/, 2008).
20. H.C. Lu, Rutgers University, 1997.
21. A.S. Foster, A.L. Shluger and R.M. Nieminen, *Phys. Rev. Lett.* **89**, 225901 (2002).
22. U. Brossmann, R. Wurschum, U. Sodervall *et al.*, *J. Appl. Phys.* **85**, 7646 (1999).
23. S. Mukhopadhyay, P.V. Sushko, A.M. Stoneham *et al.*, *Phys. Rev. B* **71**, 235204 (2005).

Dipole-Exchange Theory of Spin Waves in Nanowires: Application to Arrays of Interacting Ferromagnetic Stripes

Hoa T. Nguyen and M. G. Cottam

Department of Physics & Astronomy, University of Western Ontario,
London, Ontario N6A 3K7, Canada

Abstract. Calculations for the dipole-exchange spin waves in small arrays of nano-scale ferromagnetic stripes are reported using a microscopic, or Hamiltonian-based, method. The stripes are coupled via the dipolar fields that extend across the nonmagnetic spacers between elements of the arrays that are either juxtaposed horizontally or vertically with respect to the stripe width. The coupling is found to be stronger in the vertical arrangement, which is the focus of this paper. The effects on the discrete spin-wave frequencies and the spatial distribution of the modes are studied in terms of dependence on an applied magnetic field and the wave number along the stripe axis.

Keywords: Spin waves; Dipole-exchange theory; Magnetic stripes; Magnetic nanostructures; Magnetic arrays; Ferromagnets.
PACS: 75.30.Ds; 75.70.Cn; 78.67.Pt; 75.50.Cc; 78.35.+c.

INTRODUCTION

Recent advances have led to the fabrication of different shapes of ferromagnetic structures at submicron or nanometer sizes, including 'dots' (such as disks and rings) and long wires [1-3]. The latter are of particular interest because they exhibit one-dimensional (1D) translational symmetry along the wire axis, and so the magnetic excitations (or spin waves) are characterized by a wave number k along this direction. While studies of nanowires with both ellipsoidal and rectangular cross sections have been reported (see, e.g., [4,5]), the rectangular wires (or stripes) have recently received special attention, both experimentally and theoretically. This is because of basic properties such as the localization of the spin waves near the lateral edges, where the magnetization is strongly inhomogeneous [6], and their potential for device applic-ations, e.g., as magnetic sensors and in magnetic recording [3,7].

On the experimental side, Brillouin light scattering (BLS) provides a sensitive tool to study the spin waves (SW) in ferromagnetic metallic stripes (see, e.g., [2,4,8]) of Ni and Permalloy. On the theoretical side, it is challenging to account for the short-range exchange interactions and the long-range dipole-dipole interactions (since both are typically important for the SW dynamics in stripes at the small wave vectors excited in

CP1147, *Transport and Optical Properties of Nanomaterials—ICTOPON - 2009*, edited by M. R. Singh and R. H. Lipson
© 2009 American Institute of Physics 978-0-7354-0684-1/09/$25.00

BLS), as well as the inhomogeneous magnetization across the stripe width. Under some circumstances (for example, if the net magnetization is along the stripe axis of symmetry) a macroscopic, or continuum, theory can be employed provided the effect-ive SW boundary conditions at the lateral stripe edges are suitable modified [6]. How-ever, many of the experiments correspond to situations where an external magnetic field applied perpendicular to the stripe axis leads to a reorientation of the net magnet-ization. It then becomes appropriate, as we showed in recent work [9], to employ a microscopic (or Hamiltonian-based) dipole-exchange theory, which accounts success-fully for the main properties of the discrete SW modes, e.g., as seen using BLS in Permalloy stripes [8,10]. The above comparisons refer to situations where the theory is carried out for an individual stripe and is then compared with the experimental data for stripe arrays where the separation between adjacent elements is sufficiently large (e.g., 100 nm or more) that the stray dipolar coupling between stripes is negligible.

However, recently there have been some preliminary BLS studies of dense arrays of stripes, where the coupling between stripes may be important and may lead to a modification of the SW properties. This has motivated our present work in which we present calculations for the SW spectra of small arrays of coupled stripes in close proximity. Two different geometries are considered, in which the stripes are either grown side-by-side (separated by a nonmagnetic spacer), as in a patterned film, or are stacked vertically above each other (again with a nonmagnetic spacer). This represents a generalization of calculations in [9].

THEORY

We study the SW spectra in two different types of dense arrays where each element is a ferromagnetic stripe with width W and thickness L. In the BLS experiments (see, e.g., [8,11,12]) L is usually $20 - 30$ nm while W may be 150 nm or larger, so $W >> L$. The stripe geometry, along with the choice of coordinate axes and a typical backscatt-ering geometry for BLS, is shown in Fig. 1(a). In an array the stripes are either grown side by side on a flat nonmagnetic substrate with a separation s between stripes or they are grown vertically (separated by a nonmagnetic spacer of thickness s) as in Fig. 1(b). We shall show that the role of the inter-stripe dipolar coupling is quite different in the two arrangements.

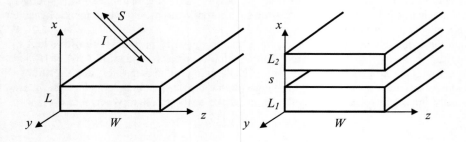

(a)   (b)

FIGURE 1. (a) Schematic of a single stripe, the choice of coordinate axes, and a typical BLS geometry with incident I and scattered S light; (b) A vertical array of two stripes with the same width but different or same thickness, separated by a nonmagnetic spacer.

The calculations for the SW spectrum in the dense stripe arrays are performed using a straightforward generalization of the microscopic dipole-exchange theory described elsewhere [9] for individual (i.e., non-interacting) ferromagnetic stripes, with the applied field H_0 here in either the longitudinal (y) or transverse (z) direction. This type of formalism, which has also been applied to other magnetic nanostructures such as cylindrical wires [13] and spheres [14], is particularly appropriate for cases where the magnetization in the array elements may be spatially inhomogeneous. It also avoids introducing assumptions for the effective pinning [6], as done in most macroscopic theories. By contrast with our recent comparison [10] of BLS experiments and theory for Permalloy stripes in widely-spaced horizontal arrays, we now extend the theory to include the coupling between stripes for both horizontal and vertical arrays.

Briefly, the modified theoretical analysis includes terms in the Hamiltonian for the dynamic and static parts of the interstripe dipolar interactions, as well as the usual intrastripe dipolar and exchange terms. This can be achieved by analogy with recent application of the microscopic theory to arrays of magnetic cylindrical nanowires [15] and spheres [14], so details will not be given here. Each stripe in the array is modelled as an infinitely long wire with thickness L and width W. The effective spins are arranged on a simple cubic lattice, with the effective lattice constant a chosen to be comparable with or less than the so-called exchange length (approximately 5.3 nm for Permalloy). The Hamiltonian H for the array has three parts, as below:

$$H = \sum_{\langle i,j \rangle} \sum_{\alpha,\beta} V_{ij}^{\alpha\beta} S_i^\alpha S_j^\beta - g\mu_B \sum_i \mathbf{H}_0 \cdot \mathbf{S}_i - \sum_i K_i (S_i^y)^2 . \tag{1}$$

Here V in the first part describes the interaction between spins $\mathbf{S}_i$ and $\mathbf{S}_j$ at lattice sites i and j, which may be in either the same or a different stripe:

$$V_{ij}^{\alpha\beta} = g^2 \mu_B^2 \left[(|\mathbf{r}_{ij}|^2 \delta_{\alpha\beta} - 3r_{ij}^\alpha r_{ij}^\beta) / |\mathbf{r}_{ij}|^5 \right] - J_{ij}\delta_{\alpha\beta} . \tag{2}$$

Following [9] the above interaction contains both the long-range dipole-dipole terms (with $\mathbf{r}_{ij}$ denoting the vector connecting sites i and j) and the short-range exchange terms J_{ij}, while α and β denote Cartesian components. The second and third parts in Eq. 1 describe the Zeeman energy of an applied field $\mathbf{H}_0$ (along either the y or z axis) and a single-ion anisotropy energy relative to the longitudinal (y) axis, respectively. The coefficient K_i may be positive or negative (for easy-axis or easy plane anisotropy, respectively); for simplicity it is assumed to have the value K_{surf} at any surface site of a stripe and the bulk value K_{bulk} otherwise.

As in [9] the steps in the theory involve first solving for the equilibrium spin configurations in the array, using an energy minimization procedure appropriate for low temperatures. Then the total Hamiltonian is re-expressed in terms of a set of boson operators, which are defined relative to the *local* equilibrium coordinates of each spin. Finally, keeping only the terms up to quadratic order in an operator expansion, we solve for the SW excitations of the interacting magnetic stripes. In general, the procedure consists of diagonalizing a $2N \times 2N$ matrix, where N is the total number of effective spins in any transverse (xz) plane of the array. Since typically N ~ 2000 or less, depending on the array size and stripe dimensions, this must be done numerically. The coupled SW modes are characterized by a wave vector component k along the longitudinal axis of the stripes. For the BLS geometry shown in Fig. 1(a), which is typical of many experiments, we have $k = 0$ because the transferred wave vector is in the transverse (z) direction. Usually we assume that the number of elements (stripes) in the array is less than about ten, otherwise the diagonalization becomes impractical.

RESULTS FOR THE LONGITUDINAL FIELD CASE

We now apply the above theory to deduce the dispersion relations for the lowest SW branches when the applied field is in the longitudinal (y) direction, taking for simplicity arrays with two or three stripes and comparing the vertical and horizontal stacking geometries. Applications are made to Permalloy $Ni_{0.81}Fe_{0.19}$ taking the mater-ial parameters to be the saturation magnetization $M_s = 0.067$ T, SW stiffness $D = 23.88$ Tnm^2, and $g\mu_B = 29.0$ GHz/T (see, e.g., [9,16]). The value of a is chosen to be 3.33 nm which is smaller than the exchange length and a fixed value 0.2 T is chosen for the applied field. For simplicity, the single-ion anisotropy coefficient K_i, which is known to be very small in Permalloy, is set equal to zero.

First we show in Fig. 2 the SW frequencies as a function of the wave number k for a symmetric array of two identical stripes stacked vertically and separated by a non-magnetic spacer of thickness $s = 50$ nm. For comparison the results for noninteracting stripes (i.e., when $s \to \infty$) are included. The initial decrease in the lowest branches as k is increased is due to competing dipolar and exchange effects, by analogy with other nanowire studies [13]. The interstripe interactions lead to a splitting of the SW frequencies, which is more apparent at small k and for the lowest branches. If the separation becomes very small value (see Fig. 3 for $s = 5$ nm), the splitting becomes much larger and there is extensive hybridization (mixing) of the modes for individual stripes. In fact, some of the frequencies in Fig. 3 become similar to those for a composite structure of thickness 20 nm.

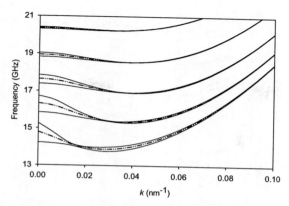

FIGURE 2. SW frequencies versus wave number k for an array of two stripes, taking $L_1 = L_2 = 10$ nm and $W = 200$ nm, stacked vertically with separation $s = 50$ nm. For comparison the dotted-dashed lines represent results for individual stripes of $W = 200$ nm and $L = 10$ nm.

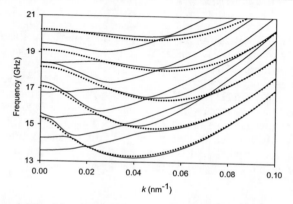

FIGURE 3. Same as in Fig. 2 but for separation $s = 5$ nm. Here the dotted lines represent results for individual stripes of $W = 200$ nm and $L = 20$ nm.

121

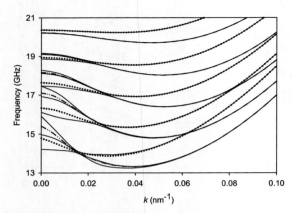

FIGURE 4. SW frequencies versus wave number k for an array of three stripes, taking $L_1 = 20$ nm, $L_2 = 10$ nm, $L_3 = 20$ nm, and $W = 200$ nm, stacked vertically with each separated by $s = 50$ nm. For compa-rison the dotted-dashed and dotted lines represent results for individual stripes of $W = 200$ nm and $L = 10$ nm and 20 nm, respectively.

Another example, this time for three Permalloy stripes stacked vertically, is shown in Fig. 4. The effects are qualitatively similar to the two-stripe arrays, but the splitting and hybridization are now more pronounced.

Next we briefly consider two interacting stripes separated horizontally, as in elem-ents of a patterned film, e.g., [10,17]. The SW frequencies versus k are shown in Fig. 5 for two Permalloy stripes of the same sizes as in Fig. 2, but taking now $s = 20$ nm. Even though s has been reduced in this example to bring the stripes closer together, it is seen that the interstripe coupling effects (such as the frequency splitting, mode hybridization, etc.) are all weaker for the case of horizontal stacking, and this is mainly a consequence of the aspect ratio L/W of the individual stripes being much less than unity.

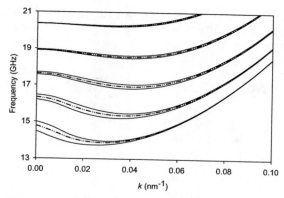

FIGURE 5. SW frequencies versus wave number k for an arrays of two stripes, taking $L = 10$ nm and $W_1 = W_2 = 200$ nm, stacked horizontally with separation $s = 20$ nm. The dotted-dashed lines represent individual stripes of L = 10 nm and W = 200 nm.

RESULTS FOR THE TRANSVERSE FIELD CASE

In this section we turn to situations where the magnetic field H_0 is applied trans-versely to the stripe length (along the z axis). By contrast to the preceding examples, this presents a more challenging (and interesting) case because the spins are now canted away from the y axis of symmetry, leading to stronger spatial inhomogeneities in both the static and dynamic parts of the magnetization. For a sufficiently large H_0 a spin-reorientation transition takes place and eventually the equilibrium direction of the spins is approximately along the z axis. The same parameters for Permalloy are used as before, but we now choose to study the dependence of the SW frequencies on H_0, taking k = 0 for simplicity.

First in Fig. 6 we provide a plot of the lowest SW frequencies versus the transverse magnetic field for an array of two identical stripes stacked vertically and separated by a nonmagnetic spacer of thickness $s = 100$ nm. Even though s is relatively large, it is seen that the field-dependence is modified by the inclusion of interstripe interactions. This is because the spin reorientation, which occurs for fields slightly less than 0.1 T in this example, is sensitive to the coupling between stripes. The behavior for individ-ual or non-interacting stripes ($s \rightarrow \infty$) is also shown for comparison; it has recently been observed in BLS and is fairly well understood [10]. A striking feature of the mode behavior is the pronounced dip in frequency for the lower SW branches, occurring as the magnetization in the bulk region of the stripes undergoes reorient-ation. The weaker dip at around 0.2 T for the lowest two (almost degenerate) branches is attributed to the reorientation of the spins at the lateral edges of the stripes. As before, splitting and hybridization effects are evident in Fig. 6.

The next example, given in Fig. 7, is for two stripes in the horizontal (side-by-side) geometry with separation $s = 50$ nm. Although the behavior is qualitatively similar to the case where the stripes are stacked vertically, the splitting and hybridization effects for the SW are less pronounced, just as in the longitudinal field case.

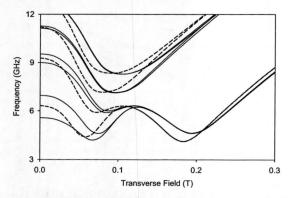

FIGURE 6. SW frequencies versus the transverse magnetic field H_0 for an array of two stripes, taking $L_1 = L_2 = 10$ nm and $W = 200$ nm, stacked vertically with separation $s = 100$ nm. For comparison the dashed lines represent individual stripes of $W = 200$ nm and $L = 10$ nm.

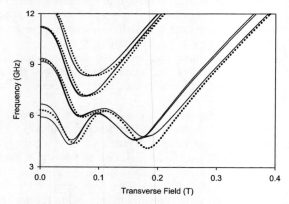

FIGURE 7. SW frequencies versus the transverse magnetic field H_0 for an array of two stripes, taking $L_1 = L_2 = 10$ nm and $W = 200$ nm, stacked horizontally with separation $s = 50$ nm. The dashed lines represent individual stripes of $W = 200$ nm and $L = 10$ nm.

CONCLUSIONS

We have presented a microscopic formalism for investigating the dipole-exchange SW modes in two different types of interacting-stripe arrays, one with the stripes stacked vertically (along the x axis in our model) and the other with the stripes stacked horizontally (along the z axis). This theoretical approach, which is convenient for studying the SW in systems with large spatial inhomogeneities in the magnetization (as in our case of the external magnetic applied in the transverse direction), is an extension of earlier work for the SW in single stripes [9,10] to incorporate the long-range dipole-dipole coupling between stripes. Numerical applications have been made to Permalloy stripes with the applied field in either of two different orientations. It is concluded that the transverse field case is of greater interest since the spin-reorientation transition has a profound effect on the SW spectrum.

Comparisons are made between SW frequencies in the two stacking geometries. We concluded that the frequency-splitting and mode hybridization of the SW depend on the separation s (becoming greater for small s) and are more pronounced in the vertical stacking case. Fortunately, more BLS data are becoming available for dense arrays to compare with our theoretical work. In the case of horizontally-stacked arrays, (patterned films) a preliminary comparison has very recently been reported [17] and further collaborative work is in progress. For the vertically-stacked arrays, while our theory is in general agreement with reported experimental work (see, e.g., [12]), it would be of interest to have more comprehensive BLS data, particularly for the transverse field case.

ACKNOWLEDGMENTS

Partial support of this project from the Natural Sciences and Engineering Research Council (NSERC) of Canada is gratefully acknowledged.

REFERENCES

1. B. Hillebrands and K. Ounadjela (eds.), *Spin Dynamics in Confined Magnetic Structures I*, Berlin: Springer, 2002.
2. B. Hillebrands and K. Ounadjela (eds.), *Spin Dynamics in Confined Magnetic Structures II*, Berlin: Springer, 2003.
3. B. Heinrich and J. A. C. Bland (eds.), *Ultrathin Magnetic Structures IV*, Berlin: Springer, 2005.
4. J. Jorzick, S. O. Demokritov, B. Hillebrands, M. Bailleul, C. Fermon, K. Yu Guslienko, A. N. Sla-vin, D.V. Berkov, N. L. Gorn, *Phys. Rev. Lett.* **88**, 047204/1-4 (2002)
5. Z. K. Wang, M. H. Kuok, S. C. Ng, D. J. Lockwood, M. G. Cottam, K. Nielsch, R. B. Wehrspohn and U. Gösele, *Phys. Rev. Lett.* **89**, 27201/1-3 (2002).
6. K. Yu. Guslienko and A. N. Slavin, *Phys. Rev. B* **72**, 014463/1-5 (2005).
7. M. L. Plumer, J. van Ek and D. Weller, *The Physics of Ultra-High-Density Magnetic Recording*, Berlin: Springer, 2001.
8. G. Gubbiotti, S. Tacchi, G. Carlotti, P. Vavassori, N. Singh, S. Goolaup, A. O. Adeyeye, A. Stash-kevich and M. Kostylev, *Phys. Rev. B* **72**, 224413/1-7 (2005).

9. H. T. Nguyen, T. M. Nguyen, and M. G. Cottam, *Phys. Rev. B* **76**, 134413/1-9 (2007).
10. H. T. Nguyen, G. Gubbiotti, M. Madami, S. Tacchi, and M. G. Cottam, *Microelectr. J.***40**, 598-600 (2009).
11. G. Gubbiotti, S. Tacchi, G. Carlotti, N. Singh, S. Goolaup, A. O. Adeyeye and M. Kostylev, *App. Phys. Lett.* **90**, 092503/1-3 (2007).
12. G. Gubbiotti, G. Carlotti, T. Ono and Y. Roussigne, *J. App. Phys.* **100**, 023906/1-7 (2006).
13. T. M. Nguyen and M. G. Cottam, *Phys. Rev. B* **72**, 224415/1-10 (2005).
14. H. T. Nguyen and M. G. Cottam, *Surf. Rev. & Lett.* **15**, 727-744 (2008).
15. T. M. Nguyen and M. G. Cottam, *J.Magn. Magn. Mater.* **310**, 2433-2435 (2007).
16. K. Yu. Guslienko, V. Pishko, V. Novosad, K. Buchanan and S. D. Bader, *J. Appl. Phys.* **97**, 10A709/1-3 (2005).
17. S. Tacchi, M. Madami, G. Gubbiotti, G. Carlotti, S. Goolaup, A. O. Adeyeye, H. T. Nguyen and M. G. Cottam, *J. Appl. Phys.* **105**, 07C102/1-3 (2009).

Microdomain Phases in Magnetic Multilayers

J. P. Whitehead[a], J. I. Mercer[b] and A. B. MacIsaac[c]

[a]*Department of Physics and Physical Oceanography*
Memorial University, St. John's, Newfoundland, Canada A1B 2W8
[b]*Department of Computer Science*
Memorial University, St. John's, Newfoundland, Canada A1B 2W8
[c]*Department of Applied Mathematics*
University of Western Ontario, London, Ontario, Canada N6A 5B7

Abstract. The magnetic properties of ultra-thin magnetic films can be fine tuned to a remarkable degree by varying the number and composition of the individual layers and the substrate. Of particular interest is the ability to vary the effective magnetic anisotropy by changing the thickness of the film. This allows experimentalists to study in some considerable detail the nature of the magnetization close to the spin reorientation transition. In this paper we review some preliminary results from a theoretical analysis and simulation studies that describe some of the key properties of microdomain phases observed near the spin reorientation transition.

Keywords: ultra-thin magnetic films, stripe domains, spin reorientation transition.
PACS: 75.70.Ak, 75.70.Kw

INTRODUCTION

Ultra-thin magnetic films (UTMFs) consist of several layers of magnetic atoms deposited on a non-magnetic substrate. By varying the number and composition of the various layers it is possible to "fine-tune" the parameters that describe the key magnetic interactions between the atoms[1-4]. In addition to their scientific interest in these materials they are also of considerable technological significance[5].

Of particular important is the ability to change the magnetic anisotropy by varying the number and composition of the layers, making it possible to create magnetic materials in which the magnetic moment are aligned perpendicular or parallel to the surface[6-15]. By carefully selecting the composition of layers the it is possible to observe a reorientation transition in which the orientation of the magnetic moments changes relative to the surface normal changes as the number of layers or temperature is varied.

In the case of systems in which the magnetisation is aligned perpendicular to the surface the magnetization exhibits a stripe pattern consisting of elongated microdomains with the direction of the magnetization in adjacent microdomains aligned antiparallel[7,9,11,13,16,17]. The characteristic width of these domains can vary significantly from a fraction of a μm down to the nm length scale and is observed to

CP1147, *Transport and Optical Properties of Nanomaterials—ICTOPON - 2009*, edited by M. R. Singh and R. H. Lipson
© 2009 American Institute of Physics 978-0-7354-0684-1/09/$25.00

depend strongly on the thickness of the film and the temperature[13,16,17]. The formation of these stripes arises as a consequence of the competition between the short exchange interaction and the long range dipolar interaction[18-23] and there are instances of good quantitative agreement between experiment and certain theoretical models[16,19].

Experiments show that in the microdomain phase the stripes can be aligned along a common axis[16] or can be disordered to form a far more complex structure with no orientational ordering[7]. The transition from a phase in which the stripes are orientationally ordered to one in which they are orientationally disordered has been observed in a number of experiments[13,17] and simulations[24,25] and studied theoretically[26,27], however, a while there exists similarities between the results of simulation, theory and experiment a complete understanding of the precise character of this transition remains elusive.

In this paper we present the results of some preliminary calculations and simulation studies which show many of the features observed experimentally. The interactions that determine the properties of such systems are the exchange and dipole interactions, and the single site magnetic surface anisotropy. The energy of the system may therefore be modeled by the following expression

$$E = -J_0 \sum_{\langle ij \rangle} \vec{S}_i \cdot \vec{S}_j - K \sum_i \left(S_i^z \right)^2 + \frac{\mu_0}{4\pi} \sum_{i \neq j} \left(\frac{\vec{\mu}_i \cdot \vec{\mu}_j}{r_{ij}^3} - 3 \frac{(\vec{\mu}_i \cdot \vec{r}_i)(\vec{\mu}_j \cdot \vec{r}_j)}{r_{ij}^5} \right) \quad (1)$$

where $\vec{S}_i$ and $\vec{\mu}_i$ denote the spin and magnetic moment vector of an atom at the site located at $\vec{r}_i$. The first term represents the exchange interaction and is given by a sum over nearest neighbours. The second and third terms represent the single site magnetic surface anisotropy and the dipolar interaction respectively.

In this paper we explore the microdomain structure of the magnetization that is observed close to the spin reorientation transition. Since the microdomain phase consists of elongated ferromagnetic domains or stripes with length scales typically in submicron range it is reasonable and computationally essential to coarse grain the system. In the present work we approximate the system as a single layer consisting of cells of area $d \times d$, where d is typically on the order of several nm. If we denote the location of the ith cell by the vector $d \times R_i$ and the direction of the net magnetization in the cell by the unit vector $\vec{\sigma}_i$ then we may approximate the energy given by Eq. 1 in terms of the coarse grained variables as

$$E = -J \sum_{\langle ij \rangle} \vec{\sigma}_i \cdot \vec{\sigma}_j - \kappa \sum_i \left(\sigma_i^z \right)^2 + g \sum_{i \neq j} \left(\frac{\vec{\sigma}_i \cdot \vec{\sigma}_j}{R_{ij}^3} - 3 \frac{(\vec{\sigma}_i \cdot \vec{R}_i)(\vec{\sigma}_j \cdot \vec{R}_j)}{R_{ij}^5} \right) \quad (2)$$

where the parameters J and g are given by

$$\vec{R}_i = \frac{\vec{r}_i}{a} \qquad J = nS^2 J_0$$

$$g = \frac{\mu_0}{4\pi} \frac{(g_L \mu_B S)^2}{a^3} \left(\frac{n^2 d}{a} \right) \tag{3}$$

where a denotes the atomic lattice parameter and n the number of layers. The relation ship between the effective anisotropy parameter κ for the coarse grained system and the microscopic parameters in Eq. 1 is somewhat more complicated and in the present work we treat it as a variable parameter.

Ground State Properties

The expression for the coarse grained energy given by Eq. 2 is known to exhibit a number of interesting features. In the limiting case $\kappa \to 0$ the ground state is a simple planar ferromagnet with a ground state energy given by $E_0 = -2J - 4.5168\,g$ [22,23]. In limit $\kappa \to \infty$, in which the spins are aligned perpendicular to the surface, the situation is somewhat more complicated. Calculations for the ferromagnetic exchange/dipolar Ising model reveal a ground state consisting of parallel stripes aligned along one of the lattice axis. The spins within each stripe are aligned ferromagnetically with the spins in successive stripes aligned antiparallel[18-22]. In the case of a square lattice the width of the stripes is given by $h_{\text{Ising}} = h_0 \exp J/4g$ and has a ground state energy given by $E_0 \approx -2J - \kappa + \left(9.034 - 8/h_{\text{Ising}} \right) g$ [21-23]. These stripes, or microdomains, arise as a consequence of the competition between the short-range exchange interaction and the long-range dipolar interaction. Such microdomain structures have been observed in a wide variety of systems[7,9,11,13,16,17].

For finite κ, previous calculations[25], for the case $J/g = 8.9$, show that there is a transition from the planar ferromagnetic state to a canted phase at around $\kappa = \kappa_c \approx 12\,g$. In the canted phase the spins acquire a spatially modulated component perpendicular to the film with a wavelength $\lambda = 2h$. As κ is increased above κ_c, the amplitude of the modulation increases until the spins in the centre of the stripes saturate to form ferromagnetic stripes separated by Bloch domain walls. Calculations show that the spins inside the Bloch walls are aligned along a common direction resulting in a net in-plane magnetization. As κ is increased further the width of the stripes increases while the domain wall separating the stripes shrinks until its width becomes comparable to the cell size d at which point the system makes a transition to a perpendicular stripe phase in which all the spins are perpendicular to the plane. At this point the in-plane magnetization goes to zero and the domain walls become Ising-like with the spins on either side of them aligned antiparallel.

One feature that characterises all of the experimental measurements close to the reorientation transition is the dramatic change in the stripe width as a function of film

129

thickness. This is attributed to the dependence of the effective anisotropy of the film on the thickness[11,16,19]. While these earlier calculations[25] shed some light on the role and nature of the microdomain phase close to the spin reorientation transition, the dependence of the stripe width h on κ is very weak and certainly does not show the strong dependence that is observed experimentally[13,16] and predicted theoretically[19]. This is due to the fact that the canted phase separating the planar ferromagnetic phase and the perpendicular stripe phase is realised only over a very narrow region range of κ.

To examine the dependence of the stripe width on the anisotropy, the ground state calculations presented in the earlier work have recently been extended to larger values of J/g, and hence larger stripe widths. The values for the equilibrium stripe width h and the transverse magnetization are plotted as a function of κ in Figure 1 for several values of J/g corresponding to the range $64 \leq h_{\text{Ising}} \leq 256$.

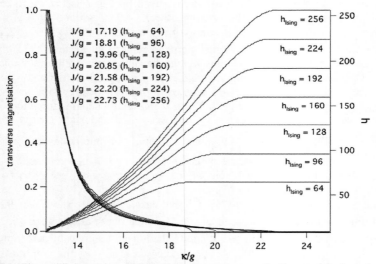

FIGURE 1. The equilibrium stripe width h and the transverse magnetization are plotted as a function of κ/g for several values of J/g.

Figure 1. shows the transition from the planar ferromagnetic phase, with a net transverse magnetization $m_{\parallel} : 1$, to the striped phase, with $h : 10$, at around $\kappa = \kappa_c : 12g$. As κ is increased above κ_c, the transverse magnetisation decreases as the spins begin to align perpendicular to the plane to form stripes with the stripe width h increasing up to a maximum value $h = h_{\text{Ising}}$, at which point the transverse magnetization goes to zero. This corresponds to the transition to the perpendicular stripe phase. While these results are consistent with the earlier calculations for J/g: 8.9, the stripe width shows a far stronger dependence on κ, as expected both from experimental and previous theoretical studies. It is interesting to note that the transverse magnetization is relatively insensitive to the particular value of the ratio J/g for $h < h_{\text{Ising}}(J/g)$.

Finite Temperature

Monte Carlo simulations show that many of the key features found in the ground state calculations persist at finite temperature[24,25]. However, simulation studies of microdomain phases obviously require a system size large enough to accommodate a sufficient number of stripes. The long-range character of the dipolar interaction, however, significantly increases the computational requirements for large systems. This presents significant a challenges in extending the earlier finite temperature simulation studies to larger values of the ratio J/g, and hence large stripe widths. However, by adapting the techniques used by experimentalists to study the variation in anisotropy, we have carried out a series of simulation studies on systems with a κ gradient. The simulations integrate the Landau-Lifshitz-Gilbert (LLG) equation and include a stochastic field to represent the effect of the thermal fluctuations.

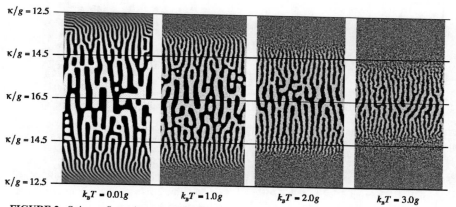

FIGURE 2. Spin configurations obtained by integrating the LLG equations on a 800x1600 lattice with $J/g = 20$ and linear κ gradient with $12.5 \leq \kappa \leq 16.5$ for $T/g = 0.01, 1.0, 2.0$ and 3.0.

The LLG equation is integrated using code developed in house and optimized for large parallel computers using shared memory[28]. Using this code and hardware we are able to study systems consisting of multiple layers, with each layer consisting of upwards of a million lattice sites. Spin configurations for a 800×1600 lattice are presented in Figure 2 for $J/g = 20$. Each of the four spin configurations in Figure 2 is the result of over 1000

CPU hours running on 64bit AMD Opterons. To achieve this final configuration we employ an annealing process consisting of an alternating field perpendicular to the lattice and a constant field along the long axis of the lattice. The fields decay exponentially and reach negligible values at approximately 20% of the total runtime. The states are sampled over the final 80% of the runtime to ensure that the system is close to equilibrium.

These preliminary results reveal a number of interesting features regarding the nature of the microdomain phase. In addition to the length scale associated with the stripes decreasing with decreasing κ, the data also show that the stripe width decreases with increasing temperature, consistent with experimental studies[13,16]. Moreover the simulations indicate a fairly distinct boundary between the planar ferromagnetic phase and the stripe phase and show that the value of κ at which this boundary occurs increases with increasing temperature consistent with observations[13].

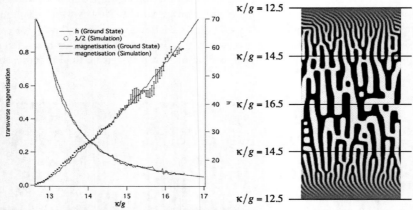

FIGURE 3. A plot of the ground state stripe width h and the transverse magnetization are plotted as a function of κ/g for $J/g = 20$, together with the corresponding results obtained from the spin configuration obtain from simulation studies for $k_BT/g = 0.01$ shown on the right.

In order to compare the results from the ground state calculations with the results of the simulations, estimates of the stripe width as a function of κ were obtained for the simulation shown in Figure 3 for $k_BT = 0.01g$. The stripe width estimates were obtained from an analysis of 2D Fourier transforms on windowed subsections, centered at each lattice site, of the spin configuration shown in Figure 3. The resultant spectra show well-defined peaks corresponding to a characteristic wavelength λ and orientation. Averaging λ over each row (constant κ) we obtain a relationship between κ and the characteristic wavelength λ. In Figure 3 we plot the transverse magnetization and $\lambda / 2$ together with the transverse magnetization and equilibrium stripe width obtained from the corresponding ground state calculations.

From the graph presented in Figure 3, we see that the estimates of the transverse magnetization and the equilibrium stripe width $h = \lambda/2$ obtained from the simulation compare very well with the corresponding results from the ground state calculations. This agreement strongly suggests that we can obtain reliable estimates of the equilibrium stripe width and transverse magnetization as a function of κ from the results of simulations. This is an important observation as it suggests that by extending the above analysis to the finite temperature simulations, we can also calculate the temperature dependence of the stripe width of the microdomain phase using this method.

CONCLUSIONS

We have compared the results of ground state calculations with those obtained from simulations for a simple model of an ultrathin magnetic film described by the coarse grained energy given in Eq. 2. Results have been presented for the ground state spin configuration for a range of values of J/g and κ/g. The results extend earlier work to much larger stripe widths and show that, as the system approaches the spin reorientation transition ($\kappa \rightarrow \kappa_c$), the transverse magnetization grows and the equilibrium stripe width becomes smaller, consistent with the results of earlier theoretical and experimental studies. As κ is increased, the thickness of the domain wall separating the stripes becomes smaller as the stripe width increases until it becomes comparable to the dimensions of the cell at which point the system makes the transition to perpendicular stripe phase.

We examined the phase behaviour of the coarse grained model as a function of κ and T for the specific case $J/g = 20$ by performing simulations on a system with a κ gradient applied along one axis, for several different temperatures. The results of the simulations show the same qualitative features as the ground state calculations. Specifically, as the system approaches the spin reorientation transition, the transverse magnetization grows and the equilibrium stripe width becomes smaller. However, the simulations show that $\kappa_c(T)$ increases with increasing temperature while the length scale associated with the stripes decreases.

The comparison between the magnetization and stripe width obtained from ground state calculations to those found from simulations for $k_B T = 0.01g$, presented in Figure 3 show very good agreement. The results from these calculations suggest that the methodology of considering a system with a gradient in one of the parameters provides a useful, accurate and computationally efficient method of studying the dependence of the magnetic state on that parameter using simulation methods. We are currently extending this method to examine the effect of an applied transverse field on the stripe width and magnetization.

ACKNOWLEDGMENTS

One of the authors (JIM) wishes to acknowledge a number of useful discussions with Dr. Saika-Voivod regarding the method used to extract the length scale of the stripes from the simulation data. The authors also wish to acknowledge funding from National Science and Engineering Research Council of Canada (NSERC), SHARCnet, the Centre for Chemical Physics (CCP) at the University of Western Ontario. Computations were performed on resources at the Centre for Material and Magnetic Simulations Laboratory (CMMS) that was funded by grants from the Canadian Foundation for Innovation (CFI) the Newfoundland and Labrador Industrial Research and Innovation Fund (IRIF) and SUN Microsystems. The authors also wish to acknowledge the use of additional computational resources and technical support provided by ACEnet.

REFERENCES

1. C. A. F Vaz, J. A. C. Bland and G. Lauhoff, *Rep. Prog. Phys.* **71**, 056501 (2008)
2. Z. Q. Qiu, J. Pearson and S. D. Bader *Phys. Rev. Lett.* **70**, 1006 (1993)
3. A. Bauer, G. Meyer, T. Crecelius, et al. *J. Mag. Mag. Mat.* 282 (2004)
4. T. O. Mentes, A. Locatelli, L. Aballe and E. Bauer *Phys. Rev. Lett.* **101**, 085701 (2008)
5. P. J. Jensen and K. H. Bennemann, *Surf. Sci. Rep.* **61**, 129 (2006)
6. D. P. Pappas, K. P. Kamper, and H. Hopster, *Phys. Rev. Lett.* **64**, 3179 (1990).
7. R. Allenspach, M. Stampanoni, and A. Bischof, *Phys. Rev. Lett.* **65**, 3344 (1990).
8. B. Scholz, R. A. Brand, and W. Keune, *Phys. Rev.* B **50**, 2537 (1994).
9. F. Baudelet, M.-T. Lin, et al. *Phys. Rev.* B **51**, 12563 (1995).
10. A. Berger and H. Hopster, *Phys. Rev. Lett.* **76**, 519 (1996).
11. H. P. Oepen, M. Speckmann, Y. Millev, and J. Kirschner, *Phys. Rev.* B **55**, 2752 (1997).
12. E. Mentz, A. Bauer, T. Gunther, and G. Kaindl, *Phys. Rev.* B **60**, 7379 (1999).
13. A. Vaterlaus, C. Stamm, et al. *Phys. Rev. Lett.* **84**, 2247 (2000).
14. R. Ramchal, A. K. Schmid, M. Farle, and H. Poppa, *Phys. Rev.* B **69**, 214401 (2004).
15. Z. Q. Qiu, J. Pearson, and S. D. Bader, *Phys. Rev. Lett.* **70**, 1006 (1993).
16. Y. Z. Wu, C. Won, et al. *Phys. Rev. Lett.* **93** 117205 (2004).
17. C. Won, Y. Z. Wu, et al. *Phys. Rev. B* **71**, 224429 (2005)
18. T. Garel and S. Doniach, *Phys. Rev. B* **26**, 325 (1982).
19. Y. Yafet and E. M. Gyorgy, *Phys. Rev. B* **38**, 9145 (1988).
20. B. Kaplan and G. A. Gehring, *J. Mag. Mag. Mat.* **128**, 111 (1993).
21. J.P. Whitehead and K. Debell, *J. Phys.: Condens. Matter* 6, L731 (1994)
22. A. B. MacIsaac J. P. Whitehead, M. C. Robinson and K. DeBell, *Phys. Rev. B* **51**, 16033 (1995)
23. K. DeBell, A. B. MacIsaac and J. P. Whitehead, *Rev. Mod. Phys.* **72**(1), 225 (2000)
24. I. N. Booth, A. B. MacIsaac, J. P. Whitehead, and K. De'Bell, *Phys. Rev. Lett.* **75**, 950 (1995).
25. J. P. Whitehead, A. B. MacIsaac and K. DeBell *Phys. Rev. B* **77**, 174415 (2008)
26. A. B. Kashuba and V. L. Pokrovsky, *Phys. Rev. Lett.* **70**, 3155 (1993).
27. A. B. Kashuba and V. L. Pokrovsky, *Phys. Rev. B* **48**, 10335 (1993).
28. J. I. Mercer, "*Computational Magnetic Thin Film Dynamics*", MSc. Thesis, Memorial University of Newfoundland, 2007.

Atomic Theory Of Phononic Gaps In Nano-patterned Semiconductors

S. P. Hepplestone and G. P. Srivastava

School of Physics, University of Exeter, Stocker Road, Exeter, EX4 4QL, UK.

Abstract. An Enhanced Adiabatic Bond Charge Model is applied to a variety of semiconductor nano-patterned structures to examine their phononic behaviour. We apply this approach to determine the locations and widths of phononic band gaps, and negative phonon group velocities, for both longitudinal acoustic and transverse acoustic modes, in the nanosized $Si(4\ nm)/Si_{0.4}Ge_{0.6}(8\ nm)[001]$ superlattice. In addition to reproducing the LA band gaps observed recently by Ezzahari and co-workers using a picosecond technique, we predict this superlattice to be a true one-dimensional phononic system in the hypersonic range. We show that embedded wires of Ge in Si can be either a one-dimensional or an optical phononic crystal and that Sn wires embedded in Si can also be a 2D phononic crystal.

Keywords: Phonon, Lattice Dynamics
PACS: 63.22Np,43.35.+d,63.20.-e,68.35.lv

INTRODUCTION

Phononic crystals [1,2,3] are the vibrational analogues of photonic crystals, and offer the possibility of novel applications for phonon engineering including phonon focusing, and sound filters [4,5,6]. A phononic crystal consists of two or more periodically arranged materials with contrasting vibrational properties. These differences arise from each material's set of interatomic force constants and masses. This results in the creation of phononic band gaps and negative refraction of phonons.

Most currently realisable phononic structures rely on solid/fluid composites on the scale of μm - mm, but recent technological advances have led to the fabrication of nanophononic solid/solid materials [7,8]. Semiconducting superlattices (SL) with nanometer scale periodicity have been fabricated [9,10], and two-dimensional (2D) and three-dimensional (3D) array systems are now feasible. SLs have been shown to be one-dimensional (1D) phononic systems [11]. Theoretical treatments of these systems have been limited to continuum models [2,12,13] and simplified atomic models [14,15].

In this work we investigate the lattice dynamics of a variety of semiconductor nano-patterned structures to examine their phononic properties by applying an Enhanced Adiabatic Bond-Charge Model (EBCM). We first apply the EBCM to obtain the full phonon dispersion relation for the Si/SiGe SL fabricated by Ezzhari *et al.* [9]. In addition to replicating the results for the longitudinal acoustic (**LA**) band measurements of Ezzahri *et al*, we show that this system is a true 1D phononic semiconducting structure. We then

CP1147, *Transport and Optical Properties of Nanomaterials—ICTOPON - 2009*, edited by M. R. Singh and R. H. Lipson
© 2009 American Institute of Physics 978-0-7354-0684-1/09/$25.00

perform calculation for a variety of semiconducting systems with 1D, 2D, and 3D periodicities examining whether band gaps in the phonon dispersion relations are created and whether these band gaps exist in one-, two-, or three-dimensions. We also discuss the negative refraction phenomenon by calculating the relevant negative group velocities of phonons in these systems.

METHODOLOGY

The adiabatic bond charge model is an accurate phenomenological theory which has been used by several groups [16,17] to successfully describe the lattice dynamical properties of semiconductor systems. This model describes the semiconductor atoms and their electronic configurations for diamond and zincblende solids as tetrahedrally-bonded ions and their bond charges. The bond charges are valence electrons distributed as point charges located along the tetrahedral bonds between the ions and their first nearest neighbours. These bond charges are allowed to move adiabatically and are assumed to have zero mass.

The dynamical matrix for solving the equations of motion of the system is constructed by invoking three types of interaction: (i) Coulomb interaction; (ii) short range central force interaction; and (iii) Keating type bond bending interaction. This leads to a set of four parameters, which can be used to describe the phonon dispersion relations of each bulk Si and Ge materials reliably. In our modified approach, the EBCM, the matrix elements for each bond within a bulk region are calculated using the parameters from the corresponding bulk material, and for interface bonds the parameters are calculated as the appropriately averaged values of the parameters for the two bulk materials. Similarly, for structures containing alloys, such as Si_xGe_{1-x}, the parameters for the matrix elements of the complete system are calculated by taking appropriately weighted parameters of the individual systems. Using this procedure, for Si_xGe_{1-x} systems of differing cell sizes with a random distribution of atoms we obtain frequency results that are in good agreement with the measured results of Brya [18].

For a systematic examination of the criteria and trends we apply the EBCM to three systems consisting of dots, wires and layers (the guests) embedded in bulk Si (the host). For dots, the atoms are arranged in a cube of size $l \times l \times l$ embedded in a Si supercell of size $L_0 \times L_0 \times L_0$. For wires, the supercell is continuous in the Z axis and consists of a embedded square of size $l \times l$ embedded in a supercell of size $L_0 \times L_0$. Similarly, for SLs, the supercell is continuous in the Y and Z directions and, for the X direction, the guest layer has width l, with the system having a total period of L_0. We define the extent of the embedded material in terms of the ratio $L_f = l/L_0$, the length fraction along one axis.

RESULTS AND DISCUSSION

A Si(4 nm)/$Si_{0.4}Ge_{0.6}$(8 nm)[100] SL was recently grown and studied by Ezzahri *et al.* [9]. Using a pump-probe picosecond technique they were able to measure gaps in the longitudinal acoustic phonon frequency range. They measured three longitudinal gaps at 283 GHz, 527 GHz, and 805 GHz. Figure 1 (inset) show the full calculated phonon dispersion relations for the SL structure used in their experiment. As can be seen in the figure, this system has several gaps in the phonon dispersion relations for the entire frequency range. The high optical frequencies (greater than 4 THz) show very flat dispersion curves (i.e. low group velocity), which is because these modes become more confined within different SL layers. In the directions perpendicular to the SL layer direction, the phonon frequency spectrum is continuous as expected.

Figure 1 shows a close up of the phonon dispersion curves along the SL growth direction for the frequency range measured by Ezzahri *et al*. The highlighted regions are the gaps in the **LA** spectra measured in their experiment, showing that there is very good

TABLE 1. The positions of the gaps, their sizes, their polarization characteristics (TA=transverse, LA=longitudinal), their zone location (ZC=zone-centre, ZE=zone-edge) and the velocities of the corresponding adjacent branches for the Si (4 nm)/Si$_{0.4}$Ge$_{0.6}$ (8 nm) superlattice.

Frequency (GHz)	Gap size (GHz)	Lower branch velocity (km/s)	Upper branch velocity (km/s)	Polarisation and Location
174.8	29.5	4.18	-4.18	TA ZE
252.4	47.8	6.00	-5.94	LA ZE
350.2	34.8	-4.18	4.03	TA ZC
495.1	39.7	-5.94	5.84	LA ZC
523.2	33.1	4.03	-3.98	TA ZE
805.0	30.0	5.84	-5.80	LA ZE

agreement between the theory presented here and experiment. Table 1 shows the gaps for both the **TA** and **LA** modes and, with Fig 1, shows that there is a clear overall gap in the [100] direction between 515 GHz and 539 GHz meaning no phonon modes can propagate (in that direction). Also, within the lower **LA** bandgap only phonons possessing a negative group velocity can propagate.

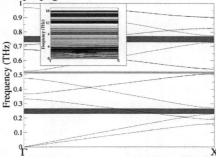

FIGURE 1. Phonon dispersion relations of a Si (4 nm)/Si$_{0.4}$Ge$_{0.6}$(8 nm)[100] superlattice. The inset shows the full range of the frequency spectrum. Highlighted are the disallowed longitudinal frequencies in the phonon dispersion range and the central highlighted region corresponds to a true 1D frequency gap in both the longitudinal and transverse frequency spectra.

When this system is compared with other Si/Si$_x$Ge$_{1-x}$ systems the following features can be observed. (i) Increasing the period of the SL causes the size and position of these gaps to vary as $\omega_{gap} \propto 1/L_0$. Hence such gaps are more easily observable on the nanometer scale, but rapidly decrease in size as the SL period becomes larger. (ii) Increasing or decreasing the length fraction from L$_f$=0.33, decreases the number and size of these gaps. This suggests that 0.33 is the optimum length fraction for phononic Si/SiGe SLs. (iii) The random disorder of the atomic makeup of the Si$_{0.4}$Ge$_{0.6}$ layer increases the number of gaps observed as opposed to an ordered Si$_{0.5}$Ge$_{0.5}$ system with silicon and germanium as the two basis atoms. This is expected as an ordered structure should provide a better crystal than that of a disordered structure.

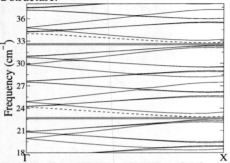

FIGURE 2. Phonon dispersion curves for the Si (27 nm)/Ge (17 nm)[100] superlattice. Highlighted are the two gap regions (in blue online) and two dispersion curves (in dashed red online).

A second type of Si/Ge[100] SL structure has been realised experimentally on the nanometer scale by Lee *et al* [10]. Figure 2 shows our calculated results for part of the phonon dispersion curves along the growth direction for the Si(27 nm)/Ge(17 nm)[100] SL. As can be seen, there are three clear gaps in the phonon spectra at 680 GHz, 826 GHz and 974 GHz, of widths 10 GHz, 3 GHz and 10 GHz, respectively. The phonon dispersion curves highlighted (dashed) in Fig. 2 are doubly degenerate and have maximum negative group velocities of -2.915 km/s and -2.018 km/s for the lower and higher branches, respectively. It can be seen that all phonon modes with frequencies between 688 GHz and 726 GHz possess negative group velocities. As described for the Si/Si$_x$Ge$_{1-x}$ structures above, the gap size and position varies as $\omega_{gap} \propto 1/L_0$ and the optimum length fraction is 0.33.

We have further studied other forms of Si/Ge systems. Embedded dots of germanium with L$_f$=0.3 embedded in bulk with period L_0=2.715 nm do not yield a true gap in the phonon density of states, but does reveal a minimum in the density of states at approximately 1021 GHz. The transverse acoustic modes show a frequency gap in folded dispersion curves between approximately 901 GHz and 1021 GHz. The corresponding folded modes possess a maximum negative group velocity of -5.0 km/s in the [100] direction. Similarly, the longitudinal branch has a gap in its frequency spectra between 1381 GHz and 1441 GHz, and its corresponding folded optical branch has a maximum negative velocity of -8.3 km/s in the [100] direction. Both of these folded branches also possess negative velocities in the [110] directions.

The phonon dispersion relations of the corresponding 2D periodic system of germanium wires of cross-section 1.95 nm × 1.95 nm embedded in a host of silicon with period of 5.43 nm (i.e. L_f=0.35) is shown in Fig. 3. This structure is close to optimum length fraction of 33% discussed previously. The highlighted region in Fig. 3 shows a clear gap at 557 GHz of size 14 GHz for both the **LA** and **TA** branches in the [110] direction. Hence, this structure (unlike the embedded dot) is a 1D phononic system. The lower of the two branches (dashed) above the gap shows a negative group velocity of up to -5.5 km/s, meaning that phonon modes lying between 564 GHz and 658 GHz propagating in the [110] direction cannot have a positive group velocity. The folded transverse and longitudinal branches in the [100] direction (highlighted) also show negative group velocities of -5.4 km/s and -8.43 km/s respectively.

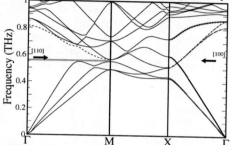

FIGURE 3. (Colour online) Phonon dispersion curves for a cubic germanium tower (3.8~nm) embedded in bulk silicon with a period of 5.743 nm. Highlighted in red and black are the transverse and longitudinal acoustic branches, respectively, and their corresponding optical folded branches in the [100] direction. Highlighted in blue in the [110] direction is the 1D phononic gap and highlighted in green is the lowest non-zero branch in the [110] direction.

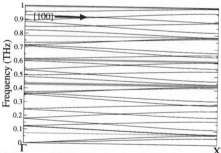

FIGURE 4. (Colour online) Phonon dispersion relations of the Si (22 nm)/Sn (11 nm)[100] superlattice. Highlighted are the gap regions (in blue) and the doubly degenerate branches possessing negative group velocity above the acoustic branches (in red).

Fig. 4 shows the phonon dispersion relations of an artificial tetrahedrally bonded semiconductor SL. This structure consists of alternating layers of Si(22 nm) and Sn(11 nm) arranged along the [100] direction. Whilst Si/Sn in a pure crystalline form is rather unrealistic to grow, the interface region being either amorphous or under heavy strain will increase the size of any gaps found rather than decrease it [13]. The figure reveals this structure to not only be a 1D phononic, but also have an acoustic gap between 68.5 GHz and 75.0 GHz. The position of this gap may be controlled by changing the period of the structure with an appropriate fabrication technique. The centre of the gap varies as $\omega_{gap} \propto 1/L_0$. It is also of note that the lowest folded branch (highlighted in red in Fig. 4) is doubly degenerate and has negative group velocities of up to -3.6 km/s. In the [100] direction (Γ-X), there are several gaps in the phonon dispersion relations as shown in the figure. These gaps continue up to the highest frequency of approximately 16 THz.

A second type of phononic gap is possible to open for most embedded structures. This gap occurs in the optical region of the phonon dispersion relations and is attributed to a reduction in the number of Si-like vibrational modes and an increase in the number of heavy-atom-like vibrational modes. This occurs for systems with a very high length fraction with very few Si atoms, and are thus dominated by the heavier atoms. In these systems, the phonon branches above and below this gap are very flat with nearly zero velocities.

For Ge wires of cross-section 5.15 nm × 5.15 nm embedded in Si with period of 5.43 nm (i.e. L_f=0.95) a large gap opens up in both the [110] and [100] directions between the frequencies 9.92 THz and 10.65 THz for all phonon modes. This gap occurs in all three degrees of freedom and hence is an absolute gap.

A cubic dot of Sn with dimensions 2.44 nm × 2.44 nm × 2.44 nm embedded in a host of Si with period 2.7 nm shows a large frequency gap of approximately 1 THz with the centre of the gap occurring at 10.2 THz. The phonon branches above and below this gap are very flat with nearly zero velocities. As discussed before the gap in the phonon density of states is due to depletion of the available silicon-like optical states and an increase in the number of tin-like states.

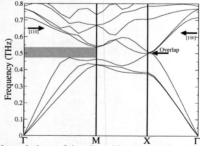

FIGURE 5. Phonon dispersion relations of tin wire with cross-section 1.95 nm×1.95 nm embedded in a host of silicon with period of 5.43 nm (i.e. L_f=0.35). Highlighted is a 60 GHz frequency gap in the [110] direction (Γ-M) and the small overlap region in [100] direction (Γ-X).

Figure 5 shows the phonon dispersion curves for a Sn wire with cross-section 1.95 nm × 1.95 nm embedded in a host of silicon with period of 5.43 nm (i.e. L_f=0.35). In the [100] direction the upper edge of the **TA** polarisation gap and the lower edge of the **LA** polarisation gap coincide at almost exactly at the zone boundary, with the difference in frequency between these two gaps being less than 10 GHz (highlighted as *overlap*). For the symmetry direction [110], the highlighted gap is much larger than in the Si/Ge case. This is due to the greater mass ratio. This gap remains consistently greater than 60 GHz between the Γ- and M-point meaning that this system is also a 1D phononic. However, unlike Ge wires embedded in Si, this gap extends across the majority of the *M-X* directions, meaning that the system is almost a 2D phononic. This shows along the 2D plane, phonon modes with a frequency 500 GHz can only be travelling in the Γ-X direction and no other. This position can be adjusted by changing the period of the structure using the previously discussed relation $\omega_{gap} \propto 1/L_0$. For phonons with a frequency between 500 GHz and 534 GHz, these modes can only possess a negative group velocity of the order of km/s.

CONCLUSIONS

In conclusion, we have applied an extended adiabatic bond charge model to a large variety of periodic semiconducting systems to explore their potential as 1D, 2D and 3D phononic systems. We have shown that several Si/X systems (where X can be Si_xGe_{1-x}, Ge, Sn) are 1D phononics and that the gap size and position varies with period as $\omega_{gap} \propto 1/L_0$. For SLs, it is possible to open several gaps in the [100] direction (the growth direction) and that these gaps allow for (some) phonon modes to propagate with very high negative group velocities (of the order of km/s). Similarly, embedded wires in silicon show large 1D phononic gaps in the [110] direction and the virtual Si/Sn system has a phononic gap in all directions in the plane except the [100] direction. However, we speculate that due to lattice strain, the reorganisation of atoms at the interface and if any of the interface atoms act similar to those in an amorphous system, then this is very likely to be a true gap in the entire 2D plane ([100]-[110]). Lastly, we have shown that for all these systems, there is a large asymmetry in the phonon dispersion relations which will have a very important effect for thermal properties such as thermal conductivity. These factors can be applied to composite semiconducting systems with important design implications for thermal management.

ACKNOWLEDGMENTS

S. Hepplestone acknowledges financial support from the Leverhulme Trust.

REFERENCES

1. M. Siglas and E. N. Economou, *J. Sound Vib.* **158**, 377 (1992).
2. M. S. Kushwaha, P. Halevi, G. Martinez, L. Dobrzynski and B. Djafari-Rouhani, *Phys. Rev B* **49**, 2313 (1994); ibid *Phys. Rev. Lett.* **71**, 2022 (1993).
3. E. L. Thomas, T. Gorishnyy and M. Maldovan, *Nature Materials* **5**, 773 (2006).
4. Z. Liu, Xixiang Zhang, Y. Mao, Y. Y. Zhu, Z.u Yang, C. T. Chan and P. Sheng, *Science* **289**, 1734 (2000).
5. M. Torres, F. R. Montero de Espinosa and J. L. Aragon, *Phys. Rev. Lett.* **86**, 4282 (2001).
6. T. Gorishnyy, C. K. Ullal, M. Maldovan, G. Fytas, and E. L. Thomas, *Phys. Rev. Lett.* **94**, 115501 (2005).
7. J.-F. Robillard, A. Devos and I. Roch-Jeune, *Phys. Rev. B* **76**, 092301 (2007).
8. A. Huynh, N. D. Lanzillotti-Kimura, B. Jusserand, B. Perrin, A. Fainstein, M. F. Pascual-Winter, E. Peronne and A. Lemaitre, *Phys. Rev. Lett.* **97**, 115502 (2006).
9. Y. Ezzahri, S. Grauby, J. M. Rampnoux, H. Michel, G. Pernot, W. Claeys, S. Dilhaire, C. Rossignol, G. Zeng, and A. Shakouri, *Phys. Rev. B* **75**, 195309 (2007).
10. S.-M. Lee, D. G. Cahill and R. Venkatasubramanian, *Appl. Phys. Lett.* **70**, 2957 (1997).
11. S. P. Hepplestone and G. P. Srivastava, *Phys. Rev. Lett.* **101**, 105502 (2008).
12. M. Siglas and E. N. Economou, *Solid State Commun.* **86**, 141 (1993).
13. J. O. Vasseur, P. A. Deymier, Ph. Lambin, B. Djafari-Rouhani, A. Akjouj, L. Dobrzynski, N. Fettouhi and J. Zemmouri, *Phys. Rev. B* **77**, 085415 (2008).
14. J. S. Jensen, *J. Sound Vib.* **266**, 1053 (2003).
15. P. G. Martinsson and A. B. Movcahn, *J. Mech. Appl. Math.* **56**, 45 (2003).
16. W. Weber, *Phys. Rev. B* **15**, 4789 (1977).
17. S. P. Hepplestone and G. P. Srivastava, *Nanotechnology* **17**, 3288 (2006).
18. W. J. Brya, *Sol. Stat. Comm.* **12**, 253 (1973).

OPTICAL PROPERTIES OF NANOMATERIALS

Study of Quantum Yield and Photoluminescence of Thiol Capped CdS Nanocrystallites

Manisree Majumder, Santanu Karan, Manik Kumar Sanyal, Aloke Kumar Chakraborty and Biswanath Mallik[*]

Department of Spectroscopy, Indian Association for the Cultivation of Science, 2A & 2B, Raja S. C. Mullick Road, Jadavpur, Kolkata-700 032, INDIA.

Abstract. For the preparation of thiol capped CdS nanocrystallites, the microwave (MW) assisted reaction of cadmium acetate with thiourea in N,N-dimethylformamide (DMF) was controlled in the presence of two capping agents, 1-butanethiol and 2-mercaptoethanol. The peak position of the absorption band of the CdS nanocrystals in DMF solution shifted towards longer wavelength with increasing duration (by repeated exposure for a fixed time) of MW irradiation, for 1- butanethiol caped CdS nanocrystals indicating growth of particle size. However, the peak position of absorption band remained nearly at the same wavelength and only the intensity of the absorption band increased with increasing duration of MW irradiation for the 2-mercaptoethanol capped CdS nanocrystals. Photoluminescence (PL) of the CdS nanocrystals in solution with 1-butanethiol as capping agent was observed to shift towards higher wavelength in the visible range of the spectrum showing a decrease in intensity with increase in the duration of exposure of MW irradiation. For the 2-mercaptoethanol capped CdS nanocrystals in solution, the photoluminescence peak remains nearly at the same position, showing a decrease in intensity, with increase in duration of exposure of MW irradiation. The CdS nanocrystals were also characterized by XRD and FTIR. The relative PL quantum yield of the CdS nanocrystallites was estimated under various experimental conditions. The formation of sulfur vacancies (surface defects) on CdS nanocrystallites in DMF for both the capping agents were indicated by emission measurements. The estimated relative PL quantum yield decreases systematically with increase in MW exposure for both 1-butanethiol and 2-mercaptoethanol capped CdS nanocrystallites.

Keywords: Cadmium sulfide, Thiol capping, Photoluminescence, Quantum yield, Sulfur vacancy.
PACS: 81.07.Bc, 71.55.Gs, 78.55.Cr, 78.67.Bf

INTRODUCTION

Attention has been paid in past for studying cadmium sulfide (CdS) nanoparticles/ nanocrystallites[1-10] and such studies relating to their photophysical properties have useful applications in various areas like optoelectronic[1], photocatalysis[5], solar energy conversion[6], photo degradation of water pollutants[7] etc. Photoluminescence (PL) of CdS nanocrystallites prepared by various methods has been studied by many researchers[3,4,8-10] and in general, the reported emission spectra of CdS nanocrystallites consist of two broad bands[3,4,8-10] (ranging 400-520 nm and 520-800 nm) peaked around 480 and 650 nm, respectively. Relatively strong emission peaked at 650 nm is attributed to radiative recombination at deep trap sites originating from lattice imperfections at the surface and the emission that peaked around 480 nm is attributed to direct recombination of electron and hole pairs at the band gap. In

CP1147, *Transport and Optical Properties of Nanomaterials—ICTOPON - 2009*, edited by M. R. Singh and R. H. Lipson
© 2009 American Institute of Physics 978-0-7354-0684-1/09/$25.00

contrast, PL spectrum with a single band/ peak was reported by some researchers[2,11] depending on the CdS nanocomposites. Bulk CdS is reported to have an emission maximum in the 500-700 nm region[12] which is due to the recombination from surface defects[13].

Previous study[4] on the preparation of 1-thioglycerol capped CdS nano-crystallites by using microwave (MW) irradiation indicated that particle growth occurs only during the continuous MW irradiation and stops when the system is cooled down[4]. To study this aspect for preparing CdS nanocrystallites with different capping agents like 1-butanethiol and 2-mercaptoethanol by using periodically interrupted MW heating for a fixed time/ duration a program was taken in our laboratory and the results of the study are discussed in this article.

EXPERIMENTAL

Cadmium acetate [$Cd(CH_3COO)_2.2H_2O$. 99.9%, Loba Chemie, India], thiourea [$C(NH_2)_2S$, GR grade, Merck, India], and 1-butanethiol [$CH_3CH_2CH_2CH_2SH$, Aldrich, USA], 2-mercaptoethanol [$HOCH_2CH_2SH$, Aldrich, USA] and N,N-dimethylformamide (DMF) [Spectrochem, India] was used as received. A microwave (MW) oven (Samsung, 2.45 GHz. max. powers 900W) was used to prepare the nanocrystallites. Solution of cadmium acetate in DMF was prepared with an appropriate amount of the capping and then thiourea was added to the solution. Two separate MW irradiation-induced chemical reaction baths were employed for the preparation of 2-mercaptoethanol and 1-butanethiol capped cadmium sulfide and were irradiated with MW continuously for 40s for one exposure. For repeated exposures, before each irradiation the irradiated solution was cooled down to room temperature 27°C. The details of the experimental procedure could be found elsewhere[14]. The nanocrystallites obtained after MW exposures for nine times were separated by centrifuging at 15,000 rpm; washing several times with pure water and then dried under vacuum were used for XRD measurements.

RESULTS AND DISCUSSIONS

Fig. 1a shows the absorption spectra of the 1-butanethiol capped CdS nanocrystals in DMF solution. The shift of the absorption peak towards longer wavelength side with the increasing number of MW exposures indicates the size growth of the CdS nanocrystallites induced by the repeated MW irradiation. From Fig.1a the presence of the tail in the visible spectral region in the absorption spectra is noticed i.e. the absorption spectra appear to be broaden out as well as grow with the increasing number of MW exposures. The presence of the tail in the visible spectral region in the absorption spectra (i.e. the broadness in the absorption spectra) is related to the size distribution of CdS nanocrystallites. The increase in intensity indicates the enhancement in the number of CdS nanocrystallites with repeated MW exposures. We could estimate[2,14] the value of radius (R_{CdS})of CdS nanoparticles prepared under 3, 4 and 5 repeated MW exposures as 2.83, 3.97 and 8.37 nm, respectively. The estimated values of R_{CdS} indicate clearly that the shift in the absorption peak towards the longer

146

wavelength region corresponds to the increase in particle size of the CdS nanocrystallites with the number of repeated MW irradiation. The increase in particle size with the MW irradiation time indicated that Ostwald ripening appeared to determine the final size of CdS nanocrystallite in the present preparation procedure[2,14] with 1- butanethiol as the capping agent.

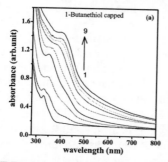 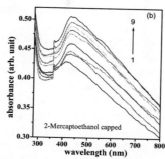

FIGURE 1. Absorption spectra of CdS nanocrystallites (a) 1-butanethiol capped; (b) 2-mercaptoethanol capped. The numbers 1→9 indicate the number of microwave exposures corresponding to the plots and the upward direction of arrows shows the gradual increase in the absorption intensity with increasing number of MW exposures.

For the 2-mercaptoethanol capped CdS nanocrystallites in DMF solution the absorption peak (Fig.1b) initially showed a very little shift towards the longer wavelength region and then showed almost no change with increasing number of MW exposures. The observed negligible change in the position of the absorption bands can be ignored. The observed broadness of the absorption spectra from Fig.1b indicates the size distribution of CdS nanocrystallites. The size of the maximum number of CdS nanocrystallites remained almost the same with repeated MW exposures.

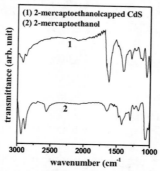

FIGURE 2. FTIR spectra of (1) 2-mercaptoethanol capped CdS; (2) 2-mercaptoethanol

The curve 1 of Fig. 2 shows the FTIR spectra recorded for the CdS nanocrystallites prepared via MW irradiation for 40s with 2-mercaptoethanol as the capping agent. This curve is similar to the FTIR spectrum of that of 2-mercaptoethanol (curve2, Fig2) except for the absence of the S-H vibration peak[15] at about 2557 cm^{-1}. The thiolates are connected to the Cd^{2+} sites on the CdS nanocrystallites surface via sulfur atoms and act as the skin of the CdS particles[15,16].

The new band near 1626 cm^{-1} could be due to the water bending of the adsorbed water molecules on the surface of the nanocrystallites and the band near 1045 cm^{-1} represents the SO_3 stretching[9]. Thus the capping of CdS nanocrystallites by 2-mercaptoethanol has been confirmed by FTIR spectroscopy. The FTIR spectrum of 1-butanethiol capped CdS nanocrystallites could not be performed as the material evaporated quickly from the sample pellets during the measurement of FTIR spectrum.

Fig.3a and b show the XRD pattern of CdS nanocrystallites capped with 1-butanethiol and 2-mercaptoethanol, respectively. The XRD pattern shows the presence of the reflection characteristics of the hexagonal phase of the nanocrystallites prepared under microwave irradiation with 1-butanethiol and 2-mercaptoethanol capping agents. For 1- butanethiol capped CdS nanocrystallites the intense and wide peaks are positioned at $2\theta = 26.6^0$, 43.8^0, and 51.2^0, which are oriented along the (002), (110), and (112) directions. For 2-mercaptoethanol capped CdS nanocrystallites the intense and wide peaks are positioned at $2\theta = 28.3^0$, and 48.2^0 which are oriented along the (101) and (103) directions. These directions are in agreement with the JCPDS file 41-1049. The size D of the prepared nanoparticles/ nanocrystallites was estimated by using the Scherrer equation to be 2.31 and 1.59nm for the 1-butanethiol and 2-mercaptoethanol capping, respectively.

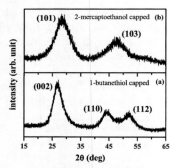

FIGURE 3. XRD patterns of thiol capped CdS nanocrystallites: (a) 1- butanethiol capped (b) 2-mercaptoethanol capped

The photoluminescence (PL) of the CdS nanocrystallites is expected to give information on their size and surface structures[6,13,14,16,18]. Fig.4a shows the emission spectra of 1-butanethiol capped CdS nanocrystals in DMF solution with the increase in the number of MW exposures. The decrease in the emission intensity and the shift in the emission peak position to the longer wavelength side with the increasing number of MW exposure can be attributed to emission from surface defects[4] and particle size growth. The dependence of PL on the photoexcitation wavelength is shown in Fig.4b. A decrease in PL intensity of 2-mercaptoethanol capped CdS nanocrystallites in DMF solution (Fig.4c) with the increasing number of MW exposures; with the emission peak remaining at the same position can be attributed to the emission from surface defects/ sulfur vacancy[6,8,19]. With the increasing photoexcitation wavelength the intensity of the emission peak was observed to increase but the emission was peaked at the same position (Fig.4d) which indicated that the CdS nanoparticle size (average) representing the PL peak position was independent of the photoexcitation wavelength[14].

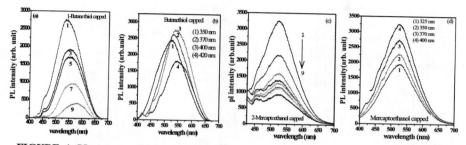

FIGURE 4. PL spectra of 1-butanethiol capped CdS nanoparticles: (a) for repeated MW exposures (numbers representing the curves are the number of MW exposures), photoexcitation at 470 nm; (b) for different photoexcitation wavelengths after first MW exposure. Emission spectra of 2-mercaptoethanol capped CdS nanoparticles: (c) for repeated MW exposures (numbers 1→9 indicate the number of MW exposures corresponding to the plots and the downward direction of arrow shows the gradual decrease in the PL intensity with increasing number of MW exposures), photoexcitation at 470 nm; (d) for different photoexcitation wavelengths after first MW exposure.

In the early stages of MW irradiation, the surfaces of the nanocrystallites are mainly covered with Cd^{2+} resulting in the formation of sulfur vacancies. Thus the sulfur vacancy giving rise to PL is formed in the early stages and the excess of Cd^{2+} is decreased in the latter stages because it is consumed to produce more CdS nanocrystallites. This change in the number of sulfur vacancies gives rise to changes in the PL intensity[4].

The relative PL quantum yield of CdS nanocrystallites (Φ_{CdS}) were estimated by integrating the area under the PL curves by the usual procedure[14.] A systematic decrease in the relative PL quantum yield of CdS nanocrystallites was noticed (Table 1) with the increase in MW exposures for both 1-butanethiol and 2-mercaptoethanol capped CdS nanocrystallites. The low quantum yield, generally less than 1%, is taken as an indication that radiationless recombination of the charge carriers is the dominating process[16].

CONCLUSIONS

The microwave (MW) irradiation/ exposure drastically influences the nucleation and growth of CdS nanocrystallites prepared by manipulating the capping agent and the time/ duration of MW irradiation (i.e number of repeated MW exposures) of the chemical bath. With the increase in the number of MW exposures in the case of 1-butanethiol as capping agent the size of the CdS nanocrystallites increases but for 2-mercaptoethanol as the capping agent the size of maximum number (average) of CdS nanocrystallites remains the same. In 1-butanethiol capped CdS nanocrystallites the size distribution and surface defects contribute to PL and in 2-

mercaptoethanol capped CdS nanocrystallites only the surface defects contribute to PL. The systematic decrease in estimated relative PL quantum yield with increase in the number of MW exposures for both 1-butanethiol and 2-mercaptoethanol capped CdS nanocrystallites indicates that the radiationless recombination of the charge carriers is the dominating process.

TABLE 1. Relative Photoluminescence (PL) QuantumYield (Φ_{CdS}) of Thiol-Capped CdS Nanocrystallites Prepared under Microwave Irradiation.

Number of Microwave Exposures	Φ_{CdS} (for 1-butanethiol)	Φ_{CdS} (for 2-mercaptoethanol)
First	0.0064	0.0023
Second	0.0036	0.0013
Third	0.0018	0.00092
Fourth	0.0013	0.00067
Fifth	0.0010	0.00068
Sixth	0.00065	0.00059
Seventh	0.00039	0.00056
Eighth	0.00034	0.00044
Nineth	0.00016	0.00042

ACKNOWLEDGMENTS

The authors express thanks to the authorities of IACS for providing working facilities. One of the authors AKC is thankful to the concerned authorities for the approval of leave and research project.

REFERENCES

1. B. A. Korgel and H. G. Monbouquette, *J. Phys. Chem.* **100**, 346-351 (1996).
2. R. He, X. Qian, J. Yin, H. Xi, L. Bian and Z. Zhu, *Colloids And Surfaces A* **220**, 151-157 (2003).
3. Y. Ohara, T. Nakabayashi, K. Iwasaki, T. Torimoto, B. Ohtani, T. Hiratani, K. Konishi and N. Ohta, *J. Phys. Chem. B* **110**, 20927-20936 (2006).
4. Y. Wada, H. Kuramoto, J. Anand, T. Kitamura, T. Sakata, H. Mori and S. Yanagida, *J. Mater. Chem.* **11**, 1936-1940 (2001).
5. C. K. Graetzel and M. Graetzel, *J. Am. Chem. Soc.* **101**, 7741-7743 (1979).
6 J. J. Ramsden and M. Gratzel, *J. Chem. Soc. Faraday Transactions I* **80**, 919-933 (1984).
7. A. Mills and G. Williams, *J. Chem. Soc. Faraday Transactions I* **85**, 503-519 (1989).
8. H. Fujiwara, H. Hosokawa, K. Murakoshi, Y. Wada, S. Yanagida, T. Okada and H. Kobayashi, *J. Phys. Chemistry. B* **101**, 8270-8278 (1997).
9. E. Caponetti, D. C. Martino, M. Leone, L. Pedone, M. L. Saladino and V. Vetri, *J. Colloid Interf. Sci.* **304**, 413-418 (2006).
10. K. S. Babu, C. Vijayan and P. Haridoss, *Mater. Lett.* **60**, 124-128 (2006).
11. H. Yang, C. Huang, X. Li, R. Shi and K. Zhang, *Mater. Chem. Phys.* **90**, 155-158 (2005).
12. Z. A. Peng and X. G. Peng, *J. Am. Chem. Soc.* **124**, 3343-3353 (2002).
13. N. Chestnoy, T. D. Harris, R. Hull and L. E. Brus, *J. Phys. Chem.* **90**, 3393-3399 (1986).
14. S. Karan and B. Mallik, *J. Phys. Chem.C* **111**, 16734-16741 (2007).
15. S. Chen, T. Ida and K. Kinamura, *Chem. Commun* 2301-2301 (1997).
16. L. Spanhel, M. Haase, H. Weller and A. Henglein, *J. Am. Chem. Soc.* **109**, 5649-5655 (1987).
17. B. D. Cullity. "Elements of X-ray Diffractions; Addition-Wesley: Reading, MA, 1978; pp 102.
18. Y. Wang and N. Herron, *J. Phys. Chem.* **92**, 4988-4994 (1988).

19 M. Kanemoto, H. Hosokawa, Y. Wada, K. Murakoshi, S. Yanagida, T. Sakata, H. Mori, M. Ishikawa and H. Kobayashi, *J. Chem. Soc., Faraday Transactions* **92**, 2401-2411 (1996).

20. E. Austin and M. Gouterman, *Bioinorg. Chem.* **9**, 281-298 (1978).

21. K. Dhara, S. Karan, J. Ratha, P. Roy, G. Chandra, M. Manassero, B. Mallik and P. Banerjee, *Chem. Asian. Journal* **2**, 1011-1100 (2007).

Hydrothermal Synthesis, Characterization and Optical Property of Single Crystal ZnO Nanorods

Prabhakar Rai, Suraj Kumar Tripathy, Nam-Hee Park, Yeon-Tae Yu *

Division of Advanced Materials Engineering and Research Centre for Advanced Materials Development, College of Engineering, Chonbuk National University, Jeonju 561-756, South Korea
* Corresponding author: Y. T. Yu (E-Mail: yeontae@chonbuk.ac.kr, Tel: +82-63-270-2288, Fax: +82-63-270-2305)

Abstract: ZnO nanorods of 100±10 nm in diameter and 900±100 nm in length were synthesized by cetyl trimethylammonium bromide (CTAB) assisted hydrothermal technique from a single molecular precursor. The influence of pH on morphology and PL property of ZnO nanorods were investigated. The phase and structural analysis were carried out by X-ray diffraction. Morphology of the nanorods was investigated by electron microscopy techniques. Optical properties were investigated by photoluminescence spectroscopy. As prepared ZnO nanorods have been single crystalline without defect and shown intense room temperature photoluminescence peak in the ultraviolet region.

Keywords: Electron microscopy, hydrothermal, luminescence, ZnO nanorods
PACS: 68.37.Hk, 81.16.Be, 78.55.Qr, 61.46.Km

INTRODUCTION

ZnO, a wide band-gap ($E_g \approx 3.37$ eV) II-VI semiconductor material possesses interesting optical, dielectric and catalytic properties that make it suitable for various industrial applications such as pigments [1], dye-sensitized solar cells [2], photocatalysts [3] and sensors [4]. However, ZnO-based materials have immense prospects for high temperature optoelectronic applications due to its high exciton-binding energy (60 meV) and high optical gain (320 cm^{-1}) at room temperature. Recently, intensive research has been focused on fabricating one-dimensional (1D) ZnO nanostructures such as nanotubes, nanowires (rods), and nanobelts (rings) due to their shape induced novel properties and potential applications [5–7]. The large surface area of the nanorods make them attractive for gas and chemical sensing, and ability to control their nucleation sites makes them candidates for micro lasers and memory arrays. Since a long time ago various physical and chemical synthetic techniques have been investigated to explore the material quality of ZnO of diverse morphology for fabrication of devices such as blue lasers and ultraviolet light emitting diodes [8]. Hydrothermal synthesis has emerged as a simple route for the processing of transition metal oxides [9]. Hydrothermal growth of ZnO crystals with variable, yet controllable, morphologies has been reported widely [10,11]. The growth habit of ZnO is determined mainly by the internal structure of the crystal, but is also sensitive to a number of external conditions such as pH, the zinc source (and its counterion), the presence (or not) of any complexing agent, the nucleation conditions (including effects due to the presence of a substrate), the extent of super-saturation, etc. Vernadou and co-workers have investigated the pH effect on the morphology of the ZnO nanostructures in aqueous chemical growth [12]. Recently cetyl trimethylammonium bromide (CTAB) assisted hydrothermal technique has emerged as an attractive technique to investigate the synthesis of zinc oxide nanostructures. Sun et al. have reported the growth of ZnO nanorods by CTAB assisted hydrothermal technique at 180^0C by using zinc powder as the source material [13]. Ni and co-workers have used a mixture of zinc acetate and CTAB to obtain zinc oxide rods in presence of KOH [14]. However to obtain regularly structured single-crystalline ZnO without defect from a single molecular precursor still remains a highly challenging issue.

Here we report a simple CTAB-assisted hydrothermal technique for the growth of well-dispersed single-crystalline ZnO nanorods from a single molecular precursor $[Zn(NO_3)_2.6H_2O]$. We investigate the possibility of controlling the nanostructures morphology by adjusting the pH of the solution. The novelty of this procedure can be characterized by the successful high yield production of single-crystalline ZnO nanorods.

EXPERIMENTAL

ZnO nanorods were grown by a simple hydrothermal reaction of $Zn(NO_3)_2.6H_2O$ (Reagent Grade, 98% Sigma-Aldrich) in presence of CTAB (Aldrich) as the promoter. In a typical procedure, $Zn(NO_3)_2.6H_2O$ (0.009mol) were dissolved in 40 ml of deionized water and pH was maintained at 7, 8 and 9 by adding ammonia followed by vigorous stirring for 1 h. A white precipitate was produced which was collected by centrifugation and washed thoroughly with deionized water. Then the precipitate was dispersed into 35ml water having 1wt%CTAB. This solution was charged in a 50 ml capacity autoclave with Teflon liner followed by uniform heating at 200°C for 10 h. The reaction mixture was heated at a rate of 5°C per minute. After the reaction the samples were allowed to cool as the natural process. After completion of the reaction, it was cooled to room temperature and powdered samples were collected by centrifugation. Powdered sample was thoroughly washed with deionized water and ethanol. Samples were dried at 80°C for 12h in presence of air. The crystal structure of the powder was studied by powder X-ray diffractometer (D/Max 2005, Rigaku), and particle size and morphology was investigated by scanning electron microscopy (SEM) and transmission electron microscopy (TEM-JEM-2010, JEOL). Photoluminescence property is studied by Cd-He laser source with 325nm excitation.

RESULTS AND DISCUSSION

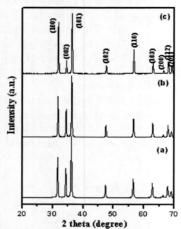

FIGURE 1. XRD pttern of as prepared ZnO nanorods at various pH.
(a) pH 7, (b) pH 8, (c) pH 9

Fig. 1 shows the XRD pattern of the as-synthesized ZnO nanorod sample. All the diffraction peaks are indexed to the hexagonal phase of ZnO (JCPDS 36-1451) and no other crystalline phases are detected. The intense peaks in the XRD pattern of the powder samples clearly show formation of hexagonal wurtzite phase of ZnO having prominent (101) plane in the sample. This is the most stable phase of ZnO.

FIGURE 2. SEM images of as prepared ZnO nanorods at various pH.
(a) and (b) pH 7, (c) and (d) pH 8,(e) and (f) pH 9

Morphology of the ZnO nanostructure was investigated by electron microscopy techniques. Fig. 2 shows the SEM images of the as-prepared ZnO nanostructure at different pH (7, 8 and 9). As can be seen, the morphology of ZnO nanostructures strongly depends on the pH of the solution. For example at pH 7 both low and high resolution images have shown the formation of well-dispersed nanorods with 100±10 nm in diameter and 900±100 nm in length (fig. 2 (a) and (b)). Increasing the pH to a value more than 8, spindle shape ZnO nanorods are formed (fig. 2 (c) and (d)). However, when pH about 9 hallow ZnO nanorods at one end were formed (fig.3 (e) and (f)). From above observation it is clear that morphology of ZnO nanostructure is controlled by NH$_4$OH. From fig.2 it is also clear although the shape of ZnO nanostructure changes their overall dimension remain the same except at pH 9. Fig. 3 shows the TEM images of the nanorods. The set of well-regulated diffraction points can be indexed to be [01−10] zone axis of the hexagonal ZnO structure, which indicates that the as prepared nanorod is single crystal [15].

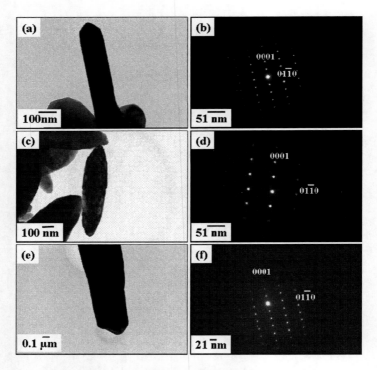

FIGURE 3.TEM images of as prepared ZnO nanorods at various pH.
(a) and (b) pH 7, (c) and (d) pH 8,(e) and (f) pH 9

Figure 4 illustrates the room temperature photoluminescence (PL) spectra of as prepared ZnO nanostructures. A sharp and strong UV emission band was recorded at around 403 nm for ZnO nanorod prepared at pH 7. By contrast, weaker UV emission band at around 399 and 413 nm was recorded for hallow shape and spindle shape ZnO nanostructure respectively. However, no defect mediated emission observed in any sample. These emission falls into violet region that is there is slight red shift in near band edge emission. Generally, the UV emission band is originated from the direct recombination of the free excitons through an exciton-exciton collision process while the visible emission is due to the impurities and structure defects in ZnO crystals [16]. The exact mechanism of violet emission is not observed however, it is consider that the violet emission comes from the transition of electron from bottom of conduction band to bottom of valance band [17]. The disappearance of the commonly encountered green emission in the PL spectrum demonstrates the absence of oxygen vacancy in our sample.

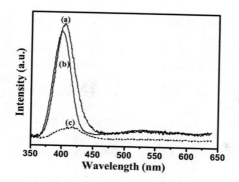

FIGURE 4. Room temperature photoluminescence of as prepared ZnO nanorods at various pH. (a) pH 7, (b) pH 8, (c) pH 9

However, change in shape of ZnO nanorod to hallow ZnO nanostructure due to change in pH can be explain as follows. Under hydrothermal condition following reaction occur at various pH,

$$NH_3 + H_2O \longleftrightarrow NH_3.H_2O \longleftrightarrow NH_4^+ + OH^- \quad (1)$$
$$Zn^{2+} + 2OH^- \longleftrightarrow Zn(OH)_2 \quad (2)$$
$$Zn(OH)_2 \longleftrightarrow ZnO + H_2O \quad (pH \text{ less than } 9) \quad (3)$$

$$NH_3 + H_2O \longleftrightarrow NH_3.H_2O \longleftrightarrow NH_4^+ + OH^- \quad (4)$$
$$Zn^{2+} + 4NH_3 \longleftrightarrow [Zn(NH_3)_4]^{2+} \quad (5)$$
$$[Zn(NH_3)_4]^{2+} + 2OH^- \longleftrightarrow ZnO + 4NH_3 + H_2O \quad (pH \text{ more than } 9) \quad (6)$$

In the formation of ZnO nanostructure below pH 9 the dehydration reaction between OH^- group and $Zn(OH)_2$ result in the ZnO seed crystal. Since growth unit incorporation in crystal lattice takes place by the dehydration reaction between OH^- ions, so the growth habit of ZnO crystal is related to OH^- ions at interface. In neutral medium, the exterior condition has smaller effect on OH^- ions at interface so the crystal growth habit is mainly affected by internal structure. However, when the hydrothermal reaction proceed above pH 8 more OH^- ions present in the solution which cause shielding of OH^- ions at interface. This shielding is larger at the (0001) and (01−1−1) facets. So that the growth rate of these facets have greater decrease than the other facets. This results into non-preferential growth of crystal. So the (0001) and (01−1−1) facets appear in case of pH more than 8. Thus due to this reason well dispersed regular shape ZnO nanorods are formed at pH 7 while spindle shape ZnO nanorods are formed above pH 8 [18].

However, above pH 9 the formation of hallow ZnO comes from the hydration of $[Zn(NH_3)_4]^{2+}$. The precursor $[Zn(NH_3)_4]^{2+}$ is related to the hydrothermal synthesis of ZnO in solution with a pH higher than 9 ammonium salt tends to bind with Zn^{2+} ions generating Zn^{2+} amino complex. They primarily adsorb on the six prismatic side planes slowing down the growth velocity of side surface. This facilitates the growth of 1D nanotower reaches certain equilibrium. After that the rate of dissolution is faster than the rate of formation (Ostwalled ripening). At this case the polar surface having the high energy will be dissolved in priority to decrease the system energy. Therefore tubular structure is formed [19]. During the formation processes, CTAB may acts as a transporter of the growth unit and modifier that leads to the orientated growth of ZnO nanorods [20].

CONCLUSIONS

High purity single-crystalline ZnO nanorods are grown by CTAB-assisted hydrothermal technique from a single molecular precursor. The influence of pH on morphology of ZnO nanorods was investigated. Optimal condition for growth of nanorods with improved crystal quality was obtained at reaction temperature of 200°C, pH of 7 and in a reaction time of 10 h. Electron microscopy studies showed the formation of well dispersed nanorods of 100±10 nm in diameter and 900±100 nm in length at pH 7. As prepared ZnO nanorods have shown a single intense room temperature photoluminescence peak in the violet region. Absence of defect mediated green luminescence peak suggests the formation of well crystalline ZnO nanorods without any impurities or structural defects.

ACKNOWLEDGEMENTS

This work is supported by Post-BK21 program from Ministry of Education and Human-Resource Development

REFERENCES

1. V. Čechalová and A. Kalendová, *J. Phys. Chem. Solids* **68**, 1096-1100 (2007).
2. A. E. Suliman, Y. Tang and L. Xu, *Solar Energy Mater. Solar Cell* **91**, 1658-1662 (2007).
3. K. M. Parida, S. S. Dash, D. P. Das, *J. Colloid Interface Sci.* **298**, 787-793 (2006).
4. T. Gao and T. H. Wang, *Appl. Phys. A* **80**, 1451-1454 (2005).
5. Z. Pan, Z. Dai, Z.L. Wang, *Science* **291**, 1947-1949 (2001).
6. X. Kong, Y. Ding, R. Yang, and Z. L. Wang, *Science* **303**, 1348-1351 (2004).
7. C.H. Liu, J. A. Zapien, Y. Yao, X. M. Meng, C. S. Lee, S.S. Fan, Y. Lifshitz, S. T. Lee, *Adv. Mater* **15**, 838-841 (2003).
8. Ü. Özgür, Y. I. Alivov, A. Take, M. A. Reshchikov, S. Doğan, V. Avrutin, S. J. Cho, and H. Morkoc, *J. App. Phys.* **98**, 041301-041404 (2005).
9. B. Liu and H. C. Zeng, *J. Am. Chem. Soc.* **125** 4430-4431 (2003).
10. B. Cheng and E. T. Samulski, *Chem. Comm.* 986-987 (2004).
11. T. Sahoo, S. K. Tripathy, Y. T. Yu, H. K. Ahn, D. C. Shin, and I. H. Lee, *Mater. Res. Bul* **43**, 2060-2068 (2008).
12. D. Vernardu, G. Kenanakis, S. Couris, E. Koudoumas, E. Kymakis, and N. Katsarakis, *Thin Solid Films* **515** 8764-8767 (2007).

13. X. M. Sun, X. Chen, Z. X. Deng, and Y. D. Li, *Mater. Chem. Phy.* **78**, 99-104 (2002).
14. Y. H. Ni, X. W. Wei, X. Ma, and J. M. Hong, *J. Crystal Growth* **283,** 48-56 (2005).
15. D. Chu, Y. P. Zeng, and D. Jiang, *Mater. Let.* **60** 2783-2785 (2006).
16. A. Umar and Y. B. Hahn, *Nanotechnology* **17**, 2174-2180 (2006).
17. Q.P. Wang, D.H. Zhang, Z.Y. Xue, and X. T. Hao, *Appl. Surf. Sci.* **201**, 123–128, (2002).
18. Wen Ju Li, Er-Wei, and Wei-Zhou Zhong, *J. Cryst. Growth* **203**, 186-196 (1999).
19. Hongwei Hou, Yi Xie, and Qing Li, *Solid State Sci.* **7**, 45–51 (2005).
20. B. G. Wang, E. W. Shi, and W. Z. Zhong, *Cryst. Res. Technol.* **32**, 659-667 (1997).

Preparation and optical characterization of PbS quantum dots

N. Choudhury[a], P. K. Kalita[b] and B. K. Sarma[a]

[a]*Department of Physics, Gauhati University, Guwahati 781014, India*
[b]*Department of Physics, Guwahati College, Guwahati 781021, India*

Abstract. PbS quantum dots are prepared on glass substrates as thin films by chemical bath deposition (CBD) method. The structure of nanocrystalline PbS is characterized by X-ray diffraction and the size of the crystallites are found to vary from 5-10 nm. The UV-visible absorption spectra show a large blue shift that increases the band gap up to 2.89 eV on decrease of molarity. Photoluminescence (PL) study indicates the size effects where PL peaks are found around 475-538 nm. IR spectra studies are made on all the films which show the relevant functional groups in addition to PbS bonding.

Keywords: quantum dots, PbS, blue shift, XRD, photoluminescence, IR spectra
PACS: 73.63.Kv, 61.05cp, 78.55.-m, 78.30.-j

Introduction

In recent years there has been growing interest in the studies of materials with nanometer sized ultrafine particles, since these particles often exhibit properties and structures different from those of the corresponding bulk materials. Reduction of sizes drastically changes the electronic, optical and structural properties of such materials[1]. In such materials, apart from phenomenon like quantum size confinement, wave-like transport and predominance of interfacial phenomenon play role in determining their properties. Among the semiconductor quantum dots, PbS is important due to its unique optical and electrical properties[2]. In the bulk stage, PbS has an energy band gap of 0.41eV at room temperature. But the band gap of PbS in nanocrystalline form can be changed from 2.8eV to as large as 5.2eV by varying the size of the grains of the material[3]. The nanoparticles of PbS and other Lead salts like PbTe and PbSe are better suited for strong confinement limit compared with other well known II-VI semiconductor nanoparticles like CdS. The radius of electrons and holes of PbS (~ 10 nm) are much larger than the corresponding values of CdSe (3 nm and 1 nm). The Bohr excitonic radius for PbS is 20 nm[4]. Out of several other methods for deposition of nanocrystalline films, Chemical Bath Deposition (CBD) is a simple method as it is relatively inexpensive and convenient for large area deposition[5, 6].

Experimental

In the present work polyvinyl alcohol (PVA) was used to act as an active and flexible medium to organize nanoclusters. For preparation of PbS quantum dots, the matrix PVA was prepared by adding lead acetate to a 2 wt% PVA solution using

CP1147, *Transport and Optical Properties of Nanomaterials—ICTOPON - 2009*, edited by M. R. Singh and R. H. Lipson
© 2009 American Institute of Physics 978-0-7354-0684-1/09/$25.00

double distilled water with constant stirring with a stirring rate at ~ 200 rpm at a constant temperature of 70°C until a transparent solution is formed. Three different concentrations of lead acetate (0.1M, 0.5M and 1.0M) were used to get three different matrix solutions. The solution was then left for 24 hours to get a transparent liquid indicating complete dissolution of lead acetate. Ammonia solution was added slowly to form the complex maintaining pH of the solution at around 11. Suitably cleaned glass substrates of size (20 mm x 35 mm) are then introduced vertically in this solution using a specially designed substrate holder. Then the equimolar solution of thiourea was added to each of the mixture solutions. After 24 hours the substrates with layers of composites of nanocrystalline PbS are taken out, rinsed with running distilled water and then dried in an oven.

X-ray powder diffraction (XRD) patterns of the products were recorded on a (Bruker AXS D8 ADVANCE) Fully Automatic Powder X-Ray Diffractometer operating at 40 KV – 30 mA with CuKα radiation (λ = 1.54 Å). SEM images were obtained on a scanning electron microscopy (Model LEO 1430 VP). UV-visible absorption was carried out using an automated spectrophotometer (HITACHI U3210) in the wavelength range 250 nm – 650 nm. PL spectra were recorded on an AB2 Luminance Spectrometer. IR spectra were recorded using PERKIN ELMER instrument series II (2400-CHNS/0).

Results and Discussion

1. XRD studies

Figure 1 depicts the XRD patterns of PbS quantum dots. From the Scherrer formula, the crystallite size of the PbS crystallines were determined and found to be within the range 5.4 – 10.1 nm. The diffraction spectra confirm the formation of fcc PbS nanocrystals. The values of crystallite sizes of the quantum dots are shown in Table 1. It is observed that crystallite size increases with increase in molarity of the lead acetate in the solution.

TABLE 1. The values of crystallite sizes and band gaps for PbS quantum dots having different molarities.

Molarity	Colour	Crystallite size (nm)		Band gap (eV) from UV
		From XRD	From SEM	
0.1 M	reddish	5.4	-	2.89
0.5 M	brown	7.1	65.5	2.53
1.0 M	black	10.1	89.0	2.30

2. SEM analysis

Scanning electron microscopy (SEM) is a convenient technique to study the surface morphology of the quantum dots. SEM images of PbS quantum dots embedded in PVA is shown in Figure 2. From this image, it can be seen that the grain size of PbS quantum dots are not uniform. The values of crystallite sizes measured from SEM are presented in Table 1. It is observed that the average grain sizes determined by SEM are comparatively larger than measured by XRD. This type of results is consistent with our earlier works[7]. This enhancement may be due to the agglomeration of nanoparticles.

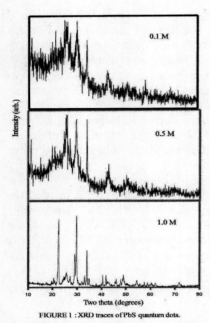

FIGURE 1 : XRD traces of PbS quantum dots.

Figure 1: XRD traces of PbS quantum dots.

Figure 2: SEM image of a PbS quantum dots for 0.5M.

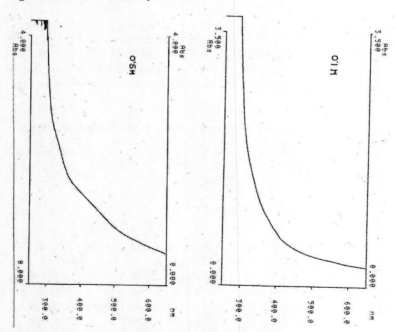

Figure 3. UV-visible optical absorption spectra of PbS samples

162

3. Absorption studies

The UV Absorption spectra of PbS/PVA solution for different molarities taken at room temperature is shown in Figure 3. It shows that the absorption peak is largely blue shifted indicating strong quantum confinement. Since the bulk absorption edge

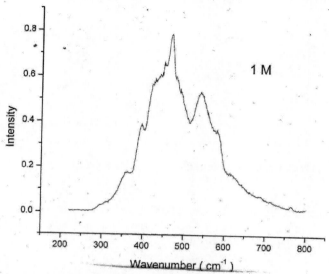

Figure 4. PI spectra of PbS quantum dots

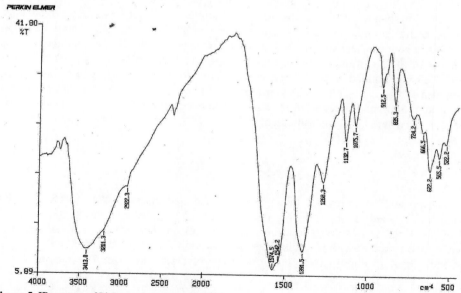

Figure 5. IR spectra of PbS quantum dots for 0.5M

TABLE 2. Vibrational assignments of PbS quantum dots with positions and intensities of absorption shown by IR spectra for a typical film (0.5M)

Positions (cm^{-1})	Intensities	Assignments
3413	Strong	OH stretching
3201 2922.3	Doublet	C-H stretching
1547.5	Strong	C-C stretching
1391.5	Medium	C-O stretching
1132.7 1057.7	Weak	SO$^-_4$ as trace
699.5 622	Medium	Pb-S stretching

for PbS is 3020 nm, so the band gap decreases with the increase of crystallite sizes.

4. PL studies

Photoluminence spectra are obtained by using AB2 Luminance Spectrometer. The observed peak from the graph (figure 4) of 1 M is 511 nm. This also supports the strong quantum confinement in the prepared PbS films.

5. IR spectra

IR spectra of PbS quantum dots for different molarities were quite similar. An IR spectrum of a typical PbS quantum dots is depicted in Fig. 5. The IR frequencies along with the vibration assignment for PbS quantum dots for 0.5 M are given in Table 2. The band at 3413 cm^{-1} is due to OH stretching vibration of water molecules. The bending vibration of water molecule appeared at 1547.5 cm^{-1}, C-C stretching and a COO- unsymmetrical stretching at 1391.5 cm^{-1} appeared which were induced by the amount of SO$_4^-$ as impurity is seen as there are small absorption around 1075.7 cm^{-1}. At 699.5 cm^{-1} and 622 cm^{-1}, there are medium to strong bands which have been assigned to Pb-S stretching. The vibration absorption peak of PbS band which should be around 430 cm^{-1} as reported by other workers could not be observed since it was beyond the extent of our measurement.

CONCLUSIONS

The present work reveals that polymer embedded quantum dot systems can be successfully used for investigating the structural and optical changes in the samples. The XRD patterns clearly show the prepared PbS quantum dots are polycrystalline having fcc structure with the crystallite sizes varying from 5.4-10.1 nm. The band gap of PbS quantum dots increases with decrease of crystallite size. The quantum size effect is also supported by UV absorption spectra as well as photoluminescence studies.

ACKNOWLEDGMENTS

One of the authors (NC) thanks UGC for providing a teacher fellowship for the period when the work was being done. The authors thank Chemistry Department, Gauhati University, Guwahati for providing UV and IR facility and IIT, Guwahati for XRD, SEM and PL data.

REFERENCES

1. R. Rosseti, J. M. Elison and L. E. Brus, *J. Chem. Phys.* **80**, 4464 (1984).
2. S. Espevik, C. Wu and R. H. Bube, *J. Appl. Phys.* **42**, 3513 (1971).
3. H. Gleiter, *ACTA. Mater* **48**, 1-29 (2000).
4. S. Chowdhury, S. K. Dolui, D. K. Avasthi and A. Choudhury, *Indian J. Phys.* **79**, 1019-1022 (2005).
5. P. Hoyer and R. Konenkamp, *Appl. Phys. Lett.* **66**, 349-351 (1995).
6. G. P. Kothiyal, B. Ghosh and R. Y. Deshpande, *J. Phys. D: Appl. Phys.* **13**, 869-873 (1980).
7. N. Choudhury and B. K. Sarma, *Indian J. Pure & Appl. Phys.* **46**, 261-265 (2008).

Dielectric Material Based Band Gap Tailoring For 1D Photonic Crystal

S. Dasgupta[a], C. Bose[b], SMIEEE

[a] ECE Dept, MCKV Institute of Engineering, 243 G.T Road, Howrah- 711204, India
reach2samrat@yahoo.co.in
[b] Dept. of Electronics and Telecommunication Engg, Jadavpur University, Kolkata-700032, India
chayanikab@ieee.org

Abstract. In this paper, we present the transmission characteristic of a 1D photonic crystal with 2-layer 10 period stack. We use the transfer matrix approach and assume light with p- polarization. The effective periodicity of the photonic crystal is modified by changing the dielectric constant of one layer with the other remaining fixed. Increase in refractive indices of either of the two layers shifts the photonic gap towards longer wavelength with an associated reduction in its width. Thus, the centre wavelength of the photonic gap is tuned to appear about 1550 nm making it useful for signal processing applications in optical communication.

Keywords: Photonic crystal, Transmittivity, Transfer Matrix.
PACS: 42.70.Qs; 78.20.Ci; 02.60.-x

INTRODUCTION

Photonic crystals, first realized by Yablonovitch and John [1,2], are periodic dielectric structures in which refractive indices of alternate layers may vary along one, two or three orthogonal directions yielding 1D, 2D or 3D crystals. In such structures, the photon confinement arising from its refractive index contrast leads to formation of allowed and forbidden energy regions for light waves traveling through it, and thereby, introduces the concept of photonic band gap. In recent years, the optical properties of photonic band gap structures are being widely investigated for their potential applications in optical communication [3-6].

Our present study deals with a 1D PBG structure, in which dielectric constant of each alternate layer is varied, keeping that of others fixed. Transmittivity of the resulting PBG structure is computed, and the transmission spectrum of light with p-polarization passing through it, is examined. The transfer matrix method is used for computation. This technique allows any extension of periodic structure with least added complexity.

CP1147, *Transport and Optical Properties of Nanomaterials—ICTOPON - 2009,* edited by M. R. Singh and R. H. Lipson
© 2009 American Institute of Physics 978-0-7354-0684-1/09/$25.00

THEORY AND MODELING

To estimate the transmittivity for a PBG structure, a general form of transfer matrix is formulated. For this purpose, a 1D PBG structure, extended along z-direction, is considered to have dielectric materials of relative permittivity ε_1, ε_2, and permeability μ_1, μ_2; the substrate and superstrate materials having the same electromagnetic properties as vacuum. A p-polarized electromagnetic field associated with a light wave moving in the xz plane is assumed to be incident on the crystal at an angle α with the z-axis.

Relations describing reflection and transmission of p-polarized wave, incident at the interface between the dielectric materials while traveling from the denser medium (medium 1 of length L_1) to rarer medium (medium 2 of length L_2) can be written as [7]

$$\frac{A_r}{A_i} = \frac{\sqrt{\mu_1/\varepsilon_1}\,\cos\alpha - \sqrt{\mu_2/\varepsilon_2}\,\cos\beta}{\sqrt{\mu_1/\varepsilon_1}\,\cos\alpha + \sqrt{\mu_2/\varepsilon_2}\,\cos\beta} \tag{1}$$

and $$\frac{A_t}{A_i} = \frac{2\sqrt{\mu_2/\varepsilon_2}\,\cos\alpha}{\sqrt{\mu_1/\varepsilon_1}\,\cos\alpha + \sqrt{\mu_2/\varepsilon_2}\,\cos\beta}, \tag{2}$$

where A denotes the complex amplitude of electric field and β, the angle of refraction.

Assuming both the materials to be non-magnetic (i.e., $\mu_1 = \mu_2 = 1$) and expressing dielectric constants in terms of refractive indices (n_1 and n_2), equations (1) and (2) can be written as

$$\left(\frac{A_r}{A_i}\right)_{12} = \frac{n_2\cos\alpha - n_1\cos\beta}{n_2\cos\alpha + n_1\cos\beta} \tag{3}$$

and $$\left(\frac{A_t}{A_i}\right)_{12} = \frac{2n_1\cos\alpha}{n_2\cos\alpha + n_1\cos\beta}. \tag{4}$$

Similarly, reflection and transmission coefficients for the interface encountered by the wave while propagating from lower to higher refractive index medium are

$$\left(\frac{A_r}{A_i}\right)_{21} = \frac{n_1\cos\beta - n_2\cos\alpha}{n_2\cos\alpha + n_1\cos\beta} \tag{5}$$

and $$\left(\frac{A_t}{A_i}\right)_{21} = \frac{2n_2\cos\alpha}{n_2\cos\alpha + n_1\cos\beta}. \tag{6}$$

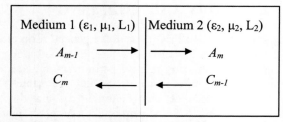

FIGURE 1. Schematic representation of forward and backward waves formed at the mth interface of two dielectric layers forming the unit cell.

The forward (A) and backward (C) waves formed at the mth interface (i.e. at $z=z_m$) between layers of higher and lower refractive index medium are schematically shown in Fig.1, and can be described, using equations (3) to (6), as

$$C_{m-1} = \frac{a_{12}}{b_{12}} A_{m-1} + \frac{d_{12}}{b_{12}} C_m e^{i\theta_2} \tag{7}$$

and $\quad A_m e^{-i\theta_2} = \dfrac{c_{12}}{b_{12}} A_{m-1} - \dfrac{a_{12}}{b_{12}} C_m e^{i\theta_2}$, $\tag{8}$

where $\quad a_{12} = n_2 \cos\alpha - n_1 \cos\beta$, $\quad b_{12} = n_2 \cos\alpha + n_1 \cos\beta$, $\quad c_{12} = 2n_1 \cos\alpha$,

$d_{12} = n_2 \cos\beta$, $\quad$ with $\theta_1 = k_{m-1} z_{m-1} \cos\alpha$, $\quad \theta_2 = k_m z_m \cos\beta$, $\quad k_{m-1} = \sqrt{\varepsilon_1}\dfrac{\omega}{c}$ $\quad$ and

$k_m = \sqrt{\varepsilon_2}\dfrac{\omega}{c}$.

Analysis Using T-matrix

To form the transfer matrix for the PBG structure, equations (5) and (6) are applied at each of the interfaces with $m=1, 2,...., M$. Assuming that the incident wave is normalized (i.e. $A_0 =1$) and suffers no further reflection after the Mth interface (i.e. $C_M = 0$), forward and backward propagating waves at each interface of an unit cell can be described in a matrix form as

$$\begin{bmatrix} A_{m-1} \\ C_{m-1} \end{bmatrix} = \begin{bmatrix} T_1 \end{bmatrix} \begin{bmatrix} A_m \\ C_m \end{bmatrix} \tag{9}$$

where T_1 is a 2 X 2 matrix with elements

$$T_{11} = \left(\frac{b^2_{12}}{c_{12}d_{12}}\right)e^{-i(\theta_1+\theta_2)}, \qquad T_{12} = \left(\frac{-a^2_{12}}{c_{12}d_{12}}\right)e^{i(\theta_2-\theta_1)},$$

$$T_{21} = \left(\frac{-a^2_{12}}{c_{12}d_{12}}\right)e^{-i(\theta_2-\theta_1)} \text{ and } T_{22} = \left(\frac{b^2_{12}}{c_{12}d_{12}}\right)e^{i(\theta_1+\theta_2)}.$$

Considering the entire PBG structure, forward and backward waves in two terminating layers can be related through transfer of matrices for intermediate unit cells, and expressed finally in a similar form as equation (9), where the resultant matrix is product of all the 2 X 2 matrices for individual unit cells.

The resultant T-matrix can finally be formulated as

$$T = \begin{pmatrix} \dfrac{1}{t} & \dfrac{r^*}{t^*} \\[2mm] \dfrac{r}{t} & \dfrac{1}{t^*} \end{pmatrix} \tag{10}$$

where r^* and t^* are complex conjugate of the reflection (r) and transmission (t) coefficients of incident light wave for the PBG structure. Obtaining transmission coefficient from equation (10), transmittivity of the structure can be readily determined using the relation

$$T_p = \left(\frac{n_2 \cos \beta}{n_1 \cos \alpha} \right)(t.t^*). \tag{11}$$

RESULTS

Transmission spectrum for light passing through a 1D PBG structure is characterized with a series of pass and stop bands. The transitions between such bands become more and more abrupt as the pair of dielectric layers is added to the structure. In principle, a PBG structure should, therefore, extend indefinitely to yield an ideal photonic crystal. The characteristic of transmission spectrum for a PBG structure, however, tends to saturate for a stack with 8-period two-layer unit cells. Therefore, in our computation, the number of periods in photonic crystal is limited to 10. In addition, normal incidence of light is assumed without sacrificing the generality of the end results.

Each dielectric layer in a unit cell of the photonic crystal described above provides optical path of length $\lambda_0/4$, where λ_0 is the free space wavelength ($\lambda_0 = 2\pi c/\omega$) and taken here as 1.55 μm. This requires that $n_1 L_1 = n_2 L_2 = \lambda_0/4$, where L_1 and L_2 are the length of the two dielectric layers in the unit cell. Fulfillment of above condition should result in a photonic band gap centered on λ_0.

The resulting transmission spectrum for light with p-polarization with normal incidence has been estimated using equations (9) and (11), and is presented in Fig.2. The spectrum exhibits a number of pass bands and stop bands of different widths. Pass bands are defined as the range of wavelengths characterized with non-vanishing transmittivity of light. Within the pass bands, transmittivity peaks periodically to unity, where such period increases with wavelength of incident light.

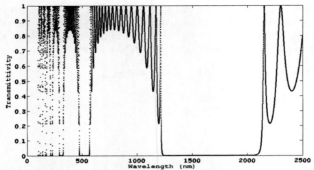

FIGURE 2. Transmittivity vs. wavelength of incident light for a 10-period PBG structure with normal incidence for λ_0=1.55 μm, n_1=3.5 and n_2= 1.5.

Stop bands in the transmission spectrum correspond to the regions of flat reflection window with zero transmitivity. Widths of both pass bands and stop bands are found to increase with wavelength of incident light. As also evident from Fig.2, pass bands and stop bands, lying in the region of smaller wavelength are not clearly distinguishable and hence, are not useful for practical applications.

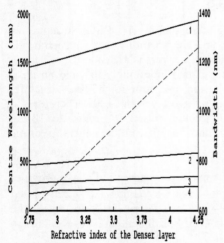

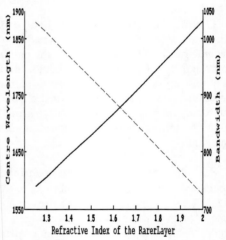

FIGURE 3. Centre wavelength of reflection window (solid lines) and bandwidth (dotted line) as a function of denser layer refractive index; labels designate different windows: 1 for the widest and 4 for the narrowest one.

FIGURE 4. Centre wavelength of reflection window (solid line) and bandwidth (dotted line) as a function of denser layer refractive index.

170

Fig. 3 depicts dependence of center wavelengths of the four widest stop bands on refractive index of the denser layer (solid lines). With increase in refractive index of the denser layer, the center wavelengths of the stop bands are found to shift linearly towards the longer wavelength region, since higher refractive index causes longer optical path offered by the dielectric layer. The widest stop band, designated as the photonic band gap, is most sensitive to change in refractive index and is only studied in details here. It is evident from Fig.3 that the width of the photonic gap increases linearly for above variation in refractive index, as it effectively increases the index contrast between the two dielectric layers.

In Fig. 4, variations of center wavelength and width of photonic gap have been presented as a function of refractive index of the rarer dielectric layer, while that of the denser layer kept unaltered. As the refractive index increases, center of the gap is seen to move upwards due to reason already explained above. But in this case, increase in refractive index reduces the index contrast between the two layers and therefore, decreases the width of photonic gap, as expected.

CONCLUSIONS

The transmission characteristic of a 10-period two-layer PBG structure is studied with variations in refractive index of the associated dielectric materials. Variation in refractive index is found to control the center of the photonic gap through modulation of optical path length, and width of the band gap through modulation of refractive index contrast between layers in each period. Our model is designed to tailor photonic gap encompassing 1.55 μm, the most important wavelength for optical communication. Accordingly, the unit cell is made of dielectric layers with refractive index 3.5 and 1.5 and of adequate widths to offer $\lambda_0/4$ optical path. To dovetail the center of photonic gap exactly at 1.55 μm, the PBG structure should be considered having pair of dielectric layers with refractive indices (i) 3.1 and 1.5, or (ii) 3.5 and 1.16. The choice will depend on the width of photonic gap required in the design. For example, alternative (ii) may be preferable for a wider photonic gap.

References

1. E. Yablonovitch, *J. Opt. Soc. Amer. B* **10**, 283-295 (1993).
2. E. Yablonovitch, *Science* **289**, 557-559 (2000).
3. Mahi R. Singh, *Phys. Rev. A* **75**, 033810 (13 pp) (2007).
4. Mahi R. Singh and R. H. Lipson, *J. Phys. B: At. Mol. Opt.* **41**, 015401 (7pp) (2008).
5. S. Noda, A. Chutinan, and M. Imada, *Nature* **407**, 608-610 (2000).
6. O. Painter, R. K. Lee, A.Yariv, A. Scheer, J. D. O'Brien, P. D. Dapkus and I. Kim, *Science* **284**, 1819-1821 (1999).
7. J. W. Haus, "Photonic band gap structures" in *Nanometer Structures: Theory, Modeling and Simulation* (Ed: A. Lakhtakia), Publisher City: New Delhi, Publisher Name: PHI, 2007, p. 45

Theoretical Modeling of Femtosecond Pulsed Laser Ablation of Silicon

Anil Kumar Singh[a] and R. K. Soni

Department of Physics, Indian Institute of Technology Delhi, New Delhi 110016, India.
[a]*Email: asbhu@rediffmail.com,*

Abstract. Femtosecond (fs) laser ablation yield of semiconductors critically depends on pulse shape in time and space domain, laser wavelength and fluence. We have carried out a theoretical study of fs-pulsed laser interaction with silicon to optimize the ablation process using two-temperature model for laser pulse (λ=800 nm), which is rectangular in time domain (pulse width = 100 fs) and uniform intensity in circular cross-section of 100 μm diameter. We have calculated ablation threshold, depth of material removal per pulse and refractive index change of silicon after irradiated by laser pulse and compared with the available experimental results. The formation of fs-laser induced periodic sub-wavelength surface grating structures on the silicon is also discussed within the frame work of carrier induced nonlinear effect.

Keywords: Laser ablation, silicon, sub-wavelength periodic grating structures.
PACS: 52.38 Mf, 72.80 Nv, 81.16 Rf

INTRODUCTION

The micro-fabrication technologies have brought an unprecedented upsurge of research interests in microscale and nanoscale fabrication of devices. Femtosecond (fs) pulsed laser processing has made tremendous impact in fabrication of sophisticated microstructures [1-2], nanostructures [3] due to precise control of heat location and depth, and realization of high heating/cooling rate. These potential applications require understanding of underlying physics and interrelation of the processes taking place in materials irradiated by fs-laser pulses. This can facilitate optimization of experimental parameters in current applications and development of contemporary technologies. In this work we have carried out the theoretical study of the fs-pulsed laser ablation process using the Two-Temperature Model (TTM) by taking into account the linear, two-photon and free-carrier absorption.

THEORY

The primary laser–semiconductor interaction process is the excitation of electrons from their equilibrium states in the valence band to conduction band by absorption of photons. The initial distribution of excited electronic states corresponds to a set of states coupled by the optical transitions. The occupation of these primary states is rapidly changed by carrier–carrier interaction processes, and a quasi-

CP1147, *Transport and Optical Properties of Nanomaterials—ICTOPON - 2009*, edited by M. R. Singh and R. H. Lipson
© 2009 American Institute of Physics 978-0-7354-0684-1/09/$25.00

equilibrium condition is established among the electrons on a time scale of about $10^{-13} s$ [4].The energy distribution of the carriers over the available states is described by the Fermi–Dirac distribution with an electron temperature (T_e), greater than the lattice temperature (T_l). The quasi-equilibrium electrons cool down on a time scale of $10^{-12} s$ by electron-phonon scattering [4]. These phonons relax predominantly by anharmonic interaction with other phonon modes. The final stage of the thermalization process is the redistribution of the phonons over the entire Brillouin zone according to Bose–Einstein distribution. After the thermalization further evolution of the system can be described in terms of thermal processes and spatial distribution of the energy can be characterized by the temperature profile. When a sufficient amount of energy is deposited in the material, the material exfoliation takes place from the surface of the target, which is known as laser ablation.

1.1 Photo-induced Free Carrier

For short laser pulse of 100 fs the carrier recombination during pulse can be neglected. Further, if the laser intensity distribution is uniform and wide enough perpendicular to the direction of propagation, diffusion in any direction can also be neglected and we can consider the problem as one dimensional. In such a case the photo-induced free carrier density and photon density will be,

$$\frac{\partial N(z,t)}{\partial t} = [\alpha + \beta h \nu I(z,t)] I(z,t) \tag{1}$$

$$\frac{\partial I(z,t)}{\partial z} = -[\alpha + \sigma N(z,t) + \beta h \nu I(z,t)] I(z,t) \tag{2}$$

For rectangular pulse in time domain we can write,

$$N(z,\tau) = \alpha Q(z,\tau) + \frac{\beta h \nu}{\tau} Q^2(z,\tau) \tag{3}$$

$$\frac{\partial Q(z,\tau)}{\partial z} = -\left[\alpha Q(z,\tau) + \frac{(\alpha \sigma \tau + \beta h \nu)}{\tau} Q^2(z,\tau) + \frac{\sigma \beta h \nu}{\tau} Q^3(z,\tau) \right] \tag{4}$$

where α is linear absorption coefficient, σ is cross-section for free carrier absorption, β is the coefficient for two-photon absorption. τ and z, are pulse duration and depth from surface, respectively, $I(z,t)$ is the photon flux at depth z from the target surface at time t, $N(z,\tau)$ and $Q(z,\tau)$ are the photo-induced free carrier and photon density at depth z from the surface at the end of incident laser pulse, respectively.

1.2 Effect of Carrier Concentration on Optical Properties of Silicon

The intensive free carrier injection leads to formation of an electron-hole plasma, which alters the optical properties of silicon.

Band gap shrinkage: If the non equilibrium carrier concentration is large enough there are exchange effects between the electrons (or holes) in the same spin state due to the spatial overlap of their wave functions. In addition, the correlations effects between oppositely charged particles and the particles of the same charge but of opposite spin also become important. The net result of these correlation effects is the reduction of bandgap energy. We use the precise expression for band gap shrinkage of [5],

$$\Delta E_g = -E_{ex}^b \frac{4.64(a_{ex}^3 N)^{1/2}}{[a_{ex}^3 N + (0.107 kT / E_{ex}^b)^2]^{1/4}} \tag{5}$$

Here $E_{ex}^b = \left(2\pi^2 m_r e^4 / h^2 \varepsilon_s^2\right)$ and $a_{ex} = (h^2 \varepsilon_s)/(4\pi^2 m_r e^2)$ are binding energy and Bohr radius of electron-hole (e-h) exciton. ε_s is dielectric constant, m_r is e-h reduced mass in silicon and N is the photo-induced free carrier density calculated by using Eqn (3).

Band filling effect: In the case of very high carrier density, the lowest energy states in the conduction band are filled and electrons from the valence band require energies greater than nominal band gap to be optically excited into the conduction band. Hence, there is a decrease in the absorption coefficient above the band gap. The reduction in interband absorption coefficient of semiconductors due to high free carrier concentration is known as the Burstein-Moss effect or band filling effect. If photon energy is E, energy in the valence band by E_a and energy in the conduction band by E_b, and then the absorption coefficient of an injected semiconductor is [2],

$$\alpha(N,E,T) = \alpha_0(E,T)[f_v(E_a) - f_c(E_b)] \tag{6}$$

where $\alpha(N,E,T)$ is the absorption coefficient at photo-induced free carrier density N.

Refractive index change: Laser pulse induced free carriers causes change in refractive index of the silicon by free carrier, band gap shrinkage and band filling effect. The laser pulse induced change in the refractive index is [6],

$$\Delta n_c(N,E,T) = \left(\frac{e^2 \lambda^2}{8\pi^2 c^2 \varepsilon_0 n}\right)\left(\frac{N}{m_e} + \frac{P}{m_h}\right) \tag{7}$$

$$\Delta n_b(N,E,T) = \frac{ch}{\pi e^2} P \int_0^\infty \frac{\Delta \alpha(N,E,T)}{E'^2 - E^2} dE \tag{8}$$

where λ is the photon wavelength, N and P are photo-induced electron and hole concentrations, respectively. m_e, and m_h are effective masses for electron and hole, respectively and n is the refractive index. Δn_c and Δn_b are free carrier induced and the combined effect of band filling and band shrinkage induced refractive index change, respectively. Photo-induced free carrier density can be calculated using Eqs (1-4) and from Eqs (7-8) one can calculate Δn_c and Δn_b. The net refractive index change in silicon caused by laser pulse would be $\Delta n = \Delta n_c + \Delta n_b$.

2. Ablation Threshold and Ablation Depth/Pulse

When the laser beam diameter is much larger than the penetration depth then the interaction of laser with semiconductor can be treated as one-dimensional two temperature model. If T_e and T_l are the electron and lattice temperatures after the laser pulse, respectively [7],

$$T_e(\tau) \approx \left(\frac{2F_a\alpha_{eff}}{C_e'}\right)^{1/2} \exp(-z/\delta), \text{ and } T_l(\tau) \approx \frac{F_a\alpha_{eff}}{C_l}\exp(-z/\delta). \qquad (9)$$

Where α_{eff} is the effective absorption coefficient of the material (by taking in to account linear, free carrier and two photon absorption). Here, z is the direction along beam axis, $C_e = C_e'T_e$ and C_l are the specific heat capacities of the electron and lattice subsystems. F_a is the absorbed laser fluence and $\delta = 2/\alpha_{eff}$ is the skin depth. The significant material evaporation occurs only when, $\rho C_l(T_l - T_m) \geq \Omega$ and putting Eqn. (9) in this condition the absorbed laser fluence and the ablation depth L per pulse is

$$F_a \geq F_{th}\exp(z/\delta) \quad \text{and} \quad L \approx \alpha_{eff}^{-1}\ln(F_a/F_{th}) \qquad (10)$$

Where ρ, T_m and Ω are density, melting temperature and latent heat of evaporation for silicon and $F_{th} = \dfrac{C_l}{\alpha_{eff}}\left[\dfrac{\Omega}{\rho C_l} + T_m\right]$ is the threshold laser fluence for evaporation.

3. Sub-wavelength Grating

When a target is irradiated by linearly polarized femtosecond pulse laser sub-wavelength grating is grown perpendicular to the direction of polarization of the incident laser. Such grating formation can be explained by interaction between the incident laser and the excited plasmon-polariton wave. The grating period is defined by the momentum conservation condition [8],

$$\Lambda = \frac{2\pi}{\sqrt{\dfrac{1}{T_e}\left(\dfrac{m_e\omega^2}{3k_B} - \dfrac{e^2N}{3\varepsilon_0 k_B}\right) - k_{ph}^2}} \qquad (11)$$

Where k_{gr}, k_{pl} and k_{ph} are grating, plasma and photon wave vectors, respectively. T_e and N are surface electronic temperature and photo-induced electron density at the end of laser pulse.

RESULTS AND DISCUSSION

The lattice temperature profile with depth from the silicon target surface was calculated using Eqn (9) and plotted in Fig.1 for different pulse energy. The rapid decrease in lattice temperature with depth at higher pulse energy is due to the dominant contribution of non-linear and photo-induced free carrier absorption. The

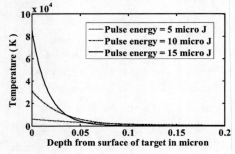 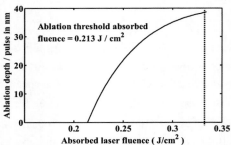

FIGURE 1. Lattice temperature variation with depth from target surface.

FIGURE 2. Variation of ablation depth /pulse with absorbed laser fluence.

lattice temperature (T_l) is not necessarily the real temperature of the silicon target. After the melting point is reached, the temperature is allowed computationally to increase in order to account for the thermal energy transferred from the incident radiation to the target. Once the lattice temperature profile with depth from the target surface is calculated, the threshold fluence for material removal from the target surface was

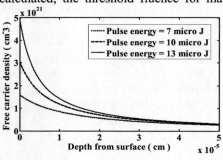

FIGURE 3. Photo-induced free carrier density profile with depth from surface

FIGURE 4. Silicon surface reflectivity as a function of incident laser pulse energy.

calculated from Eq. (10). Fig.2 shows the variation of ablation depth per pulse with absorbed laser fluence. The calculated threshold energy density for material removal is 0.213 J/cm² , very close to literature reported threshold value 0.20 J/cm² [9]. From the Fig.2 it is also clear that above the threshold fluence, the rate of material removal increases with increasing laser fluence and rate of material removal per pulse saturate for laser fluence above 0.33 J/cm². Thus for efficient material removal from silicon target surface at 100 fs pulse duration and 800 nm laser wavelength, the calculated optimum threshold is 0.33 J/cm^2.We also find that near the threshold fluence depth of material removal per pulse is ~1.5 nm thus surface structuring with few nm depth precision with fs-pulse laser is possible. Femtosecond laser pulse induced free carrier density profile at the end of pulse is calculated using Eqs. (3) and (4) and results are plotted in Fig.3.One can see from Fig. 3 that a fs-laser pulse with energy 10μJ can create electron hole plasma with carrier density ~3×10^{21} cm^{-3} which decreases by a factor of 1/e within few tens of nanometer depth from the surface. Using this laser-induced free carrier density profile we have calculated refractive index change of silicon taking into account free carrier, band filling and band shrinkage effect. The surface reflectivity variation caused by the refractive index change with incident laser pulse energy is plotted in Fig.4. Initially the surface reflectivity decreases with incident pulse energy, becomes zero at 13.2 μJ and then increases again. The calculated variation of sub-wavelength grating period with absorbed laser fluence using Eqn (11) is plotted in Fig 5. The period of sub-wavelength surface grating increases with increasing laser fluence. This is in qualitative agreement with the experimental results on grating formation in glass [8]; however, due to lack of experimental results on silicon surface gratings we could not quantitatively compare our calculated results.

TABLE 1. List of model parameters used for calculations

Parameters	Value
Wavelength (λ)	800 nm
Pulse duration (τ)	100fs
Linear absorption coefficient (α_L)	2.99×10^3 cm^{-1}
Cross-section for free carrier absorption (σ)	4.5×10^{-18} cm^{-2}
Coefficient for two photon absorption (β)	9 GW^{-1}cm
Density of silicon (ρ)	2.33 gcm^{-3}
Lattice heat capacity (C_l)	0.88 Jg^{-1}K^{-1}
Latent heat for melting (L_m)	4206 Jcm^{-3}
Latent heat of evaporation (Ω)	12.8 KJg^{-1}
Circular beam (diameter)	100μm

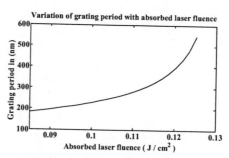

FIGURE 5. Variation of sub-wavelength grating period with absorbed laser fluence for linearly polarized, normally incident fs laser.

REFERENCES

1. S. Ameer-Beg et al., *Appl. Surf. Sci.* **127-129**, 875-880 (1998).
2. Q. Z. Zhao et al., *Appl. Surf. Sci.* **241**, 416-419 (2005).
3. Stephen Y. Chou, Chris Cheimel and Jian Gu, *Nature* **417**, 835-837 (2002)
4. D. von der Linde, K. Sokolowski – Tinten and J. Bialkowski, *Appl. Surf. Sci.* **109-110**, 1-10 (1997).
5. P. P. Paskov and L.I. Pavlov, *Appl. Phys. B* **54**,113-118 (1992).
6. Brian R. Bennett, Richard A. Soref and Jesus A. Del Alamo, *IEEE J. Quantum Electron.* **26**, 113-122 (1990).

7. B. N. Chichkov, C. Momma, S. Nolte, F. Von Alvensleben and A. Tunnermann, *Appl. Phys. A* **63**,109-115 (1996).
8. Yasuhiko Shimotsuma, Peter G.Kazansky and Jiarong Qui, *Phys. Rev. Lett*. **91**, 247405-247408 (2003).
9. Min Kyu Kim, Takayuki Takao, Yuji Oki and Mitsuo Maeda, *Jpn. J. Appl. Phys*.**39**, 6277-6280 (2000).

Nanophotonic technologies for innovative all-optical signal processor using photonic crystals and quantum dots

Y. Sugimoto[a], N. Ikeda[a], N. Ozaki[b], Y. Watanabe[b], S. Ohkouchi[c], S. Nakamura[c] and K. Asakawa[b]

[a]Nanotechnology Innovation Center, National Institute for Materials Science (NIMS), 1-2-1 Sengen, Tsukuba, Ibaraki 305-0047, Japan.
[b]Center for Tsukuba Advanced Research Alliance (TARA), University of Tsukuba, 1-1-1 Tennoudai, Tsukuba, Ibaraki 305-8577, Japan.
[c]Nano Electronics Research Laboratories, NEC Corporation, 34 Miyukigaoka, Tsukuba, Ibaraki 305-8501, Japan.

Abstract. GaAs-based two-dimensional photonic crystal (2DPC) slab waveguides (WGs) and InAs quantum dots (QDs) were developed for key photonic device structures in the future. An ultrasmall and ultrafast symmetrical Mach–Zehnder (SMZ)-type all-optical switch (PC-SMZ) and an optical flip-flop device (PC-FF) have been developed based on these nanophotonic structures for an ultrafast digital photonic network. To realize these devices, two important techniques were developed. One is a new simulation method, i.e., topology optimization method of 2DPC WGs with wide/flat bandwidth, high transmittance and low reflectivity. Another is a new selective-area-growth method, i.e., metal-mask molecular beam epitaxy method of InAs QDs. This technique contributes to achieving high-density and highly uniform InAs QDs in a desired area such as an optical nonlinearity-induced phase shift arm in the PC-FF. Furthermore, as a unique site-controlled QD technique, a nano-jet probe method is also developed for positioning QDs at the centre of the optical nonlinearity-induced phase shift arm.

Keywords: Photonic crystal, Quantum dots, Molecular beam epitaxy, All-optical switch.
PACS: 73.21.La, 78.55.Cr, 78.67.Hc, 81.15.Hi, 81.16.Dn

INTRODUCTION

An optical switching device is an intensively required optical digital signal processing device promising in the future photonic network. There are some theoretical works for optical switching devices composed by photonic crystals (PCs) [1-3]. An ultrasmall and ultrafast symmetrical Mach–Zehnder (SMZ)-type all-optical switch [4] (PC-SMZ [5]) is one of the candidates. The principle of the PC-SMZ is based on the time-differential phase modulation caused by the optical non-linear (ONL)-induced refractive index of the ONL arms. Two-dimensional photonic crystal (2DPC) waveguides (WGs) are composed of single missing line defects, while ONL-induced phase shift arms are selectively embedded with quantum dots (QDs). After these designs were successfully performed, ultra-fast all-optical switch was

CP1147, *Transport and Optical Properties of Nanomaterials—ICTOPON - 2009*, edited by M. R. Singh and R. H. Lipson
© 2009 American Institute of Physics 978-0-7354-0684-1/09/$25.00

demonstrated by using a PC-SMZ sample with 600×300 μm in size. Rise and fall times are ~2 ps, and the switching energies of the control pulsed are estimated as low as 100 fJ [6].

In the next stage, the PC-SMZ evolved into a new functional key device, i.e., an ultra-fast all-optical flip-flop (PC-FF [7]) device essential for the digital photonic network. For these two stages, two important techniques have been developed. One technique is a new simulation method, i.e., topology optimization (TO) [8, 9] of 2DPC WGs with wide/flat bandwidth, high transmittance, and low reflectivity. Another technique is a new selective area growth (SAG) method, i.e., metal-mask (MM) molecular-beam epitaxy (MBE) method of InAs QDs [10, 11]. This technique helps achieve high-density and highly uniform InAs QDs in a desired area such as an optical nonlinearity-induced phase shift arm in the PC-FF. Furthermore, as a unique site-controlled QD technique, a nano-jet probe (NJP) method [12] has also been developed for positioning a single QD at the centre of the optical nonlinearity-induced phase shift arm.

In this paper, we describe an ultrasmall and ultrafast optical flip-flop device PC-FF and two important techniques to realize this device.

CONCEPT OF OPTICAL FLIP-FLOP DEVICE: PC-FF

The PC-SMZ is switched by two control pulses, i.e., set and reset pulses, suggesting a pseudo optical flip-flop (FF) operation. However, an "on-state" of the PS-SMZ is restricted by the carrier relaxation time in the semiconductor ONL material (this decay time is ~ 100ps in this experiment). By using this result, unique techniques for changing into the normal FF device are proposed, as shown in Fig. 1(a). The PC-FF is composed of two PC-SMZs (SW-1 and SW-2) [5] that are coupled by a feedback loop for supporting optical bistability. Both PC-SMZs basically exhibit time-differential phase modulators [4]. In SW-1, two beams divided from CW light (B) generate an output of Q = 0 or 1, depending on the constructive or destructive interference resulting from the optical nonlinear (ONL) phase shift. A fraction of the output Q (A1) excites SW-2, where clock pulses (A2) exhibit ONL phase shifts and form an output (A3). As shown in Fig. 1 (a), when the set pulse is applied, the first arm of SW-1 is excited and the output is switched from the initial Q = 0 to Q = 1. This causes excitation of SW-2, and its output is added to the first arm of SW-1 as feedback. Consequently, continuous excitation of the first arm of SW-1 starts within the feedback delay time, thus maintaining the output as Q = 1. When the reset pulse is input, the second arm of SW-1 is excited, canceling out the ONL phase shift in the first arm, and the output is switched to Q = 0; this state is maintained until the set pulse arrives. A time chart corresponding to the process is shown in Fig. 1 (b).

Here, we simulated the response for the set and reset pulses with random pulse intervals [13]. Simulation is based on analyzing rate equations of carrier density and light intensity in QD nonlinear waveguides. The relation between the carrier density and the nonlinear phase shift is based on the discussion in Ref. 14. The four parameters, which are the carrier lifetime (τ), the feedback delay time (T), the duration of set and reset pulses (t_1), and the minimum interval between set and reset pulses (t_2), govern the dynamic response of the optical flip-flop.

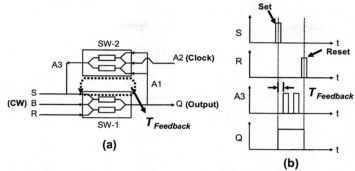

FIGURE 1. (a) Schematic diagram of PC-FF. (b) Time chart of PC-FF.

A typical measured temporal evolution of the induced nonlinear phase shift in a QD nonlinear waveguide indicates a carrier lifetime of about 50 ps [6]. The carrier lifetime was observed to be controllable in the range from less than 30 ps to over 100 ps by modifying growth conditions. When $\tau = 100$ ps, $T = 10$ ps, $t_1 = 32$ ps, and $t_2 = 100$ ps, corresponding to 10 Gb/s, clearly eye opened output is obtained as shown in Fig. 2(a). With the carrier lifetime of 100 ps, we did not successfully obtain eye opened output at 40 Gb/s. However, when the carrier lifetime τ is reduced to 25 ps with $T = 10$ ps, $t_1 = 8$ ps, and $t_2 = 25$ ps, corresponding to 40 Gb/s, clearly eye opened output is obtained as shown in Fig. 2(b). The results are very promising for achievement of the high speed optical flip-flop using the PC-FF.

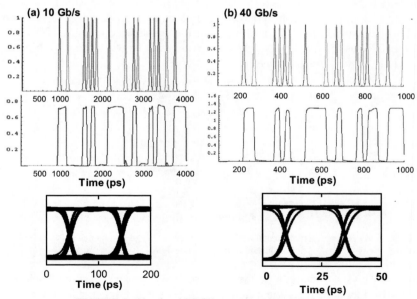

FIGURE 2. Simulated PC-FF operations for random inputs.

KEY ISSUES IN THE PC-FF TECHNOLOGY

TO design for 2DPC WG

Different from the case of a 2DPC straight waveguide, the PC-FF includes a variety of waveguides such as bend, wavelength-selective Y branch and directional coupler. Such non-straight waveguides significantly degrades bandwidth and transmittance characteristics as compared with the straight waveguide if the conventional PC design method is used. To solve this problem, a TO design method has been developed. The TO method, which has been developed for PC designs by Borel et al. [8], was applied to our PC-FF waveguides design [9].

A typical TO design example for the bend waveguide is shown in Fig. 3. The figure compares the standard and TO designs, as shown by the inset SEM photographs, respectively. The sample in the experiment was fabricated in an epitaxial heterostructure grown by MBE. A 250-nm-thick GaAs core layer was deposited on top of a 2-μm-thick $Al_{0.6}Ga_{0.4}As$ clad layer on a GaAs substrate [5]. A hexagonal-lattice pattern of air holes and a single line-defect waveguide were drawn by electron-beam (EB) lithography and transferred into the core layer surface by reactive ion beam etching (RIBE). The lower sacrificial clad layer was removed by HF etching through the air holes, creating a GaAs membrane suspended in air. The lattice constant and diameter of the air holes were set to 345 and 210 nm, respectively. The total length of the 2DPC WG sample was 1 mm. The optical properties of the fabricated PC waveguide were characterized by transmission spectra for TE polarization measured with a white light source. A spectrometer-attached InGaAs multichannel-type detector was used for the measurements. Transmission spectra for the standard (STD) and TO method are shown in Fig. 3. It is found that the transmission for the TO design is almost the same as that for the straight waveguide, thus largely improved in particular in the vicinity of the band gap range.

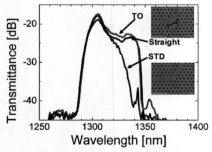

FIGURE 3. Comparison of experimental transmission spectra for the double bends PC-WGs between the standard and TO designs.

Figure 4 shows an example of the TO-designed Y-branch used at the input and set/reset pulse junction. After the TO design of the Y-junction, an asymmetric pattern for the air hole configuration is obtained, as shown in Fig. 4 (a). Figure 4 (b) shows the experimental results of the transmission spectra. For reference, a transmission spectrum of the straight 2DPC WG is included, as indicated by the dotted line curve in Fig. 4 (b). The characteristics of the measured transmission spectra for routes AC and BC are almost similar to those predicted by the calculations, as shown in Fig. 4 (b). The resultant isolations at λ_1 and λ_2 are 10 dB and 4 dB, respectively. The larger the isolation, the smaller is the cross-talk for the unwanted path in the Y-junction.

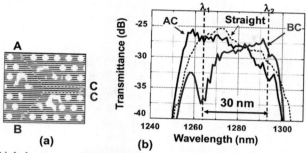

FIGURE 4. (a) Air hole patterns of TO-designed Y-junction. (b) Measured transmission spectra for the lights along routes AC and BC in the TO-designed Y-junction.

Selective area growth by MM-MBE

Figure 5 (a) shows a PC-FF pattern designed within a chip size of 300×500 μm on the GaAs substrate. In the SW-1, wavelengths for input and S/R pulses are set to 1.31 μm and 1.29 ☐μm, respectively. Based on this condition, absorption peak wavelengths for QDs should be set to 1.29 ☐μm in the SW-1 and 1.31 μm in the SW-2, respectively. This implies that QDs should be selective-area-grown with different wavelengths at different sites.

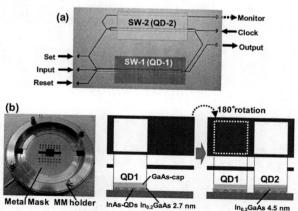

FIGURE 5. (a) Optical microscope image of a PC-FF sample fabricated on the GaAs wafer within a chip size of 300×500 μm. (b) Photograph of the rotational MM and schematic explanation of the SAG of two-color quantum dots.

For this purpose, a SAG technique has been developed by using a MM-MBE method (10, 11). InAs-QDs were grown on a GaAs substrate using the conventional Stranski-Krastanov-mode growth with 2.6-monolayer amount of InAs deposition by MBE. The SAG of two-color InAs-QDs (QD1 and QD2) was achieved by a specially developed rotational MM method. As shown in Fig. 5(b, c), the MM, which has windows (several 100 µm × several 100 µm in size) with a rotationally asymmetric pattern, was mounted on a substrate and enabled SAG of QDs during conventional MBE growth. Furthermore, by varying the composition and thickness of the InGaAs strain-reducing layers (SRLs) [11] in both the SAG areas, QDs with different absorption wavelengths can be monolithically fabricated on a wafer.

We examined the grown QDs by photoluminescence (PL) measurements. A PL intensity mapping obtained at room temperature (RT) clearly shows two high-intensity areas attributed to QD1 and QD2. Figure 6 (a) shows the spectra from QD1 and QD2 regions. The emission peaks of the spectra from the QD1 and QD2 regions appear at 1276 and 1296 nm, respectively. These results demonstrate the successful SAG of two-color QDs by the MM-MBE method. We fabricated PC WGs designed for the PC-FF on the QD-grown wafer by EB lithography and RIBE. Figure 6 (b) shows images of the fabricated PC-FF obtained using an optical microscope and the corresponding PL intensity mapping, indicating the successful fabrication of the PC-FF structure.

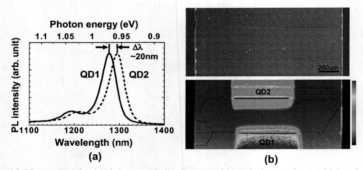

FIGURE 6. (a) PL spectra obtained from each SAG area, with emission peaks at 1276 and 1296 nm. (b) Optical microscope image of PC-FF fabricated on the SAG areas by EB lithography, RIBE, and corresponding PL intensity mapping.

Site-controlled QD technique: nano-jet probe (NJP) method

The control of the nucleation site of quantum dots (QDs) is one of the key issues in the nanoscale design of optoelectronic functional devices. We have developed a new nano-probe-assisted bottom-up technique that enables the formation of site-controlled QDs by using a specially designed atomic-force-microscope (AFM) probe, referred to as the Nano-Jet Probe (NJP) [12]. Beginning with the fabrication of regular QD arrays by using the NJP method, we vertically aligned the self-assembled InAs QDs by using the strain-induced stacking method [15], thereby producing 3D QD structures.

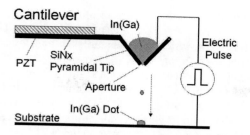

FIGURE 7. Schematic illustration of the NJP and principle of nano-dot formation.

Figure 7 shows a schematic illustration of the NJP and the principle of the nano-dot formation. This probe has a hollow pyramidal tip with a micro-aperture on its apex and a reservoir tank within the stylus. In the pyramidal tip, one of the constituents of the QDs -In and/or Ga- was charged by using an evaporation method in a high-vacuum environment. By applying a voltage pulse between the pyramidal tip and the sample under an ultra-high vacuum condition, nano-dots were reproducibly fabricated at the desired position on a GaAs substrate. As an example, an AFM image of the In nano-dot arrays fabricated by using the NJP method is shown in Fig. 8 (a). The Ga/In alloy nano-dots were also formed by changing the charged material in the pyramidal tip. The fabricated nano-dots can be converted into In(Ga)As QDs by subsequent annealing with arsenic flux. That is, the arsenic atoms are incorporated into the nano-dots and the latter are consequently converted into In(Ga)As QDs. Subsequently, self-assembled InAs QDs were grown on the converted QD arrays by using the formed QD arrays as a strain template. The photoluminescence measurements of the assembled 3D structures revealed a good crystallographic quality as shown in Fig. 8 (b).

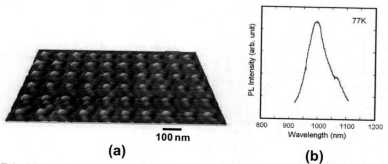

(a)   **(b)**

FIGURE 8. (a) 3D AFM image of an In nano-dot array fabricated by the NJP method.(b) Typical PL spectrum of the fabricated QDs.

SUMMARY

A 2DPC- and QD-based all-optical PC-FF switch capable of synchronizing with a high-frequency clock was designed and fabricated for developing an advanced optical digital signal processing device with high-speed capability. In order to achieve the new PC-FF device, two advanced techniques for upgrading the PC-SMZ performances have been developed. One is a new 2DPC design method, i.e., TO method, effective for WGs with wide/flat bandwidth, high transmittance and low reflectivity. Another one is a selective area growth of InAs QDs by using the MM method. Recent experimental results on these two technologies and scheme for implementing the optical flip flop operation have been shown. The formation of In (Ga) nano-dots using the NJP method and the fabrication of 3D QD structures were succeeded. These techniques have great advantage to fabricate for not only our developed PC/QD-based all-optical devices but also any optical integrated devices.

These results will pave the way for the implementation of innovative photonic devices in advanced ultrafast photonic network systems.

ACKNOWLEDGMENTS

This study was partly supported by the New Energy and Industrial Technology Development Organization (NEDO) and a Grant-in-Aid from the Ministry of Education, Culture, Sports, Science and technology (MEXT), Japan. This research was also partly supported by "Nanotechnology Network Program" of MEXT, Japan.

REFERENCES

1. M.R. Singh, *Appl. Phys. B* **93**, 91-98 (2008).
2. M. R Singh and R. H. Lipson, *J. Phys. B: At. Mol. Opt. Phys.* **41**, 015401 (2008).
3. M.R. Singh, *Phys. Rev. B* **75**, 155427 (2007).
4. K. Tajima, *Jpn. J. Appl. Phys* **32**, L1746-9 (1993).
5. Y. Sugimoto, N. Ikeda, N. Carlsson, K. Asakawa, N. Kawai and K. Inoue, *J. Appl. Phys.* **91**, 622-929 (2002).
6. H. Nakamura, Y. Sugimoto, K. Kanamoto, N. Ikeda, Y. Tanaka, Y. Nakamura, S. Ohkouchi, Y. Watanabe, K. Inoue, H. Ishikawa and K. Asakawa, *Opt. Express,* **12**, 6606-6614 (2004).
7. K. Asakawa, Y. Sugimoto, Y. Watanabe, N. Ozaki, A. Mizutani, Y. Takata, Y. Kitagawa, H. Ishikawa, N. Ikeda, K. Awazu, X. Wang, A. Watanabe, S. Nakamura, S. Ohkouchi, K. Inoue, M. Kristensen, O. Sigmund, P. I. Borel and R. Baets, *New J. Phys.* **8**, 1-26 (2006).
8. P.I. Borel, A. Harpøth, L. H. Frandsen, M. Kristensen, P. Shi, J. S. Jensen and O. Sigmund, *Opt. Express* **12**, 1996-2001 (2004).
9. Y. Watanabe, N. Ikeda, Y. Sugimoto, Y. Takata, Y. Kitagawa, A. Mizutani, N. Ozaki and K. Asakawa, *J. Appl. Phys.* **101**, 113108 (2007).
10. S. Ohkouchi, Y. Nakamura, N. Ikeda, Y. Sugimoto and K. Asakawa, *Rev. Sci. Instrum.* **78**, 073908 (2007).
11. N. Ozaki, Y. Takata, S. Ohkouchi, Y. Sugimoto, Y. Nakamura, N. Ikeda and K. Asakawa, *J. Cryst. Growth* **301-302**, 771-775 (2007).
12. S. Ohkouchi, Y. Sugimoto, N. Ozaki, H. Ishikawa and K. Asakawa, *J. Cryst. Growth* **301–302**, 726-730 (2007).
13. S. Nakamura, A. Watanabe, X. Wang, N. Ikeda, Y. Sugimoto, N. Ozaki, Y. Watanabe and K. Asakawa, "Optical flip-flop based on coupled ultra-small Mach-Zehnder all-optical switches," in *OFC/NFOEC 2008*, San Diego, USA, 2008, pp. 045-10.
14. H. Nakamura, K. Kanamoto, Y. Nakamura, S. Ohkouchi, H. Ishikawa and K. Asakawa, *J. Appl. Phys.* **96**, 1425-1434 (2004).
15. Q. Xie, A. Madhukar, P. Chen and N. P. Kobayashi, *Phys. Rev. Lett.* **75** 2542-2545 (1995).

Optical Stability and Photoluminescence Enhancement of Biotin Assisted ZnS:Mn²⁺ Nanoparticles

Ashish K. Keshari[a, b*], Vyom Parashar[a] and Avinash C. Pandey[a, b]

[a]Nanophosphor Application Centre,
[b]Department of Physics,
University of Allahabad, Allahabad-211 002, India.
Tel. /Fax: +91-532-2460675
* Corresponding author e-mail: jeevaneshk26@yahoo.co.in

Abstract: We synthesized the ZnS: Mn²⁺ nanoparticles passivated by biocompatible layer namely 'biotin' by chemical precipitation route and studied their temporal evolution for optical and photoluminescence stability. Structural analysis was carried out using high resolution transmission electron microscope. To monitor the optical and photoluminescence properties of the nanoparticles with time, we have characterized the grown product by UV-Visible and photoluminescence spectroscopy at regular interval for the period three months. Results showed that the properties of nanophosphors capped with biotin are remaining same even after three months. We found that, biotin capping will enhance the luminescence from ZnS: Mn²⁺ nanoparticles as compared to without cap particles. Absence of biotin will gradually degrade the luminescence upon aging while drastic degradation in the luminescence intensity was observed after annealing.

Keywords: ZnS:Mn²⁺; Biotin; Aging; Photoluminescence; Stability.
PACS: 61.46.Hk, 78.67.Bf, 81.07.Bc

Introduction

The development of nanophosphors of desired sizes and properties for various practical applications and its controlled growth in an organic matrix has attracted a great deal of attention in recent years [1-4]. In this way, nanophosphor get surface passivated and are stabilized against environmental attacks. In many cases, semiconductor nanoparticles are synthesized in polymers, using the chemical capping method or by embedding them in an inert matrix. Such a matrix is useful in avoiding coalescence of the nanoparticles. This class of new materials have not only provided many unique opportunities for studying physics in low dimensions but also exhibited novel optical and transport properties which are potentially useful for technological applications. In contrast to the un-doped (ZnS) nanocrystals, the impurity (Mn) states in doped semiconductor nanocrystals (ZnS: Mn²⁺) has played a special role in affecting the electronic structures and transition probabilities. These novel nanoscale materials like ZnS:Mn²⁺/Eu²⁺ [5] offer

CP1147, Transport and Optical Properties of Nanomaterials—ICTOPON - 2009, edited by M. R. Singh and R. H. Lipson
© 2009 American Institute of Physics 978-0-7354-0684-1/09/$25.00

various interesting phenomenon such as radiative life-time shortening, electroluminescence, photo-blinking effects and effective low-voltage cathodoluminescence which leads to various tremendous potential applications in areas such as solid state lighting, opto-electronics device technology, photo catalyst fabrication to biotechnology such as bio-labeling (DNA markers) and drug delivery systems [6-10].

Experimental

It is still a challenge for chemists and material scientists to find some convenient, economical, less energy consuming and environmentally friendly synthesized routes to these semiconductor nanocrystals. The surface chemistry is an efficient tool not only to organize and immobilize the NCs but also to effectively modify the emission properties. The possible of performing the manipulation of the prepared NCs by properly engineering the surface by means of biomolecules. Biological systems can control mineralization and synthesis of various nanocrystals in the exact shapes and sizes with high reproducibility and accuracy. Therefore it is logical to use biomolecules to grow monodisperse nanocrystals via bio-mineralization. The synthesis of biotin capped $ZnS:Mn^{2+}$ nanoparticles is described elsewhere [11]. Structural analysis was carried out by high-resolution transmission electron microscope (HRTEM) of FEI Tecnai G^2 20 which shows the growth of the nanoparticles is strongly arrested by the biotin. Absorbance spectra were recorded from Perkin Elmer Lambda 35 spectrometer and PL spectra were recorded with 325 nm excitation wavelength using He-Cd laser (KIMMON) and Mechelle 900 spectrograph. The effect of biotin capping is explored in detail in terms of optical stability and PL enhancement as compared to without cap particles.

Results and Discussion

To access the size and structure of the particles, we performed a high-resolution transmission electron microscopic (HRTEM) study of the biotin capped $ZnS:Mn^{2+}$ nanoparticles. It shows an abundance of nearly spherical particles (Fig. 1(a)) which have particle sizes from 3 to 6 nm. The contrast in sizes obtained from XRD spectra [11] using Scherrer's formula to that determined from HRTEM images suggesting that the prepared particles are mostly multi-domain crystallites. Fig. 1(a) clearly shows that, the particles are strongly arrested by the biotin to control the growth and agglomeration. Selected area electron diffraction (SAED) pattern of $ZnS:Mn^{2+}$ nanocrystals are shown in Fig. 1(b). As expected, the electron diffraction (ED) pattern shows a set of rings instead of spots due to the random orientation of the crystallites. The ED pattern shows three rings, which corresponds to the (111), (220) and (311) planes of the cubic phase of Mn doped ZnS and are well agreed with the standard pattern (JCPDS file # 800020).

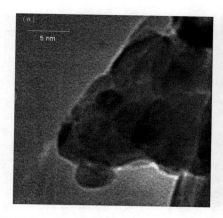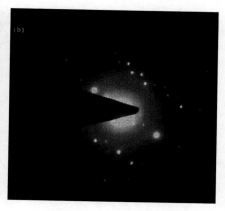

FIGURE 1. (a) High-resolution transmission electron microscope (HRTEM) image of biotin capped ZnS:Mn²⁺ nanoparticles where the particles are strongly arrested by the biotin (b) Selected area electron diffraction (SAED) pattern of Mn doped ZnS nanocrystals.

UV-Visible spectroscopy is a useful technique to monitor the growth of the nanoparticles and their optical properties due to consequence of quantum confinement of photo generated electron-hole carriers [12]. Optical absorption spectra of Mn doped ZnS nanoparticles passivated with biotin were recorded at regular interval of times up to 90 days and are shown in Fig. 2. The excitonic absorption peak is observed at 308 nm and is blue shifted as compared to the bulk band gap of 3.7 eV. The blue-shift in the excitonic absorption is caused by quantum size effect and consequently changes in the band gap of nanocrystals whereas the broadening is due to defects such as lattice distortions and vacancies, which are probably located close to the surface in the case of small particles. It is noted that the absorption edge is remains same and is not red shifted with time during the period of study that confirms the stabilization of ZnS: Mn²⁺ nanoparticles in biotin matrix even on 90th days. The curves in Fig. 2 describing the long wavelength edge of the absorption bands of nanoparticles of biotin capped ZnS: Mn²⁺ which is consequently occurs as a result of allowed direct band to band transitions. So the band gap is simply calculated from the excitonic absorption peak by using the equation $E_g = h\,c\,/\lambda$. Where h is Planck's constant, c is speed of light and λ is the peak wavelength. The band gap calculated thus, using above equation is found to be ~ 4 eV which is much larger than the bulk band gap. Examination of spectral curves shown in Fig. 2, showed that nanocrystals of ZnS:Mn²⁺ are able to absorb quanta of light with energies less than E_g. The appearance of additional absorption in the region with $h\nu < E_g$ may be attributed either due to the presence of surface states or to high concentration of defects in the crystals lattice of ZnS:Mn²⁺ nanoparticles, which underlies the lower of the two conduction bands and are capable of ionization under the influence of quanta of light with $hv < E_g$ (intra band absorption of light). For nanocrystalline ZnS, a quite good first-order approximation to the energy of the band gap, in electron volts as a function of particle radius r, in nanometers is given by the Brus equation [13]

$$r(E_g) = \frac{0.32 - 2.91\sqrt{E_g - 3.49}}{2(3.50 - E_g)} \tag{1}$$

The particle size calculated using Eq.1 corresponds to band gap of 4 eV is about 3.5 nm for all spectra and are well agreed with sizes obtained from HRTEM. Here biotin plays an important role for surface passivation of ZnS: Mn^{2+} nanoparticles to control the growth and agglomeration thereby enhance the opto-electronic properties. Since the particle sizes obtained are in the quantum confinement regime therefore band gap becomes function of particle size and it is possible to tune the visible spectrum by varying the size.

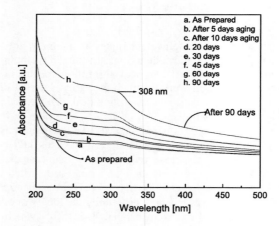

FIGURE 2. Absorption spectra of biotin capped ZnS:Mn^{2+} nanocrystals recorded at different interval of time for the period of 3 months. The excitonic absorption peak is observed at 308 nm. Figure clearly show that the absorption edge is remains same and is not red shifted with time during the period of study. So it confirms the stabilization of ZnS: Mn^{2+} nanoparticles in biotin matrix even on 90th days. Band gap is calculated from excitonic absorption peak is about 4 eV and corresponding particle size calculated using Brus equation is about 3.5 nm for all spectra and which are well agreed with sizes obtained from HRTEM.

The PL property of without cap and biotin capped ZnS: Mn^{2+} nanoparticles are also observed for the entire period and only PL spectra of initial (as prepared) and final (after

90 days) are shown in Fig. 3. The annealing effect on PL of without cap nanoparticles is also shown. The spectral features of the emission spectra were unchanged in every measurement except that in annealed sample where PL emission of Mn in wurtzite is slightly blue shifted than Mn in zinc blende ZnS which is also reported by H.E. Gamlich [14]. As seen in Fig. 3, we have found that the PL intensities are largely enhanced by surface passivation with a biocompatible layer biotin as compared to without cap particles while annealing of uncapped particles were drastically degrade the luminescence. There is no significant changes in luminescence intensity were observed in biotin capped ZnS: Mn^{2+} nanoparticles after 3 months however absence of biotin will gradually degrade the luminescence upon aging. The capping molecules (biotin) present on the ZnS:Mn^{2+} nanoparticles, actively altered the crystal field near the surface which depends on the number of defects that have been passivated on the surfaces. The biotin coating reduces unsaturated bond density and hence passivate surface, resulting in reduction in the number of surface trap sites for non-radiative recombination processes to occur, which finally enhance the luminescence intensity [11]. The increase in quantum efficiency as well as the life time shortening has also been attributed to the result from the strong hybridization of s-p electrons of the ZnS and d-electrons of Mn impurity due to confinement [15]. It has been observed that the luminescence from PVA and CA capped ZnS: Mn^{2+} nanoparticles are less than that obtained from biotin capped particles [11]. Thus stabilizing agent biotin not only enhances the luminescence intensity but play an important role in retains the luminosity of the ZnS: Mn^{2+} nanophosphors.

The incorporation of Mn^{2+} significantly enhanced the PL intensity in nanocrystalline ZnS:Mn^{2+} by several orders of magnitude thereby suggesting Mn^{2+} induced PL. Mn^{2+} ion d electrons act as efficient luminescent states which interact strongly with s-p electronic states of the host material. The spatial overlap of the host and dopant states, which varies with their distributions, influences the luminescence performance of the particles. This is important because sensors, displays and light emitting devices utilizing ZnS: Mn^{2+} nanoparticles can be better performing due to rapid energy transfer to the Mn^{2+} ions and fast recombination with the Mn^{2+} color centre.

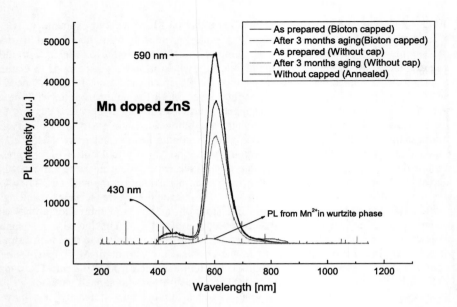

FIGURE 3. PL spectra of ZnS: Mn^{2+} nanoparticles. The spectral features of the emission spectra were unchanged in every measurement except that in annealed sample. All the samples show orange-red luminescence under UV-excitation. For 325 nm excitation wavelength of radiation from He-Cd laser, the emission at 590 nm was observed. It is clear that biotin capping will enhance the luminescence as compared to without cap particles. There is no significant changes in the luminescence intensity are observed in biotin capped particles even after 3 months aging. Absence of biotin will gradually degrade the luminescence upon aging while drastic degradation in the luminescence intensity was observed after annealing.

CONCLUSIONS

So we have concluded from our observation that capping is unavoidable and the application of biotin as a capping agent has resulted in successful stabilization of the structural and opto-electronic properties of ZnS:Mn^{2+} nanoparticles. It not only retained the size against growth and agglomeration but also protected from any environmental attack with time. It was found that biotin is an effective passivation of the unsaturated bonds on the particle surface leading to not only high luminescence efficiency but play an important role in stabilizing the luminesity of the ZnS:Mn^{2+} nanophosphors.

ACKNOWLEDGEMENTS

We thank Department of Science and Technology, India for funding the project under IRHPA in collaboration with Nanocrystals Technology, New York.

REFERENCES

1. Yuanrong Cheng, Changli Lu, Zhe Lin, Yifei Liu, Cheng Guan, Hao Lu and Bai Yang, *J. Mater. Chem.* **18**, 4062–4068 (2008).
2. Changli Lü, Junfang Gao, Yuqin Fu, Yaying Du, Yongli Shi and Zhongmin Su, *Adv. Funct. Mater.* **18**, 3070 – 3079 (2008).
3. R. Tu, B. Liu, Z. Wang, D. Gao, F. Wang, Q. Fang and Z. Zhang, *Anal. Chem.* **80**, 3458-65 (2008).
4. F. C. Liu, T. L. Cheng, C. C. Shen, W. L. Tseng and M. Y. Chiang, *Langmuir* **21** (2008).
5. Wei Chen, Alam G. Joly, Jan-Olle Malm, and Jan-Olov Bovin, *J. Appl. Phys.* **95**, 667-672 (2004).
6. Honma, S. Hirakawa, K. Yamada, and J. M. Bae, *Solid State Ionics* **118**, 29 (1999).
7. M. L. Curri, R. Comparelli, P. D. Cozzoli, G. Mascolo, and A. Agostiano, *Mater. Sci. Eng C* **23**, 285-289 (2003).
8. T. S. Phely-Bobin, R. J. Muisener, J. T. Koberstein, and F. Papadimitrakopoulos, *Synth. Met.* **116**, 439 (2001).
9. X. Zhang, A. V. Whitney, J. Zhao, E. M. Hicks, and R. P. VanDuyne, *J. Nanosci. Nanotechnol.* **6**, 1920-1934 (2006).
10. W. C. W. Chan and S. Nie, *Science* **281**, 2016 (1998).
11. A. K. Keshari, M. Kumar, P. K. Singh, and A. C. Pandey, *J. Nanosci. Nanotechnol.* **8**, 301-308 (2008).
12. P. Calandra, M. Goffredi, V. T. Liveri, *Colloids Surf., A* **133**, 69 (1998).
13. L. E. Brus, *J. Phys. Chem.* **90**, 2555 (1986).
14. H. E. Gamlich, *J. Lumin.* **23**, 73 (1981).
15. T. A. Kennedy, E. R. Glaser, P. B. Klein and R. N. Bhargava, *Phys. Rev. B* **52**, R14356 (1995).

Photoabsorption In (CdS)$_n$ Clusters On The Basis of Time-dependent Density Functional Theory

P. Dhuvad and A. C. Sharma

Physics Department, Faculty of Science, The M S University of Baroda, Vadodara 390002, Gujarat, INDIA

Abstract. We present the study of the optical properties of the small size (CdS)$_n$, n=1 to 4 clusters, using the plane wave based density functional formalism. Photoabsorption spectrum have been computed in a lower energy range using the time dependent Density Functional Theory, that uses time propagation method. The computed results are in good agreement with those obtained from another approach developed earlier known as Casida's formalism. Our results are reported for frequency range in which photoabsorption is prominent. The study can be useful for the future experimental investigations.

Keywords: TDDFT, Photoabsorption, small CdS clusters
PACS: 78.40. –q, 78.66 Hf, 73.22 -f

INTRODUCTION

Study of optical properties of very small size cluster of group II-VI semiconductors has potential applications in lasers, solar cell, sensors and biological imaging. The fundamental properties of the nano size material and atomic clusters remarkably differ from those of bulk semiconductor. Advancement in experimental methods has prompted significant theoretical efforts to understand various properties of the materials at atomic level. By changing the size and shape of clusters it is possible to modify their properties, and such modification is predominantly important for small size clusters. There has been no experimental investigations on optical properties of group II-VI semiconductor atomic clusters. However, there exists several theoretical studies on the electronic structure and optical properties of the small clusters. Photoabsorption of the small and medium size CdSe has been computed using different approaches in past. Many methods of first principle calculations have been developed. It has been reported that the simple LDA is not a good approximation compared to TDLDA.[1] First principle GW calculation is also another good approach for calculating the excitations and absorption spectra for real systems.[2] In this paper, we have used the TDDFT in two different ways one is propagation of the Kohn-Sham (KS) equation in real time and another is Casida's linear response theory. We used both approaches with the plane wave basis.

CP1147, *Transport and Optical Properties of Nanomaterials—ICTOPON - 2009*, edited by M. R. Singh and R. H. Lipson
© 2009 American Institute of Physics 978-0-7354-0684-1/09/$25.00

METHODOLOGY

There are mainly two methods to calculate the photoabsorption spectra using TDDFT. One is based on propagation of time-dependent KS equation in real time [3,4,5] ,while other calculates the linear response of the electrons to the frequency dependent external potential.[6,7] We have used both approaches and comparison of results from both approaches is made. Propagation of KS equation in real time is described by time-dependent Schrödinger equation,[8,9,10]

$$ i\hbar \frac{\partial}{\partial t}\psi_i(\mathbf{r},t) = \hat{H}\psi_i(\mathbf{r},t) \tag{1} $$

Where $\psi_i(\mathbf{r},t)$ is the one-electron wave function of the noninteracting time-dependent KS equation, and the Hamiltonian is given by

$$ \hat{H} = -\frac{\hbar^2 \nabla^2}{2m_e} + v_{ext}(\mathbf{r},t) + v_H(\mathbf{r},t) + v_{xc}(\mathbf{r},t) \tag{2} $$

which consists of the kinetic energy of the electrons, a time-dependent external potential $v_{ext}(\mathbf{r},t)$, the Hartree potential $v_H(\mathbf{r},t)$, and exchange-correlation potential $v_{xc}(\mathbf{r},t)$. The initial state at time t = 0 is perturbed with a time-dependent potential, $v(r,t) = -\hbar k_0 \chi_v \delta(t)$ and $v_{ext}(r,t) = v_{static}(r,t) + v(r,t)$ [4,5,6]. Here, $\chi_v = x,y,z;$ is polarization direction and k_0 is a momentum transferred to the system, which has to be small in order to assure that the response of the system is linear and that only dipole transitions are excited. This means that all frequencies of the system are excited with equal weight. If we define $\varphi_i(r)$ to be the ground-state KS wave functions of our system, the initial state for the time evolution at t = t^0 is $\psi_i(r,0^+) = exp(i k_0 \chi_v) \varphi_i(r)$

After this initial step, the KS wave function is propagated during a finite, but very long, time. The information of the excitations is deduced from dipole-strength function $S(\omega)$. It measures how strongly a given frequency ω excites the system and it can be written as [3,4,5]

$$ S(\omega) = \frac{2m_e\omega}{\pi\hbar e^2}\Im\sum_v \alpha_{vv}(\omega), \tag{4} $$

where dynamical polarizability $\alpha_{\mu v}$ is

$$ \alpha_{\mu v}(\omega) = \frac{e^2}{\hbar k_0}\int d\mathbf{r}\chi_v \delta n(\mathbf{r},\omega) \tag{5} $$

Details about the Casida's linear response theory are reported in Refs. 6 & 7.

The first step for calculating the cluster properties is the determination of the lowest energy configuration. Existence of several configurations with varying coordination and multiple minima in the potential energy surface enormously complicate the problem and make it almost impossible to decide by direct calculation of lowest energy configuration for cluster. For the lowest energy structure of the clusters we

195

referred to the previously reported results.[8] We used the plane augmented wave method that is implemented in the VASP package[11] within the frame work of the GGA[12], which provides better fist principle simulation as compared to ultra-soft pseudo potentials[13]. For the structure optimization we have used a super-cell of size a=25Å and the plane wave energy cut-off was set to be 247.3eV. We performed the calculations for different cell size and energy cutoff and values of cell size and energy cutoffs have been chosen accordingly. The valence electron configuration used for Cd and S are; $5s^2 4d^{10}$ and $3s^2 3p^4$ respectively. For all the structure minimization of force and energy convergence was obtained at $\sim 10^{-3}$ eV / Å and $\sim 10^{-4}$ eV respectively. The photoabsorption spectra have been computed using OCTOPUS program[14], a TDDFT real-time real-space code. The norm-conserving Troullier-Martins pseudopotentials[15] have been used through out the calculations. Total time prorogation of 25 $\hbar/eV$ was given with time step of 0.001 $\hbar/eV$. The different size clusters were kept in the spherical domain of radius 18 Å and spacing of real-space mesh was set to 0.27 Å. Linear response TDDFT allows calculation of excitation energies of finite system. Time dependent density functional response theory (TDDFRT), Casida's formulation, is implemented in OCTOPUS, which is fastest way to calculate excitation energies for small size clusters. The ground state wave function is subjected to a time-dependent perturbation modifying its external potential. The number of the unoccupied orbitals for the excitations was three times grater than the occupied orbitals. The results converge with the higher number of unoccupied orbitals.

RESULTS AND CONCLUSIONS

Our optimized structure of $(CdS)_n$, n= 1 to 4 are shown in figure1 (a)-(d) along with our computed photoabsorption spectra. Different configuration of $(CdS)_n$ has varying Cd-S bond length form 2.21 Å to 2.57 Å that is different from the Cd-S bond length, 2.519 Å, in a bulk wurtzite structure. Our optimized structure are in good agreement with those reported in Ref. 6. We have computed photoabsorption spectra of the small size $(CdS)_n$, n=1 to 4, using both the time-propagation method and Casida's linear response approach as described in methodology. Recent theoretical studies on photoabsorption shows that time-propagation results give better agreement with the experimental results as compared to those obtained form Casida's linear response technique, which exhibits some additional peaks in absorption spectra that are not seen in experiments[16]. In our study we focused on the results obtained with time-propagation method and obtained several important results. For $(CdS)_1$ molecule the prominent absorption occurs in energy range of 5 eV to 6.5 eV with additional peak of relatively lesser intensity at around 7 eV. For $(CdS)_2$ the prominent absorption takes place in energy range of 6 eV to 7 eV. $(CdS)_3$ which has a planner structure shows two absorption peaks at around 3.7 eV and 7.4 eV. The $(CdS)_4$ cluster that has three dimensional structure shows absorption in wide range of 5.5 eV to 7.5 eV. We did not find any significant variation in the behavior of the absorption spectrum by using additional unoccupied states for the computation of the excitations. This indicates that the results are well converged. In time-propagation method results are converged for the smaller time steps for propagation. We feel that our obtained results can be useful for the experimental investigations of small CdS clusters in lower energy range.

ACKNOWLEDGMENTS

Authors acknowledge with thanks the financial support from Department of Science and Technology, Government of India, New Delhi through a research project No. SR/S2/CMP-16/2005.

REFERENCES

1. M. Claudia Troparevsky, Leeor Kronik, and James R. Chelikowsky, *Phys. Rev. B* **65**, 033311 (2001).
2. Giovanni Onida, Lucia Reining and Angel Rubio, *Rev. Mod. Phy,* **74**, 601 (2000).
3. K. Yabana and G. F. Bertsch, *Phys. Rev. B* **54**, 4484 (1996).
4. K. Yabana and G. F. Bertsch, *Int. J. Quantum Chem.* **75**, 55 (1999).
5. K. Yabana and G. F. Bertsch, *Phys. Rev. A* **60**, 1271 (1999).
6. M. E. Casida, in *Recent Advances in Density Functional Methods*, edited by D. P. Chong ,World Scientific, Singapore, (1995), Part 1, p. 155
7. M. E. Casida, in *Theoretical and Computational Chemistry*, edited by J. M. Seminario, Elsevier Science, Amsterdam, (1996), Vol. 4, p. 391
8. M. Claudia Troparevsky and James R. Chelikowsky *J. Chem. Phys.* **114**, 943 (2001).
9. R. van Leeuwen, *Int. J. Mod. Phys. B* **15**, 1969 (2001).
10. M. A. L. Marques and E. K. U. Gross, *Annu. Rev. Phys. Chem.* **55**, 427 (2004).
11. G. Kresse and J. Furthmuller, *Phys. Rev. B* **54**, 11169 (1996); *Comput. Mater. Sci.* **6**, 15 (1996).
12. P. E. Blochl, *Phys. Rev. B* **50**, 17953 (1994); G. Kresse and D. Joubert, ibid. **59**, 1758 (1999)
13. Somesh Kr. Bhattacharya and Anjali Kshirsagar, *Phys. Rev. B* **75**, 035402 (2007).
14. A. Castro, H. Appel, M. Oliveira, C. A. Rozzi, X. Andrade, F. Lorenzen, M. A. L. Marques, E. K. U. Gross, and A. Rubio, *Phys. Status Solidi B* **243**, 2465 2006; http://www.tddft.org/programs/octopus
15. N. Troullier and J. L. Martins, *Phys. Rev. B* **43**, 1993(1991).
16. Michael Walter, Hannu Häkkinen, Lauri Lehtovaara, Martti Puska, Jussi Enkovaara, Carsten Rostgaard and Jens Jørgen Mortensen, *J. Chem. Phys.* **128**, 244101 (2008).

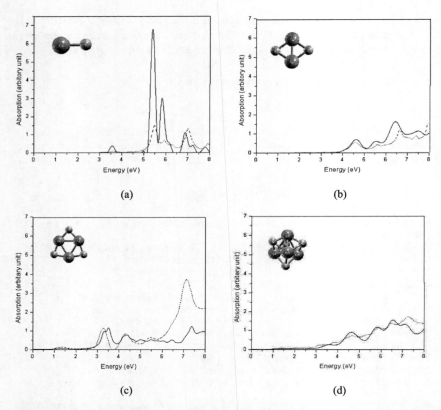

FIGURE 1. Calculated photoabsorption spectra is plotted as a function of energy for (CdS)1,(CdS)2, (CdS)3 and (CdS)4 in figures 1(a),1(b),1(c) and 1(d) respectively. Solid line corresponds to time-propagation approach while dashed line corresponds to Casida's approach. Cluster structure is shown in inset.

Synthesis of Nickel Nanomaterial by Pulsed Laser Ablation in Liquid Medium and its Characterization

R. Gopal[a], M. K. Singh[b], A. Agarwal[b], S.C. Singh[a], R. K. Swarnkar[a]

[a]Laser Spectroscopy & Nanomaterials Lab, Department of Physics, University of Allahabad, Allahabad-211002, India.
[b]Department of Physics, M. N. National Institute of Technology, Allahabad-211004, India.

Abstract. Laser ablation of nickel nanoparticles suspended in double deionized water has been studied using Nd:YAG laser (355nm) with energy 30 mJ/pulse. Produced nanoparticles are analyzed by UV-visible absorption spectroscopy at certain interval (0, 20, 40, 60 minute). The particles are characterized by Transmission Electron Microscopy (TEM), and X-ray Diffraction (XRD). TEM image shows that particles have crystal like structure and these particles are in the range of 20 nm to 50 nm.

Keywords: Laser ablation; Nanomaterials; X-ray Diffraction
PACS: 81.07.-b, 78.67.Bf,

INTRODUCTION

Nanostructure materials exhibit unique physical and chemical properties, significantly different from those of conventional bulk materials, because of their extremely small size and large specific surface area. Recently, transition metal nanomaterials have drawn considerable attention owing to their potential applications in various fields such as optoelectronics, photoelectrical, catalysis, data storage, ferro-fluids, biomedicine [1-6]. Among various types of magnetic materials, nickel has been well documented because of its excellent properties and industrial applications. Several methods have been used for the synthesis of nickel nanoparticles such as chemical [7], surfactant mediated synthesis [8], spray pyrolysis [9], sonochemical decomposition [10], and laser ablation method [11]. Among these methods pulsed laser ablation in liquid media (PLAL) is an emerging technique in material science for fabricating metal and metal-oxide nanoparticles. Several synthesis of metal/metal-oxide has been done by using PLAL [11-13]. There are many advantages of the PLAL technique which includes inexpensive equipment for controlling the ablation atmosphere, simplicity of the procedure and the minimum amount of chemical species required for synthesis compared to the conventional chemical process [14]. It involves ablating the pure metal in liquid medium that confines the plasma plume formed when the high energy laser beam interacts with the material and laser induced reactive quenching leads to the formation of nanoparticles of metal and metal oxide. This mechanism is called ''laser-induced reactive quenching'', which is essentially the same as oxidation of metal by water [13]. The size of the nanoparticles can be controlled by varying laser parameters and liquid medium. It has

CP1147, *Transport and Optical Properties of Nanomaterials—ICTOPON - 2009*, edited by M. R. Singh and R. H. Lipson
© 2009 American Institute of Physics 978-0-7354-0684-1/09/$25.00

been observed that the laser energy is one of the important parameter which affects the size of nanoparticles and the absorption spectrum [15]. The nanoparticles are obtained in a solution form which makes them easier to handle.

Here we report the synthesis of nickel nanoparticles by pulsed laser ablation in liquid medium, using powder source of micron size. This method of preparing nickel nanoparticles is not reported in the literature. The synthesized nickel nanoparticles were characterized by X- ray diffraction (XRD) for structural parameters, UV-vis spectrometry was employed to check the reduction in particle size and transmission electron microscopy (TEM) to access the precise size of the nanoparticles.

EXPERIMENTAL

The experimental setup for the synthesis of nickel nanoparticles by PLAL method is shown in Fig. 1. 25 mg of micron size nickel powder was suspended in 50 mL of double distilled water contained in a glass vial at the room temperature. The suspensions were stirred by a magnetic stirrer. Ablation of suspensions was carried out for one hour by focused output from 355 nm of Nd: YAG laser (Spectra Phys., USA) operating at 35 mJ/ pulse having 5 ns pulse duration with repetition rate of 10 Hz. The UV-vis. absorption spectra of synthesized colloidal solution of nanoparticles were acquired from Perkin Elmer Lambda 35, double beam spectrophotometer. XRD patterns of the dried powder were recorded by Thermo electron corporation ARL X'TRA X-ray diffractometer using 1.5405 Å Cu K_α line as reference. High Resolution Transmission Electron Microscope images of the nanoparticles were recorded on Technai G20 – stwin electron microscope operating at 200kV accelerating potential with 1.44 Å point and 2.32 Å line resolutions. A drop from each colloidal solution of nanoparticles was placed onto a carbon coated copper grid and was dried for TEM analysis.

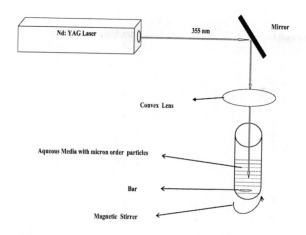

FIGURE 1. Experimental setup for laser ablation of nickel nanoparticles suspended in the deionized water.

RESULTS AND DISCUSSION

UV-vis absorption spectra of the synthesized colloidal solution at different ablation time (0, 20, 40 and 60 minute) are shown in Fig. 2. The optical absorption spectra show red shift (from 309nm to 380nm) of the optical absorption edge with the reduction in particle size [16]. In some cases however, the optical absorption edge shows a trend of blue shift also. The shift of the optical absorption edge is the competition result between blue-shift effect and red-shift effect [17].

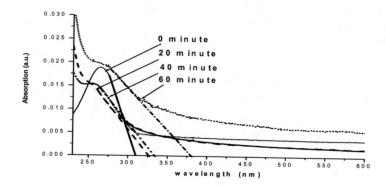

FIGURE 2. Optical absorption spectra of synthesized nickel nanoparticles at different time intervals.

The XRD spectrum of the dried powder of synthesized nickel nanoparticles after ablation for one hour is shown in Fig 3. The spectrum shows three peaks at 2 θ= 44.4, 51.8, 76.3 corresponding to the (111), (200) and (220) lattice planes (JCPDS Card No. 87-0712). No peak was observed for oxide and hydroxide, which reveal the formation of pure elemental nickel with a face centered cubic (FCC) structure. The crystallite size was calculated from XRD peaks by using Scherrer formula and it was found of the order of 38 nm.

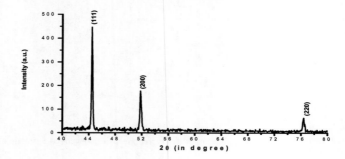

FIGURE 3. XRD spectrum of as synthesized nickel nanoparticles after the ablation for one hour.

The TEM images of the synthesized nickel nanoparticles are shown in Fig. 4 (a-b). TEM images show that the nanoparticles are in the range of 20 to 50 nm. Wu et al. [17] have synthesized nickel nanotubules (500–1000 nm) by reduction of nickel chloride using metallic zinc in the presence of ethanol amine at room temperature. Gao et al. [18] have prepared nanocrystalline nickel powder (58–102 nm) from nickel chloride solution in the presence of surfactant and hydrazine hydrate as reducing agent at pH 9–10 and temperature of 85–95 °C. In addition, Zhang et al. [19] have prepared size-controlled nickel nanocrystals (20–60 nm) by decomposition of nickel acetylacetone in a noncoordinating reagent, oleylamine. The nanoparticles obtained in this process have almost same size as we have obtained by laser ablation in liquid medium.

High magnification micrograph [Fig 4(b)] shows crystal like structure. Selected Area Electron Diffraction pattern (SAED) [inset, Fig. 4(b)] shows the superposition of the ring as well as spot-pattern which suggest the presence of both polycrystals and monocrystals.

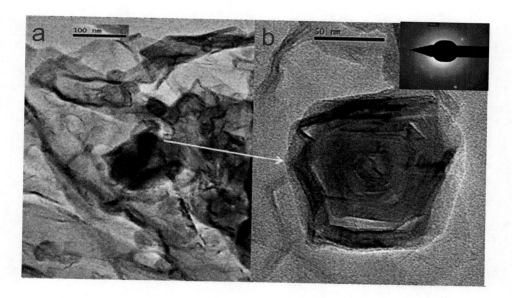

FIGURE 4. (a) TEM image of nickel nanoparticles **(b)** high magnification image of single nanoparticles and SAED pattern of nickel nanoparticles.

CONCLUSIONS

We have prepared stable nickel nanoparticles from the laser ablation of bulk nickel powder dispersed in double deionized water. The method described here may provide an alternative and simple route to prepare nickel nanoparticles from the bulk phase of metal. X-ray diffraction showed that the resultant particles are pure nickel crystalline with FCC structure. The colloids prepared in this experiment offer an opportunity to be utilized as a catalyst in some useful reactions.

ACKNOWLEDGMENTS

Authors are thankful to Prof. B.R. Mehta, IIT Delhi, for providing TEM facility and UGC, Govt. of India, New Delhi for financial assistance. One author M.K. Singh is thankful to MNNIT for providing the fellowship.

REFERENCES

1. C. Stamm, F. Marty, A. Vaterlaus, V. Weich, S. Egger, U. Maier, U. Ramsperger, H. Fuhrmann and D. Pescia, *Science,* **282**, 449-451 (1998).
2. L. Lu, M. L. Sui and K. Lu, *Science* **287**, 1463-1466 (2000).
3. H. Sakurai and M. Haruta, *Appl. Catal. A. Gen.* **127**, 93-105 (1995).
4. V.F. Puntes, K.M. Krishnan and A. P. Alivisatos, *Science* **291,** 2115-2117 (2001).
5. N. Cordente, M. Reapaud, F. Senocq, M. J. Casanove, C. Amiens and B. Chaudret, *Nano Lett.* **1**, 565-568 (2001).
6. S. Sun, H. Zeng, *J. Am. Chem. Soc.* **124,** 8204-8205 (2002).
7. V. Tzitzios, G. Basina, M. Gjoka, V. Alexandrakis, V. Georgakilas, D. Niarchos, N. Boukos and D. Petridis, *Nanotechnology* **17**, 3750-3755 (2006).
8. Y. D. Wang, C.L. Ma, X.D. Sun and H. D. Li, *Inorg. Chem. Commun.* **5,** 751-755 (2002).
9. S. L. Che, K. Takada, , K. Takashima, , O. Sakurai, , K. Shinozaki, and N. Mizutani, *J. Mater. Sci.* **34**, 1313-1318 (1999).
10. S. Ramesh, Y. Koltypin, R. Prozorov and A. Gedanken, *Chem. Mater.* **9,** 546-551 (1997).
11. R. Mahfouz, F.J. Cadete Santos Aires, A. Brenier, B. Jacquier and J.C. Bertolini *Appl. Surf. Sci,* **254,** 5181-5190 (2008).
12. S.C. Singh and R. Gopal, *J. Phys. Chem. C*, **112**, 2812-2819 (2008).
13. T. Tsuji, T. Hamagami, T. Kawamura, J. Yamiki and M. Tsuji, *Appl. Surf. Sci.* **243,** 214-219 (2005).
14. G.W. Yang, *Prog. Mater.Sci.* **52,** 648-698 (2007).
15. H. Usui, T. Sasaki and N. Koshizaki *J. Phys. Chem. B*, **110**, 12890-12895 (2006).
16. M.V. Rama Krishna and R. A. Friesner, *J. Chem.Phys.* **95,** 8309-8322 (1991).
17. P. Liu, W. Cai and H. Zeng *J. Phys.Chem. C*, **112**, 3261-3266 (2008).
18. M. Wu, Y. Zhu, H. Zheng and Y. Qian, *Inorg. Chem. Commun.* 5, 971-974 (2002).
19. J. Gao, F. Guan, Y. Zhao, W. Yang, Y. Ma, X. Lu, J. Hou and J. Kang, *Mater. Chm. Phys*. **71**, 215-219 (2001).
20. H.T. Zhang, G. Wu, X.H. Chen and X.G. Qiu, *Mater. Res. Bull.* **41,** 495–501 (2006).

Synthesis of Copper/Copper-Oxide Nanoparticles: Optical and Structural Characterizations

R. K. Swarnkar, S. C. Singh and R. Gopal

Laser Spectroscopy & Nanomaterials Lab, Physics Department (Centre for Advanced Studies)
University of Allahabad, ALLAHABAD- 211002, INDIA.
spectra2@rediffmail.com

Abstract. In the present study, we have synthesized copper/ copper-oxide nanoparticles by laser ablation of copper metal in aqueous solution of sodium dodecyl sulfate. The focused output of 1064 nm wavelength of pulsed Nd:YAG laser is used for ablation. The synthesized nanoparticles are characterized by UV-visible absorption, X-ray diffraction, transmission electron microscopy and Raman spectroscopy techniques. The synthesis of copper/ copper oxide nanoparticles are confirmed by XRD and Raman studies. The possible mechanism of nanoparticle formation is also discussed.

Keywords: Laser ablation in liquid media; Copper oxide nanomaterial; Synthesis of nanomaterial; Raman spectroscopy.
PACS: 81.07.b; 81.05.Dz; 52.38.Mf.

INTRODUCTION

Metal oxide nanoparticles (NPs) have attracted their great attention due to its tunable optical, electronic, magnetic and catalytic properties [1]. In addition, size dependent optical, electronic and vibrational properties of the nanomaterials are also observed in several semiconductors and oxides materials [2]. Copper oxide is an excellent nanoparticles (NP) system for investigating the size induced structural transformations and phase stability. Copper oxide has two phases i.e. cuprous oxide (Cu_2O) and cupric oxide (CuO). The oxides of copper are p- type semiconductor and have direct band gap. Palker et al. [3] have studied CuO NPs synthesized by chemical route and reported that smaller size CuO (< 25nm) are not stable while that of cubic and more ionic Cu_2O is formed in smaller size regime. Lin *et al.* [4] have studied electro-optical properties of copper oxide nanomaterial and found that nanofibrils of 50 nm diameter exhibits high field emission current density therefore can be used as electron source in cathode ray tubes and flat panel displays. Copper oxide is considered as an efficient catalytic agent and also a good gas sensing material [5]. There are several methods reported to synthesize copper oxide NPs like sol-gel, hydrothermal route, electrochemical [6-8] etc. Materials synthesized by these processes include significant amount of chemical contamination on the surface of synthesized nanomaterials, which limits their applications in optoelectronic and sensing device fabrication. Therefore, it is highly required to synthesize nanomaterials having chemical contaminations free surfaces for the fabrication of surface phenomenon based devices and biological applications. Lasers have open new doors in the processing of nanomaterials. Pulsed laser ablation process has several advantages over

CP1147, *Transport and Optical Properties of Nanomaterials—ICTOPON - 2009*, edited by M. R. Singh and R. H. Lipson
© 2009 American Institute of Physics 978-0-7354-0684-1/09/$25.00

other conventional routes including (a) large number of available ablation parameters for controlling the size and shape of nanomaterials, (b) produced nanomaterials have inherent stochiometry as their mother targets therefore, capability to produce nanomaterials of desired chemical composition and (c) ability of producing nanomaterials having surfaces free from chemical contamination. This technique is used by several workers to synthesized colloidal NPs of different metals and semiconductors [9, 10]. In present study, copper oxide NPs has been synthesized by laser ablation of copper metal plate in 50mM aqueous solution of SDS.

EXPERIMENTAL

Copper oxide NPs have been synthesized by pulsed laser ablation of high purity copper plate. The copper plate is placed on the bottom of glass vessel containing 20 ml aqueous solution of 50 mM SDS and irradiated with focused output of 1064 nm of Nd:YAG laser (Spectra Physics Inc. USA) operating at 40 mJ/ pulse energy, 10 ns pulse duration and 10 Hz for 30 minutes. This results a light green colored colloidal solution. The synthesized NPs have been characterized by using UV-visible absorption, X-ray Diffraction (XRD), Transmission Electron Microscopy (TEM), Selective Area Electron Diffraction (SAED) and Raman spectroscopy techniques.

The absorption spectrum of as synthesized colloidal solution of NPs has been recorded by using Perkin Elmer Lambda 35 double beam spectrophotometer. TEM image of the NPs is recorded on Technai G20 – stwin electron microscope operating at 200kV. A drop from colloidal solution of NPs is placed onto a carbon coated copper grid and dried for TEM analysis. Small amount of colloidal solution is dried at 60°C in an oven for 24 hours and collected the powder for XRD analysis. XRD pattern of synthesized powder of NPs is recorded with Rikagu, D-Max X-ray diffractometer using 1.5405 $\overset{\circ}{A}$ Cu K_{α} line. For recording Raman spectrum of synthesized colloidal NPs, 488 nm line of Ar^{+} laser (Spectra Physics Inc., USA) is used as excitation source and spectrum is recorded on computer controlled 0.5 M triple grating monochromator (Acton Research Corp., USA) with resolution of 0.03 nm having R928 PMT as detector. Spectra Sense and Grams 32 software are used for recording spectrum and spectral analysis respectively.

RESULTS AND DISCUSSIONS

UV-VIS absorption spectrum of as synthesized colloidal solution of NPs in the 50mM concentration of anionic surfactant SDS is shown in figure 1. The absorption spectrum reveals four peaks at ~ 300, 380, 650 and 800 nm. The peak at 300 nm corresponds to inter band transition from deep level electrons of valance band of Cu_2O

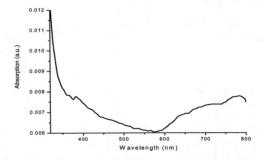

FIGURE 1. UV-VIS Absorption spectrum of as synthesized sample.

while that of peak around 380 nm is due to the band edge transition of Cu_2O. The peaks at 650 and 800 nm are attributed due to presence of CuO on the surface of synthesized material [11].

Fig.2 shows XRD spectrum of the dried powder of NPs synthesized in aqueous solution of 50mM SDS. Presence of sharp and intense diffraction peaks in the XRD spectrum shows synthesized NPs are crystalline in nature. Most of the peaks correspond to CuO. Peaks at $2\theta = 17.67°$, $22.06°$, $38.03°$ and $41°$ correspond to the (111), (202), (131) and (024) Bragg's reflection planes of monoclinic CuO (JCPDS No.011117). Peak at $2\theta = 52.6°$ correspond to the (211) plane of cubic Cu_2O (JCPDS No. 770199). XRD spectrum confirms the synthesis of copper oxide NPs. Remaining peaks are due to some impurities or defects, which may be removed after annealing.

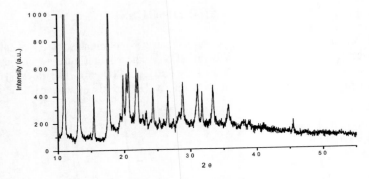

FIGURE 2. XRD spectrum of as synthesized sample.

TEM image of NPs synthesized by pulsed laser ablation of copper in aqueous media of SDS is shown in figure 3. It is evidenced from the TEM micrograph that synthesized NPs are spherical and have average diameter of 4-5 nm. SAED pattern [inset, Fig. 3] shows the superposition of the ring as well as spot-pattern which suggest the presence of both polycrystals and monocrystals.

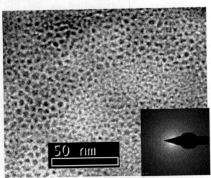

FIGURE 3. TEM image of as synthesized sample alongwith SAED pattern

Raman spectroscopy is a very useful tool for the study of phases and structures of oxide systems. The Raman shift and bandwidth change with the decreasing particle size. Raman spectrum of the NPs synthesized in 50mM SDS is shown in fig 4. The spectrum has three peaks located at around 300, 360 and 620 cm^{-1}. The strong peak at ~300 cm^{-1} and the weak peak at 360 cm^{-1} are assigned to CuO , while the peak at ~620 cm^{-1} is due to

Cu$_2$O [12]. The Raman spectrum reveals the presence of both CuO and Cu$_2$O in synthesized sample, agreeing well with the results of XRD and UV-VIS absorption.

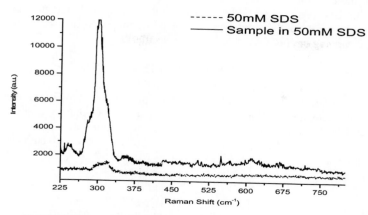

FIGURE 4. Raman Spectrum of the sample.

Yin et al. [11] have studied copper oxide NPs synthesized by chemical method, which involved metal acetates as precursors and thermal decomposition of metal acetate-surfactant complex in a hot organic solvent. They have found that the absorption spectrum of Cu$_2$O nanocrystals have peaks at 260 and 340 nm, attributed to band to band transition of Cu$_2$O. Another peak at 630 nm is due to band gap transition of CuO present at the surface of nanocrystals. Borgohain et al. [13] have found that the band gap absorption for Cu$_2$O NPs with diameter of 2.0 ± 0.5 nm occurs at 368 nm. With the red shift in absorption, particles size increases. Pedersen et al. [14] have studied oxidation of copper NPs by annealing the sample of Cu NPs at different temperatures in air environment and found that oxide of copper NPs with diameter of 3.0 ± 0.3 nm have absorption peak at 280 and 340 nm, consistent with the direct conversion of Cu to Cu$_2$O. They have also observed a broad peak at 800 nm, assigned as characteristic of CuO. Our experimental result obtained for optical property of copper oxide NPs synthesized by laser ablation in aqueous media of 50mM SDS is in good agreement with all above result. Gan et al., Balamurugan et al. and many other groups [12, 15, 16] have studied structural property of copper oxide nanomaterial by Raman spectroscopy. The structural results obtained by these research groups for copper oxide NPs synthesized by chemical routes are almost same as we have got in NPs synthesized by laser ablation in liquid media. They found Raman spectrum of copper oxide nanomaterial have strong peak at ~300 and ~360 cm^{-1} ascribed to CuO and a broad peak at 600 cm^{-1} due to Cu$_2$O.

The mechanism of the laser ablation can be explained in terms of the dynamic formation mechanism [9] given as "A dense cloud of the metal atoms (plume) is accumulated in the closed vicinity of laser spot on the metal surface during the course of the ablation. These metal atoms supersonically expand against liquid media and form clusters of atoms/ molecules after cooling. These atoms/ molecules act as seed for the growth of the NPs. Termination of the growth of the particles and their stabilization in the colloidal solution is achieved by equilibrium between production of atoms by laser ablation and capping of clusters by surfactant molecules".

CONCLUSIONS

Copper oxide NPs has been synthesized by pulsed laser ablation of copper metal plate in 50mM aqueous solution of SDS. TEM image shows that synthesized NPs are spherical and have average diameter of 4-5 nm. Formation of copper oxide NPs is confirmed by XRD and Raman spectrum which shows presence of both cuprous and cupric oxide NPs.

ACKNOWLEDGEMENTS

Authors are thankful to Prof. B.R. Mehta, IIT Delhi, for providing TEM facility,   Prof. D. Pandey, Dept. of Material Science, B.H.U., Varanasi for providing XRD facility and DRDO, New Delhi for financial assistance.

REFERENCES

1. S.C. Singh, R.K. Swarnkar and R. Gopal, *J. Nanosci. Nanotech.* **9**, 1-5 (2009).
2. B. Balamurugan, B. R. Mehta and S. M. Shivprasad, *Appl. Phys. Lett.* **79**, 3176 (2001).
3. V. R. Palkar, P. Ayyub, S. Chattopdhyay and M. Multani, *Phys. Rev. B* **53**, 2167 (1996).
4. H. Lin, C. Wang, H. C. Shih, J. Chen and C. Hsieh, *J. Appl. Phys.* **95**, 5889 (2004).
5. J. J. Zhang, J. F. Liu and Y. D. Li, *Chem. Mater.* **18**, 867 (2006).
6. H. Zhang and Z. Cui, *Mater. Res. Bull.* **43**, 1583 (2008).
7. K. Borgohain, J. B. Singh, M.V. R. Rao, T. Shripati and S. Mahamuni, *Phys. Rev. B* **61**, 11093 (2000).
8. J. Gong, L. Luo, S. Yu, H. Qian and L. Fei, *J. Mater. Chem.* **16**, 101 (2006).
9. S.C. Singh and R. Gopal, *J. Phys. Chem. C,* **112**, 2812 (2008).
10. C.H. Liang, Y. Shimizu, T. Sasaki, N. Koshizaki, *Appl. Phys. A,* **80**, 819 (2005).
11. M. Yin, C. Wu, Y. Lou, C. Bruda, J. T. Koberstein, Y. Zhu and S. O'Brien, *J. Am. Chem. Soc.,* **127**, 9506 (2005).
12. Z. H. Gan, G. Q. Yu, B. K. Tay, C. M. Tan, Z. W. Zhao and Y.Q. Fu, *J. Phys. D: Appl. Phys.* **37**, 81 (2004).
13. K. Borgohain, J. B. Singh, M. V. Rama Rao, T. Shripathi and S. Mahamuni, *Phys. Rev. B* **61**, 11093 (2000).
14. D. B. Pedersen, S. Wang and S. H. Liang, *J. Phys. Chem. C* **112**, 8819 (2008).
15. B. Balamurugan, B. R. Mehta, D. K. Avasthi, F. Singh, A. K. Arora, M. Rajalakshmi, G. Raghavan, A. K. Tyagi and S. M. Shivprasad, *J. Appl. Phys. B* **92**, 3304 (2002).
16. Y. S. Gong, C. Lee, and C. K. Yang, *J. Appl. Phys.* **77**, 5422 (1995).

Cadmium Oxide Nanostructures in Water; Synthesis, Characterizations & Optical Properties

S.C. Singh, R.K. Swarnkar and R. Gopal

Laser spectroscopy and Nanomaterials Lab, Department of Physics,
University of Allahabad, Allahabad-211002, INDIA
E-mail: spectra2@rediffmail.com

Abstract: Cadmium oxide nanoparticles are synthesized by pulsed laser ablation of cadmium metal in double distilled water, which are found stable for 24 hours. The sample is characterized by using UV-VIS absorption, scanning electron microscopy, thermo gravimetric analysis, differential thermal analysis, X-ray diffraction and fourier transform infrared techniques. XRD pattern shows that as synthesized nanopowder has maximum amount of the Cadmium hydroxide rather than its oxide, which converts into Cadmium oxide after annealing at 350°C.

PACS: 81.07.b; 81.05.Dz.
Key words: Cadmium oxide nanocrystals, Pulsed laser ablation, X-ray diffraction; Thermal analysis

INTRODUCTION

Recently, CdO nanomaterials are found a suitable candidate for optical and sensing applications due to its ionic nature coupled with wide band gap. CdO nanomaterial plays important role in the field of solar cells and front panel displays. Inspite of these applications, there is a lack of experimental as well as theoretical data on the CdO nanomaterial in comparison to other II-VI semiconductors such as ZnO, ZnS, CdS, ZnSe, CdSe etc.. However considerable attention is also paid to CdO nanomaterial due to its applicability in low voltage and shorter wavelength electro optical device fabrication such as UV LEDs and UV diode lasers [1]. Different optical and electronic properties of CdO nanomaterials are highly dependent on their size, shapes and surface structures. Therefore it is important to synthesize good quality of the CdO nanomaterials. There are several physical, chemical and mechanical approaches [2-5] for example vapor phase transport, precursor calcination, solvothermal method and mechano-chemical reaction method to synthesis of CdO nanomaterials. Materials synthesized by salt reduction, precursor calcinations and solvothermal process include significant amount of chemical contamination on the surface of synthesized nanomaterials, which limits their applications in optoelectronic and sensing device fabrication.

Lasers have open new doors in the processing of nanodimmensional materials and their characterization in current decades. Synthesis of colloidal solution of noble metal nanocrystals using pulsed laser ablation [6] of corresponding metal target in aqueous media and study of their size, shape and other properties on ablation parameters are intense field of research now a days. Synthesis of CdO nanocrystals by pulsed laser ablation and its structural, thermal and optical characterization is main theme of the present investigation.

EXPERIMENTAL

Experimental arrangement for the synthesis of colloidal solution of nanomaterials using pulsed laser ablation in aqueous media is described elsewhere [7], briefly high purity cadmium target (99.99 %, Johnston Mathey, U.K.), placed on the bottom of glass vessel containing 30 ml double distilled water, is allowed to irradiate with focused output of 1064 nm from pulsed Nd:YAG laser (Spectra Physics Quanta Ray, USA) operating at 35 mJ/pulse energy, 10 ns pulse width and 10 Hz repetition rate for an hour. As synthesized colloidal suspension has brown color and found stable for 24 hours. Solution is centrifuged at 4000 rpm and obtained residue is dried at 60°C in oven for 24 hours. Dried powder is used for Thermo Gravimetric Analysis (TGA), Differential Thermal Analysis (DTA), X-ray Diffraction (XRD) and Fourier Transform Infrared (FTIR) characterizations. Sample is annealed at 350°C for 9 hours and characterized by UV-VIS absorption and XRD techniques.

UV-visible absorption spectrum of as synthesized colloidal solution and annealed powder dispersed in methanol is recorded with Perkin Elemer Lambda 35 double beam spectrophotometer. SEM image of the as synthesized powder is recorded with JEOL- SEI scanning electron microscope. X-ray diffraction pattern of the powder dried at 60°C and powder annealed at 350°C are recorded with PAN-Alytical X-ray Diffractometer (Philips model) using Cu-K$_\alpha$ radiation (λ = 1.5406 Å). TGA and DTA of as synthesized powder are carried out on Perkin Elmer model no. 7 in the nitrogen atmosphere at the heating rate of 10°C/ minute. As synthesized powder is dispersed into KBr matrix and is palletized at 10 ton pressure. FTIR spectrum of the powder is recorded by placing the pellet in the path of the IR beam of FTIR spectrophotometer Perkin Elmer RX-1.

RESULTS AND DISCUSSIONS

Figure 1 (a) depicts TGA and (b) DTA spectra of as synthesized powder dried at 60°C in oven. TGA spectrum of the sample illustrates two steps of the mass decay. In the first step from 337°C- 400°C, there is almost 18% weight loss in the initial weight, which is almost equal to the loss of one water molecule from each of the $Cd(OH)_2$ molecules. There is a sharp endothermic peak at 373°C in the DTA spectrum corresponding to the first step of the TGA, which confirms that sample absorbs heat at this temperature and transition of $Cd(OH)_2$ into CdO takes place. There is almost 33 % weight loss of sample observed in the temperature range of 475 - 640° C, but there is no any phase transition of CdO in this temperature range as shown by DTA.

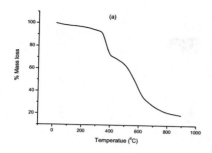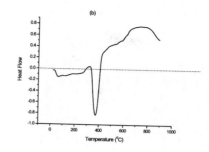

FIGURE 1. (a) TGA spectrum (b) DTA spectrum of as synthesized sample

XRD patterns of (a) as obtained powder and (b) after annealing at 350°C for 9 hours are illustrated in Figure 2. The XRD peaks for as synthesized powder at 2θ = 29.50°, 35.18° and 49° are assigned as [100], [101] and [102] miller indices corresponding to hexagonal symmetry of $Cd(OH)_2$, while that of peaks at 2θ = 33° and 38° are attributed for [111] and [200] indices for the cubic symmetry of CdO nanocrystals (Figure 2 a). Peak at 2θ = 48° corresponds to the unreacted cadmium metal. XRD data of sample annealed at 350°C for 9 hours (Figure 2b) illustrates that almost all the peaks corresponding to $Cd(OH)_2$ disappears and CdO peaks become dominant. Peaks at 2θ = 33°, 38.32° and 55.3° correspond to (111), (200) and (220) Bragg's reflection planes of cubic CdO respectively (JCPDS card No. 5-640). Sharp and intense peaks prove the synthesis of good quality crystalline CdO NPs with 15.9 nm average diameter calculated by using Scherrer's formula $D = (0.9\lambda / \beta \cos\theta)$; where D is the crystalline diameter, λ is X-ray wavelength, β is FWHM at a selected 2θ. Transition of $Cd(OH)_2$/ CdO nanocomposite material into pure CdO NPs at 350 °C, as evidenced by XRD data, strongly supports our TGA and DTA observations.

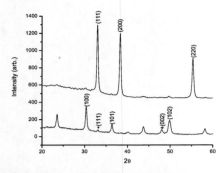

FIGURE 2. XRD spectra of (a) as synthesized and (b) annealed sample

Figure 3 shows UV-VIS absorption spectrum of (a) as-synthesized colloidal solution of NPs and (b) colloidal solution of NPs obtained by dispersing powder annealed at 350°C in methanol. As synthesized solution of NPs illustrates a wide absorption peak centered at 300 nm with a long tail towards a higher wavelength side. The absorption peak lies in between the surface plasmon of Cd metal nanocrystals (243 nm) and that of CdO nanocrystals (344), therefore it may be due to the Cd(OH)$_2$ nanocrystals, which is supported by XRD and DTA analyses.

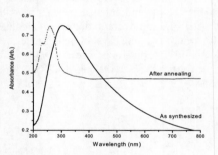

FIGURE 3. UV-VIS Absorption spectra

FIGURE 4. SEM Image of as synthesized sample

Surface morphology of as synthesized dried powder is studied by using SEM and illustrated in Figure 4. SEM image shows that the powder has granular morphology and these are highly agglomerated. The agglomeration is quite obvious as any surfactant is not used.

Figure 5 reveals the FTIR spectrum of as synthesized powder dispersed in KBr. There are poorly resolved shoulders in the IR spectrum at 565 and 715 cm^{-1}, due to the characteristic vibrations of Cd-O. Since the amount of CdO is very small as compared with Cd(OH)$_2$, the IR spectrum of sample shows band for –OH group. Therefore the peaks at 857, 1425.36, 1603 and 3400 cm^{-1} correspond to vibrations of stretching, bending and their overtones of H-OH molecules.

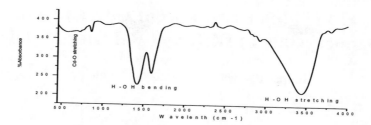

FIGURE 5. FTIR spectrum of as synthesized nanoparticles

CONCLUSIONS

The study demonstrates a new method for the synthesis of CdO/Cd(OH)$_2$ nanocrystals by pulsed laser ablation of cadmium metal plate in water. XRD pattern reveals that as synthesized NPs have maximum amount of Cd(OH)$_2$ which convert in to CdO after annealing at 350°C, also verified by DTA data. This method can provide an alternative, pollution free, way for the synthesis of oxides and hydroxide nanocrystals of other metals. Nanocrystals synthesized by this method have chemical contamination free surfaces, which can be used for biological applications.

ACKNOWLEDGEMENTS

Authors are thankful to Material Research Centre, IISc. Bangalore for TGA and DTA characterization, NCEMP, Allahabad University for XRD, Prof. S.B. Rai, Physics Department, Banaras Hindu University, Varanasi for IR facilities and DRDO, New Delhi for financial support.

REFERENCES

1. Y. Dou, R. G. Egdell, D. S. L. Law, N. M. Harrison and B. G. Searle, *J. Phys.-Condens. Mat.* **10**, 8447 -8458 (1998).
2. T. J. Kuo and M. H. Huang, *J. Phys. Chem. B*. **110**, 13717- 13721 (2006).
3. X Liu, C. Li, S. Han, J. Han and C. Zhou, *Appl. Phys. Lett.* **82**, 1950- 1952 (2003).
4. W.Q. Wang, G.Z. Wang and X. P. Wang, *Int. J. Nanotechnol.* **4**, 110- 118 (2007).
5. H. Yang, G. Qui, X. Zhang, A. Tang and W. Yang, *J. Nanopart. Res.* **6**, 539-542 (2004).
6. S. C. Singh and R. Gopal, *Physica E* **145**, 724 (2008).
7. S. C. Singh and R. Gopal, *J. Phys. Chem. C* **112**, 2812-2819 (2008).

Optical Properties Of Hybrid Composites Based On Highly Luminescent CdS and ZnS Nanocrystals In Different Polymer Matrices

Shipra Pandey[2] and Avinash C. Pandey[1, 2]

[1]Nanophosphor Application Centre. [2]Physics Department,
University of Allahabad, Allahabad, India.
Email- shipra.pandey04@gmail.com

Abstract. The luminescent CdS and ZnS nanocystals embedded in different polymer matrices like poly-vinyl alcohol (PVA), poly-vinyl pyridine (PVP) polymer matrices have been prepared by simple wet chemical precipitation method. The thin films of nanocomposites are deposited on glass substrate by spin-coating. The optical absorption spectra show a clear blue shift in absorption edge, such that the band gap calculated from the absorption spectra are higher than those calculated for the bulk. Photoluminescence spectra show defect related emission at different wavelength in visible range.

Keywords: nanocomposite, nanocrystal, polymer.
PACS: 82.35.Np, 81.05.Qk, 61.46.Hk

INTRODUCTION

II-VI nanocrystals (NCs)/organic polymer composites[1-3] or in silica matrix[4] are receiving considerable attention in recent years for fundamental study due to their size-dependent properties and great potential for many applications such as nonlinear optics, sensors, electroluminescent devices[5] lasers [6] biomedical tags [7], photochemical cells heterogeneous photo catalysis and single electron transistors [8-14]. The polymer matrices provide good processability of organic materials, better solubility and control of the growth and morphology of the nanoparticles. In nanocomposites, organic polymer can not only stabilize the nanoparticle in a solid matrix, but also effectively combine the peculiar features of organic and inorganic components [15]. Due to all these properties the use of polymer is an important method for the synthesis of semiconductor nanoparticles.

The organic-inorganic polymer nanocomposites combining both the properties of inorganic and organic materials offer unique mechanical, thermal and optical features and can apply in versatile areas. Various approaches have been employed to prepare nanoparticles/nanocomposites. Conventionally, polymerization of monomers and formation of inorganic nanoparticles were performed separately and then the inorganic nanoparticles were mechanically mixed to form composites [16]. However, sometimes it is difficult to disperse nanoparticles into the polymeric matrix homogeneously owing to

CP1147, *Transport and Optical Properties of Nanomaterials—ICTOPON - 2009*, edited by M. R. Singh and R. H. Lipson
© 2009 American Institute of Physics 978-0-7354-0684-1/09/$25.00

the easy agglomeration of nanoparticles and the high viscosity of polymers. Therefore, more attention has been paid to the *in situ* synthesis of inorganic nanoparticles in polymer matrices. Reports have been presented by many research groups to embed nanocrystals in polymeric matrix. Lu et al. [17] prepared ZnS NCs/polythiourethane via ultraviolet radiation initiated free radical polymerization. Yao et al [18] reported polymer-controlled growth strategy to prepare CdS nanowires/polymer composite films using ethylenediamine as the reaction medium and PVA as the polymer–controller matrix by the solvothermal method. Alivisataos et al [19] constructed light emitting diodes from cadmium selenide nanocrystals and semiconducting polymer p-paraphenylene vinylene (PPV) hybrid. Pich et al. [20] explored the preparation of ZnS nanocomposites with the method of ultrasonication. Shen et al [21] prepared a PbS/polystyrene nanocomposite with lead methylacrylate as the precursor.

In this paper, we report the synthesis of CdS and ZnS nanocrystal composites in PVA and PVP polymer matrices by simple wet chemical method. The key step in our preparation method is initial capping of nanocrystals in the presence of polymer and then dispersion of capped nanocrystal in thick polymer solution by long term ultrasonication. Thus the properly dispersed nanocrystals in polymer matrices have been prepared without using any other coupling/capping agent.

EXPERIMENTAL DETAILS

Chemicals: Polyvinyl alcohol (PVA), Polyvinyl pyridine (PVP), Cadmium Acetate $Cd(CH_3COO)_2.2H_2O$, Zinc Acetate $Zn (CH_3COO)_2.2H_2O$, Sodium Sulphide (Na_2S). The chemicals were used as obtained without further purification.

Sample preparation: Firstly we prepared the polymeric solution by dissolving 2 gm polymer (PVA or PVP) in 25 ml distilled water at ambient conditions. Then we added drop-wise 25ml aqueous solution of 0.1M cadmium acetate $Cd(CH_3COO)_2.2H_2O$ (99.99 %) in the above prepared polymeric solution. After 10 min, the 50ml aqueous of 0.1M sodium sulfide flakes (Na_2S, 99.99%) was also added in the previous solution. The solution turned yellowish immediately due to the formation of cadmium sulfide in the polymeric (PVA or PVP) matrix. Thus we obtained 100 ml polymer capped CdS nanocrystal solution. Now this polymer capped CdS was mixed with large amount (6gm) of polymer (PVA or PVP) to obtain the required CdS:PVA or CdS:PVP nanocomposite. Same process had been repeated for the ZnS:PVA and ZnS:PVP nanocomposite. The solution was put for ultrasonication for 10 hour. The thin films of as prepared nanocomposites were deposited on glass substrate by spin coating. The glass substrates were ultrasonically cleaned by nitric acid and acetone for 30 min before spin coating process and then dried.

Instrumentation: All the characterization has been done at room temperature. The optical absorption spectra of the nanocomposites have been recorded using Perkin

Elmer Lambda 35 UV-visible spectrometer and the luminescence studies were performed on Perkin Elmer LS 55 spectrophotometer (PL) have been used for photoluminescence study.

RESULTS AND DISCUSSIONS

Particle size and band gap study

The optical absorbance spectra of CdS and ZnS nanocomposites in different polymer matrix are shown in figure 1 and figure 2. The shift in the absorbtion edge is different in each case. In CdS case, the absorption edge is different for CdS, CdS:PVP and CdS:PVA. The spectrum exhibits the absorption edge at ~473nm for CdS, ~441 for CdS:PVP and ~437nm for CdS:PVA, which are considerably blue, shifted relative to the absorption of bulk CdS (515nm). It is known that the semiconductor nanoparticles have unique size dependent chemical and physical properties [22-27].When the size of these nanocrystals become smaller than the exciton radius, a remarkable quantum size effect leads to increase in band gap with blue shift in the absorption onset.

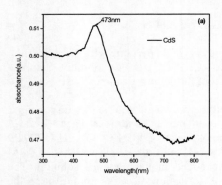 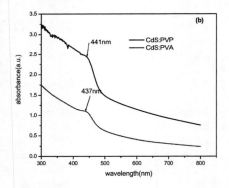

FIGURE 1. Absorption spectra of (a) CdS nanocrystal and (b) CdS nanocomposites.

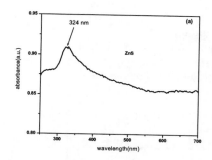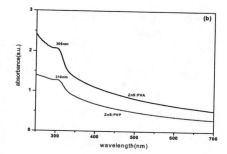

FIGURE 2. Absorption spectra of (a) ZnS nanocrystal and (b) ZnS nanocomposites.

The blue shift of the absorption edge indicates that the particles formed are in the quantum confinement regime, indicating quantum confinement effect [28]. The calculated value of band gap of CdS, CdS:PVP and CdS:PVA corresponding to absorption edge are 2.6eV,2.7eV and 2.8eV respectively.

With these band gap values, the calculation of particle size was done using formula [29]:

$$E_g(QD) = E_{g,0} + \frac{\hbar^2\pi^2}{2m_{eh}R^2}$$

where E_g is the band gap of the nanoparticles, $E_{g,0}$ is the band gap value of the bulk material (2.4eV), and m_e^* (0.19 m_e in CdS), m_h^* (0.9 m_h in CdS) [30] are the effective masses of the electrons and holes respectively. R is the radius of the particle.

$$m_{eh} = m_e^* m_h^* / (m_e^* + m_h^*)$$

The estimated particle size are 8 nm for CdS, 6.6 nm CdS:PVP and 4.8 nm for CdS:PVA nanocomposites.

In the case of ZnS, the shift in absorption edge are also observed which are different in different polymer composites, the spectrum exhibits the absorption peak at ~324 nm for ZnS, at ~310nm for ZnS:PVP and at ~306 nm ZnS:PVA respectively, which are also blue shifted relative to the absorption peak of bulk ZnS (342nm) arising due to quantum confinement effect.

The calculated band gaps for corresponding absorption edge are 3.80eV, 3.98eV and 4.0eV. With the band gap value, the particles radius is calculated, using formula [31]:

$$R(E_g) = \frac{0.32 - 2.9\sqrt{(E_g - 3.49)}}{2(3.50 - E_g)}$$

where, E_g is the band gap of nanocrystal.

The estimated particle size is 4.2 nm for ZnS, 3.6 nm for ZnS:PVP and 3.5 nm for ZnS:PVA respectively. The calculated particle size and band gap of CdS, CdS:PVP, CdS:PVA, ZnS, ZnS:PVP and ZnS:PVA nanocomposites are shown in table 1.

Table 1. The calculated particle size, band gap of CdS, CdS:PVP, CdS:PVA, ZnS, ZnS:PVP and ZnS:PVA nanocomposites.

Sample	absorption edge (nm)	band gap (eV)	calculated particle size (nm)
CdS	473	2.6	8.0
CdS:PVP	441	2.7	6.6
CdS:PVA	437	2.8	4.8
ZnS	324	3.8	4.2
ZnS:PVP	310	3.9	3.6
ZnS:PVA	306	4.0	3.5

Photoluminescence Study

Figure 3(A) shows the photoluminescence spectra of the CdS nanocrystal and their nanocomposites CdS: PVA and CdS:PVP under 390nm excitation wavelength. As a typical semiconductor, CdS nanocrystal exhibits the interesting optical properties. When the CdS nanocrystals are embedded in different polymer matrices, the composites show the enhanced characteristic optical properties of CdS nanocrystals. The photoluminescence spectra show the characteristic emission for CdS nanocrystals at around 481nm and 526nm in the visible range. The emission at 481nm is attributed to the band edge emission, while the 526nm emission is related to the electron hole recombination at deep traps because of sulfur vacancies.

The Figure 3(B) shows the photoluminescence spectra of ZnS nanocrystal and nanocomposite under 300nm excitation wavelength. The characteristic emission of ZnS nanocrystal and its nanocomposite are nearly 417nm and 445nm respectively. The peaks appearing at 417nm and at 445nm are attributed to the sulfur vacancies in the lattice. This emission results to the recombination of photo generated charge carriers in shallow and deep traps.

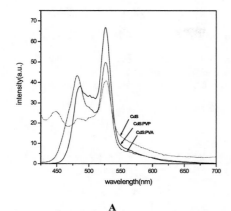

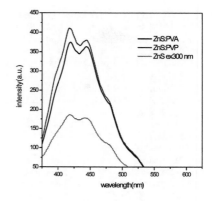

A   **B**

FIGURE 3. Photoluminescence spectra of (A) CdS nanocrystal and nanocomposites, (B) ZnS nanocrystal and nanocomposites.

From the above figure, the enhancements in emission peaks are observed when the nanocrystals embedded in the different polymeric matrices. The nanocomposites exhibit better optical properties in comparision to their bare nanocrystals.

CONCLUSIONS

We have synthesized the different nanocrystals embedded in different polymeric matrices by wet chemical method and investigated the optical properties of organic inorganic nanocomposites. Incorporation of CdS and ZnS nanocomposite in different polymer matrices exhibit better optical properties. The nanocomposites show enhanced emission in visible range in comparison to bare nanocrystal. Absorption spectra show the significant absorption in nanocomposites. The adopted process provides a simple and effective route for the production of polymer based nanocomposites for optical and electronic devices.

ACKNOWLEDGEMENTS

The authors are very grateful to all the scientific members of Nanophosphor application centre, University of Allahabad, Allahabad. This work was financially supported by DST, India.

REFERENCES

1. A. K. Keshari, M. Kumar, P. K. Singh and A. C. Pandey, *J. Nanosci. Nanotechnol.* **8,** 1 (2007).

2. P. K. Sharma, R. K. Dutta, A. C. Pandey, S. Layek and H. C. Verma, *J. Magn. Magn. Mater.* doi:10.1016/j.jmmm.2009.03.043. (2009).
3. P. K. Sharma, M. Kumar, P. K. Singh, A. C. Pandey and V. N. Singh, *IEEE Trans. Nano, article in press* (2009).
4. P. K. Sharma, R. K. Dutta, M. Kumar, P. K. Singh and A. C. Pandey, *J. Lumin.* **129**, 605 (2009).
5. N. Tessler, V. Medvedev, M. Kazas, S. Kan and U. Banin *Science* **295**, 1506 (2002).
6. V. L. Klimov, A. A. Mikhlailowsky, S. Xu, A. Malko, J. A. Hallingsworth, C. A. Leatherdale, H. J. Eisler and M. G. Bawendi *Science* **290**, 314 (2000).
7. M. Han, X. Gao, J. Z. Su and S. Nie, *Nat. Biotechnol.* **19**, 631 (2001).
8. A. P. Alivisatos, *Science* **271**, 933 (1996).
9. G. Fasol, *Science,* **280**, 545 (1998).
10. H. Zhao and E.P. Douglas, *Chem Matter.* **14**, 1418 (2003).
11. K. Murakoshi, H. Hosokawa. M. Saitoh, Y. Wada T. Sakata H. Mori, M. Satoh and S. Yanagida *J. Chem. Soc. Faraday Trans.* **94**, 579 (1998).
12. U. Resch, A.Eychmurller, M. Haase and H. Weller *Langmiur* **8**, 2215 (1992).
13. B. Liu, H. Li, C. H. Chew, W. Que, Y. L. Lam, C. H. Kam, L. M. Gan and G. Q. Xu, *Mater. Lett.* **51**, 461 (2001).
14. J. Y. Kim, H. M. Kim, D. H. Shin, and K. J. Ihn, *Macromol. Chem. Physic.* **207**, 925 (2006).
15. W. Chen., A. G. Joly, J. O. Malm, J. O. Bovin and S. Wang *J. Phys. Chem. B* **107**, 6544 (2003).
16. B. Z. Tang, Y. Geng, J. W. Y. Lam, B. Li, X. Jing, X. Wang F. Wang, A. B. Pakhomov and X. X. Zhang, *Chem. Mater.* **11**, 1581 (1999).
17. C. L. Lu, Ze Cui, Y. Wang, Z. Li and B. Yang, *J. Mater. Chem.* **13**, 2189 (2003).
18. J. X. Yao, G. L. Zhao, D. Wang and G. R. Han, *Mater. Lett.* **59**, 3652 (2005).
19. V. L. Colvin, M. C. Schiamp and A. P. Alivisatos, *Nature* **379**, 354 (1994).
20. A. Pich, J. Hain, Y. Lu, V. Boyko. Y. Prots and H. J. Adler, *Macromolecules* **38**, 6610 (2005).
21. M. Y. Gao, Y. Yang, F.L. Bian and J. C. Shen, *J. Chem. Soc. Chem. Commun.* 2779 (1994).
22. R. S. Kane, R. E. Cohen and R. Silbey, *Langmiur*, **15**, 39 (1999).
23. L. Qi, H. Colfen and M. Antonietti, *Nano Lett.* **1**, 65 (2001).
24. R. Gangopadhyay and A. De, *Chem. Mater.* **12**, 608 (2000).
25. Y.Yang, H. Chen and X. Bao, *J. Cryst. Growth* **252**, 251 (2003).
26. G. Carrot, S. M. Scholz, C. J. G. Plummer, J. G. Hilborn and J. L. Hedrick, *Chem. Mater.* **11**, 3571 (1999).
27. J. Kuczynski and J. K. Thomas, *J. Phys. Chem.* **89**, 2720 (1985).
28. Hongmei Wang, Pengfei Fang, Zhe Chen and Shaojie Wang, *Appl. Sur. Sci.* **253**, 8495-8499 (2007).
29. V. I. Klimov, *Semiconductor and Metal Nanocrystals* Marcel Dekker, Inc, New York. pp. 1-2
30. B. Saraswathi Amma and K. Ramakrishna Manjunatha Pattabi, *J. Mater. Sci.-Mater. El.* **18**, 1109-1113 (2007).
31. J. F. Suyver, S. F. Wuister, J. J. Kelly and A. Meijerink *Nano Lett.* **1**, 8 (2001).

Optical Characterization Of Chemically Deposited Nanostructured CdS Films

Y. C. Goswami and Archana Kansal

Institute of Technology and Management, Sithouli Gwalior M.P. INDIA 474001

Abstract. Newly modified hot chemical deposition method was used to grow Cadmium sulfide films. Substrates were kept at relatively higher temperature than the bath using local heating. The bath was consisting of aqueous solutions of Cadmium chloride, Thiourea and complexed by TEA. The Ph of the bath was maintained around 8-10 by adding ammonia solution. The soda lime glass slides were used as substrates. Good thick films were obtained few minutes. Air annealing was used to study the effect of heat treatment on quality of the films. All films were analyzed using optical spectrophotometer. The step like nature in transmission spectra and band gap curves could be due to discrete energy levels, which exist in nanomaterials. Blue shift is observed in samples. Band gap shift from higher value to lower value suggest that films are either of thickness of few nanometer range and/or grain size is of the nanometer range. This paper includes details about new modified dipping technique and optical, structural studies of these films.

Keywords: CdS films, Semiconductors, Nanostructure, solar cells
PACS: 78.20.-e; 62.23.St; 84.60.Jt

INTRODUCTION

Cadmium sulphide (CdS) is a direst band gap II-VI semiconductor [1]. Cadmium sulphide (CdS) thin films play a very important role in photovoltaic technology and optoelectronic devices. CdS films are regarded as a promising material for heterojunction-thin film solar cells such as cadmium telluride (CdTe), copper indium diselenide ($CuInSe_2$), indium phosphide (InP) and copper sulphide (Cu_2S). These heterojunction thin-film solar cells operate in the efficiency range of 14% to 16%. CdTe/CdS heterojunction solar cells with efficiency of about 16% have been reported [2]. Due to the high material cost involved there is a major drive toward developing polycrystalline compound semiconductors, especially in the form of thin films. CdS thin films has been obtained by several methods such as electrodeposition, vacuum evaporation, sputtering, screen printing, photochemical deposition, CBD, spray pyrolysis and sol gel[3-10], However the basic problem with CdS is to obtain uniformity over a large area and stoichiometry. Chemical bath deposition (CBD) is one of the conventional methods adopted to produce chalcogenide thin-film solar cells using CdS n-type layers. The chemical bath technique appears to be a relatively simple, inexpensive method to prepare a homogeneous film with controlled composition and useful for large area industrial applications. Researchers showed that the CdS layers prepared by solution growth increase their solar energy conversion remarkably [11,12]. CBD is a process to

CP1147, *Transport and Optical Properties of Nanomaterials—ICTOPON - 2009*, edited by M. R. Singh and R. H. Lipson
© 2009 American Institute of Physics 978-0-7354-0684-1/09/$25.00

achieve high quality films which are obtained by adjusting the PH, Temperature and reagent concentrations. However depending on the deposition conditions like pH of the solution, temperature, stirring and time of deposition the quality as well as stiochiometry of the films differ. In this paper we have discussed the modified CBD setup where substrates are kept at relatively higher temperature than solution and its optical characterization.

EXPERIMENTAL DETAILS

CdS films were grown by newly modified hot chemical deposition method Experimental Set up for Chemical Bath Deposition is shown in Fig.1. The bath was consisting of aqueous solutions of cadmium chloride, thiourea and complexed by TEA. The PH of the bath was maintained in the basic region (around 8-10) by adding ammonia solution. The soda lime glass slides were used as substrates. The bath temperature was kept at 50-55°C. Glass and mica slides cleaned with chromic acid, soap and distill water were used as substrates. These Substrates were kept at relatively higher temperature than the bath using local heating. The bath was continuously stirred using magnetic stirrer. Samples were pulled out from solutions at different time intervals.

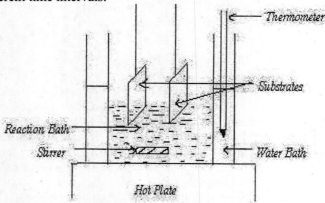

FIGURE 1. Experimental Set up for Chemical Bath Deposition of CdS films

To study the effect of heat treatment on quality of the films the post growth heat treatment was given to some of the samples by annealing in air at 300-400°C. For annealing, the samples were mounted on a horizontally placed hot plate and the plate heated to the required temperature. The transmission data of as grown and annealed samples were obtained by single beam UV-Vis spectrophotometer 119 Systronics. The optical transmission of CdS films on glass support was measured at near normal incidence in the 300-800 nm wavelength ranges. Optical band gap energy (Eg) was determined graphically after extrapolation of the plot at α =0 using the standard expression for direct transiotion between two parabolic bands $(\alpha \, hv)^2$ =A(hv-E_g)[13].

RESULTS AND DISCUSSION

Good yellow thick films were obtained few minutes. Since the substrate were kept at relatively higher temperature than solution so the maximum deposition was occur only on substrates than anywhere else.

Fig.2 show the Optical transmission spectra for chemical bath deposited CdS films before and after annealing .The pattern of interference films suggests that films are adherent and have enough uniform thickness. The corresponding $(\alpha \, hv)^2$ versus hv plot whose intersect on the hv axis give the direct band gap is shown in the Fig.3. These figures clearly show a shift in the band gap of the CdS after annealing . The reduction in the band gap shows that the grain size increase on annealing [14, 15]. This clearly shows that the films are nanocrystalline as bulk material does not exhibit this change in gap with grain size. The transmission and band gap curve also show a step like nature in the region where absorption occurs. This step like nature could be due to the discrete energy levels and step like density of states for nanomaterials instead of quasi-continuous levels in the conduction and valance band for bulk materials.

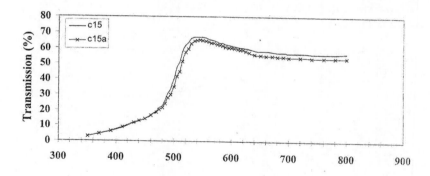

FIGURE 2. Transmission curves for chemical bath deposited CdS thin films (a) as grown (b) annealed.

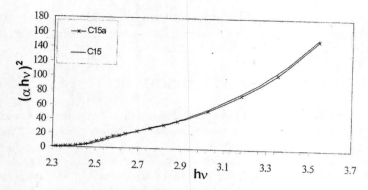

FIGURE 3. $(\alpha h\nu)^2$ vs hν graph for chemical bath deposited CdS thin films.

Fig 4 shows the optical transmission spectra for three CdS samples with different growth time of 15, 30 and 75 minutes. The corresponding $(\alpha\ h\nu)^2$ versus hν plots are shown in Fig. 5. The transmission spectra as well as the intercepts on the Fig 4 clearly shows that the band gap shifts from higher values to lower values as the time duration increases. This suggests that the shift in the bandgap could be due to an increase in the grain size (which is in the nanorange) and/or and increase in the film thickness (Which could be in the nanometer range) this shifty suggests that the samples are nanocrystalline. It is also clear from the graph that absorbance increases with time this suggests that the changes in transmittance arise actually due to increase in thickness.

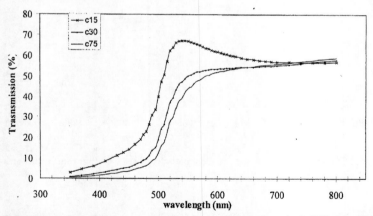

FIGURE 4. Transmission curves for chemical bath deposited CdS thin films grown in different time intervals.

226

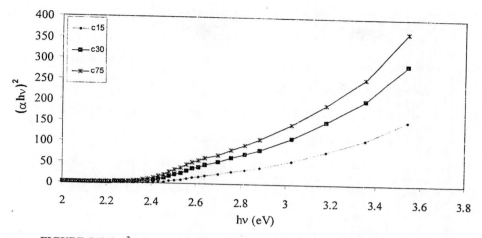

FIGURE 5. $(\alpha h\nu)^2$ vs hν graph for chemical bath deposited CdS thin films at different time intervals.

CONCLUSIONS

The CdS films were prepared by newly modified hot chemical deposition method in which substrates were kept at relatively higher temperature than the bath using local heating. Bath was maintained at lower temperature that makes this method less hazardous than other. The step like nature in transmission and band gap curves could be due to discrete energy levels which exist in nanomaterials. Blue shifts are observed in samples. Band gap shift from higher values to lower value suggest that films are either of thickness of few nanometer range and/or grain size is of the nanometer range.

ACKNOWLEDGEMENTS

Authors are thankful to Department of Electronics and Instrumentation, ITM for providing characterization facility. They are also thankful to Shri Ramashankar Singhji Chairman ITM Universe for providing financial assistance, all kind of cooperation and encouragement to the project.

REFERENCES

1. V. Popescu, E. M. Pica and I. Pop, and R. Grecu, *Thin Solid Films* **349**, 67-70 (1999).
2. J. Britt and C Ferekides, *Appl. Phys. Lett.* **62**, 2851 (1993).
3. J. Torres and G. Gordillo, *Thin Solid Films* **207**, 23 (1962).
4. J. Santanamaria, *Sol. Cells* **28**, 31 (1990).

5. C. Nascui, V. Lonecu, E. Indrea and I. Bratu, Mat. Lett. **32** 73 (1997).
6. J. G. Ibanez, O. Solorza and E. Gomez, *J. Chem. Educ.* **68** 872 (1991).
7. A. M. Andriesh, V. I. Verlan and L. A. Malahoca, *J. Optoelectron. Adv. Mater.* **5**, 817 (2003).
8. Y. Hashimoto, N. Kohara, I. N. Nishitani, and T. Wada, *Sol. Energ. Mat. Sol. C.* **50**, 71 (1998).
9. A. V. Feitosa, M. A. Miranda, J. M. Sasaki and M. A. Araujo Silva, *Braz. J. Phys.* **34**, 656-658 (2004).
10. M. Zelner, H. Minti, R. Reisfeld, H. Cohen and R. Tenne, *Chem. Mater.*, 9 (11),(1997), 2541
11. W. E. Devaney, W. S. Chen and R. A. Micvkelsen, *IEEE Trans. Electron. Dev.* **37**, 428 (1990).
12. Ramanujam Kumaresan, Masaya Ichimura, Ken Takahashi, Kazuki Takeuchi, Fumitaka Goto and Eisuke Arai, *Jpn J. Appl. Phys.* **40**, 3161 (2001).
13. P. Rajaram, Y. C. Goswami, S. Rajagoplan and V. K. Gupta, *Mat. Lett.* **54** 158-163 (2002).
14. D. V. Petrov, B. S. Santos, G. A. L. Pereira, and C. de Mello Donega, *J. Phys. Chem. B* **106**, 5325-5334 (2002).
15. N. Venkatram, D. Narayana Rao and M. A. Akundi, *Opt. Express* **13**, 867 (2005).

Color Tunability and Raman Investigation in CdS Quantum Dots

Ashish K. Keshari* and Avinash C. Pandey

*Nanophosphor Application Centre, Department of Physics,
University of Allahabad, Allahabad-211 002, India.
Tel. /Fax: +91-532-2460675
* Corresponding author e-mail: jeevaneshk26@yahoo.co.in*

Abstract: CdS quantum dots (QDs) with improved luminescence properties places it among active area of research for their exploitation in appealing application in next generation opto-electronic devices and in photonics. We present here the tunability of emission spectra in CdS QDs in whole visible spectrum i.e. from violet to red region by proper choice of the synthesis temperature and photoluminescence excitation wavelength. Different luminescence behavior is observed at low temperature and at higher excitation energy. Raman spectra show no blue shift in longitudinal optical (LO) phonon bands due to phonon confinement with synthesis temperature. There is noticeable asymmetry in the line shape indicating the effect of phonon confinement which confirms the small crystallites of good quality.

Keywords: CdS; Quantum dots; Semiconductors; Photoluminescence; Raman spectra
PACS : 61.46.Hk, 78.67.Bf, 81.07.Bc

INTRODUCTION

The development of faster and smaller opto-electronic devices demands a continuous decrease of the element sizes and resulted in remarkable progress in electronics, data processing and communication techniques. The aim of this trend is not only to increase the integration level, but mainly, to increase the operation speed. Commercial requirements for miniaturized microelectronic devices provide strong motivation for exploring the synthesis of nanoscale systems using bottom-up techniques. II-VI nanocrystals (NCs) are receiving considerable attention for fundamental studies, as an example of zero-dimensional quantum confined material and for their exploitation in appealing applications in opto-electronics and photonics [1]. The NC based emitters can be used for many purposes such as optical switches, sensors, electro-luminescent devices [2], lasers [3] and biomedical tags [4]. Nanometer size quantum dots exhibit a wide range of electrical and optical properties that depends sensitively on the size of the nanocrystals and are of both fundamental and technological interest. It is, therefore, possible in principal to manipulate the properties of the nanomaterials for specific application of interests by designing and controlling the parameters that affect their properties. There is great departure in optical properties from their bulk counterparts [1, 5-6] due to quantum size effects in nanocrystals which leads to tunable blue shifts in both optical absorption

CP1147, *Transport and Optical Properties of Nanomaterials—ICTOPON - 2009,* edited by M. R. Singh and R. H. Lipson
© 2009 American Institute of Physics 978-0-7354-0684-1/09/$25.00

and emission spectra with decreasing nanocrystal size. Indeed it is possible to synthesize differently sized NCs emitting from blue to red and up to near IR with a very narrow band width [7-8]. The optical properties get modified dramatically due to the confinement of charge carriers within the nanoparticles. Similar to the effects of charge carriers on optical properties, confinement of optical and acoustic phonons leads to interesting changes in the phonon spectra. For this reason the control and improvement of the luminescence properties of quantum dots (QDs) has been a major goal in synthetic 'nanochemistry' and related preparative procedures [9]. The recombination processes of photogenerated carriers in semiconductor nanocrystals are important to their applications in opto-electronic devices. The PL technique is widely used to investigate both radiative and non-radiative transitions of carriers in nanocrystals. The band edge PL emissions were attributed to various recombination mechanisms such as the donar-acceptor pair emission [10] and the recombination luminescence of shallow traps [11-12] and excitons [12-13]. The CdS is an important semi-conducting material that has attracted much interest owing to their unique electronic and optical properties, and their potential applications in solar energy conversion, photoconducting cells, nonlinear optics and heterogeneous photocatalysis [14-16].

EXPERIMENTAL

In this work, we report a study of the temperature dependent PL spectra of CdS semiconductor NCs for different excitation wavelength. The color tunability of this semiconductor nanoparticle with proper choice of the synthesis temperature and excitation wavelength is one of the most attractive investigations. Details Raman investigation were also performed. The CdS nanocrystals at four different set temperatures viz. 4, 25, 35 and 45^0C were synthesized by chemical precipitation route. The synthesis procedure was reported elsewhere [17] by excluding the capping agent PVA from the reaction mechanism. It is observed that color of the precipitate obtained is changes from light green to orange-red as we change the Rx temperature from 4 to 45^0C.The materials were characterized by Photoluminescence and Raman spectroscopy immediate after the synthesis. PL spectra of all the prepared samples were recorded from Perkin-Elmer LS-55 spectrophotometer at 230, 325, 375 and 425 nm excitation wavelengths from Xenon lamp respectively. Raman spectra were recorded from Renishaw micro-Raman setup described elsewhere in more details [18]. For Raman excitation, radiation of wavelength 514 nm from Ar ion laser was used.

RESULTS AND DISCUSSION

The Raman spectra of as prepared CdS samples were recorded as shown in Fig. 3. The Raman signals are strong for CdS-like longitudinal optical (LO) phonon mode located at

297 and 594 cm^{-1} with overtone progression for all synthesis temperatures and in agreement with what expected by the two-mode behavior of the lattice vibrations in CdS alloy. There is no shift in the phonon band are observed with reaction (Rx) temperature however Raman intensity becomes stronger as we increase the Rx temperature. Both first and second harmonics of the LO CdS phonon can be observed by Raman spectroscopy. There is noticeable asymmetry in the line shape indicating the effect of phonon confinement. An asymmetric broadening of the Raman peak caused by the small crystal size can be observed as well. This shows that the semiconductor crystalline quality is fairly good in comparison with that reported [19]. The broadening of the half-widths compared with the bulk material could have two reasons: (1) the size distribution of the quantum dot (QD) and the presence of defects, (2) the contribution of phonon confinement. The phonon confinement only contributes to a slightly asymmetric line shape, which results from the relaxation of the k=0 conservation momentum rule. In view of these results the Raman spectra shown in Fig. 3 indicate a better crystal quality for QDs considering the half-widths and overall intensity. Shiang *et al.* [20] pointed out a connection between the width of the overtone lines and the vibrational relaxation. In accordance with this model a relaxation mechanism dominated by the decay of LO phonons into acoustic phonons can be concluded for our samples. A dephasing mechanism would result in a square dependence between half-widths and the order of the phonon bands.

Similar to the effects of charge confinement on the optical spectra, confinement of optical and acoustic phonons leads to the interesting changes in the phonon spectra. Confinement of the acoustic phonons may lead to the appearance of new modes in the low-frequency Raman spectra, whereas optical phonon line shapes develop marked asymmetry [21-22]. The spectra in Fig. 3 exhibit strong but broad peak at ~ 297 cm^{-1} corresponding to the LO phonon mode. This peak has also slightly shifted to lower frequency compared to the LO mode of bulk CdS (305 cm^{-1}). In bulk crystals, the eigenstate is a plane wave and the wave vector selection rule for first-order Raman scattering requires $q \approx 0$. However, the confinement of the phonon to the volume of the nanocrystals results in the relaxation of the conservation of crystal momentum in the process of creation and decay of phonons. This relaxation of the $q \approx 0$ selection rule results in the additional contribution of the phonon with $q \neq 0$ which causes the asymmetric broadening and low frequency shift of first-order LO-Raman peak.

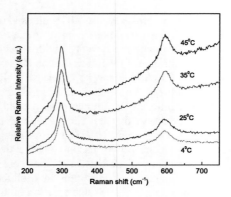

FIGURE 1. Raman spectra of CdS nanoparticles for different synthesis temperature. The wavelength of excitation laser line is 514 nm.

The photoluminescence spectra (PL) of the CdS nanoparticles synthesized at different temperatures viz. 4, 25, 35 and $45^{O}C$ were recorded at excitation wavelength of 230, 325, 375 and 425 nm. The selected characteristics spectra are shown in Fig 2. It is seen that there is a pronounced change in the characteristics and intensity of the spectra for a particular combination of reaction temperature and excitation wavelength. There is a degradation in intensity is observed in almost all emission either with increasing reaction temperature or with increasing excitation wavelength while the general features of the excitonic emission remains unaltered. The different emission features observed at different condition are listed in the table1. The characteristic band edge green emissions are observed to be varying from 512 nm to 532 nm explicitly depends upon reaction temperature and excitation wavelength. The high energy band in the green region also referred to as band edge luminescence is related to various radiative mechanisms [10-11]. Decreasing the excitation energy results in a blue shift and broadening the band edge luminescence perhaps due to band filling. The temperature dependence of the exciton emission (band edge) energy is expected to follow the band gap of bulk material. The similar dependence may also be observed in the recombination processes of the shallow traps. Thus it is difficult to distinguish whether the band edge emission originates from the recombination of excitons or shallow traps. The recombination of shallow traps is dominant at weak excitation intensities and saturates at higher excitation intensities [23], where the Luminescence of excitons becomes dominant. The PL intensities of the band edge luminescence at different excitation wavelength decreases rapidly with increasing reaction temperature. Although the PL spectra of grown CdS generally depends on the growth characteristics, the band edge luminescence is possibly composed of recombination luminescence of shallow traps or defect level of interstitial sulfur [24], localized excitons and intrinsic excitons. The PL bands placed at lower energies indicate another type of defect such as Cd or S vacancies, interstitial Cd etc. In general, the variation in the energy gap with temperature is believed to results from the following two mechanisms: (1) lattice dilation, which causes a linear effect with temperature, and (2) temperature dependant electron-phonon interactions.

The spectra generated by the higher excitation energies at 4°C display two strong, broad peak centered at 420 nm and 645 nm plus weak green emission at 525 nm (Fig 2(a)). Realizing that our CdS nanoparticles sample may not be strictly mono-dispersed. The low intensity of the spectrum for 325, 375 and 425 nm excitations indicate that the CdS nanoparticles are not sufficiently excited at this excitation energy: A few reports [25] on the effect of change in the excitation wavelength on PL spectra of CdS and CdSe nanoparticles are reported. Rodrigues et al.[26] has reported the band edge emission to be strongly dependent on the excitation photon energy. This idea was first reported by Tews et al.[27] to explain the dependence of exciton spectra on excitation photon energy. The observation of these selectively excited PL depends very much on the size distribution of nanocrystals. If the distribution is very broad, a large number of particles of different sizes will always be excited. Hence a broad PL spectrum with no distinct features will be observed independent of photon energy. But, if the distribution is extremely narrow, the emission peak will always be occur at the same energy determined by unique crystallite size. In the intermediate case of distribution which is not too broad or too narrow, appropriate excitation energy can excite several nanocrystals simultaneously producing PL spectrum which contains more than one peak. So the present sample may not be strictly mono-dispersed.

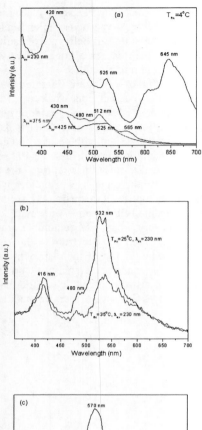

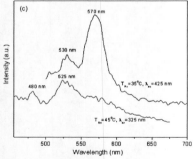

FIGURE 2. PL spectra of CdS nanoparticles recorded for different synthesis temperature at different excitation wavelengths of radiation from a Xenon lamp.

TABLE 1. The different emission features of CdS nanoparticles are observed for different synthesis temperature at different excitation wavelength. Corresponding visible color band are also shown.

Reaction temperature (^{O}C)	Excitation wavelength (nm)	Emission band (nm)	Visible region (Color)
4	230	420	Violet
		525	Green
		645	Orange-red
	375	430	Violet
		480	Blue
		512	Green
	425	525	Green
		565	Yellow
25	230	416	Violet
		480	Blue
		532	Green
35	230	416	Violet
		480	Blue
		532	Green
	425	530	Green
		570	Yellow
45	325	480	Blue
		525	Green

A lot of efforts have been spent to study the luminescence of CdS nanoparticles. Liu *et al.* reported that there were two emission bands. One is the green emission peak at 552 nm, the other is the broad red emission at 744 nm [28], Xu *et al.* found there were two luminescence peaks at 680 nm and 760 nm (IR), which were distributed to the formation of the sulfur vacancies and Cd-S composite vacancies respectively [29]. Moore *et al.* believed that Q-CdS showed the band edge PL peak centered at 450 nm [30].

CONCLUSIONS

Raman spectra show no blue shift in LO phonon bands due to phonon confinement with synthesis temperature. Line shape asymmetry due to phonon confinement confirms the prepared CdS nanoparticles are of small crystallites of good quality. Emission spectra of CdS QDs could be easily tuned in whole visible spectrum simply by proper choice of the synthesis temperature and PL excitation energy.

ACKNOWLEDGEMENTS

We thank Department of Science and Technology, India for funding the project under IRHPA in collaboration with Nanocrystals Technology, New York.

REFERENCES

1. A. P. Alivisatos, *Science* **271**, 933 (1996).
2. N. Tessler, V. Medvedev, M. Kazes, S. Kan and U. Banin, *Science*, **295**, 1506 (2002).
3. V.L. Klimov, A.A. Mikhailowsky, S. Xu, A. Malko, J. A. Hallingsworth, C.A. Leatherdale, H. J. Eisler and M.G. Bawendi, *Science* **290**, 314 (2000).
4. M. Han, X. Gao, J. Z. Su and S. Nie, *Nat. Biotechnol.* **19**, 631 (2001).
5. A.P. Alivisatos, *J. Phys. Chem. B* **100**, 13226 (1996).
6. Review articles on colloidal nanocrystals *Acc. Chem. Res.* **32**, 387 (1999).
7. X. Zhong, M. Han, Z. Dong, T.J. White, and W. Knoll *J. Am. Chem. Soc.* **125** 8589 (2003).
8. D. Battaglia, and X. Peng, *Nano Lett.* **2**, 1027 (2002).
9. L. Qu and X. Peng, *J. Am. Chem. Soc.* **124**, 2049 (2002).
10. N. Chestnoy, T.D. Haris, R. Hull and L.E. Brus, *J. Phys. Chem.* **90**, 3393 (1986).
11. A. Eychmuller, A. Hasselbarth, L. Katsikas and H. Weller, *J. Lumin.* **48&49**, 745 (1991).
12. M. O'Neil, J. Marohn and G. McLendon, *J. Phys. Chem.* **94**, 4356 (1990).
13. B.G. Potter Jr. and J.H. Simmons, *Phys. Rev.* **B37**, 10838 (1988).
14. K. Hu, M. Brust and A. J. Bard, *Chem. Mater.* **10**, 1160 (1998).
15. L. E. Brus, *J. Phys. Chem.* **90**, 2555 (1986).
16. A. Henglein, *Chem. Rev.* **89**, 1861 (1989).
17. A. K. Keshari, M. Kumar, P. K. Singh, and A. C. Pandey, *J. Nanosci. Nanotechnol.* **8**, 301-308 (2008).

18. B. Schreder, T. Schmidt, V. Ptatschek, U. Winkler, A. Materny, E. Umbach, M. Lerch, G. Muller and L. Spanhel, *J. Phys. Chem. B* **104**, 1677 (2000)
19. Y. A. Vlasov, V. N. Astratov, O. Z. Karimov, A .A. Kaplyanskii, V. N. Bogomolov, and A. V. Prokofiev, *Phys. Rev.* **B 55**, 13357 (1997)
20. J.J. Shiang, S.H. Risbud, A.P. Alivisatos, *J. Chem. Phys.* **98**, 8432 (1993)
21. I. H. XCampbell and P .M. Fauchet, *Solid State Commun.* **58**, 739(1986)
22. P. Nandakumar, C. Vijayan, M. Rajalakshmi, A. K. Arora and Y .V .G. S. Murti, *Physica* **E11**, 377 (2001)
23. J. Zhao, K. Dou, Y. Chen, C. Jin, L. Sun, S. Huang, J. Yu, W. Xiang and Z. Ding, *J. Lumin.* **66&67**, 332 (1996)
24. O. Vogil, I. Reich, M. Garcia-Rocha and O. Zelaya-Angel, *J. Vac. Sci. Technol.* A **15**, 282 (1997)
25. S. Okamotu, Y. Kanemitsu, H. Hosukawa, K. M. Koshi and S. Yanagida, *Solid State Commun.* **105**, 7 (1998)
26. P.A.M. Rodrigues, G. Tammulaitis, P. Y. Yu and S. H. Risbud, *Solid State Commun.* **94**, 583 (1995)
27. H. Tews, H. Venghaus and P.J.Dean, *Phys. Rev.* **B19**, 5178 (1979)
28. B. Liu, G.Q. Xu, L.M. Gan, C.H. Chew, W. S. Li, Z. X. Shen, *J. Appl. Phys.* **89**, 1059 (2001)
29. G.Q. Xu, B. Liu, S.J. Xu C. H. Chew, S.J.Chua, L.M. Gana, *J. Phys. Chem. Solids* **61**, 829 (2000)
30. D.E. Moore and K. Patel, *Langmuir* **17**, 2541 (2001)

Quantum regression formula and luminescence spectra of two coupled modes under incoherent continuous pumping

Elena del Valle*, Fabrice Laussy† and Carlos Tejedor*

*Dep. de Fisica Teorica de la Materia Condensada, Universidad Autonoma de Madrid, Spain
†School of Physics and Astronomy, University of Southampton, United Kingdom

Abstract. We study the quantum regression formula for two coupled dissipative modes in the steady state under incoherent continuous pumping. We analyze the equations for one and two-time correlators, needed to compute the spectra of emission of the system, for two coupled harmonic oscillators (linear model), on the one hand, and two coupled two-level systems, on the other hand. We present a comparison between them, on the basis of fully analytical results.

Keywords: strong coupling,quantum dots,microcavities,decoherence,luminescence spectra
PACS: 42.50.Ct, 78.67.Hc, 42.55.Sa, 32.70.Jz

Introduction

The aim of this text is to make a comparative study of the coupling between two modes, in the cases where they are both bosonic (harmonic oscillators [HO] with field operators a and b) on the one hand, or fermionic (two-level systems [2LS] with field operators α and β) on the other hand. The first case is the so called *linear model* [1], that we used in Ref. [2] to analyze the physics of strong coupling of a quantum dot in a microcavity in presence of pump and decay. The second case is its simplest extension, that can still be solved fully analytically, in stark contrast with the widely used Jaynes-Cummings model [3] that, under incoherent pumping, must be solved numerically [4]. Both models converge in the so-called *linear regime* which is the limit of low excitation with less than one particle in the system, and depart as pumping is increased with Bose-enhancement on the one hand vs. Pauli blocking on the other. We overview the general theory to compute exactly single and two-time correlators using the *quantum regression formula* (QRF) [5], from which we obtain averages and luminescence spectra. We analyze how the two models give rise to two different notions of Strong-Coupling (SC).

Master equation, QRF and the spectra of emission

The direct coherent coupling between two modes in the same point of space is described by the well known Hamiltonian (in the Rotating Wave Approximation):

$$H = \omega_a a^\dagger a + \omega_b b^\dagger b + g(ab^\dagger + a^\dagger b), \tag{1}$$

CP1147, *Transport and Optical Properties of Nanomaterials—ICTOPON - 2009*, edited by M. R. Singh and R. H. Lipson
© 2009 American Institute of Physics 978-0-7354-0684-1/09/$25.00

(with a, b replaced by α, β for the fermionic case throughout). The two coupled modes under study are damped due to residual coupling with the environment. We assume that they are excited in an incoherent continuous way. The flow of incoming and outgoing particles are given by the rates P_a, P_b and γ_a, γ_b, respectively. The coherent Hamiltonian dynamics is constantly interrupted by the incoherent and random arrival or departure of particles. The steady state (SS) is a statistical mixture of many quantum states and, therefore, it is given by a density matrix that follows a *master equation*. We define the Liouvillian $\mathscr{L}^c$ that acts on the density matrix through the jump operator c as $\mathscr{L}^c = 2c\rho c^\dagger - c^\dagger c\rho - \rho c^\dagger c$ and consider the general master equation

$$\frac{d\rho}{dt} = i[\rho, H] + \frac{\gamma_a}{2}\mathscr{L}^a\rho + \frac{\gamma_b}{2}\mathscr{L}^b\rho + \frac{P_a}{2}\mathscr{L}^{a^\dagger}\rho + \frac{P_b}{2}\mathscr{L}^{b^\dagger}\rho. \tag{2}$$

All together, these elements can be put in the form of a total superoperator $\mathscr{L}$ that allows to write the master equation as $d\rho/dt = \mathscr{L}\rho$. In what follows, we consider that these parameters can take any value as long as they drive the system to a steady state.

The *luminescence spectrum* $s_a(\omega) = \langle a^\dagger(\omega)a(\omega)\rangle$ is the mean number of a-particles in the system with frequency ω. It is proportional to the intensity of particles emitted at this frequency. In the SS, that we take as the origin of time, $t = 0$, it reads $s_a(\omega) = \frac{1}{2\pi}\Re\int_0^\infty\langle a^\dagger(0)a(\tau)\rangle e^{i\omega\tau}d\tau$. The QRF provides a method to compute any two-time correlator from a master equation of the form (2). As explained in textbooks [6], one must find a set of operators $C_{\{\eta\}}$ and the *regression matrix* M that satisfy $\mathrm{Tr}(C_{\{\eta\}}\mathscr{L}\Omega) = \sum_{\{\lambda\}}M_{\{\eta\lambda\}}\mathrm{Tr}(C_{\{\lambda\}}\Omega)$ for a general operator Ω. Then, the equations of motion for the two-time correlators (for $\tau \geq 0$) read $\frac{d}{d\tau}\langle\Omega(0)C_{\{\eta\}}(\tau)\rangle = \sum_{\{\lambda\}}M_{\{\eta\lambda\}}\langle\Omega(0)C_{\{\lambda\}}(\tau)\rangle$.

In the general problem of two coupled modes, a, b, we refer with the label $\{\eta\} = (m, n, \mu, \nu)$ to the two-time correlator $\langle\Omega(0)C_{\{\eta\}}(\tau)\rangle$ with $C_{\{\eta\}} = a^{\dagger m}a^n b^{\dagger\mu}b^\nu$. The two-time correlators are grouped in *manifolds* $\mathscr{N}_k$, where k is the minimum number of particles that should be in the system (regardless of the regression matrix) so that the correlator can be nonzero. If $\Omega = a^\dagger$, the correlators will have as initial condition the SS mean values $\langle C_{\{m+1,n,\mu,\nu\}}\rangle$. We can find them also applying the QRF with $\Omega = 1$ and the new set of operators $\{\tilde\eta\} = (m+1, n, \mu, \nu) \in \mathscr{N}_k$. The final result for the correlator of interest is always of the form:

$$\langle a^\dagger(0)a(\tau)\rangle = \sum_p (l_p^a(t) + ik_p^a(t))e^{-i\omega_p\tau}e^{-\frac{\gamma_p}{2}\tau}, \tag{3}$$

where weights, l_p^a, k_p^a, frequencies ω_p and effective decay rates γ_p, are all real. The spectrum reads:

$$s_a(\omega) = \frac{1}{\pi}\sum_p\left[l_p^a\frac{\frac{\gamma_p}{2}}{\left(\frac{\gamma_p}{2}\right)^2 + (\omega - \omega_p)^2} - k_p^a\frac{\omega - \omega_p}{\left(\frac{\gamma_p}{2}\right)^2 + (\omega - \omega_p)^2}\right], \tag{4}$$

which is a sum of many peaks, that we label with p, each with two contributions: a Lorentzian and a dispersive part. ω_p and γ_p are the line positions and broadenings. They originate from the energy level structure and uncertainties, whose skeleton is the Hamiltonian eigenstates, but that can be greatly distorted by decoherence. As such, they

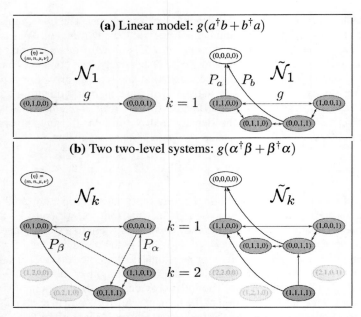

FIGURE 1. Chain of two-time and one-time correlators, $\mathcal{N}_k$ and $\tilde{\mathcal{N}}_k$, respectively, linked in the QRF by the incoherent pump (in green/blue) and the coherent coupling (red) for the LM (a) and coupled 2LSs (b).

are independent of the channel of detection (s_a or $s_b = \langle b^\dagger(\omega)b(\omega)\rangle$). Coefficients l_p and k_p depend on the single-time mean values $\langle C_{\{\tilde{n}\}}\rangle$ and, therefore, on the channel of emission that determines $\Omega = a^\dagger$ or $b^\dagger$. They select the lines that actually appear in the spectra, and with which intensity, depending on the channel of emission and the quantum state of the system.

As a first approximation to the problem, we consider the uncoupled bosonic (fermionic) field, a (α), in which case the spectrum of emission is simply a Lorentzian broadened by an effective bosonic (fermionic) decay rate:

$$\Gamma_a = \gamma_a - P_a = \frac{\gamma_a}{1+n_a} \quad \text{(bosons)}, \quad \Gamma_\alpha = \gamma_\alpha + P_\alpha = \frac{\gamma_\alpha}{1-n_\alpha} \quad \text{(fermions)}, \quad (5)$$

with $n_a = \langle a^\dagger a \rangle$. The nature of the particles is clear from the bosonic, $1+n_a$ (or fermionic, $1-n_\alpha$) factors, resulting in a narrowing (broadening) of the linewidth with the pump as the number of particles increases. From another point of view, n_a diverges if $P_a \rightarrow \gamma_a$ while n_α saturates to 1.

The linear model

In the LM, both modes a, b are bosonic, and an analytic solution is possible for correlators and spectra. The general regression matrix for the linear problem is given by the rules:

$$M_{\substack{mn\mu\nu \\ mn\mu\nu}} = i\omega_a(m-n) + i\omega_b(\mu-\nu) - (m+n)\frac{\gamma_a - P_a}{2} - (\mu+\nu)\frac{\gamma_b - P_b}{2},$$

$$M_{\substack{mn\mu\nu \\ m-1,n-1,\mu\nu}} = P_a mn, \quad M_{\substack{mn\mu\nu \\ m,n,\mu-1,\nu-1}} = P_b\mu\nu, \quad M_{\substack{mn\mu\nu \\ mn-1,\mu\nu+1}} = -ign,$$

$$M_{\substack{mn\mu\nu \\ m+1,n\mu-1,\nu}} = ig\mu, \quad M_{\substack{mn\mu\nu \\ m-1,n\mu+1,\nu}} = igm, \quad M_{\substack{mn\mu\nu \\ mn+1,\mu\nu-1}} = -ig\nu, \tag{6}$$

and zero everywhere else. However, in order to compute $\langle a^\dagger(0)a(\tau)\rangle$, we only need the subset of correlators $\{0,n,0,\nu\} \rightarrow \{n,\nu\}$ with a regression matrix defined simply by:

$$M_{\substack{n\nu \\ n\nu}} = -i(n\omega_a + \nu\omega_b) - n\frac{\Gamma_a}{2} - \nu\frac{\Gamma_b}{2} \quad \text{and} \quad M_{\substack{n\nu \\ n+1,\nu-1}} = M_{\nu n\nu-1,n+1} = -ig\nu. \tag{7}$$

Furthermore, the links between correlators in the LM are self truncated by manifolds of excitations and it is enough to consider the first one, $\mathcal{N}_1 = \{\langle a^\dagger(0)a(\tau)\rangle, \langle a^\dagger(0)b(\tau)\rangle\}$. In Fig. 1(a) we can see a scheme of this finite set of two-time correlators ($\mathcal{N}_1$, left) and the one-time mean values associated ($\tilde{\mathcal{N}}_1 = \{n_a, n_b, n_{ab}, n_{ab}^*\}$, right). The thick red arrows indicate which elements are linked by the coherent (SC) dynamics, through the coupling strength g, while the green/blue thin arrows show the connections due to the incoherent pump P_a/P_b. The sense of the arrows indicates which element is "calling" which in its equations. The self-coupling of each node to itself is not shown. This is where $\omega_{a,b}$ and $\Gamma_{a,b}$ enter from the diagonal elements of the regression matrix. Higher manifolds (not plotted), that include higher order correlators, increase their dimension as $k(k+1)$ and $(k+1)^2$, respectively. A manifold k is only linked directly to $k-1$ in this model and this is why we can solve it exactly. But in this system there is no saturation for the excitation, it can be pumped until divergence (the equivalent of $P_a \rightarrow \gamma_a$ in the single HO). The parameters for which the system reaches a SS are those satisfying $\Gamma_a + \Gamma_b > 0$ and $4g^2 > -\Gamma_a\Gamma_b$. In Fig. 2 we can see that the spectra (in a dashed red line) is simple: two *Rabi peaks* splitted in the SC regime (in thin red lines). SC is only determined by $4g > |\Gamma_a - \Gamma_b|$, therefore the pump, usually overlooked in the literature on SC, is as crucial as the decay.

Two coupled two-level systems

The most general set of operators one can create with two 2LS is restricted by Pauli blocking ($\sigma_1^\dagger \sigma_1^\dagger = 0$) and only correlators with indices $m, n, \mu, \nu \in \{0,1\}$, exist. The

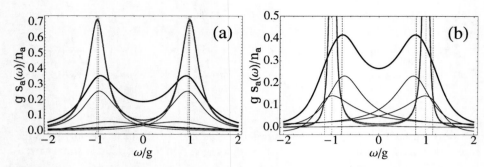

FIGURE 2. Normalized spectra of emission $s_a(\omega)/n_a$ for the LM (dashed red) and $s_\alpha(\omega)/n_\alpha$ for the coupled 2LS (solid black) in SC. Their composing respective 2 and 4 peaks [Eq. (4)], are plotted in thin lines with corresponding colors. In the example (a), SC manifests as two peaks in both models ($\gamma_a = g$ and $P_a = P_b = \gamma_b = 0.1g$). In example (b), the dressed modes of the 2LSs undergo a second splitting and the resulting doublet is distorted ($\gamma_a = P_b = 0.4g$, $P_a = 0$ and $\gamma_b = 0.1g$).

regression matrix is defined by

$$M_{\substack{mn\mu v \\ mn\mu v}} = i\omega_\alpha(m-n) + i\omega_\beta(\mu - v) - \tfrac{\Gamma_\alpha}{2}(m+n) - \tfrac{\Gamma_\beta}{2}(\mu + v),$$

$$M_{\substack{mn\mu v \\ 1-m,1-n,\mu v}} = P_\alpha mn, \quad M_{\substack{mn\mu v \\ mn,1-\mu,1-v}} = P_\beta \mu v,$$

$$M_{\substack{mn\mu v \\ m,1-n,1-\mu,v}} = 2ig(v-\mu)(1-n)(1-\mu),$$

$$M_{\substack{mn\mu v \\ 1-m,n,\mu,1-v}} = 2ig(n-\mu)(1-v)(1-m),$$

$$M_{\substack{mn\mu v \\ 1-m,n,1-\mu,v}} = ig[m(1-\mu)+\mu(1-m)],$$

$$M_{\substack{mn\mu v \\ m,1-n,\mu,1-v}} = -ig[n(1-v)+v(1-n)], \tag{8}$$

and zero everywhere else. For the computation of the spectrum, we need two more correlators from manifold $\mathscr{N}_2$ than in the LM (compare (a) and (b) in Fig. 1). Thanks to Pauli blocking (causing saturation), the number of correlators that exist is finite and we can also solve this system analytically. The correlators for the linear regime are shown in green, the same as for the LM, but we added in blue the second manifolds, $\mathscr{N}_2$ and $\tilde{\mathscr{N}}_2$, that were not needed then. As in the LM, the one-time correlators in $\mathscr{N}_1$ can be obtained independently, and with the same LM regression matrix (changing to the fermionic parameters and sign). Their expressions are, therefore, the counterpart of those in the LM, without any restriction for the parameters that always lead to a SS. On the other hand, the correlator $\{1,1,1,1\}$, that is the probability of double excitation, belongs to $\tilde{\mathscr{N}}_2$, and finds its expression separately, in terms of n_α and n_β:
$\langle \alpha^\dagger \alpha \beta^\dagger \beta \rangle = (P_\beta n_\alpha + P_\alpha n_\beta)/(\Gamma_\alpha + \Gamma_\beta)$. The results of the two models for the mean values in $\mathscr{N}_k$ diverge exactly at this point, when it comes to the dynamics of the states and it is not simply a matter of total average populations n_a/n_α. Also the two-time correlators in $\mathscr{N}_k$ are in general dissimilar.

242

Another important feature of this system in contrast with the LM, is that the pumping mechanism is of the same nature than the decay, because they are both limited thanks to saturation. The master equation is symmetrical under exchange of the pump and the decay ($\gamma \leftrightarrow P$) when the two levels of both 2LS are inverted. When either pump or decay are absent, the results for the spectra and two-time correlators are equivalent and simple, but when they are both driving the SC dynamics, the dressed structure is greatly affected. In Fig. 2(a) we can see an example where the SC spectra (solid black line) has the same structure than the LM: symmetric peaks (4 instead of 2) sitting on two splitted positions (marked with vertical dashed lines). In the example (b), there is a new second splitting of the four peaks that distorts the final doublet. The 2LSs and LM give qualitatively different results.

The definition of SC is in general also more complicated for the 2LSs than for the LM, in the presence of both pump and decay. The coupling strength g gets renormalized by $\sqrt{8(\gamma_\alpha P_\beta + \gamma_\beta P_\alpha)/(\Gamma_\alpha + \Gamma_\beta)}$ that, together with Γ_α/g and Γ_β/g, determines the regions and types of coupling. In the LM, the parameters Γ_a/g and Γ_b/g are enough to determine the coupling regime.

Conclusions

We have compared the SC regime of two HOs with that of two 2LSs, in the SS under incoherent continuous pumping. We have derived their respective QRFs that lead to analytical expressions for their spectra of emission. We have shown the links that the regression matrix establishes between correlators (Fig. 1) and the lineshapes of the luminescence spectra that would be experimentally observed (Fig. 2). The models converge in the linear regime and hold numerous similarities for a wide range of parameters (like one or two peaks in the emission). However, for some configurations, the 2LSs undergo a second SC transition characterized by four splitted peaks in the spectra. This differs largely with the LM emission, always composed of two peaks in SC.

ACKNOWLEDGMENTS

This work has been supported by the Spanish MEC under contracts Consolider-Ingenio2010 CSD2006-0019, MAT2005-01388 and NAN2004-09109-C04-3 and by CAM under contract S-0505/ESP-0200. EdV acknowledges support of the FPU (Spanish MEC).

REFERENCES

1. J. J. Hopfield, *Phys. Rev.* **112**, 1555 (1958).
2. F. P. Laussy, E. del Valle, and C. Tejedor, *Phys. Rev. Lett.* **101**, 083601 (2008).
3. E. Jaynes, and F. Cummings, *Proc. IEEE* **51**, 89 (1963).
4. E. del Valle, F. P. Laussy, and C. Tejedor, *arXiv:0812.2694* (2008).
5. M. Lax, *Phys. Rev.* **129**, 2342 (1963).
6. H. J. Carmichael, *Statistical methods in quantum optics 1*, Springer, 2002, 2 edn.

Optical properties of excimer laser nanostructured silicon wafer

Prashant Kumar[a], M. Ghanashyam Krishna[a, b], A. K. Bhattacharya[b]

[a]School of Physics, University of Hyderabad, Hyderabad, India-500046
[b]Department of engineering science, University of Oxford, Oxford, UK

Abstract. KrF excimer laser nanostructuring of [311] single crystal silicon surface is reported for laser fluence above ablation threshold for silicon. Laser irradiation of silicon surface gives rise to growth of nano-sized grains. Cooling time between the shots is an important factor in determining the nanostructures, apart from the number of shots and the laser fluence. Such nanostructured silicon surfaces become highly absorbing and photoluminescent and these peaks can be tuned by laser dressing parameters.

Keywords: Excimer laser nanostructuring, photoluminescence
PACS: 68.55. A-, 81.07.Bc, 81.16.-c.

INTRODUCTION

Quantum confinement is known to lead to application of Si for optical devices since it leads to photoluminescence in the visible range[1-5], allows tunability with size, and increases the quantum efficiency. Nanocrystals that are unpassivated i.e., with dangling bonds or other surface defects have their luminescence quenched via non-radiative recombinations involving these defect sites. Laser fluence used during porous silicon preparation is a key factor[6] in determining the subsequent photoinduced evolution of the photoluminescence spectra. The post-preparation evolution results from the combination of at least two effects. One of them is ruled by the size changes of the silicon nanostructure due to photo-oxidation, and dominates for samples prepared under low fluence. On the other hand, for samples prepared under high fluence the post-preparation evolution is dominated by a quenching effect, resulting from photoinduced dangling bonds generation in the hydrogen-rich surface of the nanostructure.

In the current paper, morphological, reflectance and photoluminescence variation is reported as a function of laser dressing parameters for KrF Excimer laser irradiation of single crystal [311] silicon surface is reported.

EXPERIMENTAL

Ultrasonically cleaned surfaces of [311] silicon were irradiated using an Excimer laser (CompexPro 201F of Lambda Physik) for laser fluence of 2-5 J/cm^2. The effects of variation in laser fluence, number of shots and rep rate were studied for the possible effects on morphology, reflectance and photoluminescence. Two types of experiments were carried out in this study: (1) At an energy density of 2 and 3 J/cm^2, at repetition rates of 1-10Hz and for a total of 100-500 shots in succession. The effects of this kind of irradiation are shown in figures 1, 3 and 4. (2) The silicon surface was also irradiated at a given energy density and number of shots. The surface was allowed to relax and after a gap of 24 hours it was again irradiated at the same energy density and total number of shots. For example the AFM image in figure 2(c) corresponds to the silicon surface irradiated with 4 shots at 2 J/cm^2 and again with 4 shots at the same energy density but after a gap of 24 hours. The objective of these experiments was to understand whether the self assembly took place during the irradiation process or after the irradiation has been completed. The results indicate that there are significant changes in nanostructures, in both cases. Evidently, though the nature of these changes is, different the self assembly is taking place during the irradiation process itself and not thereafter. Scanning probe microscope (SPA 400 of SII Inc. Japan) was employed for imaging at nanoscale. UV-VIS-NIR spectrophotometer was used to measure reflectance. Laser scanning confocal microscope was used for photoluminescence measurements.

RESULTS

At 2 J/cm^2, 100 shots and 1 Hz rep rate the formation of closely spaced silicon nanoparticles (100-150 nm dia) is observed as shown in Fig 1(a). At a higher laser fluence of 3 J/cm^2 and at higher rep rate, surface features are quite different as shown in Fig 1. In case of higher value of rep rate (10Hz), fragmentation of particles has been observed (Fig 1 (d)).

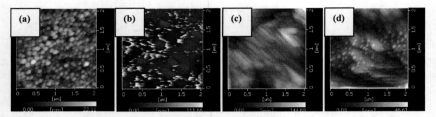

Figure 1. AFM images of laser nanostructured silicon surfaces for (a) 2 J/cm², 100 shots, 1 Hz., (b) 2 J/cm², 500 shots, 1 Hz (c) 3 J/cm², 100 shots, 1 Hz and (d) 3 J/cm², 100 shots, 10 Hz.

Detailed investigations were carried out on the effect of cooling time on the evolution of nanostructures. These experiments were carried out by irradiating the surface with a total of 4 and 8 shots in two different ways (1) all laser shots irradiated in sequence with reprate of 1 Hz without further time gap and (2) laser shots irradiated half numbers of shots in one batch and rest half number of shots in next batch. Time gap between the batches were kept 24 hours. An interesting observation was that it is not only the number of shots and laser fluence which are important in determining the kind of nanostructures but also the cooling time between the shots as shown in Fig 2.

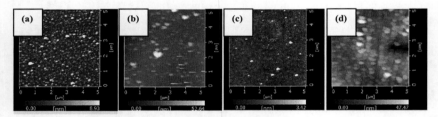

Figure 2. AFM image for excimer laser nanostructured silicon wafer surface for (a) 2 J/cm², 2 shots irradiated twice one after the other after few hours, (b) 2 J/cm², 4 shots at a time. (c) 2 J/cm², 4 shots irradiated twice one after the other after few hours, and (d) 2 J/cm², 8 shots at a time.

Specular reflectance percentage reduced with the laser fluence and also with number of shots as shown in Fig 3. At 3 J/cm² and few hundred shots, silicon surface becomes highly absorbing (98%).

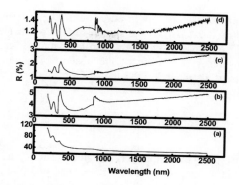

Figure 3. Optical reflection spectra for nanostructured silicon wafer surface for (a) untreated surface, (b) 2J/cm^2, 100shots, (c) 2J/cm^2, 500shots, (d) 3J/cm^2, 100 shots.

Photoluminescence study (as shown in Fig 4) of excimer laser irradiated silicon surface reveals that PL peak position (in energy units) which for bulk is 2.22 eV; is blue shifted to 2.45 eV for 2 J/cm^2 and 100shots at 1 Hz rep rate and further blue shifted to 2.73 eV for 3 J/cm^2 and 100 shots at 1 Hz rep rate.

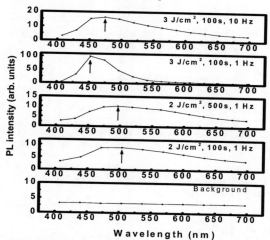

Figure 4. Photoluminescence spectra for laser nanostructured silicon wafer surface for (a) background (untreated wafer) (b) 2 J/cm^2, 100s, 1 Hz, (c) 2 J/cm^2, 500s, 1 Hz, (d) 3 J/cm^2, 100s, 1 Hz and (e) 3 J/cm^2, 100s, 10 Hz.

The fact that photoluminescence is observed would suggest that there are nanostructures, not resolved by the AFM, on the silicon surface of the size required to cause confinement effects. The high energy involved in laser induced nanostructuring would also lead to defects. Hence the origin of the observed PL can be traced to combination of these effects. It should be noted that the origin of PL in silicon is a subject of much debate [1-7]

CONCLUSIONS

Excimer laser nanostructuring of [311] single crystal silicon surface has been carried out and morphological reconstruction, specular reflectance and photoluminescence has been studied in detail. Cooling time between the shots has been found to be important apart, from number of shots and laser fluence in determining the surface nanostructures. Photoluminescence from the laser nanostructured surfaces has been observed.

ACKNOWLEDGEMENTS

A project fellowship for PK under the UGC-CAS programme is acknowledged. Facilities provided under the UPE, DST-ITPAR and Centre of Nanotechnology are also acknowledged.

REFERENCES

1. X. Y. Lin, K. X. Lin, R. H.Yao, W. H. Shi, M. Y. Li, C. Y. Yu, Y. P. Yu, H. Y. Liang and Y. P. Xu, *Chin. Phys. Lett.* **16**, 670, 1999.
2. V. Kabashin and M. Meunier, *Mat. Sci. and Engg. B* **101**, 60, 2003.
3. E. A. Boer, Ph.D. thesis "Synthesis, Passivation and Charging of Silicon Nanocrystals", California Institute of Technology, Pasadena, California 2001.
4. E. Edelberg , S. Bergh and R. Naone, *Appl. Phys. Lett.* **68**, 1415, 1996.
5. M. Ray, K. Jana, N. R. Bandyopadhyay, S. M. Hossain, D. Navarro-Urrios, P. P. Chattyopadhyay and M. A. Green, *Sol. Stat. Comm.* **149**, 352, 2009.
6. R. R. Koropecki, R. D. Arce, A. M. Gennaro, C. Spies and J. A. Schmidt, *J. Non-Crystal. Sol.* **352**, 1163, 2006.
7. S. M. Prokes, *J. Mater. Res.* **11**, 305, (1996).
8. P. D. J. Calcott, K. J. Nash, L. T. Canham, M. J. Kane and D. Brumhead, *J. Lumin.* **57**, 257, (1993).
9. G. Cullis, L. T. Canham and P. D. J. Calcott, *J. Appl. Phys.* **82**, 909, (1997).
10. F. Ozanam, J.-N. Chazalviel, and R. B. Wehrspohn, *Thin Solid Films* **297**, 53, (1997).
11. M. Chang and Y. F. Chen, *J. Appl. Phys.* **82**, 3514, (1997).
12. W. Kolasinski, M. Aindow, J. C. Barnard, S. Ganguly, L. Koker, A. Wellner, R. E. Palmer, C. Field, P. Hamley and M. Poliakoff, *J. Appl. Phys.* **88**, 2472, (2000).

Fiber- Optic pH Sensor Based on SPR of Silver Nanostructured Film

Rajib Saikia[a], Mridul Buragohain[a], Pranayee Datta[a], Pabitra Nath[a], Kishor Barua[b]

[a]Dept. of Electronics Science, Gauhati University, Guwahati-14, Assam, India.
[b]Dept. of Physics, Tezpur University, Tezpur-784028, Assam, India.

Abstract. Surface Plasmon Resonance (SPR) nanosensor has become an increasingly exploited technology for detection and analysis of chemical and biological compounds. Silver (Ag) nanostructured film fabricated by thermolysis method is shown to exhibit a strong Localized Surface Plasmon Resonance (LSPR) at wavelength around 400 nm. The spectral position of LSPR is sensitive to its local environment and also the nanoparticle size and shape. In this paper, we will demonstrate optical sensing of pH using nanosilver coated fiber optic technique based on the principle of LSPR.

Keywords: Nano particle, Localized Surface Plasmon Resonance, Sensor.
PACS: 87.85 RS.

INTRODUCTION

The localized surface plasmon resonance (LSPR) of plasmonic nanoparticles has been attracting great attention due to the potentiality in sensing and spectroscopic application. The plasmonic nanoparticles such as silver and gold exhibit a unique optical property, which strongly absorb and scatter light around LSPR wavelength due to the collective oscillation of the free electrons in resonance with the illuminating light field. The spectral position of LSPR is sensitive to its local environment such as refractive index of the surrounding medium [1, 2, 3], temperature, pressure, humidity, pH etc. In this paper, we demonstrate that pH sensor could be realized through shift in the localized surface plasmon resonance (LSPR) extinction maximum (λmax.) of Ag nanoparticles. Thermal reduction method followed by *Liu. et. al* [4] for the fabrication of Ag nano in PVA matrix has been modified in the present investigation for application in the field of optoelectronics. Among different polymers, polyvinyl alcohol (PVA) is the most promising because of its unique properties. PVA is highly soluble in water and biologically friendly. Its easy process ability and high transmittance (Khanna et.al 2005) [5] make it a good matrix of polymer films. It acts both as a reducing agent as well as a stabilizer. It is non corrosive in nature, so suitable for electronic and optoelectronic application. In the present work, a pH sensor is constructed from low cost fiber optic and

CP1147, *Transport and Optical Properties of Nanomaterials—ICTOPON - 2009*, edited by M. R. Singh and R. H. Lipson
© 2009 American Institute of Physics 978-0-7354-0684-1/09/$25.00

optoelectronic components including a blue light emitting diode and a photodiode. The fabricated Ag-PVA nanocomposite is coated on U-bent and tapered U-bent fiber having bent radius of 2.5 mm by dip coating method. For comparison of the sensitivity of the nanocomposite with that of the bulk material we have repeated the experiment with the two fibers having bulk coating.

EXPERIMENT AND RESULTS
Fabrication of Ag -PVA Nanocomposite Films

Polymer -metal nanocomposites can be obtained by two approaches, viz. in situ and ex situ techniques. In the in situ methods, metal particles are generated inside a polymer matrix of a metallic precursor dissolved in the polymer. In the ex situ approach, nanoparticles are first produced by soft-chemistry routes and then dispersed into the polymeric matrices. In our work the first method is employed.

In chemical route, polymer matrices play the central role in controlling the particle size. Physically the matrices provide some gaps in the polymer chain. These gapes are of nm range. During chemical reaction to produce small conducting nano particle, the matrix is mixed with the reactants and then stirred. As soon as the nano structure is produced, it immediately enters into the gap provided by polymer matrix. Once the particle enters the gap, it can neither come out nor can enhance in size. During sample preparation, the particle size can be varied by controlling-

(i) Temperature of the reaction.
(ii) Stirring rate of the reactants and matrix
(iii) Concentration of reactants and matrix.
(iv) The molecular ratio of reactants to matrix.

The silver nanoparticles have been generated in the PVA matrix by thermal reduction process. In the present work, silver nitrate $AgNO_3$ (MERCK Specialties Pvt. LTD., Mumbai) and PVA (LOBA CHEMIE Pvt. LTD, Mumbai) of analytical grade purity were used without further purification. The synthesis of Ag –PVA nano composite is a three step process. In the first step $AgNO_3$ solution of different concentration (10 m mol/L to 70 m mol/L) were prepared by dissolving $AgNO_3$ powder in distilled water. The mixture is stirred at room temperature in magnetic stirrer.

In the second step, 4 wt % PVA solutions are prepared in distilled water. The mixture is stirred at 60 °C for 3 hrs. in magnetic stirrer.

In third step, $AgNO_3$ solution and PVA solution is mixed in 1:1 ratio and the reaction mixture is heated at different temperatures (100 °C to 150 °C) and also for different time of heating. The solution is then kept overnight for stabilization. The sample is then cast over cleaned glass slide for further experimentations.

Characterization of Nanocomposite Films
Absorption Spectra Characterization

For absorption spectra characterization we have prepared number of samples by varying the concentration of AgNO₃ from 10 m mol/L to 70 m mol/L, keeping temperature and heating time constant. Again keeping heating time and concentration constant, we have varied the heating temperature from 100 ^{0}C to 150 ^{0}C. It has been found that, for heating temperature of 130 ^{0}C, AgNO₃ concentration of 40 m mol/L and heating time of 10 minutes, sharp SPR peak of Ag occurs at 456.2 nm. For others, prominent SPR resonance peak has not been found. Using Mie Scattering theory the corresponding particle size was found to be 68 nm. To reduce the particle size further, we varied the heating time keeping concentration and temperature constant at 40 m mol/L and 130 ^{0}C respectively. The results obtained are shown in table1 and Fig.1

TABLE 1. Concentration and temperatures are kept constant at 40 m mol/L and 130 ^{0}C respectively

Sample No.	Heating time (minutes)	λmax (nm)	Particle radius (nm)
1	10	456.2	68
2	15	457.2	68.5
3	20	407.2	41
4	25	408	42.5
5	30	460	69

UV-VIS spectrometry confirmed formation of silver nanoparticles prepared in situ approach because silver particle can show an intense absorption peak around 400 nm originating from Surface Plasmon Resonance (SPR) of nano sized silver nanoparticles (Table1)[6].The observed UV-VIS absorption spectra for different samples shows(Fig.1) LSPR at different wavelengths. The UV-VIS absorption shows that, as heating time increases, the wavelength of absorption is blue shifted up to ~ 25 minutes and then red shifted beyond 25 minutes of heating time. The particle size is calculated from Mie Scattering theory, performed with the software "MiePlot 3501" [7]. The standard deviation used in the calculation was 1 %. The particle size variation with heating time is shown in fig.2. This observed characteristic calls for further experimental as well as theoretical investigations. Although two SPR peak are observed for sample no. 2 and 5 which indicates non spherical shape of the particles [3], only the peak of maximum absorption is adopted for estimating particle size.

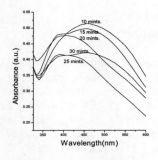

FIGURE 1. UV-vis spectra of the sample 1 to 5

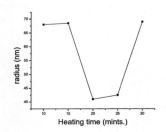

FIGURE 2. Heating time vs particle size.

Scanning Electron Microscopy (SEM) Characterization

SEM photograph (WL= 300 nm, EHT=10KV, WD=13mm, SIGNAL A=SE1) for sample no.3 and 4 are given in fig. 3(A) and 3(B). The minimum particle -diameter for the sample no. 3 and 4 from SEM characteristics has been found to be ~ 75 nm and 88 nm respectively and the radii estimated from Mie Scattering theory are 41 and 42.5 nm respectively. Therefore the experimental and theoretical observations are in good agreement.

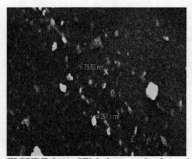

FIGURE 3(A). SEM photograph of sample no.3

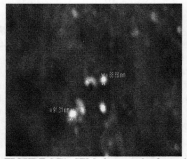

FIGURE 3(B). SEM photograph of sample no.4

Optical Fiber pH sensor
Preparation of Ag – nano coated pH Probe

FIGURE 4. Photograph of the tapered U-bent fiber probe.

To prepare the pH probe, a PMMA (polymethyl methyl acrelyte) fiber of core diameter 980/1000 micron and numerical aperture (NA) 0.5 were used. The refractive indices of the core and the cladding were 1.48 and 1.401 respectively. The cladding was removed from a small portion of the central region (sensing region) of the fiber by rubbing it with the help of a razor. To make the fiber U-shape, it was heated at temperature of 100 °C and then slowly bent to diameter 2.5 mm of the sensing region. This bare portion of the fiber is then converted into a taper by rubbing it with the help of a razor then by heating at 100 °C and then pulling it to one end of the fiber while the other end is fixed. The prepared U-bent and the tapered U-bent (Fig.4) fiber were cleaned with acetone and then dried in air for few minuets. The sensing region of the two fibers was coated with the fabricated Ag-PVA nanocomposite of minimum particle size (sample no.3 of table1, size 41 nm). To observe the performance of the probe with bulk material the nano coating is removed with acetone and then coated with bulk Ag-PVA composite.

Experimental Set Up For pH Sensor Realization

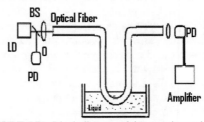

FIGURE 5(A). Photograph of the experimental set up.　**FIGURE 5(B).** Block diagram of the experimental set up.

The optical source used in this experiment is intensity stabilized 4 mW diode laser (LD) source of wavelength 670 nm (Fig. 5(B). To avoid any fluctuation in the LD, a 50:50 beam splitter is used at the input end of the optical fiber, so that a part of the light signal propagated in a direction orthogonal to the beam passing through the optical fiber can be used as reference level for the present study. Light from the beam splitter (BS) is collimated by means of microscopic objectives (O) on to the input face of the fiber. The intensity modulated beam coming out from the sensing region of the fiber is then refocused to the detector by means of another microscopic objective lens (focal length=2.5 mm). For stable light-coupling any possible relative movement between the second objective and the fiber was avoided. The diode laser source, the objectives, fiber stand and the detectors are mounted on the same bread boards so that any external mechanical disturbance of the beam launcher affects the fiber as well, thereby minimizing variations in the power coupled into the fiber. The detector is a photodiode (PD) whose output is amplified by a preamplifier circuit. A digital multimeter is used for measuring the detector output. In this experiment, water (liquid) with different pH (from 3.6 to 9.3) is used which changes the intensity of the light beam from the sensitive region of the fiber probe due to shift in LSPR position of Ag nanoparticles.

Sensitivity of the Sensor

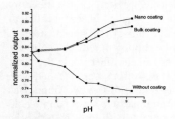
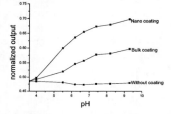

FIGURE 6(A). Sensitivity of the U-bent fiber.   **FIGURE 6(B).** Sensitivity of the tapered U-bent fiber.

To study the sensitivity of the fiber-optic pH sensor based on LSPR of Ag nanstructured film, experimental investigations have been carried out with U-bent and tapered U-bent fiber of same characteristics (Fig. 6(A) and 6(B)). Here the sensitivity factor is defined as the change in normalized output for a given change in pH value. For both the cases we have investigated the sensitivity of the probe for without coating, bulk coating and nanocomposite coating. For without coating U-bent probe, normalized output decreases with increase in pH. So it can't be useful for higher pH. But the same probe with bulk and nano coating, normalized output increases with increase in pH. The sensitivity factor in the pH range from 3.6 to 9.3 is 0.06 for bulk coating while it is 0.08 for nano coating. In case of tapered U-bent, for without coating probe the sensitivity is negligible. The sensitivity factor in the same pH range i.e. from 3.6 to 9.3 for bulk coated and nano coated are 0.11 and 0.21 respectively.

For each probe the sensitivity is more for nano coating in comparison with bulk and without coating probe. It is due to the shift in LSPR extinction maximum (λmax) which is caused by local refractive index change at different pH environment.

In the present work, it was observed that pH sensor probe with tapered U-bent is more sensitive than U-bent probe. Tapering of the fiber directly influences the number of internal reflections occurring at the beam through the tapered part of the fiber and the angle of incidence of the internal reflections while maintaining the high coupling efficiency of large fiber end face. This results in an increase in penetration depth without significantly decreasing the light transport efficiency. Consequently the sensitivity of the tapered probe sensor is enhanced by more than one order of magnitude [8].

CONCLUSIONS

The present investigation is summarized as- (i) UV-VIS spectrometry confirms formation of Ag nanoparticles in PVA matrix by simple thermal treatment, (ii) The wavelength corresponding to maximum absorption (λmax) and hence the particle size can be controlled by controlling the heating time, (iii) The LSPR of Ag nanoparticles in PVA matrix is highly sensitive to pH of the environment and (iv)The Ag nano coated tapered U-bent fiber is more sensitive to pH of the environment than U-bent fiber with the same coating of Ag nano.

To observe the effect of particle size on the sensitivity of the sensor and hence on the LSPR of Ag nanoparticles, it is proposed to carry out investigations of the tapered U-bent fiber probe coated with the sample no. 1 to 5 (table1).

ACKNOWLEDGMENTS

The authors would like to acknowledge IIT, Guwahati, Assam (India) for SEM and UV-VIS observations, Dept. of Science and Technology, Govt. of India under FIST programme for the computer lab facility provided in the Department of Electronics Science, Gauhati University, Assam, India.

REFERENCES

1. T. R. Jensen, M. L. Duval, K. L. Kelly, A. A. Lazarides, G. C. Schatz and R. P. Van Duyne, *J. Phys. Chem. B* 103, 2394-2401 (1999).
2. M. Himmelhaus and H. Takei, *Sensor Actuators B* **63**, 24-30 (2000).
3. D. R. Lide, CRC *Handbook of Chemistry and Physics* 78th edition, 1997, published by Boca Raton, FL: CRC Press.
4. Liu Shuxia, He Junhui, Xue Jianfeng and Ding Wenjun, *Journal of Nanoparticle Research*, published online in Springer Science (2007).
5. P. K. Khanna, N. Singh, S. Charan, V. V. V. S. Subbarao, R. Gokhale and U. P. Mulik, *Mater. Chem. Phys.*, **93**, 117-121 (2005).
6. Yingwei Xie, Ruqiang Ye and Honglai Liu, *Colloid Surface A* **279**, 175-178 (2006).
7. The "Mieplot" software is available from Philip Laven, available at: www.philiplaven.com/mieplot.htm.
8. B.D. Gupta and C.D. Singh, *Fiber Integrated Opt.* **13**, 433-443 (1994).

Resonant Photons in Nanophotonic Quantum Well Heterostructures

Mahi R. Singh and Joel D. Cox

Department of Physics and Astronomy, The University of Western Ontario, London, Ontario, Canada
N6A 3K7.

Abstract. In this work, we propose to study the resonant photonic states in a new class of photonic crystal heterostructures, which can be used to make smaller and faster optoelectronic devices. Photonic quantum well (PQW) heterostructures are made from two different photonic crystals A and B. The structure is denoted as A/B/A/B/A, where the core crystal A is sandwiched between two B crystals. If crystals A and B are chosen so that the upper band edge of A lies within the gap of B, the resulting band structure acts as a double potential barrier for photons. Here we have obtained an expression for the transmission coefficient of the PQW heterostructure using the transfer matrix method, and have found that there are resonant states within the well. We have shown that the number of resonant states can be controlled by varying parameters of crystals A and B, such as their lattice constants, indices of refraction and thicknesses. It is anticipated that the resonance states described here can be used to make new types of photonic switching devices and quantum computers.

Keywords: Photonic crystal heterostructures, resonant tunneling, photonic devices, photonic quantum well, nanophotonics.
PACS: 42.70.Qs, 73.21.–b, 78.67.De, 42.50.Ct, 42.25.Gy

INTRODUCTION

Photonic crystals have attracted much attention during the past two decades due to their numerous potential applications in communications and computing [1, 2]. Considerable effort has been placed into finding ways to build smaller and faster optoelectronic devices using photonic crystals. To that end, photonic crystal heterostructures provide a method for turning raw photonic crystals into functional devices [3, 4].

Photonic crystals have a band gap in their photonic dispersion relation, which prevents photons with energy lying within the band gap from propagating through the crystal structure [1, 2]. By joining two or more photonic crystals in a single structure, a photonic crystal heterostructure is formed. Photonic crystal heterostructures have been used to develop devices such as resonant cavities demonstrating very high quality factors [3, 5-7], as well as high-transmission photonic waveguides [3, 8-11]. Photonic crystal heterostructure devices such as these may be used to develop new types of all-optical switching devices, photonic integrated circuits and quantum computers [3, 5-12].

CP1147, *Transport and Optical Properties of Nanomaterials—ICTOPON - 2009,* edited by M. R. Singh and R. H. Lipson

In this paper we study the resonant photonic states formed in a photonic quantum well heterostructure made from the two photonic crystals A and B, each possessing a different band structure. The heterostructure is denoted as A/B/A/B/A, where at the core a layer of crystal A is sandwiched between two crystals B (see Figure 1). By choosing crystal parameters of A and B so that the conduction band of crystal A lies within the band gap of crystal B, a quantum well is formed. Photons with energies between the conduction bands of crystals A and B will propagate freely in crystal A, while crystal B acts as an energy barrier [3, 4, 13]. As in the case for an electron in an electronic potential barrier [13, 16], there is some chance that the photon will tunnel through the barrier (crystal B) and become confined within the quantum well (crystal A). In order for this to occur, the photon must have an energy level corresponding to one of several bound states within crystal A.

Using the transfer matrix method an expression for the transmission coefficient of the heterostructure is calculated as a function of photon energy. From this expression we have found the resonant photonic states, which have interesting properties. Resonance occurs when a photon tunnels through crystal B and occupies a bound state within crystal A for a finite period of time, before tunneling back through a layer of crystal B again. Thus photon has occupied a *quasi-bound* or *resonant* state within the photonic quantum well. We have found that by increasing the width of crystal A, which corresponds to widening the photonic quantum well, the number of available resonant states increases. We have also shown that varying the difference between the conduction bands of A and B alters the height the photonic potential barrier, controlling the number of resonant states. Photons at a resonant energy undergo perfect transmission through the photonic quantum well heterostructure, and appear as narrow peaks approaching unity in the transmission spectra.

It is anticipated that the resonance states described here can be used to make new types of all-photonic switching devices and optical computers.

THEORY

We consider two photonic crystals A and B, each composed of dielectric spheres arranged periodically in a background material, which is taken as air ($n = 1$). Lattice constants of crystals A and B are L_A and L_B, respectively. Similarly, the radii of the spheres in crystals A and B are given by a_A and a_B, while the length of the air gaps between the spheres are b_A and b_B. The refractive indices of the spheres in each crystal are denoted as n_A and n_B. The band structure of isotropic photonic crystals such as these has been calculated by John and Wang [17]. According to this model the dispersion relation for a photonic crystal is

$$\cos(k_i L_i) = \frac{(n_i+1)^2}{4n_i} \cos\left[\frac{\varepsilon_k(2n_i a_i + b_i)}{\hbar c}\right] - \frac{(n_i-1)^2}{4n_i} \cos\left[\frac{\varepsilon_k(2n_i a_i - b_i)}{\hbar c}\right]. \quad (1)$$

Here k_i is the Bloch wave vector for the ith crystal (i.e. $i = A$ or B) and ε_k is the photon energy. To simplify the energy band calculations for crystals A and B, we choose parameters that satisfy $2n_ia_i = b_i$ [17]. This condition eliminates the cosine in the second term on the right hand side of (1) for both crystals.

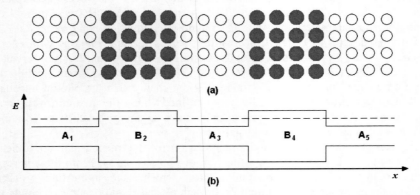

FIGURE 1. Schematic diagram of the photonic quantum well heterostructure. (a) Cross-sectional view of the crystal arrangement in the heterostructure. The circles represent the scattering spheres in each crystal, which have different sizes and/or indices of refraction. (b) Energy band structure of the quantum well. The dashed line represents the energy of a resonant state.

We consider transverse electric field (TE) waves propagating along the direction of crystal growth (x-direction) through photonic crystals A and B. The TE waves propagate through the photonic crystals in the form of Bloch waves, which are denoted as k_A and k_B. From (1), band gaps appear in crystals A and B. The energy bands above and below a band gap are called the conduction and valence bands, respectively. The valence and conduction band edges of crystal A are denoted as ε_v^A and ε_c^A; similarly we have ε_v^B and ε_c^B for crystal B. It is considered here that the conduction band edge of crystal A lies between the valence and conduction band edges of crystal B, such that $\varepsilon_v^B < \varepsilon_c^A < \varepsilon_c^B$. Due to this condition a quantum well is formed in the heterostructure, where the height of the barriers corresponds to $\Delta\varepsilon = \varepsilon_c^B - \varepsilon_c^A$ (see Figure 1).

We consider a photonic quantum well (PQW) structure grown along the x-direction where the thicknesses of crystals A and B are d_A and d_B, respectively. Transverse electric field (TE) waves are propagating in the structure in the form of Bloch waves. Due to the band structures of crystals A and B, the TE Bloch waves traveling along the x-direction propagate freely within crystal A and exponentially decay in crystal B.

The PQW structure may be denoted in numbered form as $A_1/B_2/A_3/B_4/A_5$ in order to clarify calculations (see Figure 1). Here we use the transfer matrix method [3, 18], which matches the electric field and its first derivative at the interfaces between adjacent layers in order to relate the incident and reflected amplitudes in each layer. The amplitudes in layer A_1 are related to those in B_5 by the matrix equation

$$\begin{pmatrix} \alpha_1 \\ \beta_1 \end{pmatrix} = P^{12} P^{23} P^{34} P^{45} \begin{pmatrix} \alpha_5 \\ \beta_5 \end{pmatrix}, \qquad (2)$$

where α_j and β_j are the amplitudes of the incident and reflected plane waves (respectively) in the jth layer, and $P^{j,j+1}$ is the transfer matrix relating the incident & reflected amplitudes between adjacent layers. For the system considered here there are only four interfaces between crystals A and B, so the index j runs from 1 to 4. The transfer matrix at the jth interface is given as [18]:

$$P^{j,j+1} = \begin{pmatrix} \dfrac{(k_j + k_{j+1})\exp[i(-k_j + k_{j+1})x_j]}{2k_j} & \dfrac{(k_j - k_{j+1})\exp[i(-k_j - k_{j+1})x_j]}{2k_j} \\ \dfrac{(k_j - k_{j+1})\exp[i(k_j + k_{j+1})x_j]}{2k_j} & \dfrac{(k_j + k_{j+1})\exp[i(k_j - k_{j+1})x_j]}{2k_j} \end{pmatrix} . \quad (3)$$

Here x_j is the location of the jth interface on the x-axis. Note that k_j and k_{j+1} correspond to the Bloch wave vector in crystal A or B, depending on the value of j. Thus we have $k_j = k_A$ in layers 1, 3 and 5 and $k_j = k_B$ in layers 2 and 4.

After performing the matrix multiplication in (2), the transfer matrix equation reduces to

$$\begin{pmatrix} \alpha_1 \\ \beta_1 \end{pmatrix} = P \begin{pmatrix} \alpha_5 \\ \beta_5 \end{pmatrix}, \quad (4)$$

where $P = P^{12}P^{23}P^{34}P^{45}$ is the total transfer matrix relating the incident & reflected wave amplitudes in layer A_1 to those in layer A_5. The transmission coefficient T is then easily obtained from the (1, 1) entry in the matrix P as [16, 18]

$$T = |\alpha_5 / \alpha_1|^2 = \frac{1}{|P_{11}|^2}. \quad (5)$$

The resulting expression for t is a function of the photon energy, ε_k, and is plotted between ε_c^A and ε_c^B to determine the resonant states of the PQW.

RESULTS AND DISCUSSION

Parameters used for crystals A and B were taken similar to those in recent experimental measurement [5-7]. Here $n_A = n_B = 1.45$, $L_A = 420$ nm and $L_B = 410$ nm. The other crystal parameters are obtained by imposing the condition that $2n_i a_i = b_i$, and from the definition $L_i = 2a_i + b_i$. The conduction band edges of crystals A and B were calculated from (1) as $\varepsilon_c^A = 1.3947$ eV and $\varepsilon_c^B = 1.4287$ eV. These energies correspond to photon wavelengths between 868.5 and 889.7 nm. The width of the layers of crystal B was taken as $d_B = 5L_B$, i.e. five layers of dielectric spheres. It was found that if d_B becomes too large then the transmission coefficient becomes vanishingly small for the range between the conduction band edges of A and B, which corresponds to no resonant states within the PQW. This is expected, as incident photons cannot tunnel through the barrier if it becomes too thick.

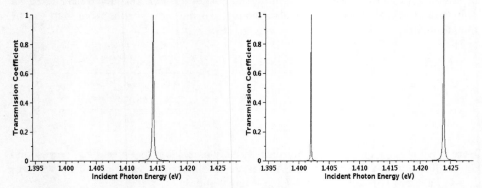

FIGURE 2. Plots of transmission coefficient vs. incident photon energy for a PQW system with $d_B = 5L_B$ (= 2.05 μm). Well widths are (left) $d_A = 15L_A$ (= 6.30 μm), (right) $d_A = 25L_A$ (= 10.50 μm). The incident photon energy is plotted from the conduction band edge of crystal A to that of crystal B.

From the transmission spectra of the PQW heterostructure given in Figure 2, we can identify distinct resonant states. Note that as the well width is increased, the number of resonant states also increases, as is the case for an electronic quantum well system [13, 16].

By adjusting the parameters of crystal B, the conduction band edge ε_c^B can be increased. This increases the barrier height $\Delta\varepsilon = \varepsilon_c^B - \varepsilon_c^A$, which increases the number of quasi-bound states within the PQW. In Figure 3 below we take $L_B = 370$ nm, while keeping all other parameters as before. The conduction band edges of crystals A and B are now $\varepsilon_c^A = 1.3947$ eV and $\varepsilon_c^B = 1.5832$ eV, which corresponds to photon wavelengths between 783.8 and 889.7 nm.

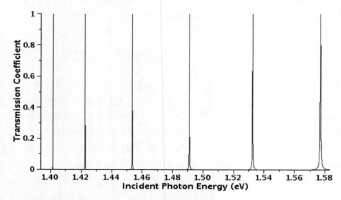

FIGURE 3. Plot of transmission coefficient vs. incident photon energy for a PQW system with $d_B = 5L_B$ (= 2.05 μm) and $d_A = 25L_A$ (= 10.50 μm). The incident photon energy is plotted from the conduction band edge of crystal A to that of crystal B.

From Figure 3 we can see that the number of resonant states has increased simply by changing the lattice constant of crystal B. By changing the parameters of crystal A or B and/or varying their thicknesses in the PQW heterostructure, the number of resonant states can be controlled.

Here we have used the transmission matrix method along with the dispersion relation for isotropic photonic crystals given in [17] to calculate the transmission spectra of a PQW heterostructure. From the transmission spectra the resonant frequencies for various heterostructures are easily obtained. Qualitatively, we have found that these results are in good agreement with other theoretical studies on PQW heterostructures, which used alternative methods for obtaining their transmission spectra [4, 13-15]. The resonant states described here have also been demonstrated experimentally through the fabrication of photonic crystal resonant cavities, several of which have been shown to exhibit very high-quality factors [3, 5-7].

From the transmission spectra presented here we have shown that the PQW heterostructure is ideal for selecting electromagnetic modes of specific frequencies [3]. The number of resonant states can be controlled by choosing the appropriate crystal parameters (such as lattice constant or index of refraction) for photonic crystals A and B. Additionally, the number of resonant states can be varied by changing the width of the PQW itself (i.e. the central crystal A). It is anticipated that the resonant states described in this paper can be used to develop new types of photonic-switching devices and optical computers [3, 5, 6, 11, 12].

ACKNOWLEDGMENTS

M.R.S. is thankful to the Natural Sciences and Engineering Research Council of Canada (NSERC) for financial support in the form of a research grant.

REFERENCES

1. E. Yablonovitch, *Phys. Rev. Lett.* **58**, 2059 (1987).
2. S. John, *Phys. Rev. Lett.* **58**, 2486 (1987).
3. E. Istrate and E. H. Sargent, *Rev. of Mod. Phys.* **78**, 455 (2006).
4. E. Istrate and E. H. Sargent, *J. Opt. A* **4**, S242-S246 (2002).
5. B. S. Song, S. Noda, T. Asano and Y. Akahane, *Nat. Mater.* **4**, 207-210 (2005).
6. B. S. Song, T. Asano and S. Noda, *J. Phys D* **40**, 2629 (2007).
7. K. Srinivasan, P. Barclay, M. Borselli and O. Painter, *Phys. Rev. B* **70**, 081306(R) (2004).
8. A. Mock, L. Lu and J. D. O'Brien, *Optics Express* **16**, 9391 (2008).
9. S. Y. Lin, E. Chow, S. G. Johnson and J. D. Joannopoulos, *Opt. Lett.* **25**, 1297-1299 (2000).
10. J. Salzman and O. Katz, *Phys. Stat. Sol. (c)* **1**, No. 6, 1531–1536 (2004).
11. A. Chutinan and S. John, *Phys. Rev. E* **71**, 026605 (2005).
12. M. R. Singh, *Applied Physics B* **93**, 91 (2008);
13. Y. Jiang, C. Niu and D. L. Lin, *Phys. Rev. B* **59**, 9981-9986 (1999).
14. X. Hu, P. Jiang and Q. Gong, *J. Opt. A* **9**, 108-113 (2007).
15. L. An and G. P. Wang, *Chin. Phys. Lett.* **23**, 388-391 (2006).
16. J. H. Davies, *The Physics of Low-Dimensional Semiconductors: An Introduction.* Cambridge University Press (1998).
17. S. John and J. Wang, *Phys. Rev. B* **43**, 12772 (1991).
18. H. Yamamoto, *Appl. Phys. A* **42**, 245-248 (1987).

Dielectric Properties of ZnO Nanoparticles Synthesized by Soft Chemical Method

Ramna Tripathi and Akhilesh Kumar

Department of Physics, Govt. P.G. College, Rishikesh-249201, Deharadun, Uttarakhand, India

Abstract: II-VI semiconductor nanoparticles have recently attracted a lot of attention due to the possibility of their application in various devices. In the present paper, nanoparticles of ZnO are prepared by soft chemical method and thioglycerol was used as capping agent, which controls the agglomeration of particle size. XRD and SEM technique was used for characterization. For the study of dielectric properties, complex permittivity (ε' and ε'') and loss tangent (tan δ) with frequency is analyzed in the frequency range 50 Hz-1 MHz and temperature range 293K-383K. The frequency-dependence maxima of the imaginary part of impedance are found to obey Arrhenius law with activation energy 1.05 eV. The frequency-dependent electrical data are analyzed in the framework of conductivity formalisms.

Keywords: A. Zinc oxide, B. Chemical synthesis; C. Impedance spectroscopy; D. Dielectric properties
PACS Nos. 61.46.Df; 61.46.Hk; 81.07.Wx

INTRODUCTION

Zinc oxide is a II-VI semiconductor has large band gap (E_g =3.37 eV) with higher exciton binding energy (60 meV). It is widely used in a number of applications like photo catalysis [1], gas sensors [2], varistors [3], and low-voltage phosphor material [4] and so on. ZnO probably has the richest family of nanostructure among all the semiconducting materials, both in structures and in properties due to its unique properties [5, 6]. Impedance spectroscopy is one of the powerful tools for characterization of electrical properties of ZnO. AC impedance spectroscopy allows measurement of the capacitance and loss tangent (tan δ) and/or conductance over a frequency range at various temperatures. From the measured capacitance and tan δ, four complex dielectric functions can be computed: impedance (Z^*), electric modulus (M^*), permittivity (ε^*), and admittance (Y^*). Studying dielectric data in the different functions allows different features of the materials to be recognized.

Though, various researchers have studied the electrical properties of ZnO, but a complete systematic study of various dielectric parameters in a wide frequency and temperature range is still lacking. In this paper we have studied systematically the dielectric properties of ZnO synthesized by a soft chemical method in the temperature range from 293 K to 383 K and in the frequency range from 100 Hz to1 MHz by impedance spectroscopy.

CP1147, *Transport and Optical Properties of Nanomaterials—ICTOPON - 2009,* edited by M. R. Singh and R. H. Lipson

EXPERIMENT

Different solutions were prepared by dissolving 0.2725 g of $ZnCl_2$ (1/10 M, 20ml), 0.4 g NaOH (1/10 M, 100 ml) in methanol keeping the variable concentration of thioglycerol in methanol. . Thioglycerol was used as the capping agent, because it arrests the agglomeration of ZnO nanoparticles during precipitation or to achieve the stability and avoid the coalescence. Thioglycerol (TG) solution was slowly added to NaOH solution while it was continuously stirred. The resulting solution was stirred for one hour before adding $ZnCl_2$ solution to it. After three hours of constant stirring a milky white solution of zinc oxide was obtained. The white powder of ZnO is palletized into a disc using poly vinyl alcohol (PVA) as binder.

The dielectric measurement of the sample of thickness 2.06 mm and diameter 10.41 mm was carried out using gold electrodes by an LCR meter (Hioki) in the frequency range from 42 Hz to 1 MHz and in the temperature range from 293 K to 383 K. The temperature was controlled with a programmable oven. All the dielectric data were collected while heating at a rate of 1 ^{0}C min^{-1}.

RESULTS AND DISCUSSION

The relation of angular frequency ω ($=2\pi v$) with dielectric constant (ε') and dielectric loss (tan δ) at various temperature was described in Fig. 1. The nature of dielectric permittivity related to free dipoles oscillating in an alternating field may be described in the following way. At very low frequencies ($\omega<<1/\tau$, τ is the relaxation time), dipoles follow the field and the real part of dielectric constant $\varepsilon' \approx \varepsilon_s$ (value of the dielectric constant at quasi static fields). As the frequency increases (with $\omega<1/\tau$), dipoles begin to lag behind the field and ε' slightly decreases. When frequency reaches the characteristic frequency ($\omega=1/\tau$), the dielectric constant drops (relaxation process). At very high frequencies ($\omega>>1/\tau$), dipoles can no longer follow the field and $\varepsilon' \approx \varepsilon_\infty$ (high-frequency value of ε'). In Fig. 1(b), the increase in the maximum value of tan δ with the increase in temperature and frequency indicate that the no. of charge carriers increases by thermal activation. The width of loss peaks in Fig. 1(b) cannot be accounted for in terms of monodispersive relaxation, but points indicate towards the possibility of a distribution of relaxation times.

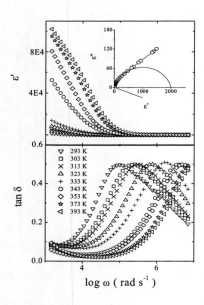

FIGURE 1. Frequency (angular) dependence of the ε' and tan δ at various temperatures and complex Argand plane plots of ε'' against ε' shown in the inset for ZnO.

One of the most convenient ways to check the polydispersive relaxation is complex Argand plane plots of ε'' against ε', usually called Cole-Cole plots [7]. The parameter α can be determined from the location of the centre of the Cole-Cole circles, of which only an arc lies above the ε'-axis. The complex dielectric constant in such situations is to be described by the empirical relation [8] given as

$$\varepsilon^* = \varepsilon' - j\,\varepsilon'' = \varepsilon_\infty + \frac{(\varepsilon_s - \varepsilon_\infty)}{1 + (j\omega\tau)^{1-\alpha}} \qquad (1)$$

where ε_s and ε_∞ are the low and high frequency values of ε'. α is a measured of the distribution of relaxation times, which can be determined from the location of the centre of Cole-Cole circle of which only an arc lies above the ε'-axis. We have plotted ε'' against ε' in the inset of Fig. 1(a) at 313 K. It is evident from this plot that the relaxation process differs from the monodispersive Debye process ($\alpha = 0$) and found to be 0.44 for ZnO. This value of α from Cole-Cole plot confirms the polydispersive nature of dielectric relaxation of ZnO.

The conductivity can be expressed as $\sigma\,(\omega) = \omega\varepsilon_0\varepsilon''$, here σ is the real part of the conductivity and ε'' is the imaginary part of dielectric constant. The frequency spectra of the conductivity for ZnO are shown in Fig. 2 at different measuring temperatures. The very basic fact about AC conductivity in complex solids is that σ is an increasing function of frequency. The real parts of conductivity spectra can be explained by the power law [9] define as

$$\sigma = \sigma_{dc}\left[1+\left(\frac{\omega}{\omega_H}\right)^n\right] \quad (2)$$

where σ_{dc} is the DC conductivity, ω_H is the hopping frequency of the charge carriers, and n is the dimensionless frequency exponent. The experimental conductivity spectra of ZnO is fitted with σ_{dc} as variable keeping in mind that the value of parameter n are weakly temperature dependent. The best fit of the conductivity spectra are shown in Fig. 2 by solid lines.

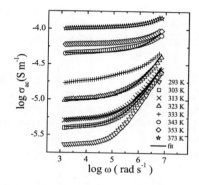

FIGURE 2. Frequency dependence of the conductivity (σ) for ZnO at various temperatures where the symbols are the experimental points and the solid lines represent the fitting.

We have plotted the frequency (angular) dependence of the imaginary parts of complex impedance (Z″) of ZnO at various temperatures in Fig. 3. The temperature affects strongly the magnitude of resistance. At lower temperature, it is evident that the position of the peaks in Z″ plot shifts to higher frequency with increasing temperature. The spectrum of Z″ at each temperature exist one relaxation peak, where peak frequency ω_m increases with increasing temperature and follows the Arrhenius law with activation energy 1.05 eV as shown in the inset of Fig. 3. Also the magnitude of Z″ decreases with increasing temperature. Fig. 3 indicates the spreading of the relaxation time, this would imply that the relaxation is temperature dependent and there is apparently not a single relaxation time, thereby relaxation process involve with there own discrete relaxation times depending on the temperature.

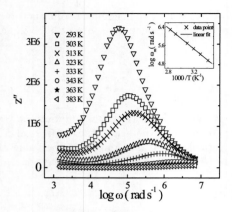

FIGURE 3. Frequency (angular) dependence of the Z'' at various temperatures and the Arrhenius plot of Z'' in the inset for ZnO.

If we plot the $Z''(\omega,T)$ data in scaled coordinates, i.e., $Z''(\omega,T)/Z''_m$ and $\log(\omega/\omega_m)$, where ω_m corresponds to the frequency of the peak value of Z'' in the Z'' versus $\log\omega$ plots, the entire data of imaginary part of impedance collapses into one master curve, as shown in Fig. 4. The scaling behavior of Z'' clearly indicates that the relaxation shows the same mechanism at various temperatures.

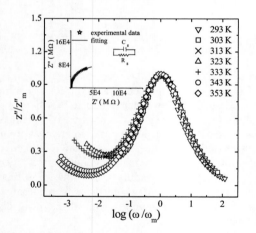

FIGURE 4. Scaling behavior of Z'' at various temperatures for ZnO

The inset of Fig. 4 shows the complex plane impedance plots of ZnO, plotting the imaginary part Z″ against the real part Z′ at 343 K. To analyze, the impedance data are usually modeled by an ideal equivalent electrical circuit comparing of resistance (R) and capacitance (C) [10]. The grain circuit consists of parallel combination of grain resistance (R_g) and grain capacitance (C_g). The equivalent electrical equation can be represented by

$$Z^* = Z' + Z'' = \frac{1}{R_g^{-1} + j\omega C_g} \qquad (3)$$

$$Z' = \frac{R_g}{1 + (\omega R_g C_g)^2} \qquad (4)$$

$$Z'' = R_g \left[\frac{\omega R_g C_g}{1 + (\omega R_g C_g)^2} \right] \qquad (5)$$

CONCLUSIONS

The frequency-dependent dielectric dispersion of ZnO nanoparticles synthesized by soft chemical method is investigated in the temperature range from 293 K to 383 K. The increasing dielectric constant and loss tangent with decreasing frequency is attributed to the conductivity which is directly related to an increase in mobility of localized charge carriers. The frequency-dependence maxima of the imaginary part of impedance are found to obey Arrhenius law with activation energy. Analyses of the real and imaginary part of complex impedance with frequency were performed assuming a distribution of relaxation times as confirmed by Cole-Cole plot as well as the scaling behavior of impedance spectra.

ACKNOWLEDGEMENTS

Authors are thankful to Uttarakhand state council of science and technology (U-COST) for its financial support and Author is also thankful to Prof. T. P. Sinha, Department of Physics, Bose Research Institute, Kolkata for lab support.

REFERENCES

1. J. R. Harbour and M. L. Hair, *J. Phys. Chem.* **83**, 652 (1979).
2. P. Mitra, A. Chatterjee and H. Maiti, *Mater. Lett.* **35**, 33 (1998).
3. T. K. Gupta, *J. Am. Ceram. Soc.* **73**, 1817 (1990).
4. A. V. Dijken, E. A. Mulenkamp, D. Vanmaekelbergh and A. Meijerink, *J. Lumin.* **90**, **1** 123 (2000).
5. S. Hotchandani and P. V. Kamat, *J. Electrochem. Soc.* **113**, 2826 (1991).
6. S. Sakohapa, L. D. Tickazen and M. A. Anderson, *J. Phys. Chem.* **96**, 11086 (1992).
7. K. S. Cole and R. H. Cole, *J. Chem. Phys.* **9**, 341 (1941).
8. A. Dutta and T. P. Sinha, *Int. J. Mod. Phys. B* **21**, 2965-2978 (2007).

9. A. K. Jonscher, Dielectric Relaxation in Solids (Chelsea Dielectric Press, London, 1983).
10. A. Dutta, C. Bharti, T. P. Sinha, Mater. Res. Bull. **43**, 1246-1254 (2008).

Silver Nanoparticles Dispersed in Poly(methyl methacrylate) Thin Films: Spectroscopic and Electrical Properties

Manisree Majumder, Aloke Kumar Chakraborty and Biswanath Mallik[*]

Department of Spectroscopy, Indian Association for the Cultivation of Science,
2A & 2B, Raja S. C. Mullick Road, Jadavpur, Kolkata 700 032, INDIA
* Corresponding Author: e-mail-spbm@mahendra.iacs.res.in

Abstract. Silver nanoparticles of different sizes were prepared following chemical route and using silver nitrate as starting material. The nanoparticles were dispersed in Poly(methyl methacrylate) (PMMA) matrix and thin films were prepared. The PMMA films with dispersed silver nanoparticles were characterized by UV–Vis absorption spectroscopy and FESEM. The electrical measurements were carried out by using surface type cells. The absorption spectra of the films have shown plasmon peaks at different wavelengths with the variation in size of the silver nanoparticles. The film with dispersed silver nanoparticles showing plasmon peak at higher wavelength manifested increasing trend in dark current values with time at room temperature. On the contrary, the decreasing dark current values with time were observed in the case of the film showing plasmon peak at lower wavelength under similar measuring conditions. The photocurrent profile of the PMMA film with dispersed silver nanoparticles as a function of photoexcitation wavelength showed dependence on the position of the plasmon bands.

Keywords: Plasmon band, silver nanoparticles, dark current, photocurrent, thin film, PMMA.
PACS: 73.50 Pz, 81.07 Bc, 81.07 Nb, 78.66 Qn.

INTRODUCTION

In many different areas of science and technology metal nanoparticles play important roles. Metal nanoparticles are utilized in various technological fields such as: biological labeling[1], optoelectronics[2], information storage[3], photography[4] etc. The intrinsic properties of a metal nanoparticle are mainly determined by its size, shape, composition, crystallinity, and structure (solid versus hollow)[5]. For the purpose of utilizing the metal nanoparticles in some specific areas of use for better results one could control (in principle) any one of these parameters to fine-tune the properties of the nanoparticles[5]. It is well known that the size and shape of nanomaterials play a crucial role in their various applications[6]. Efforts have been made for the controlled syntheses of these materials[7]. The use of an inert solid matrix like poly(methyl methacrylate) (PMMA), a well known polymer, facilitates many investigations in spectroscopic and optoelectronic areas. Now-a days, it has been a practice to modulate the properties of polymeric materials, especially optical and electrical properties, by proper formulation of polymeric nanocomposites in which nano-sized particles are dispersed in the polymer network[8]. Earlier studies in our laboratory[9] on electrical conductivity (measured in a sandwich structure) of thin solid films of PMMA with dispersed silver nanoparticles showed anomalous hysteresis in current–temperature characteristics during heating and cooling cycles. Dark current in the PMMA films with dispersed silver nanoparticles has been observed to be higher than the

CP1147, *Transport and Optical Properties of Nanomaterials—ICTOPON - 2009*, edited by M. R. Singh and R. H. Lipson
© 2009 American Institute of Physics 978-0-7354-0684-1/09/$25.00

corresponding current in the PMMA films without silver nanoparticles due to the creation of conduction paths by the silver nanoparticles/ nanoclusters. Also, a decrease in photocurrent under illumination of light was observed due to the destruction of conduction paths by the illumination of light from a mercury lamp[10]. Ferrocene-dispersed PMMA thin films, prepared from the mixture of solvents, benzene and chloroform have shown significant changes in the current versus time profile depending on the amount of chloroform molecules present in the films[11]. Plasmonic metal nanoparticles have great potential for chemical and biological sensor applications, due to their sensitive spectral response to the local environment of the nanoparticle surface and ease of monitoring the light signal due to their strong scattering or absorption[12]. In the present work, we report the dependence of electrical behavior of the silver nanoparticles on the surface plasmon resonance band position due to the various size of the nanoparticles (confined in the PMMA inert matrix).

EXPERIMENTAL

Solutions were prepared : (i) 75mg of poly(methyl methacrlylate) [PMMA, average MW 15,000, Aldrich, USA] in 6ml chloroform (Spectroscopic grade, SRL, India); (ii) 350mg of silver nitrate (GR grade, Merck, India) in 15 ml of distilled water ; (iii) 507mg of sodium hydroxide (purified, E.Merck, India) in 15 ml distilled water. Then, soln.(iii) was added to vigorously stirred soln. (ii) and stirring continued until precipitation appeared. The precipitate was collected, washed with distilled water several times, dissolved in 25 ml diluted ammonia solution (25% water solution, GR grade, Merck, India) and was vigorously stirred. Soln.(i) was added into the stirring solution of ammonia and few drops of formic acid (Merck, Germany) were added in to it until the solution separated into two distinct layers. The lower layer was separated by passing the whole solution through a suitably designed glass pipe attached with a stop cock at lower end. The extreme upper portion of this separated solution (containing silver nanoparticles and PMMA) was used for characterization purpose and making spin coated thin film[10]. The sizes of the silver nanoparticles were controlled by varying the parameters involved in chemical synthesis and thereby considering temperature, humidity etc., during the experiment. The spectroscopic studies of the films at room temperature (300K) in the range of 200 to 800 nm were carried out by an UV-Vis spectrophotometer, Model UV-2401 PC (Shimadzu, Japan). Dark and photo currents of the films (following heating and cooling treatments) were measured (electrometer, Model 617, Keithley Inst., Cleveland, OH, USA) with silver paste contacts and 5V bias in vacuum using a surface type cell placed in a specially designed conductivity chamber fitted with a quartz window for light incidence. The mentioned electrometer was used for 5V bias supply and the studies of the I-V characteristics. Temperature was controlled by a temperature controller (Model: Eurotherm 2404, UK). A Xenon lamp (Spectral Energy, USA) was used as the light source. The maximum irradiated power reaching the sample cell was about 17.12 mW / cm^2. FESEM (Model: JSM – 6700F, JEOL, Japan) was used to record the scanning electron micrograph images of the PMMA films with and without silver nanoparticles.

RESULTS AND DISCUSSION

Absorption spectra of two similar thin solid films of PMMA with dispersed silver nanoparticles are shown in Fig.1. Fig.1a shows strong absorption band peaked around 428 nm having larger sized silver nanoparticles embedded in PMMA matrix while Fig.1b represents that of dispersed smaller sized silver nanoparticles showing the absorption band peaked around 414 nm. Both bands are due to surface plasmon resonance absorption (referred to as plasmon band) of the silver nanoparticles dispersed in PMMA matrix.

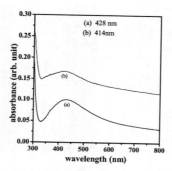

FIGURE 1. Absorption spectra for PMMA films with dispersed silver nanoparticles showing plasmon band peaked around: (a) 428nm, (b) 414nm.

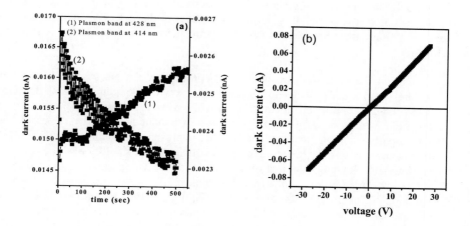

FIGURE 2. (a) Plots of dark current vs time with bias 5V for silver nanoparticles dispersed PMMA films showing plasmon bands at : (1) 428 nm (right scale), (2) 414 nm (left scale); (b) I-V characteristic for silver nanoparticles dispersed PMMA film showing plasmon band at 414 nm.

The dark current measured at room temperature 300K with 5V bias voltage in the surface cell of the PMMA film with dispersed silver nanoparticles of different sizes are shown in Fig.2a. The larger sized silver nanoparticles having surface plasmon peak around 428 nm shows increasing trend with increasing time as shown in Fig.2 (curve1). On the contrary, decreasing trend in dark current values with increasing time was observed from Fig.2 (curve 2) under identical measurement conditions in PMMA films with dispersed silver nanoparticles of smaller sizes, showing the surface plasmon peak around 414 nm.

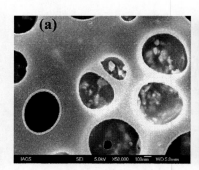 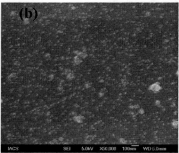

FIGURE 3. FESEM image of PMMA film (a) with dispersed silver nanoparticle (magnification: 50,000), (b) without silver nanoparticles (magnification: 50,000).

FESEM images of PMMA film with and without dispersing silver nanoparticles are shown in Fig. 3a and b, respectively. The Fig.3a shows the existence of various pores created possibly during the process of dispersing silver nanoparticles and in Fig.3b the image represents irregular geometrical structures of the inert matrix of PMMA.

During the measurement of electrical properties, the electric field is applied across the sample cell. The electric field is known to manipulate and assemble particles/ nanoparticles on an electronic chip with microelectrodes[13]. In presence of an electric field there is a possibility of formation of conduction path by the assembled silver nanoparticles in the pores/ voids. Due to the presence of the pores in the PMMA films, it is obvious that for larger silver nanoparticles the possibility of the creation of conduction paths is better than that of the smaller silver nanoparticles dispersed in the PMMA films. Also, the PMMA polymer matrix is known to possess structural reorientation[14]. Under the action of electric field, the observed increase in dark current with the passage of time in case of larger silver nanoparticles dispersed in PMMA film may be due to formation of time dependent better conduction paths. But in the case of smaller nanoparticles, the decrease in dark current may be due to the creation of weaker conduction paths with the passage of time. Therefore, the increase in dark conductivity (dark current) of the film containing the dispersed silver nanoparticles in PMMA matrix having surface plasmon band around 428 nm may be attributed to the conductive phase formed by dispersed silver nanoparticles (of larger size) in polymer matrix[15]. The dark I-V characteristic of some of the PMMA films with dispersed silver nanoparticles were studied in our laboratory in the voltage range of 0-30V for both forward and reverse bias and it was found to be ohmic in nature (Fig.2b).

The photocurrent in the surface cell of the PMMA films with dispersed silver nanoparticles showing absorption band positions around 428 and 414 nm were measured with respect to the increasing photoexciting wavelength of light incident from Xenon lamp. Although the photocurrent values are small, such current profiles have been considered as the photocurrent profile (for having an idea of their behaviour). These profiles at room temperature 300K with 5V bias are presented in Fig.3a for 428 nm and in Fig.3b for 414 nm band positions, respectively. The film containing larger sized nanoparticles (band position 428 nm) showed the photocurrent vs. wavelength profile of irregularly overall decreasing trend (Fig.4a) with slight wavy nature.

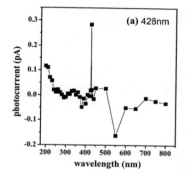 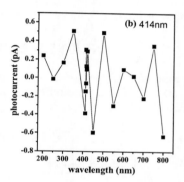

FIGURE 4. Plot of photocurrent with photoexciting wavelength for silver nanoparticles dispersed PMMA film showing plasmon band at (a) 428 nm, (b) 414 nm.

Comparatively, in case of the photocurrent vs. wavelength profile (Fig.4b) for the film with smaller sized silver nanoparticles (band position 414 nm) the change in photocurrent with photoexcitation wavelength shows larger fluctuations. In fact, Fig.4b shows more pronounced wavy trend of the photocurrent with the increasing wavelength of exciting light. It is known that illumination of light destroys the conduction paths created by the metal nanoparticles/nanoclusters[10]. The size of the nanoparticles seems to be a factor for creation of conduction path (in the silver nanoparticle dispersed PMMA film under study) as well as fluctuating behavior / instability in photocurrent. Fig. 4(a) and (b) show such an effect under such photoexcitation. The more pronounced wave like profile of photocurrent in the case of smaller size of dispersed silver nanoparticles in PMMA matrix is not clearly understood. However, factors like size of the dispersed silver nanoparticles, manipulation of silver nanoparticles in the PMMA matrix by the electric field[13], structural reorientation of the PMMA matrix[14], surface morphology of the PMMA films (i.e., presence of pores/ voids) and photoinduced changes in PMMA films are possibly responsible for controlling the more pronounced wavy nature of photocurrent observed in Fig. 4b. Further investigations are in progress.

CONCLUSIONS

PMMA film with dispersed silver nanoparticles representing plasmon band at higher wavelength around 428 nm is accounted for larger silver nanoparticles and the band around 414 nm for smaller ones. The FESEM images of the PMMA matrix/ thin films with and without dispersed silver nanoparticles show different surface morphologies. The electrical properties like dark and photoconductivity are observed to depend drastically on the position of the plasmon bands, i.e., size of the silver nanoparticles dispersed in PMMA matrix. Other controlling factors for the observed changes in electrical properties have been attributed to the manipulation of silver nanoparticles in the PMMA matrix by the electric field, structural reorientation of the PMMA matrix, surface morphology of the PMMA films (i.e., presence of pores/ voids). Photoinduced changes are additional factors in the case photoconductivity.

ACKNOWLEDGMENTS

The authors express thanks to the authorities of IACS for providing working facilities. One of the authors AKC is thankful to the concerned authorities for the approval of leave and research project.

REFERENCES

1. S. R. Nicewarner-Peña, *et al.*, *Science* **294**, 137-141 (2001)
2. P. V. Kamat, *J. Phys. Chem. B* **106**, 7729-7744 (2002)
3. C. B. Murray, S. Sun, H. Doyle, T. Betley, Mater. *Res. Soc. Bull.* **26**, 985 (2001).
4. D. M.K. Lam, B. W. Rossiter, *Sci. Am.* **265**, 80 (May 1991).
5. Yugang Sun and Younan Xia; *Science* **298,** 2176-2179 (2002) .
6. a) M. A. El-Sayed, *Acc. Chem. Res.* **34,** 257 (2001).
    b) A. C. Templeton, W. P. Wuelfing, and R. W. Murray, *Acc. Chem. Res.* **33,** 27-36 (2000).
    c) L. N. Lewis, *Chem. Rev.,* **93,** 2693-2730 (1993).
    d) Y. Cui, Q. Wei, H. Park, and C. M. Lieber, *Science* **293**, 1289-1292 (2001).
    e) P.-A. Brugger, P. Cuendet, and M. Gratzel, *J. Am. Chem. Soc.* **103**, 2923-2927 (1981).
7. Zhen Xie, Zhiyong Wang,Yanxiong Ke, Zhenggen Zha, and Chao Jiang; *Chemistry Letters* **32**, 686-687 (2003).
8. H.C. Ling, W.R. Holland, H.M. Gordon, *J. Appl. Phys.***70**, 6669-6673 (1991).
9. A.Thander, B. Mallik, *Chem. Phys. Lett.* **330,** 521-527 (2000).
10. D. Basak, S. Karan, B. Mallik, *Solid State Commun.***141**, 483–487 (2007).
11. D. Basak and B. Mallik; *Synth. Met.* **156,**176-184 (2006).
12. Kyeong-Seok Lee and Mostafa A.El-Sayed, *J. Phys. Chem. B* **110**, 19220-19225 (2006).
13. O. D. Velev, K. H. Bhatt, *Soft Matter* **2**, 738-750 (2006).
14. Chunxia Chen, Janna K.Maranas, Victoria Garcia-Sakai, *Macromolecules* **39**, 9630-9640 (2006).
15. N. L. Singh, A. Qureshi, A. K. Rakshit and D. K. Avasthi, *Bull. Mater. Sci.* **29,** 605–609 (2006).

Study of H$_2$S Sensitivity of Pure and Cu Doped SnO$_2$ Single Nanowire Sensors

Vivek Kumar[a], Shashwati Sen[b], K. P. Muthe[b], N. K. Gaur[a], R. K. Singh[c], Bhushan Dhabekar[d] and S. K. Gupta[b]*

[a]Department of Physics, Barkatullah University, Bhopal, M. P.- 462026, INDIA
[b]Technical Physics & Prototype Engineering Division, [d]Radiological Physics & Advisory Division,
Bhabha Atomic Research Centre, Mumbai,400085, INDIA
[c]School of Basic Sciences, MATS University, Raipur, C.G. – 492002, INDIA
*Corresponding author: e-mail: drgupta@barc.gov.in

Abstract. Pure and Cu doped SnO$_2$ nanowires and nanobelts were grown in a tubular furnace under argon atmosphere at ambient pressure. X-ray diffraction showed that Cu gets incorporated into the SnO$_2$ lattice. Isolated single nanowire sensors have been fabricated for detection of H$_2$S gas at room temperature. Cu doping was found to reduce the sensitivity of these sensors in contrast to results reported for Cu doped SnO$_2$ thin films. Photoluminescence studies indicate that the reduction is sensitivity may be due to reduced defect density.

Keywords: SnO$_2$, nanowire, gas sensor, H$_2$S.
PACS: 62.23.Hf; 07.07.Df

INTRODUCTION

Tin dioxide (SnO$_2$) is an n-type semiconductor with wide band gap ($E_g \sim 3.6$ eV) that has been extensively used for detection of various oxidizing and reducing gases [1, 2]. These sensors can detect many of the toxic gases and therefore have poor selectivity. They also need and high operating temperatures for detection. For practical applications, it is desirable to improve selectivity and reduce operating temperature. Various dopants have been added to improve selectivity. For example, doping of SnO$_2$ thin films with CuO is known to improve its selectivity for the detection of H$_2$S gas [1-4]. Nanostructured materials enable operation of sensors at lower temperatures and therefore, much attention has been paid to the synthesis of SnO$_2$ nanostructures such as nanoparticles [5], nanowire [6], and nanobelts [7, 8] etc. Single nanowires or nanobelts have also been employed for fabrication of miniature sensors of some oxide materials [9].

In the present study, we report the growth of pure and Cu doped SnO$_2$ nanowires for their application as H$_2$S gas sensor. The study has been carried out to (a) investigate if copper can be doped in SnO$_2$ nanowires and if it leads to improvement in sensitivity as reported for CuO doping in SnO$_2$ thin films, (b) reduce operating temperature due to higher reactivity of nano-materials and (c) enable fabrication of

CP1147, Transport and Optical Properties of Nanomaterials—ICTOPON - 2009, edited by M. R. Singh and R. H. Lipson
© 2009 American Institute of Physics 978-0-7354-0684-1/09/$25.00

miniature single wire sensors. The results showed that the sensitivity of single wire sensors decreases on doping with in contrast to results reported on thin films [2].

EXPERIMENTAL DETAILS

Pure SnO$_2$ nanowires/belts were grown by thermal evaporation of tin powder at atmospheric pressure in a tubular furnace. The powder was placed in an alumina boat and growth was carried out under flowing high purity argon gas (with ~0.1% oxygen) at a temperature of 900°C for 1 hour. Tin powder was found to evaporate and white color, wool like material consisting of nanowires and nanobelts was seen to accumulate on the walls of alumina boat. Growth of Cu doped SnO$_2$ nanowires/belts was similarly carried out by using a mixture of tin and copper (5% by weight) powders. All these samples were characterized by scanning electron microscopy (SEM) (TS 5130 MM TESCAN), energy dispersive X-ray analysis (EDX) and X-ray diffraction (XRD) carried out by employing a Cu K$_\alpha$ radiation source. The photoluminescence spectra of the samples were obtained by Hitachi made (F-4500) fluorescence spectrophotometer.

For the fabrication of isolated single nanowire sensors, a drop of very dilute suspension of nanowires in methanol was placed on a glass substrate. A metal mask (for contacts deposition) with 12 µm width was aligned to a single nanowire using an optical microscope. The alignment was made possible by long length of nanowires (100-500 µm) and solution sticking to wires by surface tension. Gold pads with 12 µm spacing were thermally evaporated and electrical contacts were made by soldering silver wires to them. Sensitivity of these single wire sensors (to H$_2$S gas) was measured at different temperatures using a measurement setup that has been described elsewhere [10].

RESULTS AND DISCUSSION

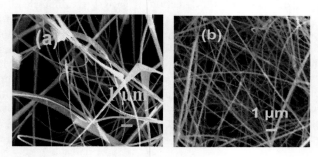

FIGURE 1. SEM images of (a) pure and (b) 5% Cu doped SnO$_2$ nanostructures.

Fig. 1 shows typical scanning electron microscopy (SEM) images of powders obtained using pure and 5% Cu doped SnO$_2$ materials. The samples consist of nanowires and nanobelts with diameter of nanowires and thickness of nanobelts in 10-100 nm range. The length of nanowires is seen to be hundreds of micrometers and doping with Cu is found to reduce their diameter. The growth of nanowires may be explained using vapor-solid (VS) growth mechanism. The source material used in our experiments is elemental tin and it melts into liquid droplets at temperature of 900°C (m. p. 232°C, b. p. 2260°C). During heating, the Sn vapors generated from the powder combine with oxygen to form SnO vapors. It is well known that SnO is metastable and decomposes into SnO_2 and Sn. The decomposition of SnO results in the precipitation of SnO_2 nanoparticles, which are deposited on the walls of the alumina boat. These nanoparticles then act as nucleation sites and initiate the growth of SnO_2 nanowires/belts via the VS growth mechanism [11]. High resolution transmission electron microscopy (HRTEM) analysis (Fig. 3) showed good crystalline nature of nanowires and their growth along [101] direction. EDX analysis of copper doped samples showed presence of ~ 1.5% (atomic fraction) copper indicating that copper is doped into nanowires [2]. Fraction of copper incorporated in SnO_2 lattice was found to be nearly independent of fraction of Cu in starting material (in 5-20% range) indicating that it is limited by solubility of Cu in nanowires.

XRD of pure and Cu doped SnO_2 nanowires (shown in Fig. 3) could be indexed to tetragonal phase with lattice parameters close to that of SnO_2. Least square fitting of all peaks showed that lattice constants (a and c) and unit cell volume reduce on doping from 4.740 Å, 3.1885 Å and 71.6397 Å^3 for undoped samples to 4.7394Å, 3.1881Å and 71.6132 Å^3 respectively for doped SnO_2 samples. Lattice parameters reduction is attributed to smaller size of Cu ion (0.57 Å) compared to that of Sn ion (0.69 Å). These results are similar to those reported for Co doping [12] of SnO_2 nanowires and indicate that Cu gets incorporated in SnO_2 nanowire lattice in contrast to results of earlier studies on thin films where it is reported that CuO and SnO_2 are immiscible [2].

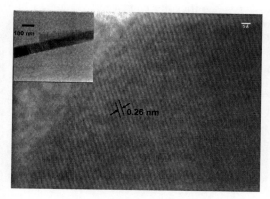 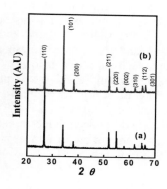

FIGURE 2. HRTEM image of a SnO_2 nanowire. Lattice spacing of 0.26 nm corresponds to (101) planes. Inset shows enlarged view of a nanowire.

FIGURE 3. XRD spectra of (a) pure and (b) 5% Cu doped SnO_2 nanowires

SEM of typical single wire sensor is shown in Fig. 4. We have measured the response (towards H_2S gas) of pure and Cu doped sensors at room temperature as a function of gas concentration and typical results are shown Fig. 5. Here, the response is defined by $S = R_a / R_g$ (where R_a and R_g are the resistances of the sensor in air and gas respectively). The response of pure SnO_2 nanowires is found to be nearly 1000 times more than that of doped wires. Dynamic response and recovery characteristics of sensors were measured for 50 ppm of gas and the typical results for a pure SnO_2 sensor are shown in Fig. 6. From these characteristics, response time defined as the time taken to attain 90% of decrease in resistance and the recovery time defined as the time taken by the sensor to return back to 10% of its original resistance were determined. The response and recovery times were found to be nearly 7 and 2 min, respectively. Good sensitivity, response and recovery times indicate that SnO_2 single wires can be employed as H_2S gas sensor operable at room temperature.

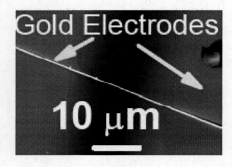

FIGURE 4. SEM image of a single SnO_2 nanowire between gold electrodes.

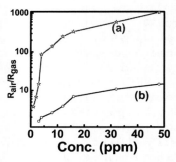

FIGURE 5. Typical response (at room temperature) as function of H_2S gas concentration for (a) pure and (b) Cu doped SnO_2 nanowire sensors.

The lower sensitivity of Cu doped single nanowires is in contrast to the results reported for thin films where the sensitivity is significantly enhanced by addition of Cu to SnO_2 [2]. A reason for this difference could be that CuO is immiscible with SnO_2 in polycrystalline material and CuO and SnO_2 are believed to form separate grains and random p-n junction responsible for high sensitivity. On the other hand, Cu is incorporated in SnO_2 nanowire lattice yielding a homogeneous material. The reported maximum sensitivity of pure and doped SnO_2 thin films for H_2S is at temperature of nearly 150°C [2-4]. Therefore, single nanowire sensors were also investigated for H_2S detection at 150°C and results are shown in Fig 7. The sensitivity was found to decrease with increase in operating temperature and pure SnO_2 nanowire shows a response of only 1.6 times at 50 ppm H_2S gas. These results are in contrast to that of thin films.

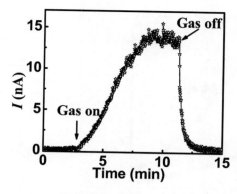

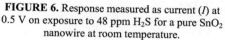

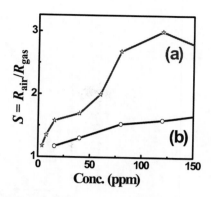

FIGURE 6. Response measured as current (*I*) at 0.5 V on exposure to 48 ppm H₂S for a pure SnO₂ nanowire at room temperature.

FIGURE 7. Response as function of concentration at operating temperature of 150°C for; (a) pure (b) 5% Cu doped SnO₂ nanowire sensors.

To understand the effect of Cu doping on SnO_2 nanowires we have recorded the photoluminescence (PL) spectra of pure as well as Cu doped SnO_2 nanowires and the results are shown in Fig 8. PL is a suitable technique to determine the defect density such as presence of oxygen vacancies in the materials which are helpful in understanding its gas sensing behaviour. It is seen that the PL emission spectrum of the pure SnO_2 nanowires is dominated by a broad yellow emission band with maximum intensity around 560 nm as reported earlier [13]. On the other hand, PL emission of Cu doped sample is significantly reduced. Broad PL band between 400–600 nm for one dimensional SnO_2 nanostructures has been attributed to Sn or O vacancies formed during the growth process inducing trapping states in the band gap. Thus Cu doping is found to decrease defects density in SnO_2. As oxygen from atmosphere is preferentially adsorbed at defect sites and interaction of H_2S with this oxygen is responsible for sensor response, a reduction in defect density may lead to reduction in sensitivity. Further investigations are required to fully understand the changes induced in SnO_2 nanowires by Cu doping.

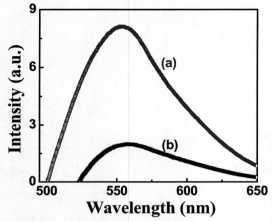

FIGURE 8. Room temperature PL spectra of (a) pure and (b) Cu doped SnO_2 nanowires.

CONCLUSIONS

In summary, the present study shows that isolated single nanowire of pure SnO_2 can be employed as room temperature H_2S gas sensor with high sensitivity as well as fast response and recovery. It is found that Cu may be doped in SnO_2 nanowire lattice in contrast to immiscibility of CuO and SnO_2 in polycrystalline form. Cu doping is found to decrease the sensitivity of these nanowire sensors which has been attributed to decrease in defect density in SnO_2 nanowire on Cu doping as revealed by PL spectroscopy.

ACKNOWLEDGEMENTS

One of the authors, Vivek Kumar, is grateful to UGC Delhi for financial support and TFDS/TPPED, BARC, Mumbai for providing experimental facilities.

REFERENCES

1. W. Yuanda, T. Maosong, H. Xiuli, Z. Yushu, D. Guorui and W. Yuanda, *Sens. Acts. B* **79,** 187 (2001).
2. V. R. Katti, A. K. Debnath, K. P. Muthe, M. Kaur, A. K. Dua, S. C. Gadkari, S. K. Gupta and V. C. Sahni, *Sens. Acts. B* **96,** 245 (2003).
3. A. Khanna, R. Kumar and S. S. Bhatti, *Appl. Phys. Lett.* **82,** 4388, (2003).
4. A. Chowdhuri, V. Gupta, K. Sreenivas, R. Kumar, S. Mozumdar and P. K. Patanjali, *Appl. Phys. Lett* **84,** 1180 (2004).
5. A. M. Mazzone, *J. Phys.: Condens. Matter* **14,** 12819 (2002).
6. A. Kolmakov, Y. Zhang, G. Cheng and M. Moskovits, *Adv. Mater.* **15,** 997 (2003).
7. J. Q. Hu, X. L. Ma, N. G. Shang, Z. Y. Xie, N. B. Wong, C. S. Lee and S. T. Lee, *J. Phys. Chem. B* **106,** 3823 (2002).

8. Z. W. Pan, Z. R. Dai, Z. L. Wang, *Science* **291**, 1947 (2001).
9. A. Kolmakov, D. O. Klenov, Y. Lilach, S. Stemmer and M. Moskovits, *Nano Lett.* **5**, 667 (2005).
10. S. Sen, K. P. Muthe, N. Joshi, S. C. Gadkari, S. K. Gupta, Jagannath, M. Roy, S. K. Deshpande, J. V. Yakmi, *Sens. Actuators B* **98**, 154 (2004).
11. T. Gao, T. Wang, *Materials Research Bulletin* **43**, 836 (2008).
12. A. Bouaine, N. Brihi, G. Schmerber, C. Ulhaq-Bouillet, S. Colis and A. Dinia *J. Phys. Chem. C* **111** 2924 (2007).
13. S. Luo, J. Fan, W. Liu, M. Zhang, Z. Song, C. Lin, X. Wu and P. K. Chu, *Nanotechnology*, **17**, 1695 (2006).

ZnO nanoparticle synthesis in presence of biocompatible carbohydrate starch

Priya Mishra,[a] Raghvendra S. Yadav [a] and Avinash C.Pandey[a,b]

[a]Nanophosphor Application Centre, University of Allahabad, Allahabad-211002, India.

[b] Physics Department, University of Allahabad-211002, India.

Corresponding author E-mail: priya_mishra15@rediffmail.com. Mob: +91-0532-

2460675, Fax: +91-0532-2460675.

Abstract. ZnO nanocrystals were successfully prepared through a simple and non-toxic route under ultrasonic irradiation using different amounts of starch. The nanocrystals were characterized by means of X-ray diffraction (XRD), photoluminescence spectroscopy (PL) and E-SEM (enviornmental mode). The X-ray diffraction analysis suggested that the ZnO nanocrystals were of wurtzite structure and ESEM showed flower-like morphology. Four different concentrations of Starch was used and pH was also varied to see the effect on optical property.

Keywords: ZnO, nanocrystal, starch, biocompatible.
PACS: 81.07.-b, 82.35.Pq.

INTRODUCTION

Among inorganic II-VI semiconductors, ZnO with a bandgap of 3.37eV and high exciton binding energy of 60meV [1] has many potential applications ranging from optoelectronic to bio-imaging [2-5].Recently much work has been done to modify the surface of nanocrystals depending on the application, for example water soluble surface is required for application as biological label. Polysaccharides such as cellulose, chitin, and starch constitute an important class of biopolymers among them starch is a renewable carbohydrate polymer which can be obtained from a wide variety of crops. These protective biocompatible polymers can effect particle size, optical property as well as morphology in combination with sonochemical method. In this work we have used starch and taken its four different amounts 0.025g, 0.05g, 1.00g and 2.00g. Ultrasonic method is known to form novel materials, ultrasonic irradiation causes acoustic cavitation, in which there is formation, growth and collapse of bubbles in liquid medium causing very high temperature and pressure [6]. Starch undergoes gelatinization in presence of water and high temperature. During ultrasonication due to cavitation phenomenon very high temperature and pressure is reached this provides ideal condition for starch to gelatinize along with formation of ZnO nanocrystals. On cooling, ZnO nanocrystals formed have plenty of hydroxyl groups of starch which acts as a capping agent.

CP1147, *Transport and Optical Properties of Nanomaterials—ICTOPON - 2009,* edited by M. R. Singh and R. H. Lipson
© 2009 American Institute of Physics 978-0-7354-0684-1/09/$25.00

EXPERIMENTAL

Chemicals

The zinc acetate, sodium hydroxide, starch and glacial acetic acid were from E. Merck Ltd., Mumbai, 400018, India. These chemicals were directly used without special treatment.

Sample Preparation

10 ml 0.1M aqueous solution of Zinc Acetate and 10 ml 1M aqueous solution of sodium hydroxide was added to 25 ml alcohol followed by starch. The amount of starch was taken 0.025g, 0.05g, 1.00g and 2.00g in four different sets of experiment. The whole solution was then kept in ultrasonic bath (33 kHz, 350 W) for 3 hours. Solution was then centrifuged and precipitate washed twice with double distilled water and then with alcohol. White precipitate was then kept in vacuum oven for drying. In another set of experiment keeping all the conditions same and taking 1.00g starch pH was varied using glacial acetic acid. Change in photoluminescence property was checked for pH= 3 and pH= 7 and was compared with that of original solutions which had pH=14.

Instrumentation

The crystal structure of ZnO nanoparticles were characterized by X-ray diffraction (XRD, Rigaku D / MAX- 2200 H/PC, Cu K_α radiation). E-SEM (environmental mode) image was taken on Quanta 200FEG (FEI company). The photoluminescence study was carried out on Perkin-Elmer LS 55 spectrometer.

RESULT AND DISCUSSION

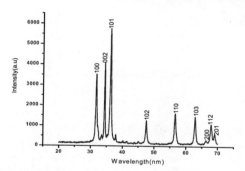

FIGURE 1. XRD pattern of ZnO nanocrystals synthesized under ultrasonic irradiation in presence of biocompatible carbohydrate starch

Figure1 shows XRD pattern of ZnO nanocrystals prepared in presence of starch (0.025g) under ultrasonication, it can be seen that all the peaks belong to hexagonal (JCPDF Card File No. 36145) ZnO, also peaks belonging to any other phase are not detected indicating that there are no impurities. Broad peaks indicate smaller crystal size.

Morphology Study

Figure 2 shows E-SEM image of ZnO nanocrystal capped with 0.025g starch under ultrasonication showing flower like morphology. There are many flat nanorods clumping together and radially coming out giving the appearance of a flower, their length varies from several hundred nanometers to 1 μm and width is in hundred nanometers, while average size of whole flower is about 1.5 μm. Due to cavitation phenomenon, occurring during ultrasonication very high temperatures and pressures [7,8,9] are reached, these conditions help in simultaneous gelatinization and formation of ZnO nanocrystals. During ultrasonication of starch in presence of water leaching of amylose occurs [10,11,12,13], amylose forms coil around ZnO nanoparticles forming nanorods which finally clump together giving the shape of flower.

FIGURE 2. ESEM image of ZnO nanocrystals showing flower like morphology

Photoluminescence Study

Figure 3(a) shows photoluminescence emission (λ_{exc}=325 nm) spectrum of ZnO nanocrystals synthesized in presence of starch with amounts 0.025g, 0.05g, 1.00g 2.00g and ultrasonic irradiation time 3 h. Several emission bands are present, with three emissions common to all the ZnO nanocrystals, weak ultraviolet emission at 415nm, blue-green emission at 480nm and green emission at 525nm. Ultraviolet emission is generally believed to be due to recombination of excitons [14,15] green emission is due to singly ionized oxygen vacancy in ZnO which results from recombination of a photon generated hole with single ionized charge state of this defect [16] and blue green emission arises from electron transition from level of ionized oxygen vacancies to the valence band. Guo L et.al.[17] have shown enhanced U.V photoluminescence and reduced green emission in PVP capped ZnO nanoparticles. This result has been attributed to nearly perfect surface passivation of ZnO nanoparticles by PVP. N.Vigneshwaran et.al.[18] have reported weak green emission from ZnO nanoparticles stabilized by starch which is due to few surface defects. In our work as we increased the amount of starch, green emission at 525nm was reduced, and defects were removed considerably which is in well agreement with previously reported results and this is attributed to good defect removal of ZnO nanocrystals by starch. It is notable, that in comparison to nanorods and nanowires [19] the UV emission exhibits a red shift of about 31nm. Findings show that the band-edge emission peak shift from 384 nm might be because of band-edge emission variations caused by native defects concentration in the different nanostructures [20]. Photoluminescence spectrum of ZnO, as shown by previous results, is sensitive to particle shape, size, temperature, preparation method [21,22,23,24]. Fig.2(b) shows effect of pH for two values pH=3 and pH=7 and compared with original solutions pH=14. As pH was lowered to 7 band-edge emission got quenched, defects got removed and photoluminescence intensity was increased considerably. Three emissions became more prominent with UV emission at 414 nm, blue-green emission at 470 nm and green emission at 520 nm. When pH was further lowered to 3, all the emission peaks were same however, photoluminescence intensity was slightly decreased it may be due to variation of amounts of defects with variation of pH.

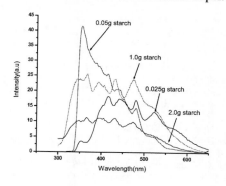

FIG.3. (a)

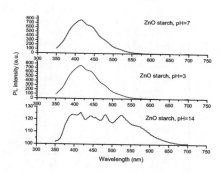

FIG.3 (b)

FIGURE 3. (a) Photoluminescence spectrum of ZnO nanocrystals synthesized by sonochemical method in presence of 0.025g, 0.05g, 1.00g and 2.00g of starch and Fig. 3(b) photoluminescence spectra of ZnO nanocrystals synthesized in presence of 1.00g starch at pH=14 (pH of original solution), pH=7 and pH=3.

CONCLUSION

In conclusion, nanocrystals of ZnO with flower-like morphology were synthesized by sonochemical method. Starch which acts as capping agent has been able to control the morphology and also influence the photoluminescence property of synthesized nanocrystals. There are plenty of hydroxyl functional groups present in starch which make these nanocrystals water dispersible; also these capped nanocrystals can be further linked to drug for potential application in biomedical area.

ACKNOWLEDGEMENTS

The authors are very grateful and also wish to express their gratitude to all the scientific members of Nanophosphor Application Centre, University of Allahabad, Allahabad, India. This work was financially supported by DST, India.

REFERENCES

1. E.M Wong, P.C Searson, *Appl.Phys.Lett.* **74**, 2939 (1999).
2. J.K Jaiswal and S. M. Simon, *Trends Cell Biol.* **14**, 497 (2004).
3. X. Gao, L Yang, J.A Petros, F. F Marshall, J. W Simon and S.Nie, *Curr. Opin. Biotechnol.* **16**, 63 (2005).
4. X Gao , Y. Cui ,R.M. Levenson , L.W.K Chung, S. Nie, *Nat. Biotechnol.* **22**, 969 (2004).
5. M. Ozkan, *Drug Discov. Today.* **9**, 1065 (2004).
6. K.S.Suslick, VCH. Weinheim, *Ultrasound: its chemical, physical and biological effects.* 1988.
7. K.S Suslick, G.J Price, *J. Ann. Rev. Mater. Sci.* **29**, 295–326 (1999).
8. K.S Suslick, D.A. Hammerton, R.E.Cline, *J. Am. Chem. Soc.* 108, 5641–5642 (1986).
9. V.G Pol, R. Reisfeld, A. Gedanken. *Chem. Mater. 14*, 3920–3924 (2002).
10. T. A. Waigh, M.J. Gidley, B.U. Komanshek, A.M. Donald, *Carbohydr. Res. 328*, 165–176 (2000).
11. P.J. Jenkins, A.M.Donald, *Carbohydr. Res.* **308**, 133–147. (1998).
12. N. J. Atkin, R. M. Abeysekera, S.L Cheng, A.W. Robards, *Carbohydr. Polym.* **36**, 173–192 (1998).
13. W. A. Atwell, L.F. Hood, D.R. Lineback, E. Varriano-Marston, H.F. Zobel, *Cereal Foods World.* **33**, 306–311 (1998).
14. J. Zhang, L.Sun, J.Yin, H.Su, C. Liao, C. Yan, *Chem.Mater.* **14**, 4172 (2002).
15. M.H.Huang, Y.Y.Wu, H.N.Feick, N.Tran, E.Weber, P.D. Yang, *Adv.Mater.* **13**, 13 (2001).
16. S. Monticone, R. Tufeu, A.V. Kanaev, *J.Phys.Chem. B.* **102**, 2854 (1998).
17. L.Guo, S. Yang, C. Yang, P. Yu, J. Wang, W. Ge, G.K.L Wong, *Appl.Phys.Lett.***76**, 2901-2903 (2000).
18. Nadanathangam Vigneshwaran, Sampath Kumar, A.A. Kathe, P.V. Varadarajan and Virendra Prasad, *Nanotechnology.* **17**, 5087-5095 (2006).
19. Z.Q. Li, Y.J. Xiong , Y. Xie , *Inorg. Chem.* **42**, 8105 (2003).
20. A.B. Djurisic, Y.H. Leung, *Small.* **2**, 944 (2006).
21. L. Guo, Y.L. Ji, H.B.Xu, P. Simon, Z.Y. Wu, *J.Am.Chem.Soc.* **124**, 14864 (2002).
22. J.L. Yang, S.J. An, W.I. Park, G.C. Yi, W. Choi, *Adv. Mater.* **16**, 1661 (2004).
23. Z.Q. Li, Y. Ding , Y.J. Xiong , Q. Yang, Y.I Xie , *Chem. Eur. J.* **10**, 5823 (2004).
24. J.F, Coley, L. Stecker, Y. Ono, *Nanotechnology.* **16**, 292 (2005).

Influence of Solvent on Size and Properties of ZnO Nanoparticles

Bikas Ranjan[a], M. Kailasnath[b], Nishant Kumar[a], P. Radhakrishnan[a], Shivanand Achari[c] and V.P.N. Nampoori[b]

[a]Centre of Excellence in Lasers and Optoelectronic Sciences, [b]International School of Photonics, [c]School of Environmental Sciences, Cochin University of Science and Technology, Kochi 682022, India.

Abstract. The nanosized zinc oxide powder was synthesized in ethylene glycol, methanol and toluene solvents using zinc oxalate and oxalic acid as precursors without using any surfactant. A variety of techniques like UV-VIS absorption spectroscopy, X-ray diffraction (XRD), photoluminescence (PL), Fourier transform infrared spectroscopy (FT-IR) and Scanning electron microscopy (SEM) were used to carry out structural and spectroscopic characterizations of the nanoparticles. The size and optical properties were investigated against used initial solvent and it was found that the solvent plays a key role in determining the size and property of zinc oxide nanoparticles.

Keywords: Optical properties; Photoluminescence; Crystal structure
PACS Nos: 78.67.Bf; 78.55.Et; 61.05.cp

1. INTRODUCTION

ZnO semiconductor nanoparticles are very important in the ongoing research activity across the world. In the past decade, synthesis of low-dimensional semiconductor nanostructures has been of great interest with the development of novel optical, electronic, magnetic, and catalytic materials. As the semiconductor particles exhibit size-dependent properties like scaling of the energy gap and corresponding change in the optical properties, they are considered as the front runners in the technologically important materials. It attracts tremendous attention due to its interesting properties like wide direct band gap of 3.37eV at room temperature and high exciton binding energy of 60meV and high melting temperature (2248K). ZnO has unique electrical and optical properties, such as low dielectric constant, high chemical stability, good photoelectric and piezoelectric behaviours. In order to obtain high-quality ZnO with better industrial performance, it is widely used in a number of applications like photocatalysis, gas sensors, varistors, low-voltage phosphor material and so on.

In this paper we report a simple, economical and moderately low temperature (450 ^{0}C) solution route for the preparation of nano powder. Precipitation of the intermediate (zinc oxalate) phase was obtained by the reaction of zinc acetate and oxalic acid in organic

CP1147, Transport and Optical Properties of Nanomaterials—ICTOPON - 2009, edited by M. R. Singh and R. H. Lipson
© 2009 American Institute of Physics 978-0-7354-0684-1/09/$25.00

solvents. The decomposition of phase intermediate gives rise to monodispersed ZnO nano particles. Structures were examined using XRD and SEM. In this paper issue of synthesis methods, structural and optical properties are summarized below.

2. EXPERIMENTAL DETAILS

In this paper nanosized ZnO powder was prepared using zinc acetate [$Zn(CH_3COO)_2 \cdot 2H_2O$] and oxalic acid [$H_2C_2O_4 \cdot 2H_2O$] in ethylene glycol, methanol and toluene solvents. Using ethylene glycol as the solvent, initially, to maximize the yield of the intermediate zinc oxalate phase, various experiments were performed by reacting aqueous concentrations of zinc acetate in the range of 0.008–0.1 M with 0.012–0.18 M solution of oxalic acid at room temperature. It was found that the reaction between 0.1 M solution of zinc acetate and 0.15 M solution of oxalic acid gave maximum yield of intermediate product as compared to other ratios of the precursors. Therefore, this ratio of concentration of the precursor solution was used for rest of the work reported here. Similar experiments were performed using ethylene methanol and toluene as solvent. There were two main steps in the preparation of nano sized ZnO: a. Preparation of Zinc oxalate b. Heat treatment of Zinc Oxalate.

$$Zn(CH_3COO)_2. 2H_2O + H_2C_2O_4.2H_2O — ZnC_2O_4.2H_2O + 2CH_3COOH + 2H_2O$$

Oxalic acid was mixed slowly with zinc oxalate under constant slow stirring with the help of magnetic stirrer for 12 hours at room temperature, and then we keep the solution under centrifuge so that intermediate Zinc oxalate can be precipitated totally. The precipitate obtained was filtered and washed with acetone for at least three times to remove impurities, if any, and dried at 120°C to remove presence of excess water. Ethylene Glycol has a dielectric constant $\varepsilon = 37$ in comparison to Methanol ($\varepsilon = 33.0$) and Toluene ($\varepsilon = 2.0$-2.4). Then the effect of different solvents on the properties and morphology of nano- size ZnO powders produced was studied.

$$ZnC_2O_4 .2H_2O + ½ O_2 — ZnO + 2CO_2 + 2H_2O$$

ZnO was prepared by the decomposition of phase intermediate synthesized zinc oxalate at 450°C according to above formula. The decomposition of phase intermediate gives rise to monodispersed ZnO nano particles.

The thermal study of zinc oxalate synthesized in different solvents was carried out using Thermo Gravimetric Analyzer (TGA-DTA) up to 1000°C in air at the heating rate of 10 °C/min. ZnO prepared by decomposition of zinc oxalate approximately at 430°C was employed for the powder XRD studies. A scanning electron microscope (SEM) was used to study the surface morphology of the zinc oxide powder. FTIR was used to observe the stretching of metal oxygen bond. UV–VIS absorption spectra and Photoluminescence (PL) spectra were recorded at room temperature.

3. RESULTS AND DISCUSSIONS

The nano-size ZnO powder was prepared using zinc acetate and oxalic acid for three different solvents viz ethylene glycol, methanol and toluene. For different solvents the yield of prepared zinc oxalate and zinc oxide obtained from zinc acetate and zinc oxalate respectively, is tabulated in Table 1.

TABLE 1. Yield of zinc oxalate from Zinc acetate and Zinc oxide from Zinc oxalate.

Solvents	% Yield of zinc oxalate	% Yield of zinc oxide
Ethylene Glycol	55.98	34.10
Methanol	43.95	43.39
Toluene	44.72	47.01

Thermal Analysis

TGA showed a weight loss in two steps and correspondingly DTA showed two endothermic peaks at temperatures given in Fig.1a.

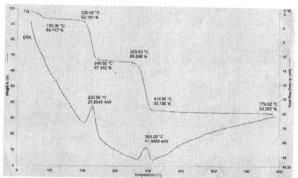

FIGURE 1. The TGA/DTA curves for decomposition of zinc oxalate synthesized in ethylene glycol

$$ZnC_2O_4 + 2H_2O + \tfrac{1}{2}O_2 \longrightarrow ZnO + 2CO_2 + 2H_2O$$

Dehydration (excess water removal) of zinc oxalate is endothermic in nature. Further decomposition of anhydrous zinc oxalate is also an endothermic process. The above TGA/DTA study also confirms the purity and phase of zinc oxalate. Ethylene Glycol mediated zinc oxalate showed a weight loss in two steps at 246.62°C and 415.00°C, respectively. The corresponding DTA showed endothermic peaks at 230.00°C and 393.00°C and is due to the removal of excess ethylene glycol solvent from zinc oxalate. The DTA peak at 393°C is due to the decomposition of zinc oxalate to ZnO. Methanol showed weight loss at 190.00°C and 430.00°C and the corresponding DTA endothermic peaks are at 160.82°C and 402.56°C temperature due to the decomposition of zinc oxalate.

When toluene was used as solvent it showed a weight loss in two steps at 150.00°C and 415.31°C and we got the corresponding DTA peaks at 139.80°C and 396.94°C. Using TGA/DTA data, the zinc oxalate obtained from various solvents was decomposed at 450°C and used for further study.

XRD Analysis

From the X-ray diffraction pattern it is found that the ZnO particles are crystalline. Decreased peak intensities in ethylene glycol and toluene indicated that ZnO prepared in methanol is more crystalline.

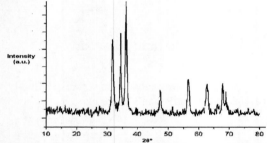

FIGURE 2(a). XRD spectra of zinc oxide powder prepared in Ethylene glycol.

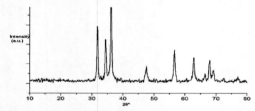

FIGURE 2(b). XRD spectra of zinc oxide powder prepared in Methanol.

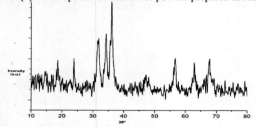

FIGURE 2(c). XRD spectra of zinc oxide powder prepared in Toluene.

The crystallite sizes calculated from FWHM using Scherrer's formula [4] for the (100), (002) and (101) XRD peaks and the average size of ZnO was calculated using the Scherer formula and is tabulated below.

TABLE 2. Average size of ZnO prepared in different solvent

Solvents	Ethylene Glycol	Methanol	Toluene
Size of ZnO	18 nm	21 nm	28 nm

X-ray diffraction patterns of ZnO prepared in Ethylene glycol, methanol and toluene solvents shows a hexagonal wurtzite structure and the lattice constant values obtained were in good agreement with the reported values [5].

Particle morphology by SEM

The SEM micrographs clearly show micro structural homogeneities and remarkably different morphologies for ZnO prepared in different solvents. Fig (a)-(c) represents scanning electron micrographs of ZnO nano particles prepared in ethylene glycol, methanol and toluene respectively. An agglomeration of particles was observed in all the cases due to prolonged reaction time. Organic solvents have better dispersing ability. The lower particle size with different morphology shows that solvents are playing a key role in controlling the nucleation and crystal orientation.

FIGURE 3(a). SEM micrograph of nano ZnO powder prepared in ethylene glycol.

291

FIGURE 3(b). SEM micrograph of nano ZnO powder prepared in methanol.

FIGURE 3(c). SEM micrograph of nano ZnO powder prepared in toluene

The particles obtained with ethylene glycol mediated samples are most ultrafine owing to its better dispersing ability as compared to others. These observations lead us to the conclusion that the selection of the solvent is a key factor for obtaining high quality ZnO nanocrystals.

Optical properties

Fourier Transform Infrared Spectroscopy

Figure 4(a) represents the FTIR of ethylene glycol, methanol and toluene mediated ZnO. The band near to highest wave number is due to the vibration of water. The vibrating mode of water OH group indicates presence of small amount of water absorbed on the ZnO nanocrystal surface. The middle band corresponds to the OH bending of water and final band is attributed to the stretching of Zn-O band.

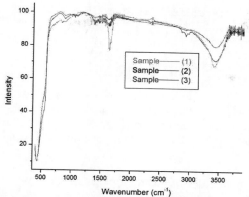

FIGURE 4(a). FTIR spectra of ZnO nano powder prepared in different solvent.

The IR band for Zn-O bonding corresponding to the ZnO nano particle is observed at 500 cm^{-1} and it matches with already reported value [5]. Furthermore this peak is shifted to 432.65 cm^{-1}, 433.74 cm^{-1} and 436.78 cm^{-1} for the ethylene glycol, methanol and toluene mediated ZnO respectively. This indicates the influence of solvents on synthesis of nanosized ZnO and also reveals that smaller particles lose their long-range order and prefer the local order with octahedral symmetry. Presence of wavenumbers 1630 cm^{-1} and 3438 cm^{-1} states bending and bond stretching of –OH molecules of water. It conform presence of water molecule in Zinc oxide powder.

UV Absorption Spectra

Absorption spectra were taken by UV-VIS spectrometer data system Jasco (V 570). ZnO powder was dispersed in, Ethylene Glycol, Methanol and Toluene and their absorption spectra were recorded.

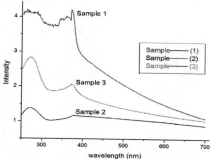

FIGURE 4(b). UV-spectrum of nano ZnO powder (hexagonal phase) prepared in: (1) Ethylene Glycol; (2) methanol (3) and Toluene solvents.

TABLE 3. Peak absorption wavelength and bandgap of nano sized ZnO prepared in different solvents

Solvents	Absorption wavelength peaks (nm)		Band gap energy(eV)
Ethylene Glycol	268.49	376.77	3.60
Methanol	271.44	379.78	3.53
Toluene	278.90	373.82	3.50

Compared with bulk ZnO, the blue shift in wavelength is observed in the ZnO nanostructures, that is due to size effect. The optical bandgap (Eg) is found to be size dependent and there is an increase in the band gap of the semiconductor with the decrease in the particle. The bandgap from the absorption spectrum is written in the table above.

Photoluminescence Spectroscopy

Fig 5(a) shows the PL spectra of ZnO nanoparticles synthesized in ethylene glycol, methanol and toluene solvent. Spectrum is taken by fluorescence spectrophotometer (varian cary Eclipse).

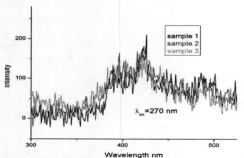

FIGURE 5(a). Fluorescence spectra of ZnO nano colloids at an excitation wavelength 270 nm.

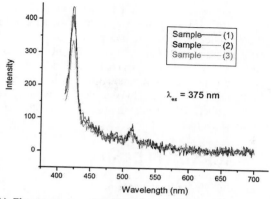

FIGURE 5(b). Fluorescence spectra of ZnO nano colloids at an excitation wavelength of 375 nm.

From the absorption spectra shown in fig.4(b), it was found that ZnO has broadband of absorption and excitation peak values are near to 270 nm and 375 nm for all the three samples, so fluorescence study was performed at two wavelengths (λ= 270 and 375 nm). When ZnO nanoparticles were excited by wavelength λ =270 nm fluorescence spectra of nano-ZnO collides having intensity maximum at wavelength 390 and 425 nm. This corresponds to UV and blue emission of ZnO nanoparticles. When ZnO nanoparticles were excited by 375 nm, fluorescence spectra of nano-ZnO collides are having intensity peaks at wavelengths 422 and 511 nm. The intensity of the emission at 422nm is very high compared to that of 511nm. Similar observation was also made by earlier work [2]. Emission at 422nm is due to band to band transition while that at 511nm is due to surface state effects. Excitation by 375nm favours emission due to band to band transition as observed at 422nm.

4. CONCLUSIONS

ZnO nanoparticles have received broad attention due to their versatile performance in electronics, optics and photonics. Compared to other methods, ZnO nanoparticles were prepared very fast and easily. TGA showed a weight loss in two steps correspondingly two endothermic peaks at two temperatures. The first endothermic peak is due to removal of water/ ethylene glycol and the second peak is due to decomposition of zinc oxalate. XRD results showed that the obtained ZnO nanoparticles were composed of hexagonal wurtzite phase with very good crystallinity. The particle size was obtained using Scherer's equation. From the pattern it is found that the ZnO particles are crystalline. Decreased peak intensities in ethylene glycol and toluene indicated that ZnO prepared in methanol is more crystalline. SEM results show that samples of ZnO nanoparticles prepared in different solvents are having different surface morphologies. The different morphology reveals that solvents are playing a key role. The particles obtained with ethylene glycol mediated samples are most ultrafine owing to their better dispersing ability as compared to others. FTIR shows the presence of ZnO bonding present in the sample. The IR band peak is shifted with respect to bulk ZnO. This indicates the influence of solvents on the synthesis of nanosized ZnO. The optical bandgap (E_g) is found to be size dependent and there is an increase in the band gap of the semiconductor with the decrease in the particle size. Compared with bulk ZnO, blue shift in wavelength is observed in the ZnO nanostructures, due to size effect. PL spectrum also shows blue emission of ZnO nanoparticles due to quantum size effect.

ACKNOWLEDGMENTS

Authors would like to thank UGC for financial assistance.

REFERENCES

1. K.G. Kanade, B. B. Kale, R. C. Aiyer and B. K. Das, *MRS Bulletin* (2006)
2. Litty Irimpan, V. P. N. Nampoori and P. Radhakrishnan, *J. Appl. Phys.* **102**, 063524 (2007)
3. Ranjani Viswanathan, Sameer Sapra, B. Satpati, B. N. Dev and D. D. Sarma, *J. Mater. Chem.* 14, 661- 668 (2004).
4. Litty Irimpan, A. Deepthy, Bindu Krishnan, V.P. N. Nampoori and P. Radhakrishnan, *Appl. Phys. B*, **90** 547-556 (2008)
5. N. Faal Hamedani and F. Farzaneh, *J. Sci. Islamic Republic of Iran*, **17**231-234 (2006).

Formation of ZnS nanostructures in SiO$_2$ matrix by RF co-sputtering

Shiv P. Patel[a], Lokendra Kumar[a], A. Tripathi[b], Y. S. Katharria, V.V. Sivakumar[b], I. Sulania[b], P. K. Kulariya[b], and D. Kanjilal[b]

[a]*Department of Physics, University of Allahabad, Allahabad-211002, India*
[b]*Inter University Accelerator Centre, Aruna Asaf Ali Marg, New Delhi-110067, India*

Abstract: Formation of nanorings of ZnS co-sputtered with SiO$_2$ on n-type Si (100) substrates is reported. A subsequent annealing in inert atmosphere at temperature 800 °C for 2 hrs was required for development of the crystallinity and the formation of nanorings. The thickness and refractive index of the films estimated by ellipsometry measurements were found to be 180 nm, and 1.59, respectively. X-ray diffraction measurements were performed to study the crystallinity of these nanostructures. It was found that the nanostructures are in wurtzite-hexagonal phase. The sizes of the nanoparticles were estimated to be 54 nm. Atomic force microscopy (AFM) and scanning electron microscopy (SEM) were carried out to see the exact shape and size of the nanostructures. Formations of nanorings on the surface were seen in AFM and SEM studies. The outer and inner diameters of the rings were 135 nm, and 65 nm, respectively. Photoluminescence (PL) studies were carried out to see the optical properties of these nanostructures.

Keywords: Nanorings, Annealing, AFM, SEM etc.

PACS: 07.79.Lh, 07.79.Cz, 68.37.Hk, 68.37.Ps, and 62.23.St.

INTRODUCTION

Wurtzite-structured materials, such as ZnO, ZnS, GaN, and AlN, have important characteristics like the presence of polar surfaces that results from cation- or anion-terminated atomic planes [1]. Consequently, a series of novel nanostructures, such as nanobelts [2], nanospring [3], nanorings [4], nanohelices [5], and nanocombs [6], and nanopropeller array [7] have been formed. The mechanism that derives the formation of these novel configurations minimizes the electrostatic interaction energy of the polar surfaces. To date, these nanostructures have been mainly observed for a mono-composition object such as ZnO [8,9], GaN [10], rutile-structured SnO$_2$ [11] and wurtzite-structured AlN [12], as results of same formation mechanism [13]. Wurtzite-structutred ZnSe nanorings, nanowires, nanobelts, nanorods have also been reported [14].

Another member of the wurtzite family, ZnS has a direct wide band gap (3.67 eV) with a high index of refraction, and high transmittance in the visible range. It is one of the most important materials in photonics [15]. Due to the large exciton binding energy (~40 meV) which is greater than the room temperature thermal energy (~25 meV), it allows to get good photoluminescence (PL) above the room temperature because of existence of exciton. The limitation of small Bohr exciton radius (2.5 nm) for ZnS has attracted much

CP1147, Transport and Optical Properties of Nanomaterials—ICTOPON - 2009, edited by M. R. Singh and R. H. Lipson
© 2009 American Institute of Physics 978-0-7354-0684-1/09/$25.00

interest in the studies of ZnS in low dimensional scale [16-17]. Recently, Manganese doping and optical properties of ZnS nanoribbons have been reported [18] without changing the crystallography. The mass syntheses of ZnS nanobelts, nanowires, and nanoparticles have been achieved by a simple method of thermal evaporation of ZnS powders onto silicon substrates in the presence of Au catalyst [19]. Development of techniques for synthesizing different nanostructures such as nanobelts, nanowire, nanorings, nanoribbons, nanosphere of desired size and shape has been a major challenge over the years. Several physical and chemical methods [9, 18-22] have been used to prepare semi-conducting nanostructures such as ion implantation [23], co-sputtering [24], and thermal evaporation [25]. In present work, we report the formation of ZnS nanorings in SiO_2 matrix by RF co-sputtering method. These nanorings are investigated for their structural and optical properties as well as surface morphology using the techniques of X-ray diffraction (XRD), Photoluminescence spectroscopy (PL), Atomic Force Microscopy (AFM), and Scanning Electron microscopy (SEM).

EXPERIMENTAL

Thin films of ZnS in SiO_2 matrix are deposited by rf sputtering method on ultra cleaned Si (100) substrate. Before deposition, ZnS pallets were glued on SiO_2 target of predetermined area to control the composition of the film. High purity ZnS powder (99.99% from Sigma Aldrich, America) was compressed into pallets by applying about 1 ton per cm^2 pressure. The pallets were sintered at 1000 °C temperature for 1 hour in Ar atmosphere. During deposition, 150 W r.f. power was used. The distance between target and sample was about 4 cm. The deposition was carried out at room temperature. After the deposition, the films were annealed at 800 °C temperature in Ar environment for 1 hour and 2 hours duration. The thickness and refractive index of the films were measured by ellipsometry. A light source of He-Ne laser with wavelength 632.8 nm, power 2 mV and beam diameter 1.5 mm was used for ellipsometry measurement. The surface morphology of the films was studied by using atomic force microscopy (AFM) of Digital Instruments (Nanoscope IIIa) in tapping mode and field emission scanning electron microscopy (Quanta 200F of FEI). X- ray diffraction (XRD) was carried out by Bruker D8 Advance diffractometer in Bragg Brentano geometry, using Cu K_α (λ = 1.5406 A°) source operating at 40 kV and 40 mA. The step was 0.02 degree and the scanning speed was 0.25 degree/minute. The PL spectra were observed on Perkins Elmer LS 55 fluorescence spectrometer with xenon lamp as an excitation source.

RESULTS AND DESCUSSION

The thickness and refractive index of the film are found to be 180 nm, and 1.59, respectively. The XRD result of the ZnS-SiO_2 film annealed at 800 °C is shown in Fig1. There is no signature of the ZnS phase in as deposited film while the film annealed at 800 °C clearly reveals the phase of ZnS in SiO_2 matrix. The diffraction peak in the XRD spectra corresponds to the wurtzite-hexagonal phase of ZnS. The Debye-Scherrer formula $D = k\lambda / \beta \cos\theta$ is used to determine the size of nanoparticles, where D is the mean particle size, k is a geometric factor (= 0.9), λ is the wavelength of the x-ray wavelength, β is the full width at half maximum (FWHM) of diffraction peak and θ is the diffraction angle. The size of the ZnS nanoparticles from most intense peak is approximated to be 54 nm. The surface morphology of the as-deposited and annealed samples is shown in Fig.2. Fig.2 (a) shows the AFM image of the as-deposited sample and Fig.2 (b) is for sample annealed at 800 °C for 1 hour. The formation of ZnS nanorings is seen in (Fig. 2 (c)) AFM image of sample annealed for 2 hrs. The outer and inner average diameters of the ring are 135 nm and 65 nm, respectively. The average thickness of the nanoring wall is found to be 70 nm as analyzed. The formation of nanorings takes place when the sample is annealed at 800 °C for 2 hrs (Fig 2). The formation of nanorings on the surface is also confirmed by SEM analysis as shown in Fig 2(d).

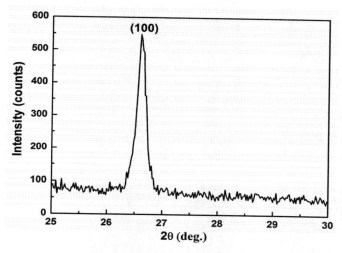

FIGURE 1. XRD pattern of ZnS-SiO$_2$ thin films annealed at 800 °C for 2 hrs.

The nanoring formation is a complex process. The growth of nanoring structures can be understood by considering the polar surfaces of the ZnS nanostructures, and is critically decided by temperature, annealing time, interaction between self-assembling particles, substrates, and surface diffusion. In as-deposited sample, no ring like structure at the surface was seen. During annealing nucleation and growth of smaller nanoparticles occur. Later agglomeration of these particles takes place to form bigger nanoparticles (clusters of nanoparticles) due to thermal-driven instability.

One possible explanation may be that bigger nanoparticles have polar charges on its top and bottom surface [8, 9]. If the surface charges are uncompensated during growth, bigger nanoparticles may tend to fold itself, to minimize the area of the polar surface. The positively charged (0001)-Zn plane (top surface) with the negatively charged (0001)-S plane (bottom surface) forms a loop with an overlapped end, thus resulting in neutralization of the local polar charges and reduction of the surface area. Agglomeration of nanoparticles occurs in secondary process leading to ring like structure [26]. The formation of nanorings depends on type of substrate, annealing temperature and time. It occurs at certain annealing temperature at which energy provided is sufficient to agglomerate these nanoparticles.

Photoluminescence spectrum was recorded with 350 nm excitation wavelength and is shown in Fig 3. From the Gaussian fit of spectrum, three peaks were obtained at 445, 482

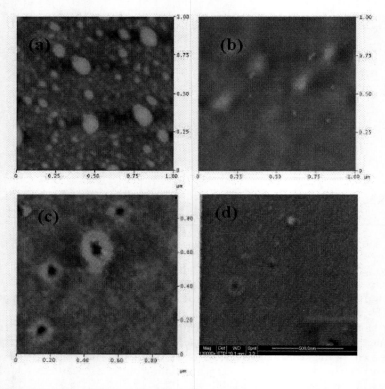

FIGURE 2. AFM image of (a) as-deposited sample (b) sample annealed at 800 °C for 1 hr (c) sample annealed at 800 °C for 2 hrs (d) SEM image of sample annealed at 800 °C for 2 hrs.

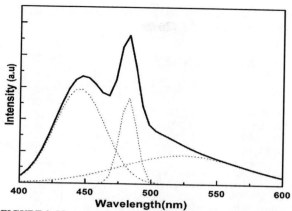

FIGURE 3. PL spectrum of ZnS-SiO₂ thin films annealed at 800 ºC for 2 hrs.

and 525 nm. The luminescence peak at 445 nm is trapped luminescence arising out of from surface states [27-28]. Emission peaks at 525 nm and 482 nm arise from the sulfur species [29] on the surface. The blue emission from ZnS-SiO₂ nanostructures is attributed to 'self-activated centers' arising from the vacancies or interstitial atoms in the lattice [30].

CONCLUSIONS

ZnS nano-rings in SiO₂ matrix were successfully synthesized by r.f. co-sputtering followed by thermal annealing in an inert atmosphere. Scanning electron microscopy and Atomic force microscopy results reveal the formation of the ZnS nano-rings. X-ray diffraction studies show that the nano-rings are in hexagonal-wurtzite phase. The mechanism that causes the formation of this novel configuration is to minimization of the electrostatic interaction energy of the polar surfaces and the agglomeration of nanoparticles in ring like structure due to thermal-induced instability in ZnS nanoparticles. The photoluminescence spectrum shows the characteristic peaks at 445, 482, and 525 nm.

ACKNOWLEDGEMENTS

The authors are thankful to Department of Science and Technology for providing XRD, AFM facility at IUAC New Delhi. The authors are grateful to Pravin Kumar, D. C. Agarwal, and Jai Prakash for their valuable help. One of the authors (SPP) would like to express his sincere thanks to IUAC, New Delhi for providing financial support through research grant to carry out the work. The authors thank the personnel at Nanophosphor

Application Centre, Allahabad University, IIT-Roorkee and Rajasthan University, Jaipur for their kind cooperation in characterization of the samples.

REFERENCES

1. Xiang Wu, Peng Jiang, Yong Ding, Wei Cai, Si-Shen Xie and Zhong Lin Wang, *Adv. Mater.* **19**, 2319-2323 (2007).
2. J.W. Pan, Z. R. Dai and Z. L.Wang, *Science* **291**,1947 (2001).
3. X. Y. Kong and Z. L. Wang, *Nano Lett.* **3**,1625 (2003).
4. X. Y. Kong, Y. Ding, R. Yang and Z.L. Wang, *Science* **303**, 1348 (2004).
5. P. X. Gao, Y. Ding, W. J. Mai, W. L. Hughes, C. S. Lao and Z. L. Wang, *Science* **309**, 1700 (2005).
6. Z. L. Wang, X. Y. Kong and J. M. Zuo, *Phys. Rev. Lett.* **91**, 185502 (2003).
7. P. X. Gao and Z. L. Wang, *Appl. Phys. Lett.* **84**, 2883 (2004).
8. Z. L. Wang, *Materials Today* **7**, 26 (2004).
9. Z. L. Wang, *J. Phys.: Condensed Matter* **16**, R829 (2004).
10. J. K. Jian, Z. H. Zhang, Y. P. Sun, M. Lei, X. L. Chen, T. M.Wang and Cong Wang, *J. Cryst. Growth* **303**, 427 (2007).
11. Rusen Yang and Zhong Lin Wang, *J. Am. Chem. Soc.* **128**, 1466 (2006).
12. Junhong Duan, Shaoguang Yang, Hongwei Liu, Jiangfeng Gong, Hongbo Huang, Xiaoning Zhao, Jili Tang, Rong Zhang and Youwei Du, *J. Cryst. Growth* **283**, 291 (2005).
13. P. X. Gao and Z. L. Wang, *J. Appl. Phys.* **97**, 044 304 (2005).
14. Y. P. Leung, W. C. H. Choy, I. Markov, G. K. H. Pang, H. C. Ong and T. I. Yuk, *Appl. Phys. Lett.* **88**, 183110 (2006).
15. R. A. Rosenberg, G. K. Shenoy, F. Heigl, S.-T Lee, P.-S. G. Kim, X.-T. Zhou and T. K. Sham, *Appl. Phys. Lett.* **86**, 263115 (2005).
16. N. Kumbhojar, V. V. Nikesh, A. Kshirsagar and S. Mahamuni, *J. Appl. Phys.* **88**, 6260 (2000).
17. S. Mahamuni, A. A. Khosravi, M. Kundu, A. Kshirsagar, A. Bedekar, D. B. Avasare, P. Singh and S.K. Kulkarni, *J. Appl. Phys.* **73**, 5237 (1993).
18. Y. Q. Li, J. A. Zapien, Y. Y. Shan, Y. K. Liu and S. T. Lee, *Appl. Phys. Lett.* **88**, 013115 (2006).
19. H. J. Yuan, S. S. Xie, D. F. Liu, X. Q. Yan, Z. P. Zhou, L. J. Ci, J. X. Wang, Y. Gao, L. Song, L. F. Liu, W. Y. Zhou and G. Wang, *J. Cryst. Growth* **258**, 225 (2003).
20. Yiguo Su, Guangshe Li, Xiangxi Bo and Liping Li, *Nanotechnology* **18**, 485602 (2007).
21. Fei Xiao, Hong-Guo Liu, Chang-WeiWang, Yong-Ill Lee, Qingbin Xue, Xiao Chen, Jingcheng Hao and Jianzhuang Jiang, *Nanotechnology* **18**, 435603 (2007).
22. Ying-Chun Zhu, Yoshio Bando, and Dong-Feng Xue, *Appl. Phys. Lett.* **82**, 1769 (2003).
23. A. Meldrum, L. A. Boatnerand C. W. White, *NIM B* **178**, 7 (2001).
24. N. Taghavinia and T. Yao, *Physica E* **21**, 96 (2004).
25. Quan Li and Chunrui Wang, *Appl. Phys. Lett.* **83**, 359 (2003).
26. Zhen Liu and Rastislav Levicky, *Nanotechnology* **15**, 1483 (2004).
27. D. Denzler, M. Olschewski and K. Sattler, *J. Appl. Phys.* **84**, 2841 (1998).
28. Soumitra Kar, Subhajit Biswas and Subhadra Chaudhuri, *Nanotechnology* **16**, 737 (2005).
29. Yiguo Su, Guangshe Li, Xiangxi Bo and Liping Li, *Nanotechnology* **18** 485602 (2007) and *references within*.
30. N. Arul Dhas, A. Zaban and A. Gedanken, *Chem. Mater.* **11**, 806 (1999).

Zinc Oxide Nano Particle Grown By Soft Solution Route at Room Temperature

Arindam Ghosh[a], Abhay A. Sagade[a], Rajesh A. Joshi[a], D. M. Phase[b] and Ramphal Sharma[a]

[a]*Thin Film and Nanotechnology Laboratory, Department of Physics, Dr. Babasaheb Ambedkar Marathwada University, Aurangabad, 431004, MS, India*
[b]*UGC-DAE Consortium for Scientific Research, University Campus, Khandawa road, Indore (M.P.) India*

Abstract. Zinc oxide thin films were deposited by simple soft solution route at room temperature. The as grown thin films of ZnO were characterized for the structural, morphological and optical properties. X-ray diffraction (XRD) analysis reveals the polycrystalline nature of ZnO thin films with hexagonal structure. The films were highly oriented along (100) and (101) planes. The broad and low intense peaks in XRD pattern indicate the presence of coarsely fine, nano crystalline grains. Transmission Electron Microscopy (TEM) micrographs shows the formation of ZnO nanocrystallites. Atomic Force Microscopy (AFM) image shows uniform deposition of ZnO thin film on the substrate. The optical band gap was found to be 3.37 eV for as grown films.

Keywords: II-VI Semiconductors, Nanocrystals and Nano particles.
PACS: R71.55 Gs, 78.67 Bf.

INTRODUCTION

Recently the study of one dimensional (1D) nanostructures like nano rods, nano wires, nano tubes etc. attracted the attention of the modern researchers, due to their physical properties and wide potential applications in the optoelectronic fields for fabricating the nano devices [1,2]. Zinc oxide (ZnO) is an important II-VI group compound due to its versatile technologically important properties like, high direct band gap of 3.33 eV, strong binding energy of 60 MeV etc. Due to these properties significant strides have been made for developing ZnO based UV blue light emitting diodes [3], solar build UV photo detectors [4], resonant tunneling devices [5] etc.

There are various methods of synthesizing ZnO thin films like physical and chemical methods. The physical methods used are vapor-liquid-solid epitaxial (VLS) mechanism [6-8], microwave plasma growth [9] but these techniques requires sophisticated instruments which are very costly while in comparison to this, the chemical techniques are simple and cost effective. One of the solution growth technique known as Successive Ionic Layer Adsorption and Reaction (SILAR) is suitable to obtain stoichiometrical ZnO thin film at room temperature even this technique provides oxygen rich deposition environment. But according to the literature

CP1147, Transport and Optical Properties of Nanomaterials—ICTOPON - 2009, edited by M. R. Singh and R. H. Lipson
© 2009 American Institute of Physics 978-0-7354-0684-1/09/$25.00

survey, there are very few reports available on the room temperature synthesis of ZnO nano particle thin films. So, in order to overcome the difficulty of high cost, we use the successive ionic layer adsorption and reaction (SILAR) method which is a simple, cost effective and less time consuming method.

EXPERIMENTAL DETAILS

The zinc oxide thin films were grown by successive ionic layer adsorption and reaction (SILAR) technique at room temperature. For the deposition purpose 0.1 M $ZnSO_4$ was used as the cationic precursor solution and H_2O_2 was used as the anionic precursor solution. For getting uniform deposition throughout the substrate triethanolamine (TEA) was added in the cationic precursor solution [10]. The pre cleaned glass substrates were immersed into the cationic precursor solution for about 20 sec. and followed by the rinsing in double distilled water for about 40 sec. for getting the uniform layer of the cations throughout the substrate. These films were then dipped into the anionic precursor solution for about 20 sec. and again rinsed into the double distilled water for about 40 sec. for removing the loosely bounded anions. The completion of such cycle produces a layer of ZnO onto the glass substrate. As the number of cycle is increased, the thickness of the film was also goes on increasing up to a certain level. The optimized thickness of the film was found to be ~380 nm. For improving the crystallinity of the films we annealed the samples at 450^0 C for about two hours in air [11].

These films were characterized by X-ray diffraction (XRD), Atomic Force Microscopy (AFM), Transmission Electron Microscopy (TEM) and the Optical absorption technique. The structure and crystallinity of ZnO thin film was studied by the X-ray diffraction (XRD) which was recorded using the Burker AXS, Germany (D8 Advanced) diffractometer in scanning range 20-60^0 (2θ) using CuKα radiation with wavelength 1.5406 Å. The optical study was done by Systronic UV-Vis spectrometer in the range 300-800 nm using Perkin Elmer, Model –lambda-25. The TEM micrographs were taken from : Tecnai 20 G2 (FEI make) unit and the monostructure of the film was obtained from the AFM images, done by using nanoscope III a provided by vecco digital instruments.

RESULT AND DISCUSSION

X-ray Diffraction

The XRD pattern of the SILAR deposited, annealed (450^0C) Zinc oxide thin film sample is shown in FIGURE 1. The low intense peaks indicate that, the film consists of coarsely fine (nano crystalline) grains. The comparative study with the JCPDS card (80-0075) reveals the hexagonal structure. The diffraction peaks were found at $2\theta = 31.70^0$, 34.40^0, 36.20^0 and 56.50^0 which attributes to the (100), (002), (101) and (110) planes respectively. The values of the lattice constants were calculated using the formula

$$\frac{1}{d^2} = \frac{4}{3}\left(\frac{h^2 + hk + k^2}{a^2}\right) + \frac{l^2}{c^2}$$

The calculated values of the lattice constants were found to be 3.26 Å and 5.27Å respectively which are very closer to the standard values. Using the scherrer's formula the calculated grain size was found to be ~ 17 nm. This significant improvement in the crystallinity is due to sintering of nanocrystals into effectively larger crystals after annealing [12]

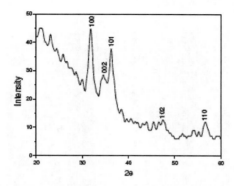

FIGURE 1. XRD pattern of the annealed ZnO thin film

Transmission Electron Microscopy

FIGURE 2.a. shows the Selected Area diffraction (SEAD) pattern and FIGURE 2.b. shows the Transmission Electron Microscopy and of the annealed ZnO thin film grown by SILAR technique at room temperature. From the SEAD pattern (FIGURE 2.a) shows a set of rings instead of the spots. This may be due to the fact of random orientation of crystals, corresponding to the diffraction from different planes of nano crystallites. The SEAD pattern confirms the (100), (101) and (102) planes respectively which is matching well in accordance to observed XRD pattern. From the TEM images (FIGURE 2b) the agglomerated form of the grains and grain size was found to be ~10 nm which is very nearer to the XRD calculation.

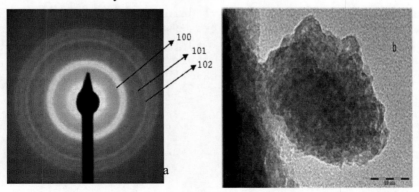

FIGURE 2. (a) SEAD pattern and (b) TEM images of the annealed ZnO thin film

Atomic Force Microscopy

Atomic Force Microscopy (AFM) was used to study surface morphology of samples. FIGURE 3 shows the AFM images of the annealed ZnO thin film with scanning area of 2µm x 2µm. In this micro graph, a fine layer of small grains confirms the uniform deposition throughout the substrate. The roughness of the sample was found to be 0.841.

FIGURE 3. AFM image of the annealed ZnO thin film

Optical Study

To study the nature of electronic transition for the annealed ZnO thin film, grown at room temperature by SILAR technique, the optical study was done. The absorption spectra (not shown here) low absorbance was found which is a well known characteristic of the ZnO thin film. From the theory of optical absorption gives the relationship between the absorption coefficient (α) and the photon energy (hv) for the direct allowed transition as

$$\alpha h v = A(h v - E_g)^n \qquad (13)$$

Where 'hv' is the photon energy, 'E_g' is the band gap and 'A' is constant having separate values for different transition. By extrapolating the linear part of the curve $(\alpha h v)^2$ as a function of hv, to the energy axis will give the energy band gap. From the FIGURE 4 the energy band gap of the annealed ZnO thin film was found to be 3.37 eV which is very close to the reported value (3.33 eV) of the bulk ZnO [14].

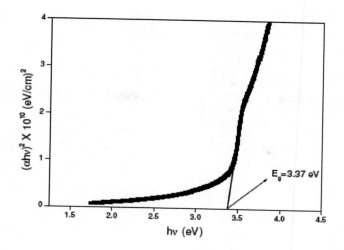

CONCLUSIONS

The nano crystalline Zinc oxide thin film engineered at room temperature by the cost effective technique, SILAR. The structural study of the film reveals that the film is nanocrystalline in nature having the polycrystalline nature with hexagonal structure. Uniform surface morphology with the fine grains covering the entire surface of the substrate is seen from the AFM images. From the optical study, the band gap was found to be 3.37 eV which is matching well with the bulk ZnO.

ACKNOWLEDGMENTS

The authors are very much thankful to UGC-DAE Consortium for Scientific Research, Indore Project no Ref. no. ESTT/DEPT/2007/9889–9967 for financial assistance to carry out the research work. We are also thankful to The Head, Department of Physics, Dr. B. A. M. University for providing the laboratory facility. In addition to this, a special thank to Dr. N. P. Lalla, and Vinay Ahire UGC–DAE Consortium for scientific research,

REFERENCES

1. J. F. Wang, M. S. Gudiksen, X. F.Duan, F. Cui and C. M. Lieber, *Science* **293**,1495-1499 (2001).
2. H. Kind, H. Q. Yan, B. Messer, M. Law and P. D. Yang, *Adv. Matter* **14**, 158-160 (2002).
3. X. L. Guo, J. H. Choi, H. Tabata and T. Kawai, *Jpn. J. Appl. Phys.* **40**, L177-L180 (2001).
4. W. Yang, R. D. Visputee, S. Choopum, R. P. Sharms, T. Venkatesan and H. Shen, *Appl.Phys. Lett.* **78**, 2787 (2001).
5. S. Krishnamoorthy, A. A. Iliadis, A. Inumpudi, S. Choopum, R. D. Viputee and T. Venkatesan, *J. Solid-State Electron.* **40**, 1633-1637 (2002).
6. P .Yang, H. Yan, S. Mao, R. Russo, J. Johnson, R. Saykally, N. Morris, J. Pham, R. He and H.J. Choi, *Adv. Funct. Mater.* **12**, 323-331 (2002).

7. B. D. Yao, Y. F. Chan and N. Wang, *Appl. Phys. Lett.* **80**, 757-759 (2002).
8. X. Wang , Y. Ding , C. J. Summers and Z. L. Wang, *J. Phys. Chem. B* **108**, 8773-8777 (2004).
9   X. H. Zhang, S. Y. Xie, Z. Y. Jiang, Z. X. Xie, R. B. Huang, L. S. Zheng, J. Y. Kang and T. Sekiguchi, *J. Solid State Chem.* **173**, 109-113 (2003).
10. S. D. Chavhan, S. V. Bagul, R. R. Ahire, N. G. Deshpande, A. A. Sagade, Y. G. Gudage and R. Sharma, *J. Alloys Compd.* **436**, 400-406 (2007).
11. A. Ghosh, N. G. Deshpande, Y.G. Gudage, R. A. Joshi, A. A. Sagade, D. M. Phase, Ramphal Sharma, *J. Alloys Compd.* **469**, 56-60 (2009).
12. R.B. Kale and C.D. Lokhande, *Mater Res Bull.* **39**,1829-1839 (2004).
13. C. H. Ku, H. H. Chiang, and J. J. Wu, *Phys. Lett.* **404**, 132-135 (2005).
14. Y. F. Chen, D. M. Bangnall, H. Koh, K. Park, K. Hiraga, Z. Zhu and T.Yao, *J. Appl.Phys.* **84**, 3912 (1998).

Isolation Of PS II Nanoparticles And Oxygen Evolution Studies In *Synechococcus Spp. PCC 7942* Under Heavy Metal Stress

Iffat Zareen Ahmad [a], Shanthy Sundaram [b], Ashutosh Tripathi [b] and K. K. Soumya [b]

[a] Department of Biotechnology, Integral University, Dasauli, Kursi Road, Lucknow-226026, India
[b] Centre for Biotechnology, University of Allahabad, Allahabad-211002, India

Abstract. The effect of heavy metals was seen on the oxygen evolution pattern of a unicellular, non-heterocystous cyanobacterial strain of *Synechococcus spp. PCC 7942*. It was grown in a BG-11 medium supplemented with heavy metals, namely, nickel, copper, cadmium and mercury. Final concentrations of the heavy metal solution used in the culture were 0.1, 0.4 and 1μM. All the experiments were performed in the exponential phase of the culture. Oxygen-evolving photosystem II (PS II) particles were purified from *Synechococcus spp. PCC 7942* by a single-step Ni^{2+}-affinity column chromatography after solubilization of thylakoid membranes with sucrose monolaurate.

Oxygen evolution was measured with Clark type oxygen electrode fitted with a circulating water jacket. The light on the surface of the vessel was $10w/m^2$. The cultures were incubated in light for 15minutes prior to the measurement of oxygen evolution. Oxygen evolution was measured in assay mixture containing phosphate buffer (pH-7.5, 0.1M) in the presence of potassium ferricyanide as the electron acceptor. The preparation from the control showed a high oxygen-evolving activity of 2,300-2,500 pmol O_2 (mg Chl)$^{-1}$ h^{-1} while the activity was decreased in the cultures grown with heavy metals. The inhibition of oxygen evolution shown by the organism in the presence of different metals was in the order Hg>Ni> Cd >Cu. Such heavy metal resistant strains will find application in the construction of PS II- based biosensors for the monitoring of pollutants.

Keywords: *Synechococcus spp. PCC 7942*, heavy metal, PS II particles, oxygen evolution, biosensor
PACS: 87

INTRODUCTION

Cyanobacteria perform oxygenic photosynthesis which requires photosystem (PS) II to catalyze the photo-oxidation of water. Biochemical and biophysical evidence has shown that PS II spans the chloroplast membrane, with a water oxidizing site on the lumenal side, a plastoquinone reductase site on the stromal side, and a set of chlorophyll and pheophytin chromophores catalyzing the photochemical reaction which separates charge between these sites. PS II is a multisubunit pigment-protein complex and comprises over twenty identified polypeptides, most of which are membrane proteins [1–3].

CP1147, *Transport and Optical Properties of Nanomaterials—ICTOPON - 2009*, edited by M. R. Singh and R. H. Lipson
© 2009 American Institute of Physics 978-0-7354-0684-1/09/$25.00

The cyanobacterium *Synechococcus spp. PCC 7942* is a single-celled photosynthetic prokaryote that is subject to a variety of environmental stressors in nature [4]. All organisms must possess mechanisms that regulate metal ion accumulation and thus, avoid heavy metal toxicity. Several resistance mechanisms exist to lessen or prevent metal toxicity. These include resistance to metals that are always toxic to the cell and serve no beneficial role, such as cadmium and mercury, and also include resistance to metals such as copper, iron, and zinc, which are toxic at high concentrations but are absolutely essential in trace amounts [5].

The photosynthetic apparatus including PS II is particularly sensitive to heavy metals and is affected by the heavy metals on both- oxidizing (donor) and reducing (acceptor) side. Based on this fact, the use of photosynthetic material and in particular the photosystem II sub-membranes fractions as the biological receptor in a biosensor provides an excellent tool for the detection of toxic metal cations [6]. Among the analytical methods used for environmental monitoring, microbial and cellular biosensors play an important role [7]. Microbial biosensors have numerous advantages in ecotoxicity testing. Microorganisms are generally cheaper to culture than higher organisms, and they can be produced in large batches, subjected to stringent quality control procedures, and freeze dried for storage. They respond rapidly to toxic compounds and indicate the bioavailability of compounds in a way that chemical analysis cannot [8]. The purpose of our study was to study the ability of *Synechococcus spp. PCC 7942* to uptake copper (II), cadmium(II), nickel(II) and mercury(II) under batch conditions.

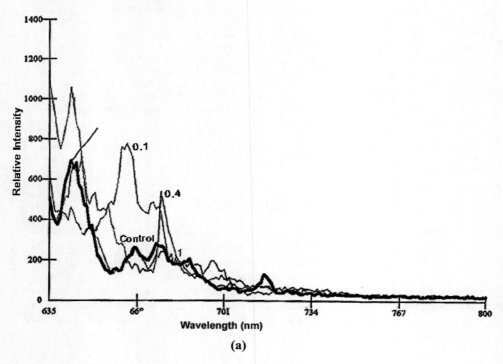

(a)

310

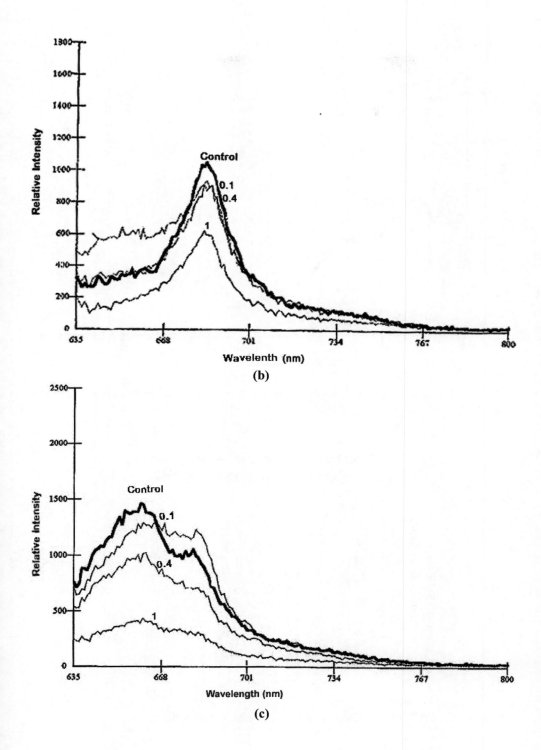

(b)

(c)

311

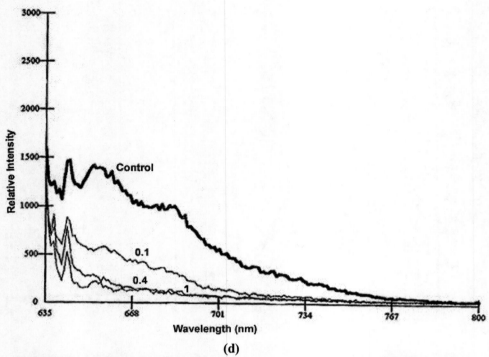

(d)

FIGURE 1. Laser induced Chl a fluorescence spectra of *Synechococcus spp. PCC 7942* under (a) copper, (b) cadmium, (c) nickel and (d) Hg stress in control and in the presence of 0.1µM, 0.4µM and 1µM metal ions.

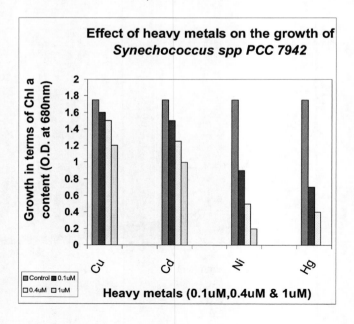

FIGURE 2. Effect of heavy metals (Cu, Cd, Ni & Hg) on the growth of *Synechococcus spp. PCC 7942* in terms of absorption spectra of Chl a at 680 nm.

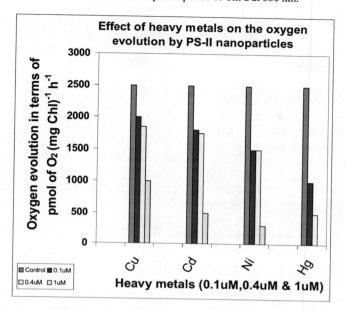

FIGURE 3. Effect of heavy metals (Cu, Cd, Ni & Hg) on the oxygen evolution activity of PS II particles of *Synechococcus spp. PCC 7942.*

MATERIALS AND METHODS

Cyanobacterial Strain and Growth Conditions

Synechococcus spp. PCC 7942 (non-heterocystous) was used in all experiments. Cyanobacterial cultures were grown with continuous aeration in liquid BG-11 medium [9] supplemented with external nitrogen source, 100mM KNO_3. The medium was buffered to pH 7.5 with 10mM NaOH. The cultures were maintained at a temperature of $25 \pm 1°C$, illuminated by white cool fluorescent tubes to receive a light intensity of approximately 100 $\mu E/m^2/s$. The cultures were kept in 14 hours light period and 10 hours dark period. Culture density for all studies was approximately 0.8 A680. All the manipulations involving the transfer of cultures in the liquid media were carried out under aseptic conditions. The growth of the cultures was measured in terms of absorption spectra at 680nm by UV-Vis spectrophotometer (Ultrospec 4000, SWIFT II - QUANT Quantification module).

Metal Ions

Cells were subjected to the divalent cations by adding a single dose of Cu^{2+} (as copper chloride), Cd^{2+} (as cadmium chloride), Ni^{2+} (as nickel chloride) and Hg^{2+} (as mercuric chloride) to give final concentrations each of 0.1 μM, 0.4μM and 1μM.

Measurement of Laser Induced Chl a Fluorescence

5ml aliquots of the cyanobacterial cultures in exponential phase were withdrawn and analyzed in terms of laser induced fluorescence spectra of Chl a at room temperature by fluorescence spectrophotometer and intensity was measured using helium-neon lamp at 632.8nm in a quartz cuvette.

Isolation and Purification of PS II Nanoparticles

In this study, the multisubunit complexes of PS-II nanoparticles were purified from a fresh water cyanobacterium, *Synechococcus spp. PCC 7942* after solubilization of the thylakoid membranes using 1.2% of sucrose monolaurate. The PS II particles were purified by a single-step Ni^{2+}-affinity column chromatography of the sucrose monolaurate extracts using histidine tagging technique. The yield of the PS II from thylakoid membranes was 6–7.2% based on Chl contents.

Oxygen Evolution Measurements

Oxygen evolution measurements of cyanobacterial cultures were performed at a light intensity of 200 $\mu E/m^2/s$ using a Clark-type oxygen electrode [10, 11] following the directions of the manufacturer, YSI Instruments Co. Inc., Yellow Springs, OH. The cultures were incubated in light for 15minutes prior to the measurement of oxygen evolution. Oxygen evolution was measured in assay mixture containing phosphate buffer (pH-7.5, 0.1M) in the presence of 2mM potassium ferricyanide as the electron acceptor.

RESULTS AND DISCUSSIONS

The effect of different toxic metals on the synthesis or inhibition of chlorophyll and its O_2 evolution capacity was seen. Laser induced fluorescence spectra of Chl a was measured in the cultures grown in the presence of heavy metals. Metals inhibit the synthesis of different pigments due to substitution of the Mg^{2+} ion from the pyrrole ring. The synthesis of Chl a was suppressed under heavy metal stress with maximum suppression at 1µM concentration with respect to control. In some cases (0.1µM & 0.4µM Cu^{2+} ion concentration), an enhancement in the synthesis of Chl a was also recorded. The thylakoids and PS II particles purified from control showed high oxygen evolving activity of 2,300-2,500 pmol O_2 (mg Chl)$^{-1}$ h^{-1} while the activity was decreased in the cultures grown with heavy metals. The inhibition of oxygen evolution was in the order Hg>Ni> Cd >Cu.

The activities of thylakoid membranes and PS II particles were in the range of 1800-2000 pmol O_2 (mg Chl)$^{-1}$ h^{-1} in the cultures grown with Cu^{2+} ion while the activity decreased to 1500-1700 pmol O_2 (mg Chl)$^{-1}$ h^{-1} with Cd^{2+} ion and decreased further to 1000-1200 pmol O_2 (mg Chl)$^{-1}$ h^{-1} with Ni^{2+} ion. The oxygen evolution activity was least in the cultures supplemented with Hg^{2+} ion, that is 400-600 pmol O_2 (mg Chl)$^{-1}$ h^{-1}.

ACKNOWLEDGMENTS

We thank Department of Science and Technology (DST), Government of India, New Delhi for financial assistance.

REFERENCES

1. N. Kamiya and J. R. Shen, *Proc. Natl. Acad. Sci.* **100**, 98–103 (2003).
2. J. Barber, J. Nield, E.P. Morris, D. Zheleva and B. Hankamer, *Physiol. Plantarum* **100**, 817–827 (1997).
3. B. Hankamer, J. Barber and E. J. Boekema, *Ann. Rev. Plant. Physiol. Plant Mol. Bio.* **48**, 641–671 (1997).
4. R. Webb and L. A. Sherman, "The cyanobacterial heat-shock response and the molecular chaperones" in *The Molecular Biology of Cyanobacteria* edited by D. Bryant, Netherlands : Springer,1994, pp. 751-767.
5. S. Silver and M. Wauderhaug, *Microbiol. Rev.* **56**, 195-264 (1992).
6. R. Rouillon, S. A. Piletsky, F. Breton, E. V. Piletska and R. Carpentier, "Biotechnological Applications of Photosynthetic Proteins: Biochips, Biosensors and Biodevices" in *Photosystem II Biosensors for Heavy Metals Monitoring*, edited by Maria Teresa Giardi and Elena V. Piletska, USA: Springer, 2007, pp. 166-174.
7. D. Frense, A. Muller and D. Beckmann, *Sensors Actuators B-Chem.* **51**, 256-260 (1998).
8. G. I. Paton, C. D. Campbell, L. A. Glover, and K. Killham, *Appl. Microbiol. Letters* **20**, 52-56 (1995).
9. M.M. Allen, *J. Phycol.* **4**, 1-4 (1968).
10. L.C. Clark, R. Wolf, D. Granger and Z. Taylor, *J. Appl. Physiol.* **6**, 189-193 (1953).
11. J.W. Severinghaus and P.B. Astrup, *J. Clin. Monit.* **2**, 125-139 (1986).

Synthesis, Characterization and Application of CdS Quantum Dot

Jumi Kakati[a] and Pranayee Datta[b]

a,b Electronic Science Department, Gauhati University, Guwhati-14, Assam

Abstract: Wide band gap semiconductor quantum dots (for example CdS) need extensive studies as they are capable of showing excitonic absorption even at room temperature. This paper reports chemical synthesis of CdS quantum dots and their characterization by XRD, TEM, UV-VIS, PL and EDS .Possible applications of the fabricated samples in the field of electronics are also investigated.

Keywords: CdS nanoparticles, Chemical synthesis, Optical absorption, Confinement.

PACS Nos: 81.07.Bc; 81.16.Be; 78.67.Bf; 72.80.Ey

INTRODUCTION

Semiconductor nanoparticles have attracted much interest during the past decades in both fundamental research and technical applications due to their unique size-dependent optical and electronic properties (1).Various techniques differing in growth conditions, size range and distribution of physical and chemical stability as well as reliability have been applied to fabricate nanocrystals(2).The size of the nano particles is controlled either by using stabilizer such as thiol,phosphates etc or restricted by using matrix like zeolite,glass, PVA etc (1).

Cadmium sulphide (CdS) is a light sensitive, direct band gap material with band gap of 2.42 eV at room temperature and Exciton Bohr radius ~3 nm, which makes possible its application in optoelectronic devices such as solar cells, photoconductors, and diode lasers (2-5).

In this paper, an attempt is made to synthesize CdS, CdS: Cu, CdS: Mn nanocrystals in PVA and to characterize the synthesized samples by XRD, UV-VIS, PL, TEM, EDS for possible applications in electronics.

EXPERIMENTAL

To prepare CdS nanocrystals embedded in PVA matrix, four different PVA solutions are prepared taking 4gms of PVA in 100 ml of distilled water, varying the stirring rate and time as given in Table-1.$CdCl_2$ solution is prepared taking 2 gms of $CdCl_2$ in 100 ml of distilled water and the solution is stirred at 60^0C for half an hour. Then Na_2S solution is prepared taking 2 gms of Na_2S in 100 ml of distilled water and the solution is stirred at 60^0C for half an hour. Then as-prepared PVA and $CdCl_2$ solutions are taken in the ratio 2:1 to which the as-prepared Na_2S solution is added drop by drop till the mixture turns

CP1147, *Transport and Optical Properties of Nanomaterials—ICTOPON - 2009,* edited by M. R. Singh and R. H. Lipson
© 2009 American Institute of Physics 978-0-7354-0684-1/09/$25.00

orange. The prepared solution is stirred with time, temperature and speed adjusted in such a way that the solution appears fully yellow. For fabrication of Cu and Mn doped CdS in PVA, $CuCl_2$ solution is prepared taking 2 gms of $CuCl_2$ in 100 ml of distilled water and then the solution is stirred for half an hour at 60^0C .Then as-prepared PVA, $CdCl_2$ solutions are taken in 2:1 ratio, to which 2 drops of as-prepared $CuCl_2$ solution are added. To this solution, Na_2S solution is added drop by drop till the colour of the solution turns black. The whole mixture is then stirred at 60^0C for 1hour.Similarly, for synthesis of CdS: Mn nanocrystals, instead of $CuCl_2$, $MnCl_2$ is taken and steps already discussed for CdS: Cu are followed.

Characterization

The prepared samples have been characterized for optical absorption in the range of 300 nm to 600 nm using HITACHI U-3210 Double beam spectrophotometer.

The photoluminescence spectra of the prepared samples have been acquired using F-2500 FL spectrometer. Photoluminescence study of the samples shows the luminescence output at around 400-500 nm for undoped samples and 400-800 nm for doped samples when excited with the optical signal ranging from 390 nm to 795 nm.

For TEM measurement, a small drop of sample is placed on a thin carbon film supported on the copper grid and photographs are taken using JEOL JEM 2100.

For XRD measurement, the diffraction patterns are recorded using Seifert XRD (3003TT).

EDS of the samples are also taken.

RESULTS AND DISCUSSION

Optical absorption spectra of CdS nanocrystals are taken at room temperature.

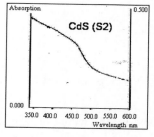 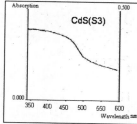 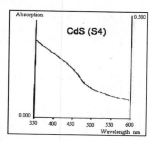

317

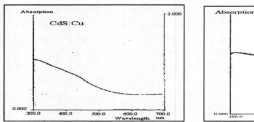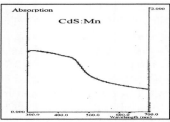

FIGURE 1. UV-VIS spectra of the prepared samples (a)S2, (b)S3, (c) S4, (d) S4:Cu, (e) S4: Mn

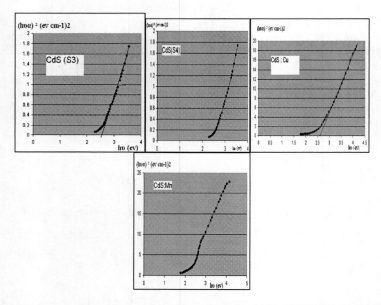

FIGURE 2. Band gap determination of the prepared samples (a)S3, (b)S4, (c)S4:Cu, (d) S4: Mn

Fig (1) shows the variation of optical absorbance of the CdS samples with wavelength. Band gap given in Table 1, has been estimated by using the Tauc's relation (Fig. 2):

$$h\upsilon\alpha = A \, (h\upsilon - E_a)^{\frac{1}{2}} \quad \text{---------------- Eq. 1}$$

where A is a constant, E_a is band gap of the nanoparticles. α is absorption coefficient. The average crystal size is calculated from the Brus equation (Table 1) (6)

$$(E_a - Eg)^{-\frac{1}{2}} = \beta d \text{ ------------------ Eq. 2}$$

where $\beta = 0.49 \text{ nm}^{-1} \text{ eV}^{\frac{1}{2}}$, Eg is the bulk band gap, d is the nanoparticle size. The photoluminescence spectra of the prepared sample are given in Fig 3.

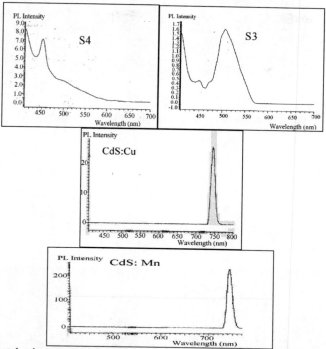

FIGURE 3. Photoluminescence spectra of the prepared samples –(a) S4, (b) S3, (c) S4:Cu, (d) S4 : Mn

The near-sharp PL peak at ~500 nm for S1, S2, S3 and S4 is due to band edge luminescence. Band edge emission may be due to low-lying dark states of the nanocrystal interior(7) .A sharp peak at 745 nm is obtained for both S4:Cu and S4:Mn.This peak is due to the surface states arising from the deep trap formed due to sulpher vacancy.

XRD pattern is given in figure 4.The size calculated from XRD results (Debye-Scherrer formula) is given in Table 1.

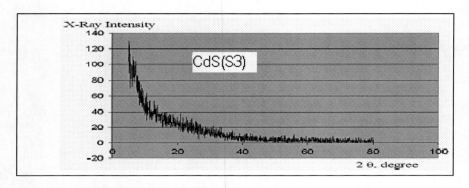

FIGURE 4: The XRD pattern of the prepared S3 nanoparticle.

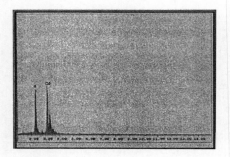

FIGURE 5. EDS of the CdS quantum dot (S3)

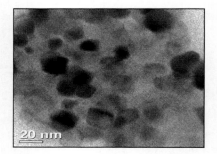

FIGURE 6. TEM of the CdS quantum dot (S3)

The composition of the CdS/PVA sample is determined by EDS spectra (Fig.5).

TEM is given in figure 6. From the TEM image, the particle size is determined to be in the range 4 to 15 nm for S3.Similar TEM results are obtained for the other samples also.

TABLE 1 : Size calculated from using XRD spectrum(Debye Scherrer formula) and UV-VIS (Brus equation)

Sample No.	Stirring Time of PVA (hr)	Stirring rate for PVA (RPM)	Temperature In Degree centigrade for the matrix while stirring	Band gap from UV-VIS (eV,nm)	Size (nm) from XRD(Debye-Scherrer formula) ; UV-VIS (Brus equation)	PL Peak position (nm)
S1	2½	900	70	2.43, 510	50.34 ; 20.41	470.0
S2	4	900	50 - 60	2.48, 500	22.65 ; 9.276	503.0
S3	2½	400	70	2.5, 496	8.1 ; 7.2	505.0
S4	3	500	70	2.5, 496	7.56 ; 6.46	451.0
S4:Cu	3	400	60	2.6, 476.9	5.32 ; 4.56	745.0
S4:Mn	3	400	60	2.43, 510.2	10.24 ; 8.33	410.0,745.0

CONCLUSIONS

Characterization results confirm that we have successfully fabricated CdS/PVA nanocomposites, both undoped and doped, by controlling various parameters and adopting chemical route. The TEM pattern displays the spherical phase of CdS nanoparticles.Size obtained from XRD is in good agreement with that from TEM. As for all the samples (exceptS4:Cu), size is greater than Exciton Bohr radius, confinement is strong. The prepared nanoparticles show blue shift in optical absorption. The photoluminescence spectra show luminescence from band edge emission as well as from trap levels caused by sulpher vacancies. Works for the applications of these samples as schottky barrier are in progress.

ACKNOWLEDGMENTS

The authors would like to acknowledge the department of Chemistry, G.U for providing PL and UV-VIS facilities and IIT Guwahati for providing XRD, TEM and EDS facilities. The authors also thank the DST, Govt. of India, for providing computational facilities in the Electronic Science Department, Gauhati University under DST-FIST programme.

REFERENCES

1. Li Cheng, Chen Xiao et. al. *Chinese Science Bulletin*, 51 No 1o May 2006.
2. Hui Zhang, Deren Yang et. al. Some critical factors in the synthesis of CdS nanorods by hydrothermal process. ELSEVIER, Materials Letters 59 (2005) 3037-3041.
3. Li-Li Ma, Hai-Zhen Sun et. al. Preparation, Characterization and photocatalytic properties of CdS nanoparticles dotted on the surface of carbon nanotubes.IOP Publishing.Nanotechonology 19 (2008) 115709 (8pp)
4. S.S. Nath, D. Chakdar, G. Gope and D K Avasthi, Characterization of CdS and ZnS Quantum Dots Prepared via A Chemical Method on SBR Latex .Journal of Nanotechnology May 12 (2008)
5. D. Mohanta, G.A.Ahmed, A.Choudhury.Spectroscopic Investigation of Carrier Confinement and Surface Phonon Detection in Polymer Embedded CdS quantum dot systems. Chinese Journal of Physics Vol: 42, No 6, December 2004.
6. Brus L 1986 J. Phys. Chem 90, p 2555.
7. R.Bhattacharya,S.Saha,Growth of CdS nanoparticles by chemical method and its characterization.PRAMANA journal of physics Vol.71,No.1,July 2008,pp.187-192

ELECTRICAL PROPERTIES OF NANOMATERIALS

Aharonov–Bohm type effects and the interference of macro–molecules

S. A. R. Horsley and M. Babiker

Department of Physics, The University of York, Heslington, York, UK, YO10 5DD

Abstract. Given that it is now possible to observe the interference of molecules with comparatively numerous internal degrees of freedom, we might wonder what role these extra degrees of freedom play in the traditional double–slit experiment. Here we seek to understand the effect of these degrees of freedom in an Aharonov–Bohm type experiment. We find, perhaps unsurprisingly that an ionised object can exhibit the Aharonov–Bohm effect in a manner which is independent of its internal structure. Meanwhile, the Aharonov–Bohm type effects involving objects with magnetic, or electric dipole moments (the Aharonov–Casher and He–McKellar–Wilkens effects, respectively) depend explicitly upon the structure. When—due to some specified dynamics of the internal variables—an object's dipole moment changes during the experiment, we find that the corresponding phase shift depends upon the time average and variance of the internal dynamics. Furthermore, given that the internal dynamics are, to an extent, dependent upon the centre of mass motion there is also a small, transient loss of coherence associated with each internal degree of freedom.

Keywords: Aharonov–Bohm effect, Aharonov–Casher effect, He–McKellar–Wilkens effect, Berry phase, macromolecule interference, atom interference, quantum coherence.
PACS: 03.65.Vf,03.75.-b,36.20-r

If we can completely eliminate the external fields from a region of space where an object is in motion ($v \ll c$) then the wave–like properties of that object may be observed through a double–slit type experiment. To see this, consider that, when we have a system of N particles in motion, in the absence of any external field, the non–relativistic Lagrangian of the system can be written as, $L = \frac{1}{2}\sum_{i=1}^{N} m_i \dot{\mathbf{r}}_i^2 - \sum_{i<j} V\left(\mathbf{r}_i - \mathbf{r}_j\right)$, where each particle has the co–ordinate, $\mathbf{r}_i$, the mass, m_i, and interacts via a potential of the form, V. Introducing the centre of mass co–ordinate, $\mathbf{R}$, and the relative co–ordinates, $\mathbf{x}_i = \mathbf{r}_i - \mathbf{R}$, this Lagrangian becomes,

$$L = \frac{1}{2}M\dot{\mathbf{R}}^2 + \frac{1}{2}\sum_{i=1}^{N} m_i \dot{\mathbf{x}}_i^2 - \sum_{i<j} V\left(\mathbf{x}_i - \mathbf{x}_j\right), \tag{1}$$

where $M = \sum m_i$ is the total mass of the system. When the action, S, is calculated from (1) it breaks up into two parts, $S = S[\mathbf{R}(t)] + S[\mathbf{x}_i(t)]$. Therefore the state of the system can take the form of the product of a centre of mass wave–function times an internal wave–function, $|\psi\rangle = |CM\rangle|INT\rangle$, and it becomes possible to observe the interference of the N particle system through a measurement of the centre of mass co–ordinate.

So far the interference of systems of 60, or even 70 atoms has been demonstrated [1, 2], and the influence of thermal emission from the system on the coherence of the centre of mass co–ordinate has been identified [3]. There has also been an interesting theoretical discussion of the mechanism for a loss of coherence into the internal degrees of freedom via the scattering from the slits [4]. In all the cases where the coherence is lost from

CP1147, *Transport and Optical Properties of Nanomaterials—ICTOPON - 2009*, edited by M. R. Singh and R. H. Lipson
© 2009 American Institute of Physics 978-0-7354-0684-1/09/$25.00

the centre of mass into the internal degrees of freedom, this occurs because an external field, V_{EXT}, is present, modifying (1) as $L \rightarrow L - V_{\text{EXT}}$. If a large molecule were to be used to demonstrate an Aharonov–Bohm type effect, then one might expect the internal degrees of freedom to provide a similar coherence 'sink'. For instance in the case of an ionised molecule—of net charge, $\sum e_i = q$—passing around an isolated region of magnetic flux, equal to Φ (c.f. [5]), the Lagrangian, (1), undergoes,

$$L \rightarrow L + \dot{\mathbf{R}} \cdot \sum_{i=1}^{N} e_i \mathbf{A}(\mathbf{R} + \mathbf{x}_i) + \cdot \sum_{i=1}^{N} e_i \dot{\mathbf{x}}_i \cdot \mathbf{A}(\mathbf{R} + \mathbf{x}_i), \tag{2}$$

and it would appear that the vector potential is responsible for entangling the internal structure and the centre of mass. However, a little consideration of the form of these 'coupling' terms shows us that they are unphysical and may be removed by the subtraction of a suitable total time derivative from the Lagrangian [6], $L' = L - d\chi/dt$. In the dipole approximation this involves the transformation,

$$L' = L - \frac{d}{dt}\left(\sum_{i=1}^{N} e_i \mathbf{x}_i \cdot \mathbf{A}(\mathbf{R})\right) = \frac{1}{2}M\dot{\mathbf{R}}^2 + q\dot{\mathbf{R}} \cdot \mathbf{A}(\mathbf{R}) + \text{internal structure}, \tag{3}$$

and the centre of mass wave–function acquires the usual phase upon passing around the region of flux, $\psi \rightarrow \psi e^{iq \oint A(\mathbf{R}) \cdot d\mathbf{R}/\hbar} = \psi e^{iq\Phi/\hbar}$. Meanwhile, in the case of the Aharonov–Casher (AC) [7] and He–McKellar–Wilkens [8, 9] (HMW) effects—where the role of the vector potential is replaced by terms involving the magnetic and electric moments, respectively—the coupling between the internal structure and the centre of mass can certainly not be removed. In the dipole approximation, these effects involve a Lagrangian of the form,

$$L = \frac{1}{2}M\dot{\mathbf{R}}^2 \begin{cases} +\dot{\mathbf{R}} \cdot \frac{1}{c^2}(\frac{1}{2}\sum_{i=1}^{N} e_i \mathbf{x}_i \times \dot{\mathbf{x}}_i) \times \mathbf{E}(\mathbf{R}) + \dots & \text{(AC)} \\[2mm] -\dot{\mathbf{R}} \cdot (\sum_{i=1}^{N} e_i \mathbf{x}_i) \times \mathbf{B}(\mathbf{R}) + \dots & \text{(HMW)}, \end{cases} \tag{4}$$

and it might be wondered how these phenomena are altered if one attempts to verify their existence with objects that include a large number of internal degrees of freedom. Here we divide the question up into two parts: firstly, how are the AC and HMW effects altered if the dipole moment is changed in some pre–specified way during the interference experiment; and secondly, the change of the dipole moment in response to the motion of the centre of mass affects the coherence of the centre of mass interference pattern.

THE AC AND HMW EFFECTS WITH CHANGING DIPOLE MOMENTS

To work out the quantum mechanics of a dipole that is changed as it passes through an electromagnetic field, we will use the path integral formalism [10, 11], where we

326

calculate the Feynman propagator, $K(\mathbf{x}_a, t_a; \mathbf{x}_b, t_b)$, through a sum over phase factors, $K(\mathbf{x}_a, t_a; \mathbf{x}_b, t_b) = \sum_{\text{paths}} e^{\frac{i}{\hbar} S[\mathbf{x}(t)]}$. In this case the advantage of using the path integral is that the Lagrangian, (4), in the situations of interest, turns out to be quadratic in the canonical variables, $\mathbf{R}$ and $\dot{\mathbf{R}}$. When the Lagrangian takes this form, the only path that counts in the path integral is the classical path of extreme action, and the propagator is therefore comparatively simple to calculate.

If the internal structure of the molecule changes in some pre–specified way, then we can take the internal dynamics in the coupling terms of (4) to be fixed functions of time; $\mathbf{d}(t) = \sum e_i \mathbf{x}_i$ and $\boldsymbol{\mu}(t) = \frac{1}{2} \sum e_i \mathbf{x}_i \times \dot{\mathbf{x}}_i$. The classical two–point action function, $S(\mathbf{R}_a, t_a; \mathbf{R}_b, t_b)$ for the motion of the centre of mass, corresponding to (4), in the AC case satisfies the two Hamilton–Jacobi equations [12]

$$\frac{\partial S}{\partial t_{a/b}} \mp \frac{1}{2m} \left(\nabla_{a/b} S \pm \frac{1}{c^2} \boldsymbol{\mu}_{a/b} \times \mathbf{E}_{a/b} \right)^2 = 0 \tag{5}$$

and in the HMW case,

$$\frac{\partial S}{\partial t_{a/b}} \mp \frac{1}{2m} \left(\nabla_{a/b} S \mp \mathbf{d}_{a/b} \times \mathbf{B}_{a/b} \right)^2 = 0 \tag{6}$$

If the interaction $\boldsymbol{\mu} \times \mathbf{E}/c^2$, or $-\mathbf{d} \times \mathbf{B}$ is uniform over the space accessible to the molecule, then the canonical momentum, $\mathbf{p}$, may be introduced as an integral of the motion, and the action for a given value of $\mathbf{p}$ may be written as [13],

$$S_{\mathbf{p}}(\mathbf{R}_a, t_a; \mathbf{R}_b, t_b) = \begin{cases} \mathbf{p} \cdot \mathbf{R}_{ba} - \frac{\mathbf{p}^2}{2m} t_{ba} + \frac{\mathbf{p}}{m} \cdot \int_{t_a}^{t_b} \frac{\boldsymbol{\mu} \times \mathbf{E}}{c^2} dt - \frac{1}{2m} \int_{t_a}^{t_b} \left[\frac{\boldsymbol{\mu} \times \mathbf{E}}{c^2} \right]^2 dt & \text{(AC)} \\ \mathbf{p} \cdot \mathbf{R}_{ba} - \frac{\mathbf{p}^2}{2m} t_{ba} - \frac{\mathbf{p}}{m} \cdot \int_{t_a}^{t_b} \mathbf{d} \times \mathbf{B}\, dt - \frac{1}{2m} \int_{t_a}^{t_b} [\mathbf{d} \times \mathbf{B}]^2 dt & \text{(HMW)}, \end{cases}$$

$$\tag{7}$$

where $\mathbf{R}_{ba} = \mathbf{R}_b - \mathbf{R}_a$ and $t_{ba} = t_b - t_a$. Substituting $\mathbf{p} = \hbar \mathbf{k}$ into (7) yields the phase for the propagator for a fixed wave–vector, $\mathbf{k}$, $K_{\mathbf{k}} = \exp(iS_{\mathbf{k}}/\hbar)$. Integrating this expression over $\mathbf{k}$–space then yields the full propagator. Such a calculation gives rise to,

$$K(\mathbf{R}_a, t_a; \mathbf{R}_b, t_b) = \left(\frac{m}{2\pi i \hbar t_{ba}} \right)^{3/2} \left\{ \begin{array}{l} \exp\left\{ \frac{i}{\hbar} \left[\frac{m}{2 t_{ba}} \mathbf{R}_{ba}^2 + \mathbf{R}_{ba} \cdot \overline{\frac{\boldsymbol{\mu} \times \mathbf{E}}{c^2}} - \frac{t_{ba}}{2m} \text{var}\left(\frac{\boldsymbol{\mu} \times \mathbf{E}}{c^2} \right) \right] \right\} \\ \exp\left\{ \frac{i}{\hbar} \left[\frac{m}{2 t_{ba}} \mathbf{R}_{ba}^2 - \mathbf{R}_{ba} \cdot \overline{\mathbf{d} \times \mathbf{B}} - \frac{t_{ba}}{2m} \text{var}(\mathbf{d} \times \mathbf{B}) \right] \right\} \end{array} \right\},$$

$$\tag{8}$$

where we note the appearance of the usual definitions of the time average and variance of $\boldsymbol{\mu} \times \mathbf{E}$ and $\mathbf{d} \times \mathbf{B}$. The result displayed in (8) represents an approximate solution to the first part of the problem we are interested in. The essence is that, when the dipole moment is altered during the experiment, there are *two* additional terms in the propagator relative to the free space case; one depending upon the vector between the start and end points of the propagation and the *time average* of the interaction term; and one depending upon the *variance* of the interaction term over the evolution time.

As an application of (8), suppose that the molecule is confined to a fixed circle of radius, R, and angular co–ordinate, θ, which is centred on a line of electric (λ) or magnetic (ξ) charge density [7, 9]. In this case, so long as the dipole moment remains oriented along $\hat{\mathbf{z}}$, and the particle passes once around the ring, $\theta_{ba} = 2\pi$ over a time period $T = t_{ba}$, then, relative to the case where there is no charged line, there is an additional phase difference [13],

$$
\frac{\Delta S}{\hbar} = \begin{cases} \frac{\mu_0 \overline{|\mu|} \lambda}{\hbar} - \frac{T}{2m\hbar} \left(\frac{\mu_0 \lambda}{2\pi} \right)^2 \mathrm{var}\,(|\mu|) & \text{(Time--dependent AC)} \\ -\frac{\mu_0 \overline{|\mathbf{d}|} \xi}{\hbar} - \frac{T}{2m\hbar} \left(\frac{\mu_0 \xi}{2\pi} \right)^2 \mathrm{var}\,(|\mathbf{d}|) & \text{(Time--dependent HMW).} \end{cases} \tag{9}
$$

The first term containing the time average of the dipole moment, in (9) is the generalisation of the usual Aharonov–Bohm type phase, while the term dependent upon the variance of the dipole moment is interpreted as an additional dynamical phase. The two expressions in (9) have the correct limiting forms if the dipole moments are constant, in which case the time average of the dipole moment becomes the static value, and the variance vanishes. One result of equation (9) is that, in contrast to the original AC and HMW proposals, if the dipole is passed around a charged line while its dipole moment is varied sinusoidally—$|\mu| \rightarrow |\mu| \cos(\omega t)$, or $|\mathbf{d}| \rightarrow |\mathbf{d}| \cos(\omega t)$—then the time average part of the phase can be made to vanish, with the dynamical part remaining.

THE LOSS OF COHERENCE DUE TO THE HMW COUPLING

At the outset we wondered whether the internal structure of a large interfering molecule could introduce a coherence 'sink' by virtue of an Aharonov–Bohm type interaction. To understand this effect we consider the HWM interaction (for simplicity), and take $N = 2$ in (4). Here we introduce a relative co–ordinate, $\mathbf{x} = \mathbf{x}_1 - \mathbf{x}_2 - \mathbf{a}$, where the constant vector, $\mathbf{a}$, is understood as the equilibrium extension of the dipole moment, and we also assume that $e_1 = e$ and $e_2 = -e$. As a model, it is assumed that the internal variable, $\mathbf{x}$ behaves as a simple harmonic oscillator, and therefore the model Lagrangian is,

$$
L = \frac{1}{2} M \dot{\mathbf{R}}^2 - e \dot{\mathbf{R}} \cdot (\mathbf{a} + \mathbf{x}) \times \mathbf{B} + \frac{1}{2} m \left(\dot{\mathbf{x}}^2 - \omega^2 \mathbf{x}^2 \right) \tag{10}
$$

where $e\mathbf{a} = \mathbf{d}_0$ is the equilibrium value of the dipole moment, and $m = m_1 m_2 / M$ is the reduced mass. Again, we interest ourselves in the calculation of the Feynman propagator for the system. As in the previous section, so long as the field is uniform in the space accessible to the centre of mass, it suffices to calculate the action associated with the classical path. To simplify the problem further, it is assumed that the dipole moment can only be extended along a single axis, $\hat{\mathbf{z}}$, $\mathbf{x} = z\hat{\mathbf{z}}$, that the field is always along $\hat{\mathbf{x}}$, and therefore that only the centre of mass motion along $\hat{\mathbf{y}}$ is of interest. Under these assumptions, (10) can be re–written in terms of only two degrees of freedom,

$$
L = \frac{1}{2} M \dot{Y}^2 - \alpha \dot{Y} (a + z) + \frac{1}{2} m \left(\dot{z}^2 - \omega^2 z^2 \right), \tag{11}
$$

with $\alpha = eB$. Through calculating the action associated with the integral of (11) along the classical path, the propagator for this system is found to be, to first order in α [6],

$$K(a;b) = \sqrt{\frac{M}{2\pi i \hbar t_{ba}}} \sqrt{\frac{m\omega}{2\pi i \hbar \sin(\omega t_{ba})}} e^{\frac{iM}{2\hbar t_{ba}} Y_{ba} - \frac{i\alpha}{\hbar} Y_{ba} \left(a + (z_a + z_b)\frac{\tan(\omega t/2)}{\omega t}\right) + \frac{i}{\hbar} S_\omega^{\text{SHO}}(z_a, t_a; z_b, t_b)}$$

(12)

where S_ω^{SHO} is the classical action associated with a simple harmonic oscillator of frequency, ω [10]. The coupling term, $\alpha Y_{ba} (a + (z_a + z_b) \tan(\omega t/2)/\omega t)$, can be understood in this approximation as the average of the dipole moment, written in terms of the oscillator co–ordinates, times the centre of mass motion [6], which is consistent with (8).

We now examine the explicit effect of the internal dynamics on some initial state of the model dipole. As the final state in the cases under consideration can now only be represented by a reduced density matrix, ρ_r, we must use the kernel of the time evolution operator—the propagator, K—to evaluate the following quantity, which is interpreted as a kind of density propagator,

$$J(1,2,0;\mathbf{R}',\mathbf{R},t) = \int dz\, K(a,0;\mathbf{R}',z,t)\, \bar{K}(b,0;\mathbf{R},z,t),$$

(13)

where the 1 and 2 indicate the variables R and z together with subscripts 1 and 2 so that the time evolution of the system can be written in terms of the initial (full) density matrix, ρ, as, $\rho_r(\mathbf{R}',\mathbf{R},t) = \int da\, db\, J(1,2,0;\mathbf{R}',\mathbf{R},t)\rho(1,2;0)$. Through substituting the expression (12) into (13) we obtain [6],

$$J(1,2,0;R',R,t) = J_{\text{FP}}(Y_1,Y_2,0;R',R,t)\, e^{-\frac{ieB}{\hbar}a[(Y'-Y_1)-(Y-Y_2)]} e^{\frac{im\omega}{2\hbar}(z_1^2 - z_2^2)\cot(\omega t)}$$

$$\times \delta\left(z_1 - z_2 + (2eB/m\omega^2 t)\sin^2(\omega t/2)\left[(Y'-Y_1)-(Y-Y_2)\right]\right)$$

$$\times e^{-\frac{ieB}{\hbar}\frac{\tan(\omega t/2)}{\omega t}[(Y'-Y_1)z_1 - (Y-Y_2)z_2]},$$

(14)

where the quantity immediately to the right of the equality is the free particle part of the expression. Take the specific case where the initial state of the system corresponds to a Gaussian distribution for the oscillator (of 'spread', σ, and extension, η), along with a superposition of two possible centre of mass position states (separated by a distance, D) on a line (i.e. in analogy with the double slit experiment),

$$\langle Y,z,0|\psi\rangle = \frac{1}{\sqrt{2}}\left[\delta(Y-D/2) + \delta(Y+D/2)\right]\left(\frac{\alpha}{\pi}\right)^{1/4} e^{-\frac{\sigma}{2}(z-\eta)^2},$$

(15)

We form the initial density matrix from this expression—$\rho(a,b,0) = \langle 1,0|\psi\rangle\langle\psi|2,0\rangle$—and place it under an integral sign with J. We then obtain the following expression for the probability density of the centre of mass,

$$\rho_r(R,R,t) = \left(\frac{M}{2\pi\hbar t}\right)\left[1 + e^{-D^2\chi(t)}\cos\left(\frac{MD}{\hbar t}R - \frac{eBD}{\hbar}\Lambda(t) + O(B^2)\right)\right]$$

(16)

where it is noticed that Λ represents the classical expression for the time average of the displacement of an oscillator which is initially extended a length η away from

329

equilibrium, $\Lambda(t) = a + \eta \frac{\sin(\omega t)}{\omega t}$, and we have the positive–definite quantity,

$$\chi(t) = \frac{e^2 B^2}{4\omega^2 t^2} \left(\frac{\sigma}{m^2 \omega^2} (1 - \cos(\omega t))^2 + \frac{\sin^2(\omega t)}{\sigma \hbar^2} \right). \tag{17}$$

The factor, χ, illustrates a kind of exchange of coherence between the internal oscillator and the centre of mass due to the spread of the dipole moment. When ωt is small, $\cos(\omega t) \sim 1 - \omega^2 t^2 / 2$, and this factor is maximised. We might imagine that, for N internal degrees of freedom this maximum might be of the order, $\chi_{\text{MAX}} \sim N \frac{e^2 B^2}{2\hbar m \omega}$, and if we take $N \sim 100$, $e \sim 10^{-19}$, $B \sim 1$, $m \sim 10^{-27}$ and $\omega \sim 10^{13}$, then χ could take the maximum initial value of $\chi \sim 10^{12}$, which is more than significant! However, $\omega^2 t^2$ is only small for less than a single oscillation of the internal degree of freedom, and practically this is no time at all: χ very rapidly decays (as $1/(\omega t)^2$) from this maximum value to zero, and the loss of coherence becomes insignificant.

CONCLUSIONS

We have shown that, when an Aharonov–Bohm type effect is performed with a molecule that might include a large number of internal degrees of freedom, then there is a distinction to be made between the effect where a charge passes around an isolated region of flux, and those effects involving the motion of multipole moments in electromagnetic fields. In the latter cases, the internal degrees of freedom cannot be separated out from the centre of mass motion, and this results in an—albeit small, and short lived—loss of coherence in the interference pattern. Furthermore, if the internal dynamics are somehow manipulated during the motion of the system, then in some simple cases, the observed phase shift depends upon the time average and variance of the internal motion. One of us (SARH) wishes to thank the EPSRC for financial support.

REFERENCES

1. M. Arndt, O. Nairz, J. Vos-Andreae, C. Keller, G. van der Zouw, and A. Zeliinger, *Nature* **401**, 680 (1999).
2. L. Hackermüller, S. Uttenthaler, K. Hornberger, E. Reiger, B. Brezger, A. Zeilinger, and M. Arndt, *Phys. Rev. Lett.* **91**, 090408–2 (2003).
3. L. Hackermüller, S. Uttenthaler, K. Hornberger, B. Brezger, A. Zeilinger, and M. Arndt, *Nature* **427**, 711 (2004).
4. M. Hillery, L. Mlodinow, and V. Bužek, *Phys. Rev. A* **71**, 062103 (2005).
5. D. Bohm, and Y. Aharonov, *Phys. Rev.* **115**, 485 (1959).
6. S. A. R. Horsley, and M. Babiker, *Phys. Rev. A* **78**, 012107 (2008).
7. Y. Aharonov, and A. C. Casher, *Phys. Rev. Lett.* **53**, 319 (1984).
8. X. G. He, and B. H. J. McKellar, *Phys. Rev. A* **47**, 3424 (1993).
9. M. Wilkens, *Phys. Rev. Lett* **72**, 5 (1994).
10. L. S. Schulman, *Techniques and Applications of Path Integration*, Dover, New York, 2005.
11. H. Kleinert, *Path Integrals in Quantum Mechanics, Statistics, Polymer Physics and Financial Markets*, World Scientific, Singapore, 2006.
12. L. D. Landau, and E. M. Lifshitz, *Mechanics*, Butterworth–Heinemann, Oxford, 2004.
13. S. A. R. Horsley, and M. Babiker, *Phys. Rev. Lett.* **99**, 090401 (2007).

Enhanced Bio-molecular Sensing Capability of LSPR, SPR-ATR Coupled Technique

N. Kamal Singh[1], Abdullah Alqudami[1], S. Annapoorni[1©], Vineet Sharma[2] and K. Muralidhar[2]

[1]Department of Physics and Astrophysics, University of Delhi, Delhi 110 007, India
[2]Department of Zoology, University of Delhi, Delhi 110 007, India
© - corresponding author

Abstract. This paper gives a comparative study of two of the SPR techniques i.e. Localised surface plasmon resonance (LSPR) and Surface plasmon resonance-Attenuated total reflection (SPR-ATR) technique and coupling of the two techniques for bio-molecular sensing applications. Silver nanoparticles prepared by using Exploding wire technique and silver thin films prepared by Thermal evaporation were used. Different concentrations of Bovine serum albumin (BSA) were taken for our studies. For LSPR studies, UV-Visible spectrometer was used and a home-made prism coupled ATR-SPR set-up was used for the study of SPR of the thin films. From the comparative studies of our system ATR-SPR method is comparatively a better tool for the optical bio-sensing. Again when bio-molecules tagged with nanoparticles are used, the sensitivity is found to be enhanced abruptly. So the use of both nanoparticles and thin film in the ATR-SPR method gives the best results which could be used as an excellent tool for sensing of bio-molecules and as an effective biosensor.

Keywords: surface plasmon resonance (SPR), localized surface plasmon resonance (LSPR), SPR-ATR technique.
PACS: 73.20.Mf, 07.07.Df

INTRODUCTION

Noble metal nanoparticles exhibit strong absorption band that are absent in the bulk metals. This absorption band, which is known as the surface plasmon resonance (SPR), results in strong absorption and creates an enhanced local electromagnetic field near the surface of the nanoparticles. Another phenomenon known as attenuated total reflection (ATR) – SPR is one of the most sensitive methods for sensor applications. In the ATR-SPR technique, the resonance occurs at the metal dielectric interface. For the SPR to occur, the dielectric function of the metal film (very thin as compared to the wavelength of the light) should have a large negative real part at the particular chosen wavelength. Surface plasmon resonance (SPR) therefore occurs in the visible range of the electromagnetic spectrum for the so called free electron-like metals such as silver and gold. A fast decaying evanescent field is developed at the interface between the metal film and the dielectric. This evanescent field interacts with the materials at the close vicinity of the thin metal film. SPR based sensors are known to detect refractive index (RI) changes smaller than 10^{-5} at the close vicinity of the metal film which influence the surface plasmon resonance (SPR) absorption peak position [1]. These interesting optical properties of noble metal nanoparticles have opened up several areas of research in materials science and biotechnology. These have led to a new area of nanobiotechnology wherein targeted drugs, gene delivery, biosensors etc.

CP1147, Transport and Optical Properties of Nanomaterials—ICTOPON - 2009, edited by M. R. Singh and R. H. Lipson
© 2009 American Institute of Physics 978-0-7354-0684-1/09/$25.00

are designed and developed [2-6]. One of the most attractive research fields of the combined nanobiotechnology is the study of the conjugation of biomolecules to nanoparticles of noble metals and their interactions. Of all the noble metals, silver and gold have received the most attention in connection with biomolecular conjugates because of their biocompatible nanoparticle-related surface chemistry [7-13].

In fact, the intrinsic optical absorption and fluorescence of silver nanoparticles prepared by electro-exploding wire (EEW) technique [14, 15] make these nanoparticles excellent candidates for bio-molecular labeling. The present study is focused on tracing probable changes in the plasmonic behaviour of silver nanoparticles prepared by EEW technique that might arise due to functionalization of biomolecules. The ATR-SPR technique was also employed for a comparative study with the LSPR technique. Bovine serum albumin (BSA), the simple but widely studied protein, was selected for this particular study which will be extended further to different biomolecules. By using the biomolecules functionalized with silver nanoparticles in the ATR-SPR technique, some unexpectedly enhanced results which may be attributed to the coupling of LSPR and ATR-SPR were observed. Haes AJ et. al. has given a brief comparative study of the LSPR and propagating SPR (ATR-SPR) techniques for bio-molecular sensing applications. Possibility of unifying the two SPR techniques for a better sensor was also suggested [16]. To our knowledge, this is the first attempt to couple Propagating SPR with LSPR for the detection of functionalized nanoparticles with a view to enhance the sensitivity for further use in biosensors.

EXPERIMENTAL DETAILS

Materials

Silver wires (0.25 mm dia.; 99.998 %, Alfa Aesar), Silver plate (20, 20, 1) mm dim.; (99.998%, Alfa Aesar), Double distilled water, BSA (Biomolecules were obtained from SRL; India) and Prisms of same refractive index (1.517).

Preparation of Silver Nanoparticles

Prior to performing the experiments, silver wires and plate were cleaned using emery paper followed by acetone. The silver wires each of 20 mm length were exploded on a silver plate in double distilled water using the electro-exploding wire (EEW) set-up described elsewhere [17]. These wires have been exploded by bringing the wire into sudden contact with the silver plate. Both the electrodes (wire and plate) are subjected to a potential difference of 36 V DC supplied from three 12 V batteries connected in series. The obtained nanoparticles remain in the water medium as colloids for further treatments.

Preparation of Silver Nanoparticles/ Biomolecules Complexes

Different amounts of BSA (10, 20, 50, 100, 400, 500, 1000, and 2000) μg were dissolved in 1.0 *mL* of double distilled water. Each 1.0 *mL* volume containing BSA was then mixed with 3.0 mL of a homogenous suspension of silver nanoparticle and was gentle stirred for 3 hrs.

Similarly for 2.0 nanomoles of BSA, the solution was made. All experiments were carried out at 4.0°C, and under the same conditions. In all the experiments, 3.0 mL of silver nanoparticles suspension was diluted with 1.0 mL of double distilled water and left stirred for 3 hrs as a control sample

Preparation of Silver Thin Films/ Silver Nanoparticles/Biomolecules Complexes

Thin silver films of 49.4 nm were coated on several prisms but of same refractive index (1.517) by thermal evaporation of silver wires (99.998 %, Alfa Aesar) using HIND-HIVAC coating unit capable of producing a vacuum of 5×10^{-5} torr. Pure solution of 2.0 nanomoles of BSA was prepared in double distilled water. The silver coated prisms were kept dipped in the solution of pure biomolecule for 3 hrs. The silver film attached with biomolecules was washed repeatedly with double distilled water to remove the excess biomolecules. Another silver coated prism was kept dipped for 3 hrs in the suspensions of silver nanoparticles functionalized with 2.0 nanomoles of BSA. Same process was carried out to remove the excesses on the surface of the films.

Characterization

Atomic force microscopy (AFM) images were collected using a Nano-R™ AFM (Pacific Nanotechnology) in the contact mode. The images were processed using NanoRule+ (version 1.10) software. The AFM samples of silver nanoparticles and BSA functionalized silver nanoparticles were prepared by drying small drops from both the solutions on cleaned glass substrates. Samples for transmission electron microscopy (TEM) imaging were prepared by drying small drops from the solutions on carbon coated copper grids. The micrographs were recorded using a JEOL JEM 2000EX transmission electron microscope (TEM).
UV-2510PC spectrophotometer was used to record UV-Visible absorption spectra for water based suspensions. A home-made prism coupled ATR-SPR set-up [18] using a Perkin Elmer Light Chopper (Model - 197) and a Stanford Research Systems Lock in Amplifier (Model – SR830 DSP) so as to minimize the background noise have been used to record the ATR-SPR signal.

RESULTS AND DISCUSSIONS

FIGURE 1 shows the 2d topographic images for both pure silver nanoparticles and BSA functionalized silver nanoparticles. The control sample of silver nanoparticles in FIGURE 1(a) shows spherical aggregated particles with spherical shapes. The Z profile shows that silver nanoparticles have a mean edge of about 50 nm. Image of BSA functionalized silver nanoparticles are shown in FIGURE 1(b). It is observed that the sizes of the functionalized silver nanoparticles seem to be smaller than the non-functionalized ones, which might be attributed to the dissociation of aggregated silver nanoparticles during the stirring / adsorption process.

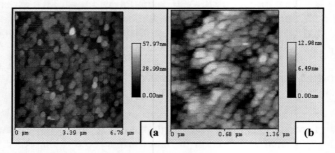

FIGURE 1. AFM 2D topographic images for: (a) silver nanoparticles; (b) BSA (5μg/mL) functionalized silver nanoparticles

As it was observed from earlier AFM images, the functionalized nanoparticles seem to be smaller in size than those non-functionalized particles as also shown from the TEM images in FIGURE 2. Moreover, the functionalized nanoparticles are not having perfect spherical shapes and clean surfaces. These might be attributed to the presence of adsorbents on its surfaces.

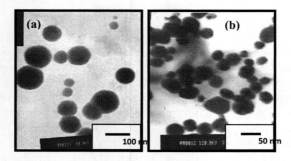

FIGURE 2. TEM images of; (a) silver nanoparticles; (b) BSA (5μg/mL) functionalized silver nanoparticles.

The UV-Visible spectra were recorded in the wavelength range 200-700 nm using quartz cells. The reference solution was double distilled water in all spectra with appropriate base line correction. FIGURE 3 shows the normalized UV-Visible spectra of BSA functionalized silver nanoparticles for different BSA concentrations. From these spectra we observe the disappearance of the LSPR signal with the increase in concentration but no significant shift in the SPR signal is observed as such. Indeed the LSPR nanosensors have modest refractive index sensitivity (2×10^2 nm RIU^{-1}) [19].

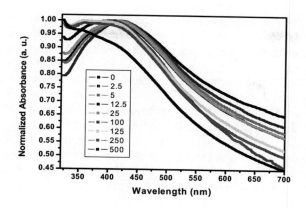

FIGURE 3. UV-Visible spectra for different concentrations of BSA.

It is well known that the ATR-SPR technique can be used to sense changes in the refractive index in the order of 2×10^6 nm RIU^{-1}. The propagating SPR has comparatively a longer decay length of the fast decaying evanescent electromagnetic field [1]. In conventional ATR-SPR experiments, the SPR is observed as a sharp decrease in the intensity of the reflected light at a certain angle of incidence. The SPR absorption peak angle, bandwidth and reflectance intensity can be modified by introducing nanoparticles on the surface of the metal film [20].

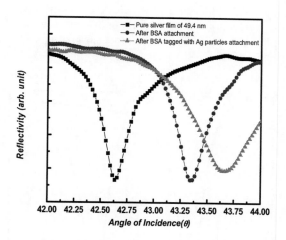

FIGURE 4. ATR-SPR of silver thin film functionalized with pure BSA and with BSA tagged with silver nanoparticles

In the present study, three sets of samples were used. These are pure silver thin film, silver thin film functionalized with pure BSA and silver thin film functionalized with BSA tagged with silver nanoparticles. FIGURE 4 shows the recorded SPR signals of the silver thin films before and after the attachment of BSA respectively. It is observed that there is an appreciable shift in the SPR peak when the biomolecules are attached. It is also observed that when silver nanoparticles tagged with the biomolecules were attached, the reflectance curve becomes broader and the shift in the SPR absorption peak is enhanced tremendously as can be seen from TABLE 1. When we used only LSPR technique, the sensing capability was insignificant. Interestingly by coupling the LSPR to the ATR-SPR technique, an abrupt enhancement was observed as compared to the ATR-SPR technique alone. It may be due to the reason that, when a metallic roughness feature is introduced to the metal film, it develops a site for non-radiative SPR to radiative electromagnetic mode conversion. Further, when the roughness feature increases, large perturbations are introduced to the SPR which in turn modifies the reflectance curve. The main features in the dramatic changes that are observed are multiple scattering phenomena or coupling of LSPR and ATR-SPR techniques.

TABLE 1. Magnitude of SPR shift due to functionalization with BSA and BSA tagged with silver nanoparticles.

	$\Delta\theta$ after functionalizing with BSA	$\Delta\theta$ after functionalizing with BSA tagged with silver nanoparticles
BSA	0.716°	1.037°

In an attempt to decipher the dielectric of the BSA, simulation was carried out using Fresnel's equation, a three layer model of Prism/Metal/Medium. The result of the simulations is given in FIGURE 5. Using the above simulations the dielectric constant of the BSA was determined.

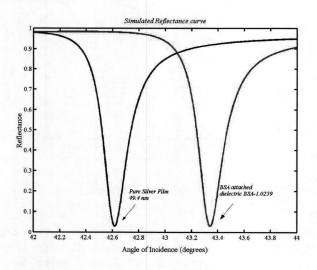

FIGURE 5. Simulated graph for estimating the value of the biomolecules using Prism/Metal/Medium Three-Layer Model.

The reflectance curve peak matching with the observed peak point was taken to be the dielectric constant of the bio-molecule. From our estimation the dielectric of the BSA attached at the interface with complex environment is found to be 1.0259.

CONCLUSIONS

Silver nanoparticles, before and after functionalization with BSA, were characterized using AFM and TEM. Images show that the sizes of BSA-adsorbed silver nanoparticles seem to be smaller than pure ones. Hence BSA acts as a stabilizing agent avoiding the aggregation of the silver nanoparticles. The effect on the LSPR and ATR-SPR were studied from which we conclude that the ATR-SPR technique is more sensitive. The sensitivity of ATR-SPR technique was found to be enhanced significantly after the biomolecules tagged with silver nanoparticles were used. This may be attributed to the coupling of LSPR and propagating SPR. Attempts were made to find dielectric constant of BSA attached on the surface for the particular specific concentration used. It would be further interesting to investigate the coupling with different BSA concentrations and the work is in progress.

ACKNOWLEDGEMENTS

We would like to acknowledge the Department of Science and Technology (DST), India for the funding through the project (SR/S5/NM-52/2002) from the Nanoscience and Technology Initiative programme. Mr. N. Kamal Singh acknowledges University Grant Commission, India for providing Junior Research Fellowship (JRF) through Joint CSIR-UGC Junior Research Fellowship (JRF) and Eligibility for Lectureship-National Eligibility Test (NET) held on 19.06.2005.

REFERENCES

1. Linda S. Jung, Charles T. Campbell, Timothy M. Chinowsky, Mimi N. Mar and Sinclair S. Yee, *Langmuir* **14**, 5636-5648 (1998).
2. A. Sharma and K. R. Rogers, *Meas. Sci. Technol.* **5**, 461-472 (1994).
3. Ch. M. Niemeyer, *Angew. Chem. Int. Ed.* **40**, 4128-4158 (2001).
4. W. J. Parak, D. Gerion, T. Pellegrino, D. Zanchet, C. Micheel, S. C. Williams, R. Boudreau, M. A. L. Gros, C. A. Larabell and A. P. Alivisatos, *Nanotechnology* **14**, R15 (2003).
5. S. G. Penn, L. Hey and M. J. Natan, *Curr. Opinion Chem. Bio.* **7**, 609-615 (2003).
6. C. C. You, A. Verma and V. M. Rotello, *Soft Matt.* **2**,190-204 (2006).
7. S. Mandal, A. Gole, N. Lala, R. Gonnade, V. Ganvir and M. Sastry, *Langmuir* **17**, 6262-6268 (2001).
8. S. H. Choi, S. H. Lee, Y. M. Hwang, K. P. Lee and H. D. Kang, *Radiat. Phys. Chem.* **67**, 517-521 (2003).
9. K. Naka, H. Itoh, Y. Tampo and Y. Chujo, *Langmuir* **19**, 5546-5549 (2003).
10. P. R. Selvakannan, S. Mandal, S. Phadtare, R. Pasricha and M. Sastry, *Langmuir* **19**, 3545-3549 (2003).
11. K. Aslan, J. Zhang, J. R. Lakowicz and C. D. Geddes, *J. Fluorescence* **14**, 677-679 (2004).
12. T. Li, H. G. Park, H.-S. Lee and S.-H. Choi, *Nanotechnology* **15**, S660 (2004).

13. J. L. Burt, C. G. Wing, M. M. Yoshida and M. J. Yacama´n, *Langmuir* **20**, 11778-11783 (2004).
14. A. Alqudami and S. Annapoorni, *Pramana J. Phys.* **65**, 815-819 (2005).
15. A. Alqudami and S. Annapoorni, *Plasmonics* **2**, 5-13 (2007).
16. J. H. Amanda and R. P. Van Duyne, *Anal. Bioanal. Chem.* **379**, 920-930 (2004).
17. P. Sen, J. Ghosh, P. Kumar, A. Alqudami and Vandana, 2003 (Indian Patent 840/Del/03); 2004 (PCT International Appl. No PCT/IN2004/000067); 2004 (International Publication No. WO 2004/112997); 2007 (US Patent Appl. No 20070101823).
18. R. P. H. Kooyman, *"Surface Plasmon Resonance, Instrumentation"* Academic Press 2302 (1999).
19. Michelle Duval Malinsky, K. Lance Kelly, George C. Schatz, and Richard P. Van Duyne, *J. Am. Chem. Soc.* **123**, 1471-1482 (2001).
20. L. A. Lyon, D. J. Pena and M. J. Natan, *J. Phys. Chem. B* **103**, 5826-5831(1999).

Effect of Substrate Temperature on Structural, Electrical and Optical Properties of Wurtzite and Cubic $Zn_{1-x}Mg_xO$ Thin Films Grown by Ultrasonic Spray Pyrolysis

Ajay Kaushal and Davinder Kaur[*]

Department of Physics and Center of Nanotechnology, Indian Institute of Technology Roorkee, India
Author to whom correspondence should be addressed; email: dkaurfph@iitr.ernet.in

Abstract. We report the synthesis of stable wide band gap hexagonal $Zn_{1-x}Mg_xO$ (x = 0.3) and cubic $Zn_{1-x}Mg_xO$ (x = 0.9) thin films at different substrate temperatures in the range from 300^0C to 750^0C on quartz substrates using inexpensive ultrasonic spray pyrolysis technique. The influence of varying substrate temperature on structural, electrical and optical properties of $Zn_{1-x}Mg_xO$ films was systematically investigated. X-ray diffraction (XRD) measurements show that the cubic film grow along the [100] direction while the hexagonal films grow along [001]. AFM images of the films deposited at optimized substrate temperature clearly reveal the formation of nanorods of hexagonal $Zn_{1-x}Mg_xO$. The optical measurement reveals the variation in optical absorption edges and band gap energies indicating the possibility of the phase transition or phase separation with increase in substrate temperature. The tuning of the band gap was obtained up to 3.78 and 6.06eV with corresponding increase substrate temperature in wurtzite and cubic $Zn_{1-x}Mg_xO$ thin films. The electrical resistivity was found to decrease with increasing deposition temperature. All the films show improvement in crystallinity with increase in deposition temperature.

Keywords: Photonic Band Gap materials; structural transitions; nanowires and nanorods.
PACS: 42.70.Qs; 64.70.Nd; 81.16.Be; 61.46.Km

INTRODUCTION

The wide band gap semiconductor ZnO and its ternary alloys have gained substantial interest in the research community for their potential application in optoelectronic devices such as solar cells [1, 2], liquid crystal display [3] and heat mirrors [4]. Alloying ZnO film with MgO or CdO potentially permits the band gap to be controlled between 2.8 and 4eV and higher, which facilitates band gap engineering. The transparent ZnMgO films are more desirable in application field of the ultraviolet optoelectronic devices as a large proportion of UV luminescence can be absorbed and utilized if ZnMgO films are used as transparent electrodes in the solar cells. It is known that ZnMgO is the solid solution consisting of ZnO and MgO. ZnO has wurtzite (a = 3.24A°and c = 5.22A°) crystalline structure, while MgO has NaCl-type cubic (a = 4.24A°) structure with wider bandgap [5]. Due to

CP1147, *Transport and Optical Properties of Nanomaterials—ICTOPON - 2009*, edited by M. R. Singh and R. H. Lipson
© 2009 American Institute of Physics 978-0-7354-0684-1/09/$25.00

structural dissimilarity between ZnO and MgO, $Zn_{1-x}Mg_xO$ alloys are of either hexagonal ($x \leq 0.49$) or cubic ($x \geq 0.5$) crystals. According to phase diagram of the ZnO-MgO binary system, [6] the thermodynamic solid solubility of MgO in ZnO is limited to only 4 atom%, and the unit cell retains its hexagonal structure. Zn-related species can be easily desorbed at higher growth temperature since Zn-species have a higher vapor pressure than that of Mg [7]. Therefore; it is interesting to see the bandgap tuning in ZnMgO films at different growth temperatures as higher growth temperature would leads to increase of Mg/Zn ratio in ZnMgO films. Various physical and chemical vapor deposition techniques such as magnetron sputtering, pulsed laser deposition, molecular beam epitaxy (MBE), spray pyrolysis and sol-gel [8-12] have been used to deposit ZnMgO thin films. In order to meet the industrial needs for the commercially available ZnMgO devices, the easier and cheaper deposition methods for the ZnMgO film should be developed. Recently, ultrasonic spray pyrolysis (USP) is considered to be a very useful method because of the simplicity of facilities and the low cost of the raw materials. In the present study we have investigated the influence of substrate temperature on structural, electrical and optical properties of ZnMgO thin films synthesized by ultrasonic spray pyrolysis.

EXPERIMENTAL

$Zn_{1-x}Mg_xO$ films with two different Mg content of x=0.3 and 0.9 were grown on quartz substrates by using ultrasonic spray pyrolysis. The detailed deposition procedures have been described elsewhere [13]. Two kind of aqueous solution, $Zn(NO_3)_2 \cdot 6H_2O$ and $Mg(NO_3)_2 \cdot 6H_2O$, were chosen as the source of zinc and magnesium respectively and 0.2molar solution were used by dissolving them in de- ionized water and methanol (2:1). Air is used as carrier gas. The aerosol was fed through the nozzle tube at a rate of 3 mL/minute which kept at 4cm above substrate. The ultrasonic frequency was kept at 1.7 MHz. The rate of deposition was controlled by the carrier gas flow rate, substrate temperature as well as precursor concentration. The substrate temperatures were varied in the range from 300 to 750°C. The crystalline quality was investigated by XRD using Bruker D8 Advance x-ray diffractrometer with Cu K_a (λ = 1.5418Å) source. The transmission spectra of the films were characterized by Cary 5000 UV-VIS-NIR spectrophotometer. The surface morphology has been investigated by using an atomic force microscopy (AFM) field emission electron microscopy (FESEM).

STRUCTURAL PROPERTIES

Figure 1(a & b) shows the XRD pattern of the hexagonal $Zn_{1-x}Mg_xO$ (x=0.3) and cubic $Zn_{1-x}Mg_xO$ (x=0.9) films respectively. XRD pattern indicate the growth of hexagonal films with preferred c-axis orientation along (002) diffraction peak.

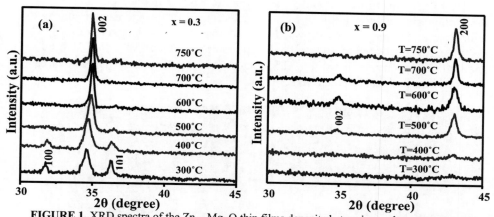

FIGURE 1. XRD spectra of the $Zn_{1-x}Mg_xO$ thin films deposited at various substrate temperatures with Mg content x = 0.3 and x = 0.9 respectively

(Figure 1a). However, the (100) and (101) peaks at lower substrate temperature of 300°C and 400°C were observed significantly. It is considered that the decrease of the adatom mobility of the species arriving on the substrate at a relatively low deposition temperature leads to the increase of the (100) and (101) peaks. From XRD spectra of $Zn_{1-x}Mg_xO$ thin films with Mg contents of x = 0.9 (figure 1b), it has been found that the films deposited at lower substrate temperatures (T_s=500°C, 600°C and 700°C) shows wurtzite (002) orientation along with (200)-cubic peak. The presence of (002)-wurtzite peak along with (200)-cubic peak at lower substrate temperature indicates the coexistence of two phases. However, appearance of only (200)-cubic peak has been found for higher substrate temperature of 750°C which indicate a structural phase transition from mixed to single cubic phase. The observed phase transformation from mixed phase to single cubic can be attributed to the stability of cubic phase at higher substrate temperature of 750°C. The intensity of (002) reflection in hexagonal $Zn_{1-x}Mg_xO$ (x=0.3) films was found to increase with increase in substrate temperature which result in decrease of FWHM of the film. The values of FWHMs calculated from (002) diffraction peak of hexagonal $Zn_{1-x}Mg_xO$ (x=0.3) films were found to decrease with increase in substrate temperature from 0.48(T_s=300°C), 0.45(T_s=400°C), 0.43(T_s=500°C), 0.38(T_s=600°C), 0.27(T_s=700°C) to

0.22(T_s=750°C) indicating the increase of particle size with improved crystallinity of the deposited films [14]. The particle size calculated using Scherrer's formula [15] $d = 0.9\lambda / B \cos\theta_B$, where λ, θ_B and B are the X-ray wavelength (1.5418Å), Bragg's diffraction angle and line width at half maximum respectively. The particle sizes found to increase from 17nm to 34nm with increase in substrate temperature from 300°C to 750°C (Table-I). The improvement of the crystalline quality of the thin films is mainly due to the increase of the adatom mobility of the atomic and molecular species on the substrate with the increasing substrate temperature. Figure 2(a- e) shows the AFM images for hexagonal $Zn_{1-x}Mg_xO$ (x=0.3) films deposited at different substrate temperatures of 300, 400, 500, 600, and 700°C respectively. From the AFM images it is clear that grain size increases with increase in deposition temperature.

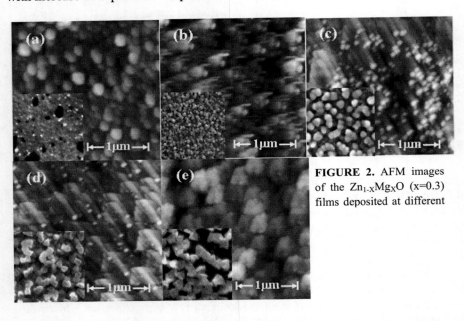

FIGURE 2. AFM images of the $Zn_{1-x}Mg_XO$ (x=0.3) films deposited at different

TABLE 1. Calculated crystallite size and optical band gap

T_s (°C)	D (nm)	E_g (eV)	
	x = 0.3	x = 0.3	x = 0.9
300	17	3.42	5.06
400	18	3.48	5.13
500	19	3.54	5.19
600	22	3.61	5.30
700	30	3.66	5.36
750	34	3.73	6.06

AFM images of films deposited at substrate temperature (T_s) of 500°C, shows vertically aligned nanorods like structure with an average grain size of about 90nm (figure 2c) which were in agreement with corresponding FESEM results shown in inset of figure 2(a-e). It can be clearly seen that with an increase in temperature to 400°C, oriented nanorod like structures starts nucleating which grow laterally with increase in substrate temperature to 500°C (figure 3(c)). Further increase in substrate temperature to T_s= 600 and 700°C (figure 3(d & e)) increases the grain size and diminishes the formation of ZnMgO nanorods. The average roughness of the films increases with increase in substrate temperature. This is because as the substrate temperature increases, grain size increases due to the increased amount of adatom mobility.

OPTICAL PROPERTIES

Figure 3a shows the transmittance spectra of both wurtzite (x=0.3) as well as cubic (0.9) $Zn_{1-x}Mg_xO$ nanocomposite thin films recorded at room temperature. The spectra shows that the average optical transmittance increases with the increase of substrate temperature and was noted transmittance up to 90%. The increase in optical transmittance was attributed to improvement of the crystalline quality of the $Zn_{1-x}Mg_xO$ thin films with increase in substrate temperature. The variation of the transmittance (T) with wavelength (λ) can be expressed as $T = A\exp(-4\pi\kappa t / \lambda)$, where κ is the extinction coefficient and is given

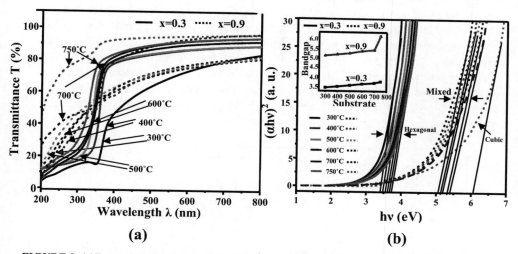

(a)   **(b)**

FIGURE 3. (a) Transmission spectra and **(b)** $(\alpha h\upsilon)^2$ vs. $h\upsilon$ plots of deposited $Zn_{1-x}Mg_xO$ thin films. Inset in (b) shows the calculated bandgap variation with substrate temperature.

by $\kappa = \lambda\alpha/4\pi$. A is approximately unity and α is the absorption coefficient. The absorption edge of the spectra shifts towards higher energy (lower wavelength) as the substrate temperature increased. The blue shift of the absorption edge in the optical transmission spectra indicates band gap expansion. The absorption coefficient (α) in this case is given by $\alpha = C(h\upsilon)^{-1}(h\upsilon - E_g)^{1/2}$, where C is a constant and $h\upsilon$ is the photon energy. The E_g is the optical band gap which is determined by plotting $(\alpha h\upsilon)^2$ verses $h\upsilon$ and extrapolating the linear portion to $(\alpha h\upsilon)^2 = 0$. As the substrate temperature increases, the UV absorption position of both the wurtzite and cubic ZnMgO films shows a blue shift indicating a band gap expansion as shown in figure 3b. Thus E_g is a function of substrate temperature. It has been reported earlier that the Mg/Zn ratio of ZnMgO films increases with substrate temperature because Zn related species can be desorbed more easily than Mg at high growth temperatures [7]. Thus increase in Mg/Zn ratio with increase in substrate temperature leads band gap tuning due to large band gap of MgO (6.67eV) than that of ZnO (3.37eV). The inset shown in the figure summarizes the band gap relations in both hexagonal as well as cubic ZnMgO thin films with substrate temperature. The optical band gap was found to tune from 3.42eV to 3.73eV in wurtzite $Zn_{1-x}Mg_xO$ (x=0.3) and from 5.06eV to 6.06eV in cubic $Zn_{1-x}Mg_xO$ (x=0.9) films with increase in substrate temperature from 300 to 750°C (Table I). An abrupt discontinuity was observed with an increase in substrate temperature from 700°C to 750°C in cubic ZnMgO films which could be attributed to the structural phase transition from mixed phase to cubic phase. It has been found that with an increase in growth temperature, the UV absorption position of the $Zn_{1-x}Mg_xO$ shows a blue shift indicating a band gap expansion. This may be attributed to the fact that films deposited at higher substrate temperature results in the decrease of total length of grain boundaries due to increase in grain size and show more ordered structure which increases the energy required for absorption and hence increases the transmittance, results in blue shift with increase in growth temperature.

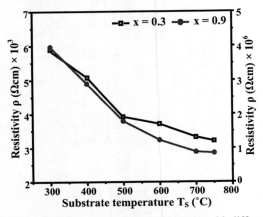

FIGURE 4. Variation of resistivity of $Zn_{1-x}Mg_xO$ thin films with different substrate temperature

ELECTRICAL PROPERTIES

Figure 4 plots the variation of resistivity as a function of substrate temperature of both wurtzite $Zn_{1-x}Mg_xO$ (x=0.3) as well as cubic $Zn_{1-x}Mg_xO$ (0.9) thin films. The resistivity of the deposited films was found to decrease with increase in substrate temperature from 300°C to 750°C which could be due to increase in crystallinity of the films with increase in substrate temperature and is in agreement with XRD results. The value of resistivity of wurtzite ZnMgO films were found to be of the order of 10^3 Ω cm whereas values of resistivity of cubic ZnMgO films were of the order of 10^5 Ωcm. It has been found a change in linearity of the decrease in resistivity as the substrate temperature is increase from 500^0C to higher substrate temperatures. This change in linearity could be attributed to increase in Mg/Zn ratio at higher temperatures.

CONCLUSIONS

$Zn_{1-x}Mg_xO$ thin films were deposited by Ultrasonic spray pyrolysis technique at various substrate temperatures. The XRD results shows single-wurtzite phase along (001) and single-cubic phase along (100) at 750°C whereas mixed wurtzite as well as cubic phases were found for films with higher Mg composition (x=0.9) deposited at lower growth temperatures. The optical bandgap of $Zn_{1-x}Mg_xO$ thin films expanded as the substrate temperature increased due to easy desorption of Zn than Mg at higher growth temperature, results in the increase of Mg composition in the $Zn_{1-x}Mg_xO$ thin films. AFM images of the films deposited at optimized substrate temperature clearly reveal the formation of nanorods of hexagonal $Zn_{1-x}Mg_xO$. These results indicate that the properties of thin $Zn_{1-x}Mg_xO$ thin films depend remarkably on substrate temperatures.

AKNOWLEDGEMENTS

The financial support provided by Ministry of Communications and Information Technology (MIT), India under Nanotechnology Initiative Program with Reference no. 20(11)/2007-VCND is highly acknowledged.

REFERENCES

1. S. Major and K. L. Chopra, *Solar Energy Mater.* **17**, 319-327 (1988).
2. S. J. Pearton, D. P. Norton, Y. W. Heo, and T. Steiner, *Prog. Mater. Sci.* **50**, 293-297 (2005).
3. J. Lan and J. Kanicki, *Mater. Res. Soc. Symp.* **424**, 347-350 (1997).
4. K. L. Chopra, S. Major and D. K. Panday, *Thin Solid Film* **102**, 1-46 (1983).
5. R. D. Shannon, Acta Crystallogr., Sect. A: Cryst. Phys., Diffr., Theor. Gen. Crystallogr. **32**, 751-756 (1976).
6. E. R. Segnit and A. E. Holland *J. Am. Ceram. Soc.* **48**, 412-416 (1965).

7. S. Choopun, R.D. Vispute, W.Yang, R.P.Sharma, and T. Venkatesan, *Appl. Phys. Lett.* **80** 1529 (2003).

8. B. Yang, A. Kumar, P.Feng and R. S. Katiyar, *Appl. Phys. Lett.* **92**, 233112-233114 (2008).

9. M. X. Qiu, Z.Z.Ye, H.P.He, Y.Z.Zhang, X.Q.Gu, L.P.Zhu, and B.H.Zhao, *Appl. Phys. Lett.* **75**, 980-982 (1999).

10. T. Makino, Y. Segawa, M. Kawasaki, A. Ohtomo, R. Shiroki, K. Tamura, T. Yasuda, and H. Koinuma, *Appl. Phys. Lett.* **78**, 1237-1239 (2001).

11. X. Zhang, X. M. Li, T. L. Chen, C. Y. Zhang, and W. D. Yu, *Appl. Phys. Lett.* **87**, 092101-092103 (2005).

12. K. Samanta, S. Dussan, P. Battacharya, and R. S. Katiyar, *Appl. Phys. Lett.* **90**, 261903-261905 (2007).

13. P. Singh, A. Kumar, Deepak, and D. Kaur, *J. Crystal Growth* **306**, 303-310 (2007).

14. J. W. Kim, H. S. Kang, J. H. Kim, and S. Y. Lee, *J. Appl. Phys.* **100** 033701-033706 (2006).

15. B. D. Cullity and S. R. Stock, *Elements of X-ray diffraction,* 3rd. ed.

Molecular Beam Epitaxy Growth of Iron Phthalocyanine Nanostructures

A.K. Debnath*, S. Samanta, Ajay Singh, D.K. Aswal, S. K. Gupta and J.V. Yakhmi

Technical Physics and Prototype Engineering Division, Bhabha Atomic Research Centre, Trombay, Mumbai-400085, India
*debnath@barc.gov.in

Abstract. FePc films of different thickness have been deposited by molecular beam epitaxy (MBE) as a function of substrate temperature (25-300°C) and deposition rate (0.02 -0.07nm/s). The morphology of a 60 nm alpha-phase film has been tuned from nanobrush (nearly parallel nanorods aligned normal to the substrate plane) to nanoweb (nanowires forming a web-like structure in the plane of the substrate) by changing the deposition rate from 0.02 to 0.07 nm/s. We propose growth mechanisms of nanoweb and nanobrush morphology based on the van der Waals (vdW) epitaxy. For air exposed FePc films I-V hysteresis was observed at 300 K and it is attributed to surface traps created by chemisorbed oxygen.

Keywords: Iron phthalocyanine nanostructures, van der waals epitaxy, molecular beam epitaxy

PACS: 61.46.Km, 68.37.Hk, 68.55.J-, 73.61.Ey, 81.07.Bc

INTRODUCTION

For last several years, motivated with the success of MBE grown inorganic (semiconductor and metal) films and heterostructures, efforts were made to grow epitaxial organic films on crystalline/glassy substrates of inorganic materials [1,2] The growth mechanisms for organic films could be considerably different and complicated due to low symmetries of the molecular crystal structures and polymorphism, internal molecular degrees of freedom, kinetic barriers, and a delicate balance between the two types of non-covalent interactions at the interface, that is, molecule-molecule and molecule-substrate. These issues are still under active investigations. While understanding the mechanism of organic epitaxy on inorganic substrates is a key issue, the impeding nanoscience and nanotechnology requires controlled organic nanostructures grown on suitable substrates. The aim of the present work is to investigate the growth mechanism of nanostructured FePc films and to study their transport properties. We demonstrate, for the first time, that by tuning these growth parameters one can reproducibly grow a-axis oriented α-FePc phase nanobrush (nearly parallel nanorods aligned normal to the substrate plane) and nanoweb (nanowires forming a web-like structure in the plane of the substrate) structures. The

CP1147, *Transport and Optical Properties of Nanomaterials—ICTOPON - 2009*, edited by M. R. Singh and R. H. Lipson
© 2009 American Institute of Physics 978-0-7354-0684-1/09/$25.00

results indicate that nanoweb and nanobrush grow on glass substrate via van der Waals (vdW) epitaxy. Interestingly, this films show I-V hysteresis, which vanishes after vacuum annealing.

EXPERIMENTAL

FePc thin films of different thickness were grown on glass substrates by MBE system and the details of the MBE system and deposition conditions are discussed in our recent papers [3,4] The deposition rate (r_d) could be controlled between 0.02 and 0.07nm/s, by varying the effusion cell temperature from 350 to 450°C. The grown films were characterized by following techniques. (1) The grazing incidence (angle 0.1°) x-ray diffraction (GIXRD) was carried out using CuKα radiation (Seifert-XRD 3003TT) in out-of-the-plane geometry. (2) The morphology of the films was recorded using scanning electron microscopy (SEM) (VEGA MV2300T/40, Tescan). (3) The vibrational spectra of the samples was recorded using the diffuse-reflectance Fourier transform infrared (FTIR) spectroscopy (VERTEX 80V, Bruker). (4) The room temperature I-V measurements were carried out using a Keithley make 6487 picoammeter / voltage source.

RESULTS AND DISCUSSION

Effect of growth temperature
Fig.1 shows the SEM images of 60 nm FePc films deposited at r_d = 0.07 nm/s at different substrate temperatures (T_s). No features were seen for Ts = 25°C indicating that the film was amorphous (Fig.1a). This amorphous character of the room temperature grown film was due to the amorphous nature of the substrate. In order to confirm this we have grown films at 25°C on single crystal sapphire substrate and granular nature of the films was observed as shown in Fig.1f. Amorphous nature of the room temperature grown films on glass substrate has been reported by Khrishnakumar et. al. [5]. At a substrate temperature of 50°C, the film morphology consists of densely packed grains of an average size of ~70 nm. At 100°C, the FePc grains were seen to get elongated in the plane of the substrate. At 200°C, long FePc nanowires (~100 nm diameter and upto 2 µm length), lying horizontally in the plane of the substrate (Fig. 1d) were seen to grow, forming a nanoweb like structure. However, further increasing substrate temperature to 300°C, the length of nanowires decreases drastically to <500 nm. There are two possible reasons for this effect: (i) increased desorption of molecules and (ii) the transformation from α- to β-FePc phase at elevated temperatures.

FIGURE 1. SEM photographs of FePc thin films (nominal thickness 60 nm) grown on glass substrate at (a) 25°C, (b) 50°C, (c) 100°C, (d) 200°C, (e) 300°C and (f) on sapphire substrate at 25°C using a deposition rate of 0.07 nm/s.

For low thickness (≤ 60 nm) growth of the crystals can occur in a particular direction but as thickness increases nanorods grow in all directions and coalescence to each other, which results in random grain growth [4]. To study the molecular stacking, crystallinity and phase formation, we have recorded GIXRD and FTIR spectra, and the results are presented in Fig. 2.

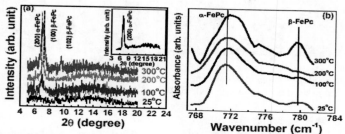

FIGURE 2. (a) Grazing incidence X-ray diffraction patterns and (b) FTIR spectra recorded for 60 nm (nominal thickness) FePc films deposited at different substrate temperatures. Inset shows the grazing incidence X-ray diffraction pattern of a 60 nm FePc film grown on sapphire substrate at 25°C.

It was seen that the film deposited on glass at 25°C is amorphous in character while that deposited on sapphire at the same temperature was crystalline as depicted from the inset of Fig.2a, This result supports the inference drawn from the SEM images. For films grown upto at a temperature of 200°C, presence of a single peak appears at $2\theta = 6.96°$, which corresponds to the (200) peak of the α-FePc phase. Films were polycrystalline but highly oriented as evident from the presence of only (200) peak in the entire spectrum which indicates that a-axis is preferentially orientated normal to the substrate plane, i.e. the edge-on stacking of the FePc molecules. The value of a-lattice parameter computed from the XRD data was 2.54 nm, which is in agreement with the reported values [6]. However, for films grown at 300°C, two additional peaks at $2\theta = 7.01°$ and $9.6°$ have been observed in the GIXRD spectrum, which have been attributed to the (100) and (102) peaks of the β-FePc phase [7]. Thus, at growth temperatures $\geq$300°C, the preferentially oriented α-FePc transforms into β-FePc phase. A further confirmation to the GIXRD results has been obtained from the FTIR data, as shown in Fig. 2(b). It is known that the characteristics peak of ν(C-N) stretching vibration for α and β-phase occur at around 773 and 780 cm⁻¹, respectively [8,9]. It is evident from Fig. 2(b) that upto a growth temperature of 200°C, the films consists of pure α-FePc phase. However, at growth temperature 300°C, appearance of an additional peak at 780 cm⁻¹ indicates that a partial transformation from α- to β-phase takes place. From these studies, it is evident that the nanoweb structure of pure α- FePc is formed for a film of nominal thickness of 60 nm deposited at 200°C.

Effect of lower deposition rate

The SEM images of 60 nm FePc films prepared at a deposition rate of 0.02 nm/s are shown in Fig. 3. It can be seen that the morphology of the films is radically different that that obtained at a deposition rate of 0.07 nm/s (Fig. 1). At low substrate temperature (e.g. 50°C), FePc nucleates in the needle-tip like structures pointing out perpendicular to the substrate plane. As the substrate temperature was increased to 100°C, the vertical growth of the needle-like tips was apparent.

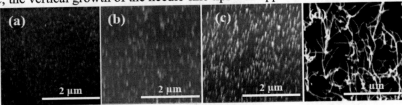

FIGURE 3. SEM photographs of 60 nm FePc thin films grown in-situ at the rate of 0.02 nm/s at different substrate temperatures (a) 50°C, (b) 100°C, (c) 200°C and (d) 300°C.

At 200°C, these needles further grow to form well defined vertical nanorods and this morphology resembles to that of nanobrush (Fig.3c). However, as the growth temperature was further increased to 300°C, growth of nanobrush was not observed, instead nanoweb structure is seen to grow, which consists of mixed α-and β-FePc phases. Thus at 200°C, for a film of 60 nm nominal thickness, the α-FePc phase morphology can be tuned from nanoweb to nanobrush by changing the growth rate from 0.07 to 0.02 nm/s. These experiments were repeated several times, and the morphology was highly reproducible. The GIXRD, FTIR data recorded for the nanobrush samples and schematic of 3D growth process of nanobrush and nanoweb structures are shown in Fig. 4, It is seen that the vertical nanorods are also a-axis (Fig.4a) oriented perpendicular to the substrate plane and have α-FePc phase (Fig.4b).

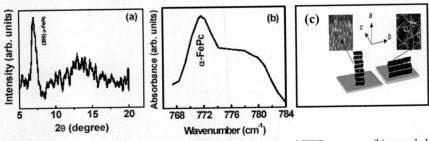

FIGURE 4. Grazing incidence X-ray diffraction pattern (a) and FTIR spectrum (b) recorded for FePc nanobrush sample and (c) Schematic of the 3D growth process of FePc nanobrush (left : the growth direction is along a-axis) and nanoweb (right : the growth direction is along b-axis)

Growth mechanisms

The most fundamental difference between growth of inorganic and organic films is that the organic molecules are 'extended objects' and thus have internal degrees of freedom. The orientational degrees of freedom allows molecules to adsorb at the surface of the substrate in 'lying-down' or 'standing-up' configurations that may lead to altogether different growth mechanism. Similarly, the vibrational degrees of freedom can influence the interaction of molecule with the surface and the thermalization upon adsorption and subsequent diffusion. Since glass was used as a substrate material in the present case, the inert character of surfaces i.e. free of unsaturated bonds indicates that the molecules get adsorbed via van-der Waals interaction. Similarly, FePc is closed-shell molecules and therefore do not have any dangling bonds. However, molecule-molecule interactions are greatest when the plate-like molecules are face-to-face (π-stacking) rather than edge-to-edge (van-der Waals interaction). Thus, the nucleation depends on the competition between the two types of interaction i.e. molecule-molecule and molecule-substrate interactions. If molecule-molecule interaction is stronger than molecule-substrate interaction, then the nuclei will form in "standing-up' configuration. Otherwise, the nuclei will form in 'lying-down' configuration. However, the growth of nuclei will depend on which of the molecule-molecule interaction, i.e. face-on or edge-on, is dominant.

It may be noted that r_d of 0.02 and 0.07 nm/s was achieved by heating the effusion cell to 350 and 450°C, respectively. Therefore, at r_d = 0.07 nm/s, molecules arrive at the substrate surface with a relatively higher kinetic energy, allowing them to make a faster diffusion across the surface of the substrate and this facilitates the face-on interaction. Therefore, nuclei grow predominantly along b-axis (Fig.4c-right side), leading to the formation of nanowire like structure at low coverage At high coverage, these nanowires grow further and merge together to form nanoweb structure. On the other hand, at r_d = 0.02 nm/s due to lower kinetic energy, molecules preferentially attach to the existing nuclei in the edge-on configuration, leading to a growth along a-axis (Fig.4c-left side). Therefore, vdWE along a-axis results in the growth of nanorods, and hence, to a nanobrush morphology. The proposed mechanism explains the GIXRD data that both nanoweb and nanobrush have same crystallographic orientation (i.e. a-axis perpendicular to the substrate plane).

Current (I) - Voltage (V) characteristics

The MPc reported to be strongly susceptible to the oxygen [10]. To understand the role of chemisorbed oxygen on the charge transport, following experiments were conducted. The films (100 nm) were annealed at 200°C for 30 min in a vacuum chamber (base vacuum $\sim 10^{-6}$ torr) and were cooled down to room temperature. The I-V measurements were performed under following conditions (i) under $\sim 10^{-6}$ torr of vacuum and (ii) after exposing the films to the atmosphere for an hour. The results are shown in Fig. 5. It is seen that the I-V characteristic for vacuum annealed film exhibits ohmic behavior and does not show any hysteresis; while hysteresis appears on air exposed films. We would like to add that the hysteresis vanishes at temperature $\leq$ 250°C. Thus the presence or absence of the hysteresis in FePc films is directly related to the chemisorbed oxygen. Interestingly, the vacuum annealed films have larger conductivity then the atmosphere-exposed film. This implies the chemisorbed oxygen creates the surface traps, which localizes the charge.

351

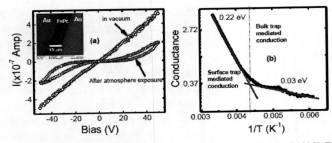

FIGURE 5. (a) *I-V* characteristics recorded at 300 K for vacuum annealed (at 200°C) FePc film under vacuum and after exposing to the atmosphere and (b) Temperature dependence of the conductance recorded for FePc film. The inset (Fig.4a) shows the SEM micrograph of the planar Au/FePc/Au structure, Au electrodes are separated by 15 μm.

The conductivity of air exposed FePc films was measured as a function of temperature, and the obtained results are presented in Fig. 5b. A change in conduction mechanism is evident at temperature ~ 250 K. Fitting of the temperature dependence of conductance with the thermally activated hopping model: $\sigma = \sigma_0 \exp(\Delta E/kT)$ reveals two activation energies, $\Delta E_1 = 0.22$ eV above 250 K and $\Delta E_2 = 0.03$ eV below 250 K. Presence of two activation energies indicates the possibility of two different types of charge trap states in the films: (i) surface traps arising due to the chemisorbed oxygen and (ii) bulk traps due to impurity/structural defects. The value of activation energy (ΔE_1) above temperature 250 K is very close to the reported value of 0.26 eV for metal phthalocyanine–O_2 interaction [10]. Thus at temperature above 250 K, the chmisorbed oxygen induced surface traps govern the charge transport. However, at low temperature bulk traps determine the charge transport.

CONCLUSIONS

We have reproducibly grown α-FePc nanoweb and nanobrush on glass substrates using MBE. The α-FePc morphology, at a growth temperature of 200°C, can be tuned from nanobrush to nanoweb by changing the deposition rate from 0.02 to 0.07 nm/s. Mechanisms of the growth of nanoweb and nanobrush, based on the van der Waals epitaxy, have been proposed. Room temperature I-V measurements of FePc films showed deep surface traps created by chemisorbed oxygen.

ACKNOWLEDGEMENTS

This work is supported by Indo-German Cooperation Programme (Project No: IND 06/005). The authors thank Prof. C. Surger for useful discussions.

REFERENCES

1. S.R. Forrest, *Chem. Rev.* **97,** 1793 (1997).

2. D.E. Hooks, T. Fritz, M. D. Ward, *Adv. Mater.* **13**, 227 (2001).
3. A.K. Debnath, N. Joshi, D.K. Aswal, S.K. Deshpande, S.K. Gupta, J.V. Yakhmi, *Solid State Commun.* **142**, 200 (2007).
4. A.K. Debnath, S. Samanta, Ajay singh, D.K. aswal, S.K. Gupta, J.V. Yakhmi, S.K. Deshpande, A.K. Poswal, C. Surger, *Physica-E* **41**, 154 (2008).
5. K. P. Krishnakumar, C.S. Menon, *Mat. Lett.* **48**, 64 (2001).
6. C.W. Miller, A. Sharoni, G. Liu, C.N. Colesniue, B. Fruhberber, I.K. Schuller, *Phys. Rev. B* **72**, 104113 (2005).
7. T. Nonaka, Y. Nakagawa, Y. Mori, M. Hirai, T. Matsunobe, M. Nakamura, T. Takahagi, A. Ishitani, H. Lin, K. Koumoto, *Thin Solid films* **256**, 262 (1995).
8. M.J. Cook, N. B. McKeown, J. M. Simmons, A.J. Thomson, M.F. Daniel, K.J. Harrison, R.M. Richardson, S.J. Roser, *J. Mater. Chem.* **1**, 121 (1991) 121.
9. E. A. Ough, M. J. Stillman, K.A.M. Creber, *Can. J. Chem.* **71**, 1898 (1993).
10. F. Sutara, N. Tsud, K. Veltruska, V. Matolin, *Vacuum* **61**, 135 (2001).

Nanoscale Confinement of Charge – Bound States in the Presence of Triangular Voids and Protrusions

S. A. R. Horsley [a], M. Al-Amri [b], H. Godazgar [c], S. Martin-Haugh [a] and M. Babiker [a]

[a] *Department of Physics, University of York, Heslington, York YO10 5DD, England, UK*
[b] *The National Centre for Math and Physics, KACST, P O Box 6086, Riyadh 11442, Saudi Arabia*
[c] *DAMTP, Centre For Mathematical Sciences, Cambridge University, Wilberforce Road , Cambridge CB3 0WA, UK..*

Abstract. We consider the quantum states for charged particles bound to their own set of images in wedge shaped voids and protrusions of nanoscale dimensions. Localized images only exist for wedge opening angles that are whole fractions of 2π, but the potential we give here is exact for any opening angle. We outline the procedure leading to the determination of the bound state energies and eigenfunactions, but display recent results only for the ground state. We point out that at the nanoscale, the charge states so found could be useful for a number of applications, including qubits in triangular grooves for quantum information processing and in the case of protrusions in metallic nano-probes.

Keywords: Image Bound States; Qubits; Nanoscale Probes; Quantum Information.
PACS: 02.70.−c; 31.15.x; 31.30.jf; 31.50.-x; 31,50.Bc

INTRODUCTION

It is well known that a charged particle can be bound to its own image it forms in a planar surface. The quantum surface states formed in this manner at a single planar surface are akin to the one-dimensional hydrogenic states. They have been experimentally detected long ago outside the surface of liquid helium and have featured recently as candidates as quantum architecture elements (qubits) towards the realization of quantum information processing [1]. These states are essentially due to the confinement of the electromagnetic fields by the introduction of surfaces, modifying the self-energy of the charge relative to the unbounded space. The presence of two parallel planar surfaces is known to lead to novel quantum states, resembling the bound states of a hydrogen molecular ion, with bonding and anti-bonding features that change with plate separation [2]. Here we focus on the quantum states for charged particles as they are formed in the vicinity of sharply intersecting planar material surfaces forming a triangular groove. Such grooves are found naturally on the surfaces of materials, often referred to as surface roughness, of depths of a few microns, or they can be deposited in a controlled manner using deposition techniques. A charge inside such a wedge-shaped groove creates an image charge, localized for some integer fractions of π, but non-localized in general, with which the

CP1147, *Transport and Optical Properties of Nanomaterials—ICTOPON - 2009*, edited by M. R. Singh and R. H. Lipson
© 2009 American Institute of Physics 978-0-7354-0684-1/09/$25.00

charge interacts and thus form bound states, with probability distributions extending in the nanometer scale. We have sought to obtain the exact expression for the binding potential and evaluated the quantum states formed in this manner for a general wedge intersection angle. Work on the effect of electromagnetic confinement on the de-excitation rate of an excited atom in wedges has already been reported [3-5].

As we outline below, dealing with the case of a free charge (i.e. unbound to an atom) in the presence of the wedge cavity requires as an important first step the determination of the two-dimensional Coulomb potential self energy distribution due to the presence of the two surfaces, followed by the solution of the Schroedinger equation incorporating the Coulomb potential to obtain the bound state eigenfunctions and the corresponding energy eigenvalues. We have used variational methods to determine the ground state as well as the first excited states and examined their variations with the intersection angle. Remarkably, the results are obtainable for any chosen opening angle in the range (0₊ to 2π₋). The angular range (0₊ to [π/2]₋) represents triangular grooves, while the range ([π/2]₊ to 2π₋) represents protrusions. The charge states so found are thus significant for a number of applications, including qubits in triangular grooves for quantum information processing and as protrusions in metallic nano-probes.

THE BINDING POTENTIAL

The physical system under consideration is schematically shown in **FIGURE 1**. It consists of a charged particle of mass m and charge e which is capable of moving in the physical space bounded by two intersecting materials surfaces forming a wedge of opening angle α. The position vector of the particle is given in polar coordinates (r, θ); with polar angle θ and radial coordinate r.

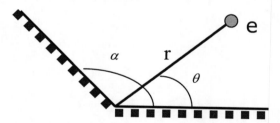

FIGURE 1: Schematic drawing showing the two material surfaces forming a wedge of opening angle α and a particle of mass m and charge e in the vicinity at the cylindrical position vector $(r, \theta, 0)$.

The bound states formed in this system are solutions of the Schrödinger equation

$$\hat{H}\psi = \left(-\frac{\hbar^2}{2m}\nabla^2 + V(r,\theta,z)\right)\psi(r,\theta,z) = E\psi(r,\theta,z) \qquad (1)$$

where $V(r,\theta,z)$ is the binding potential of the charge in the presence of the surfaces, The best derivation of this potential stems from a result by Landau and Lifshitz [6] whereby one must extract the charge self energy. The procedure for this is complicated and the details will not be given here. The required potential emerges as follows

$$V(r,\theta,z) = \frac{e}{16\pi\varepsilon_0 r}\left[\kappa(\alpha) - f(\tilde{\theta},\alpha)\right] \tag{2}$$

where $\tilde{\theta} = \theta - \alpha/2$ and the functions in the square brackets are defined as

$$f(\tilde{\theta},\alpha) = \left(\frac{\sqrt{2}}{\alpha}\right)\int_0^\infty\left[\frac{\sinh(\pi x/\alpha)}{\cosh(\pi x/\alpha) + \cos(2\pi\tilde{\theta}/\alpha)}\right]\frac{dx}{\sqrt{(\cosh(x)-1)}} \tag{3}$$

$$\kappa(\alpha) = \left(\frac{1}{\pi\sqrt{2}}\right)\int_0^\infty\left[\frac{(\pi/\alpha)\sinh(\pi x/\alpha)\sinh^2(x/2) - \sinh(x)\sinh^2(\pi x/2\alpha)}{\sinh^2(\pi x/2\alpha)\sinh^2(x/2)}\right] \tag{4}$$

$$\frac{dx}{\sqrt{(\cosh(x)-1)}}$$

From symmetry considerations, it is clear that this potential should not depend on the cylindrical coordinate z, we may consider the situation on the xy plane at z=0. We have checked by explicit calculations that the potential given in Eq.(2), gives rise to the correct potential arising from summing contribution from localized images when the opening angle α is given by $\alpha = 2\pi/n$, where $n>1$ is an integer. For illustration purposes, we display in **FIGURE 2** the variations of the potential with θ at constant radial position r for different opening angle α. The range of α shown in the figure starts at $\pi/2$, corresponding to $n=4$, ending at $\pi/4$, corresponding to $n=8$. It must be emphasized that the full potential given in Eq.(2) will yield the correct potential for any chosen angle for which there are no localized images i.e., for angles that are not integer fraction of π.

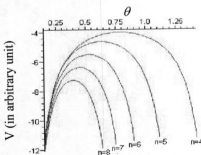

FIGURE 2: Variations of the potential with θ (horizontal axis) at constant redial position r for different opening angles α, spanning the range $\pi/2(n=4)$ to $\pi/4(n=8)$.

EVALUATION OF BOUND STATES

The Schrödinger equation given in Eq.1 with $V(r,\theta,z)$ as the full potential, given in Eq.(2) cannot be solved exactly except for the simplest case $\alpha = \pi$, which corresponds to the well known case of a charge bound to its single image in a planar surface. We are, therefore, forced to resort to the numerical evaluations to solve the Schrödinger equation Eq. (1) using the quantum variational method. The procedure is straightforward; it consists of the following steps:

(i)   Choose a physically motivated trial wavefunction $\psi(r,\theta,\{\beta\})$ where $\{\beta\}$ stands for a set of adjustable parameters.

(ii)   Evaluate $E(\{\beta\})$ defined by

$$E(\{\beta\}) = \frac{\langle\psi(r,\theta,\{\beta\})|\hat{H}|\psi(r,\theta,\{\beta\})\rangle}{\langle\psi(r,\theta,\{\beta\}|\psi(r,\theta,\{\beta\})\rangle}$$

where $\hat{H}$ is the Hamiltonian as deduced from Eq. (1).

(iii)   Minimize $E(\{\beta\})$ with the respect to the parameter set $\{\beta\}$.

The above procedure should lead to a good estimate of the energy eigenvalues and the corresponding eigenfunctions. The details will be displayed elsewhere. Here we give representative result primarily for the ground state with more detailed accounts, including results for the first excited state will be reported elsewhere.

FIGURE 3 shows the probability distribution for the ground bound state for an opening angle $\alpha = \pi/20$. It is seen that the probability distribution is concentrated near the sharp end of the wedge and decays with distance rapidly for larger radial coordinate r.

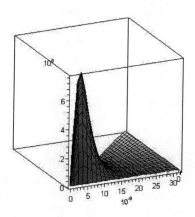

FIGURE 3: Probability distribution (arbitrary units) of the ground state for opening angle $\alpha = \pi / 20$. Distances in the horizontal plane are in nanometers.

The corresponding ground state energy eigenvalue can be read from **FIGURE 4** which shows the variations of the ground state energy with opening angle. It is seen that the ground state energy is constant for opening angles greater or equal to π and has the exact value of -0.85 eV, corresponding to ground state of a charge bound to its image in planar conductor surface. This constitutes a reassuring check of the correctness of our evaluations. What is surprising is that a constant ground state energy is obtained for protrusions (i.e., for opening angle greater than π)

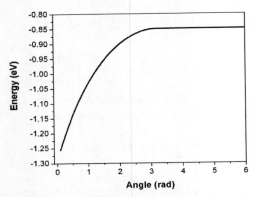

FIGURE 4: The variations of the ground state energy with opening angle for an electron in the presence of two sharply intersecting planar perfect conductors.

COMMENTS AND CONCLUSIONS

In this article we have reported an exact closed expression for the potential experienced by a charged particle due to the presence of a pair of sharply intersecting planar surfaces forming an opening angle α . This potential is general in the sense that it applies to any opening angle. It therefore includes structures for which α is a whole integer fraction of 2π , for which there is a set localized images in the conductors. It, of course, includes the case where no localized images are formed. With the potential available in this general form, we are able to consider solutions for the Schrödinger equation which yields the spectrum of bound states and their probability distributions.

We have concentrated primarily on the ground state evaluations using the quantum variational method. The main results are presented as a graph showing the variation of the ground state energy with opening angle and we have also presented the probability distribution of the ground state function for a representative opening angle. The variation of the ground state energy with the opening angle shows a strong binding for smaller angles and a constant smaller energy for opening angles $\alpha \geq \pi$. We have seen that the result for $\alpha = \pi$ obtained here using the above procedure coincides with the result obtained analytically for a charge bound to its single image inside a planar perfect conductor surface. What is quite surprising is that for angles that greater than π , where the two conductors form a protrusion, rather than a groove, the ground state energy remains constant up to 2π where the two planar conductor surfaces form a half-plane. It appears that the induced charge divides in equal amounts between the two surfaces and this remains the case no matter what the opening angle is, as long as it is greater than π. We are currently extending the treatment to evaluate the excited state energies and eigenfunctions. The results will be reported elsewhere.

ACKNOWLEDGMENTS

M A-A would like to thank KACST for financial support and MB would like to thank the Royal Society of London for a travel grant to attend ICTOPON09.

REFERENCES

1   M.J.Lea, P.G. Frayne and Y. Mukharsky, "Could we compute with electrons on helium?" in *Scalable Quantum Computers* - Eds. Samuel L.Braunstein and Hoi-Kwong Lo (Wiley-VCH, Berlin, 2000) See also *Fortsch. Phys.* **48**, 1109 (2000)
2   M. Babiker, J. Phys. C: Condens. Matter, **5**, 2137-2142 (1993)
3   S. C. Skipsey, M. Al-Amri, M. Babiker, and G. Juzeliunas. *Phys. Rev. A* **73**, 011803 (2006).
4   S. C. Skipsey, G. Juzeliunas, M. Al-Amri and M. Babiker, *Optics Comm.* **254**, 262-270 (2005).
5   G. Juzeliunas, M. Al-Amri, S. C. Skipsey, and M. Babiker, *J. Lumin.* **110**, 181-184 (2004).
6   L. D. Landau , E. M. Lifshitz, *Mechanics and Electrodynamics* (Shorter Course of Theoretical Physics, Vol 1), (Pergamon; 1st edition, June 1972)

Growth Mechanism of Zinc Oxide Nanostructures by Carbothermal Evaporation Technique

Aditee Joshi[1], S K Gupta[2]*, Manmeet Kaur[2], J.B.Singh[3], J. V. Yakhmi[2]

[1]Department of Electronic Science, University of Pune, Pune, 411007, India
[2]Technical Physics and Prototype Engineering Division ([3]Materials Science Division), Bhabha Atomic Research Centre, Mumbai 400 085, India

Abstract: Zn/ZnO nanostructures such as nanowires and nanotetrapods have been prepared by carbothermal evaporation under flowing argon and oxygen gases. Morphology of ZnO structures has been controlled by changing the time/temperature of the inlet and flow rate of the oxygen gas. Mechanism for the growth of nanowires and tetrapods has been discussed. It is seen that the nanostructure of prepared material is basically determined by conditions prevailing during formation of nuclei.

Keywords: A1. Nucleation, A1.Nanowires, A1. tetrapods, B1. ZnO
PACS: 61.46. -w; 81.05. Hd; 81.07. –b; 81.10.Bk

INTRODUCTION

Fabrication of nanometer-sized particles with desired size, composition and morphology is an important issue in the field of nanotechnology. Many techniques have been used for synthesis of one dimensional ZnO nanostructures, including "dry" methods (e.g., thermal evaporation [1], chemical vapor deposition [2]) as well as "wet" solution-phase methods [3]. Physical vapor deposition has demonstrated its versatility in obtaining a rich family of nanostructures [4,5]. In this technique, concentration of metal vapors and oxygen pressure control the formation of different nanostructures and the morphology of the products. Understanding the mechanisms by which the shape and size of nanostructures can be controlled is an important issue for their reproducible growth and this is the aim of present study. In the present work, by keeping some of the growth parameters such as the starting material, source temperature and argon gas flow rate constant, we could reproducibly produce different morphologies of ZnO nanostructures by varying the oxygen partial pressure (i.e. oxygen flow rate) and the temperature/time of inlet of the oxygen gas. Models leading to the formation of different ZnO nanostructures have been discussed.

EXPERIMENTAL

ZnO nanostructures were grown in quartz tube of 100 cm length and 6 cm diameter placed in a horizontal tubular furnace having facility for introducing gases at

CP1147, *Transport and Optical Properties of Nanomaterials—ICTOPON - 2009*, edited by M. R. Singh and R. H. Lipson
© 2009 American Institute of Physics 978-0-7354-0684-1/09/$25.00

controlled rates. Four different type of experiments (E1-E4) were performed. In experiments E1, E2 and E4, a homogeneous mixture of ZnO and graphite powders in 3:1 ratio (by weight) was used as the source material. The source material was placed in hot zone of the furnace (in an alumina boat) under flowing argon gas (500 sccm). The temperature of the furnace was raised to 1050°C at a rate of 6°C/min. While the temperature of the furnace was being raised, oxygen gas was introduced at desired temperature and flow rate. In experiments E1, E3 and E4, oxygen introduced was 4% at 1050°C after few mins of attaining maximum temperature, 10% at 700-900°C and 10% at 1000-1050°C, respectively. In experiment E2, Zn metal was heated in argon atmosphere to a temperature in the range of 600-700°C and oxygen (10%) was added at the highest temperature. The furnace was maintained at the highest temperature for 5 hrs. Samples obtained in different experiments were analyzed using SEM, EDX, XRD and TEM etc.

RESULTS

XRD, SEM and TEM results of different experiments are summarized in Figures 1-5.In experiment E1, hexagon shaped Zn polyhedra were obtained at low temperatures zone of 200-300°C as shown by XRD (Figure 1a). Spherical balls were obtained in the temperature range of 300-500°C and some of the balls at temperatures above 400°C were seen to be broken and partially hollow (Fig.2a). XRD of these balls (Fig. 1b) showed both Zn metal and ZnO peaks indicating that Zn metal balls are formed during initial evaporation and are subsequently oxidized to yield a core-shell like morphology. At temperature zone of 500-700°C, ZnO nanowires were observed to grow from these Zn/ZnO spheres (Fig. 2c). Nucleation of nanowires is clearly seen as small protrusions in Fig. 2b.

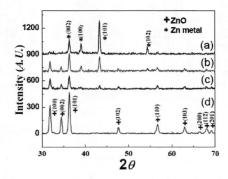

FIGURE 1. XRD patterns of: (a) E1 deposits at 200 - 300°C: zinc Polyhedra, (b) E1 deposits at 300-400°C: Zn/ZnO core/shell spheres (c) E3 deposits at 400-500°C: ZnO nanowire and (d) E4 deposits at 250- 350°C: ZnO tetrapods.

In E2 experiments, formation of hollow spheres and growth of nanowires from them was observed similar to experiment E1. In experiments E3, growth of ZnO nanowires is seen in the temperature zone of 400-500°C (Figures 3a &b) and that of

Zn metal polyhedra at ~200-250°C (Fig. 3c). XRD (Fig.1c) showed that nanowires are nearly pure ZnO. Presence of Zn polyhedra indicated that the nanowires grow in the oxygen deficient atmosphere. Transmission electron microscopy (TEM) images of nanowires (shown in Figures 4 a & b along with selected area electron diffraction pattern) shows that the nanowires have uniform diameter and are single crystalline. High-resolution transmission electron microscopy (HRTEM) images of wires and their tips are shown in Figures 4c and d. It is observed that wires grow along (0001) direction, have crystalline but somewhat rough edges and the tips are flat having defect free ZnO structure with no other material attached.

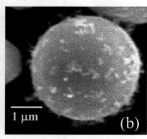

FIGURE 2. SEM micrographs of samples obtained in experiment E1, (a) hollow spheres as indicated by arrows in the temperature range of 400 - 500°C, (b) nucleation of nanowires at temperature of ~500°C and (c) growth of nanowires in temperature range of 550-700°C.

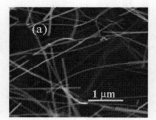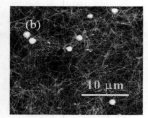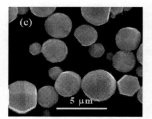

FIGURE 3. SEM micrographs of samples obtained in experiment E3, (a) ZnO nanowires in the temperature range of 400-500°C, (b) mixture of ZnO nanowires and few polyhedra at~ 400°C, (c) Zn polyhedra at temperatures of 200-300°C.

In experiment E4, formation of ZnO tetrapods was observed in temperature zone of 250-450°C. The tetrapods have pure ZnO phase as shown by XRD of Fig. 1d. TEM images of tetrapods obtained at 250°C (Fig. 5a) show that these have leg diameter in 20-50 nm range and overall size in 1-2 μm range, diameter of some legs was found to become quite narrow near its tip (Fig. 5b). Fig. 5c shows a HRTEM image of a leg of a tetrapod when its zone axis was oriented along the $[2\bar{1}\bar{1}0]$ direction. This confirmed that the growth of the leg occurred along the c-axis [6,7].Fig. 5d shows a HRTEM image of core region of a tetrapod (marked as a rectangle in the top left inset) imaged with a strong $(01\bar{1}1)$ reflection. Diffraction patterns generated using fast Fourier transform of the HRTEM images from different regions in the core confirmed the structure of the central core to be wurtzite (Fig. 5d). The diameter of legs and overall size of tetrapods was found to increase with deposition temperature [8].

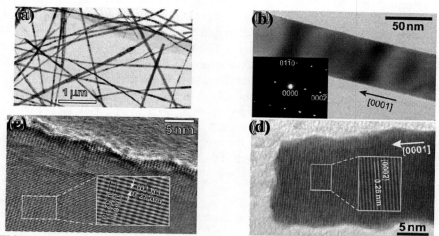

FIGURE 4. TEM images of (a) ZnO nanowires and (b) a single nanowire along with a diffraction pattern shown in the inset. HRTEM images of a wire (c) near the wall and (d) at the tip.

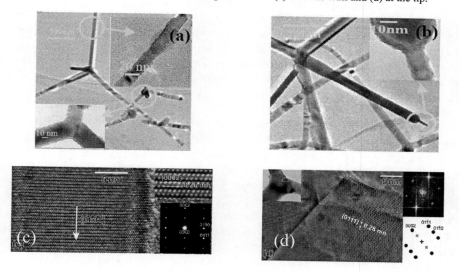

FIGURE 5. TEM images of (a) ZnO tetrapods obtained at 250°C and (b) tetrapods exhibiting further growth at the tip. (c) HRTEM image of a leg of a tetrapod when leg is oriented along the $[2\bar{1}\bar{1}0]$ zone axis. Top right inset shows magnified view of main figure and bottom right inset is diffraction pattern of the wire. (d) HRTEM image of tetrapod core in region marked with a rectangle in the left inset. Upper insets on the right shows FFT generated diffraction pattern from the region where line spacing is written in main figure and bottom inset shows schematic diagram of this pattern.

DISCUSSION

Parameters controlling the growth of different morphologies and their dependence on source/ substrate temperatures in the present study are: (a) decomposition of ZnO in the presence of graphite occurs at temperatures above 970°C [7], (b) complete oxidation of Zn occurs at $T>300$°C [9,10] and therefore in presence of Zn metal and oxygen we expect ZnO structures at $T>300$°C, and Zn/ZnO structures at $T<300$°C, (c) growth by catalytic vapor liquid solid (VLS) mechanism may occur at temperatures above 419°C (melting point of Zn) in the presence of liquid zinc droplet, (d) growth of nanocrystalline ZnO may occur by VS mechanism in presence of suitable nuclei at temperatures above ~300°C (complete oxidation temperature). In what follows we distinguish between the mechanism for nucleation and growth of ZnO structures and discuss the nucleation and growth mechanisms separately.

The results mentioned above show that the nucleation of ZnO nanowires occurs by two different mechanisms.

(1) Self catalytic vapor-liquid-solid (VLS) mechanism at temperatures above melting point of Zn when a Zn metal ball is first deposited to enable nucleation of nanowires from it [11]. This is seen in experiment E1 where nanowires emanating from Zn spheres are seen at T>500°C, but no nanowires are obtained at T<500°C. Similar growth is also seen in experiment E2 using Zn metal source.

(2) Vapor solid growth mechanism operative in the temperature range of 300-500°C (where significant oxidation rate of Zn is obtained and there is no Zn liquid droplet) as observed for experiment E3. No liquid metal droplet is formed here as oxygen flow is initiated before decomposition temperature of ZnO. Nucleation of nanowires in this case may need the presence of slight oxygen deficiency of ZnO as indicated by presence of Zn metal structures obtained at lower temperatures in the same experiment.

Nucleation of ZnO tetrapods occurs when initial deposit is significantly deficient in oxygen (inlet of oxygen at ~950-1000°C). Some Zn is evaporated before inlet of oxygen but no Zn droplets are permitted to form as occurs in experiment E1 where oxygen is introduced few minutes after attaining temperature of 1050°C. This leads to significantly oxygen deficient ZnO deposit that is believed to be conducive for growth of tetrapods in earlier studies [6,12,13]. Conditions needed for nucleation of tetrapods are therefore intermediate between those needed for nanowires by self catalytic mechanism (experiment E1) where large excess of Zn metal in the form of liquid droplet is present and vapor solid mechanism of experiment E3 where oxygen is introduced much before the evaporation from the source starts leading to the presence of mostly ZnO.

Structures obtained at temperatures below 250°C are predominantly Zn metal irrespective of deposition conditions as some Zn evaporated from source is not oxidized and deposits at lower temperatures. ZnO having lower vapor pressure, deposits at higher temperatures [9,10].

We postulate that vapor solid (VS) mechanism [14] is responsible for the growth of ZnO nanowires (nucleated by different mechanisms) as well as tetrapods. The difference of morphologies arises only due to the conditions occurring during nucleation stage. This inference is based on following points: (1) we observe flat tip (Fig. 4d) without any extraneous material indicating absence of a catalytic growth center or VLS mechanism and narrowed tip of some tetrapod legs indicating initial growth occurs on top with thinner wire that is in agreement with VS growth mechanism, (2) we obtain nanowires at T>500°C in experiment E1 with nucleation by self catalytic vapor-liquid-solid (VLS) mechanism (that occurs above 500°C only), whereas in experiments E3 and E4 nanowires or tetrapods can be grown at much lower temperatures, as nuclei have been formed by VS mechanisms, (3) in experiment E2 also, no nanowires on Zn metal liquid droplets can be grown at 500°C, as higher temperatures are necessary for nucleation by VLS mechanism in agreement with E1, (4) the structure of nanowires and legs of tetrapods is similar indicating similar growth mechanism and (5) growth conditions for nanowires and tetrapods are same in experiments E3 and E4 and are quite similar in experiments E1 and E2, and therefore the growth mechanism is expected to be similar as well.

In the present study we find that the tetrapod cores have wurtzite structure. This is confirmed by Fourier spectra taken from the tetrapod core and lattice spacings. The results are in agreement with octa-twin model as well as model of Nishio et al[15] but in disagreement with model of Shiojiri and Kaito [16]. Further detailed studies of the core are necessary to clearly distinguish between first two mechanisms and are being carried out.

CONCLUSIONS

The morphology of nanostructures during carbothermal vapor deposition technique is governed by the formation of nuclei whose nature depends on growth temperature and vapor pressures of Zn and oxygen at the time of initial deposition. Nucleation of ZnO nanowires occurs by self-catalytic nucleation mechanism or by vapor-solid mechanism. The conditions for nucleation of tetrapods are intermediate between those for nanowires by catalytic nucleation mechanism and vapor-solid mechanism. The growth of nanowires and tetrapods can occur over a wide temperature range in the presence of sufficient oxygen.

REFERENCES

1. Z.W. Pan, Z. R. Dai and Z. L. Wang, *Science* **291**, 1947-49 (2001).
2. J. J. Wu and S.C. Liu, *Adv. Mater* **14**, 215-218 (2002).
3. B. Liu, H. C. Zeng, *J. Am. Chem. Soc.* **125**, 4430-4431 (2003).
4. Z. L. Wang, X. Y. Kong, Y. Ding, P. Gao, W. L. Hughes, Y. Rusen, Y. Zhang, *Adv. Func. Mater.*, **14**, 943-956 (2004).
5. Z. L. Wang, *J. Phys.D: Condens. Matter* **16**, R829-R858 (2004).
6. F. Z. Wang, Z. Z. Ye, D. W. Ma, L. P. Zhu and F. Zhuge, *Mater. Lett.* **59**, 560-563 (2005).
7. Y.G. Wang, M. Sakurai, and M. Aono, *Nanotechnology* **19**, 245610(5pp) (2008).
8. M. Kaur, S. Bhattacharya, M. Roy, S. K. Deshpande, P.Sharma, S. K. Gupta and J. V. Yakhmi, *Appl. Phys. A.* **87**, 91-96 (2007).
9. M. S. Aida, E. Tomasella, J. Cellier, M. Jacquet, N. Bouhssira, S. Abed and A. Mosbah, *Thin Solid Films* **515**, 1494-1499 (2006).
10. J. Liu, Z. Zhang, X. Su and Y. Zhao, *J. Phys. D: Appl. Phys.* **38**, 1068-1071 (2005).
11. H.Y. Dang, J. Wang, S. S. Fan, *Nanotechnology* **14**, 738-741 (2003).

12. H. Zhang, L. Shen, S. Guo , *J. Phys. Chem. C* **111,** 12939-12943 (2007).
13. J. Singh, R. S. Tiwari , O. N.Srivastava, *J.Nanosci Nanotechno* **7** ,1783-1786 (2007).
14. B. B.Wang, J. J. Xie ,Q. Yuan,Y. P. Zhao, *J. Phys. D: Appl. Phys.***41,** 102005 (6pp) (2008)
15. K. Nishio, T. Isshiki, M. Kitano , M. Shiojiri ,*Philos. Mag. A* **76** ,889-904 (1997).
16. M. Shiojiri, C. Kaito, *J. Cryst. Growth* **52**, 173-177 (1981).

Metallic and Semiconducting [001] SiC Nanowires-An Ab-initio Study

A. Pathak, S. Agrawal and B.K. Agrawal

Physics Department, Allahabad University, Allahabad - 211002, India.
E-mail : balkagl@rediffmail.com, balkagr@yahoo.co.in

Abstract. A detailed comprehensive ab-initio study of the structural, electronic and optical properties of the unpassivated and H-passivated SiC nanowires (NWs) grown along [001] direction having diameters lying in the range, 3.35 to 15.42 Å has been made by employing the first–principles pseudopotential method within density functional theory (DFT) in the generalized gradient approximation (GGA). We investigate two types of the NWs having hexagonal and triangular cross-sections. The binding energy (BE) increases with the diameter of the NW because of decrease in the relative number of the unsaturated surface bonds. The band gap decreases with the diameter of the NW because of the quantum confinement. After atomic relaxation, appreciable distortion occurs in the nanowires where the chains of Si- and C-atoms are curved in different directions. These distortions are reduced with the diameters of the nanowires. The different nanowires reveal different electronic properties e.g., one nanowire is seen to be metallic whereas the other one is semi-metallic and the remaining nanowires are semiconducting. The optical absorption is quite strong in the ultra-violet (UV) region but is quite appreciable in the visible and infra-red (IR) regions. Thus, the nanowires may have applications in the development of the devices emitting right from infra-red to ultra-violet electromagnetic radiations.

Keywords: Semiconducting nanowires, structural stability, electronic structures, optical absorptions.
PACS: 61.46.Km, 71.20.Mq, 78.20.Ci

INTRODUCTION

The silicon carbide (SiC) bulk in its crystalline form has unique place among the IV-IV group semiconductors because of the occurrence of its vivid crystal structures. SiC crystallizes in hexagonal as well as in cubic form and it also shows polytypism [1,2]. In the various polytype structures, one finds identical layers but having different stacking sequences. In fact, these layered structures may be called as natural superlattices [3-6]. The zinc-blende 3C SiC structure has its applications as high temperature, high power, high breakdown field strength, high thermal conductivity, high saturation drift velocity, high electron mobility and high frequency electronic devices. On the other hand, the 6H SiC polytype having larger band gap can be employed for the development of the light-emitting-diodes [7]. Another form of SiC called as porous SiC can be utilized as a biocompatible material and used in bone implants.

One-dimensional (1D) SiC structures with nanometer diameters, such as nanowires, nanotubes, nanowhiskers [8], nanowire flowers [9], nanosprings [10] are attracting increasing interest due to their optical, electronic, structural and mechanical

CP1147, *Transport and Optical Properties of Nanomaterials—ICTOPON - 2009*, edited by M. R. Singh and R. H. Lipson
© 2009 American Institute of Physics 978-0-7354-0684-1/09/$25.00

properties and their potential applications ranging from probe microscopy tips to interconnections in nanoelectrical devices.

We make an ab-initio study of the SiC [001] nanowires possessing the complex wurtzite structure and consider both the unpassivated and H-passivated SiC nanowires (NWs). The diameters of the hexagon and triangle cross-sectional NWs fall in the range, 3.35 – 15.42 Å.

CALCULATIONS AND RESULTS

We have employed a soft non–local pseudopotential of Troullier and Martins[11] within a separable approximation[12] and an exchange correlation potential of Perdew et al[13] in GGA. This is generated by the FHI code [14]. We test the convergence of the wire energy after varying both the plane – wave cut off energy (ecut) and the number of the $\vec{k}$ - points in the Brillouin zone (BZ) for the two small diameter WZ-H (6,6) and WZ-(24,24) NW's. We find that the BE converges quite well for the plane wave cut off energy of 60 Ry and one $\vec{k}$ -point for the two wires of different diameters. In our future calculations, we therefore chose one $\vec{k}$ - point and a plane- wave cut off energy of 60 Ry. The interference effects between the images of a NW normal to the wire axis were removed by employing a sufficient vacuum separation of at least 10 Å between the two adjacent walls of the images.

The considered cross-sections of the presently studied unpassivated SiC NWs possessing wurtzite (WZ) crystal structure are the hexagonal and triangular ones. In the wurtzite structure normal to the c-axis, a unit cell contains four layers; each layer is formed from one type of ions and is surrounded by the layers of different ions. The atomic arrangements in two successive layers are same although, the type of ions in each layer is different. We call it a double layer. There are two different types of the double layers. The number and the configuration of the atoms in one double layer are different from those of the other layer. The structures are denoted as WZ-H (m, n) and WZ-T (m, n), respectively where m, n represent the number of atoms in each closely spaced double layer in the super unit cell. It may be pointed out that here, the numbers 'm' and 'n' do not represent the number of one types of ions and the other type of ions but instead represent the total number of both types of ions in each double layer.

The H-passivated NWs are denoted as H-WZ-H(m,n) and H-WZ-T(m,n) for the hexagon and triangle cross-sectional nanowires, respectively where the meanings of the symbols are similar to those taken for the unpassivated nanowires.

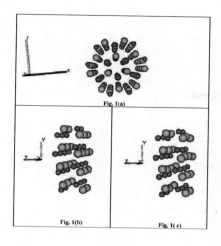

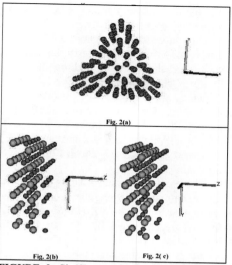

FIGURE 1. Si (C) atoms are denoted by bigger (smaller) spheres. (a) Optimized (relaxed) atomic configuration of a part of the hexagon cross-sectional WZ-H(24,24) nanowire(NW) having the wire axis normal to the plain of the paper. (b) Unoptimised and (c) optimized atomic configurations of a unit cell of the hexagon cross-sectional WZ-H(24,24) NW having [010] direction normal to the plain of the paper

FIGURE 2. Si (C) atoms are denoted by bigger (smaller) spheres. (a) Optimized (relaxed) atomic configuration of a part of the triangle cross-sectional WZ-T(50,42) NW having the [001] wire axis normal to the plain of the paper. (b) Unoptimised and (c) optimized atomic configurations of a unit cell of the triangle cross-sectional WZ-T(50,42) NW having [100] direction normal to the plain of the paper.

One may estimate the stability of a NW by defining the binding energy (BE) as,

$$BE = E_{iso} - E_{tot}/n,\qquad(1)$$

where E_{iso} is the energy of the isolated SiC molecule and E_{tot} is the energy of the unit cell of the NW containing 'n' number of the molecules. A positive BE indicates the structural stability of a NW with respect to the single SiC molecule whereas a negative value denotes the instability. In practice, the stability of a nanowire will be judged from the energy barrier height towards the other possible structural candidates which we have not addressed.

In Fig. 1, the atomic configurations of the hexagon cross-sectional WZ-H(24,24) NW in three dimensions are presented. One observes in Fig. 1(c) that in the optimized hexagon cross-sectional NW, Si and C planes are quite distorted. The atomic rows of the Si-plane lying along the [100] direction get curved in [010] direction but by different magnitudes in one Si-plane whereas the curvature is seen in the opposite direction on the other Si-plane of the unit cell. The bulk atoms remain unshifted. The six surface atoms of the Si-planes relax inward towards the NW axis whereas the three remaining surface atoms relax outward away from the NW axis. On the other hand, all the surface and bulk C-atoms relax outward but by different magnitudes. For both the C-planes, the [100] rows of the C-atoms show curvature along the NW axis having different magnitudes.

For a triangular cross-section NW [Fig. 2(c)], we find that the atoms of all the Si- and C-planes show quite small distortions in contrast to the case of the hexagon cross-sectional NWs. The variations of the different structural parameters and the BE's with the diameter of the NW are shown in Fig. 3. There are two kinds of SiC bonds having different bond lengths in a WZ structure. The Si–C bond lengths parallel to the wire (c-) axis is named as r_p whereas each of the three Si–C bond lengths which are inclined to the c-axis is named as r_i.

The electronic energy bands along with the DOS for various unpassivated NW's having hexagonal and triangular cross-sections are shown in Figs. 4 and 5, respectively. In the smallest diameter WZ-H(6,6) NW (Fig. 4), the surface states fill the energy gap and the NW becomes metallic. At $k_z=0.0$ the highest occupied 23rd state is hybridized s-p-d orbital states of both the Si and C atoms with dominant contributions from p-orbitals. This state is seen to cross the Fermi level (E_F) at a high value of k_z. The lowest unoccupied nearly doubly-degenerate (24, 25)th conduction state is comprised of the p- and d-type orbitals of the both types of the atoms having again major contributions from the p-orbitals. This state also crosses the Fermi level for a higher value of k_z. The next 26th conduction state originates from a mixture of s and p like orbitals of both types of atoms having comparable contribution of s and p orbitals.

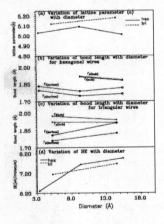

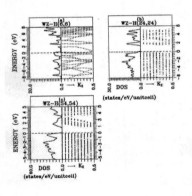

FIGURE 3. (a) Variation of the lattice parameter with the diameter of the hexagon and triangle cross-sectional nanowires. (b) Variation of Si-C bond lengths with the

FIGURE 4. Electronic structure and density electronic states for the unpassivated hexagon cross-sectional (a) WZ-H(6,6), (b) WZ-H(24,24) and (c) WZ-H(54,54) SiC nanowires.

diameters for hexagonal NW's. (c) Same as
(b) but for triangular NW's. (d) Variation of
the binding energy with the diameter (D_{NW})
for the hexagonal and triangular NW's.

   As the valence states in the vicinity of the Fermi level have small dispersion, they
give rise to high DOS. On the contrary, the low-lying surface band is quite broad which
leads to small DOS in the lower part of the conduction band.
   In WZ-H(24,24) and WZ-H(54,54) NWs (Fig. 4), the dangling bond states fill
only the upper region of the fundamental energy gap and there still remain an indirect
energy gaps of 1.62 eV and 1.16 eV, respectively. The valence states in the vicinity of the
E_F are comprised of the comparable contributions of the Si(p) and Si(d) orbitals and the
dominant contributions of C(p) orbitals. On the other hand, the conduction states are the
hybridized s-p-d-orbitals with comparable contributions of the various orbitals. Because
of the small dispersion of the states the DOS is quite high in the neighborhood of E_F as
well as in the lower part of the conduction region. The nature of the electronic structure
and DOS for the triangle cross-sectional NWs are quite different from the hexagonal
cross-sectional NWs. Here also, the surface dangling bond states do not fill completely
the band gap and an indirect small band gaps appear.
   This behaviour is in contrast to that seen earlier in the small diameter hexagonal
cross-sectional WZ-H(6,6) NW which was seen to be metallic. The lowest conduction
bands are quite flat giving rise to the high DOS. The surface conduction band lying above
these flat bands show large dispersion. The top most valence states are also show small
dispersion. The top most filled valence states are hybridized ones having dominant
contributions from Si(p), Si(d) and C(p) orbitals. The lowest conduction state is
comprised of s-p-d like orbitals from both types of the atoms.

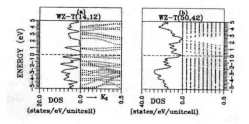

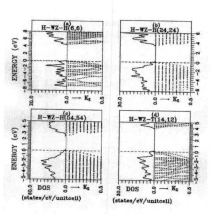

FIGURE 5. Same as for Fig.4 but for
the unpassivated triangle cross-sectional

FIGURE 6. Same as Fig. 4 but for the H-
passivated hexagon cross-sectional (a) H-

(a) WZ-T(14,12) and (b) WZ-T(50,42) SiC nanowires

WZ-H(6,6), (b) H-WZ-H(24,24) and (c) H-WZ-H(54,54) and (d) the triangle cross-sectional H-WZ-T(14,12) SiC nanowires.

The electronic structures along with the DOS's for the H-passivated hexagon and triangle cross-sectional NW's are shown in Fig. 6. Every H-passivated SiC nanowire shows a direct band gap at the origin of the BZ. For the H-passivated NWs, the gap decreases from 4.63 to 2.65 eV with the diameter of the NWs. The top most filled valence states and lowest conduction states show hybridized character having main contributions from the p and d type orbitals of Si atoms and p-like orbitals of C-atoms. In addition to the contributions of these host atomic orbitals, one also observes the comparable contributions from the s, p and d orbitals of H. The present calculation of the energy gaps for the NWs ignores the many body effects where the same site correlations are also accounted. These many body effects may shift the conduction band state to the higher energy site making the band gap larger.

The no phonon optical absorption spectra for unpassivated and H-passivated SiC nanowires are shown in Figs. 7, 8 and 9, respectively. The optical absorption for the unpassivated nanowires shows that the optical transitions involving the surface states appear in the low energy region. This low energy region absorption shifts towards the lower energy side with the diameter of the nanowire because of the decreasing energy gaps. Further, as the diameter of the nanowire increases the relative number of the surface states decreases resulting into a weaker low energy absorption in the large diameter nanowires. The strong absorption lying in the high-energy region also shifts to the low energy side again because of the decreasing electron energy gaps. Continuous absorption occurs in all the unpassivated and H-passivated NWs. For the H-passivated NWs, the commencement of the absorption shifts to the low energy region with the diameter.

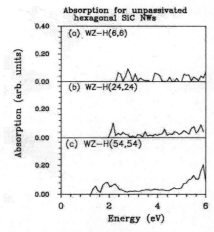

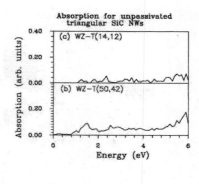

FIGURE 7. No-phonon optical absorption for the unpassivated hexagonal (a) WZ-H(6,6), (b) WZ-H(24,24) and (c) WZ-H(54,54) SiC nanowires.

FIGURE 8. Same as Fig.7 but for the unpassivated triangular (a) WZ-T(14,12) and (b) WZ-T(50,42) SiC nanowires.

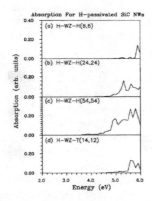

FIGURE 9. Same as Fig. 7 but for the H-passivated hexagonal (a) H-WZ-H(6,6), (b) H-WZ-H(24,24) and (c) H-WZ-H(54,54) and (d) the triangular H-WZ-T(14,12) SiC nanowires. The energy of the optical absorption has been plotted above 2.0 eV.

The present calculation has not taken into account the many body effects like the GW approximation and the excitonic effects, which may change the locations of the optical absorption peaks obtained for the various SiC nanowires. However, it has been reported that in the one dimension systems like the carbon nanotubes [15] and Ge nanowires [16], these effects may not be very significant in most of the cases because of the occurrence of the partial compensation between the self energy and the excitonic effects.

CONCLUSIONS

The facets of the hexagonal and the triangular [001] NW's are seen to be non-polar. All the optimized NWs reveal distortions where the chains of the Si- and C- atoms are curved in different directions. The surface atoms also relax. The two bond lengths, r_i and r_p are different for the atoms which lie either inside the NW or on the surface of the NW. The surface Si-C bond lengths are smaller than the bond lengths of the bulk atoms. For the thick hexagon cross-sectional WZ-H(24,24) and WZ-H(54,54) NWs, the bond lengths (r_i and r_p) for the bulk atoms are similar. The differences in the surface bond lengths r_i and r_p remain the same irrespective of the diameter of the NW. In the triangle cross-sectional NWs, the bulk r_i and r_p differ also but their difference decreases with the diameter of the NW. However, the two types of the surface bond lengths increase with the diameter of the NW.

The BE increases with the diameter of the NW because of decrease in the relative number of the unsaturated surface bonds. According to the Wulff's rule the nanowires with triangular cross-sections having sharp edges should be less stable than those having hexagonal cross-sections with rounded edges. However, Wulff's rule breaks down for the one dimensional structure at the nano scale as we find that a nanowire of small diameter of 5.26 Å with triangular cross-section has binding energy comparable to the wires possessing hexagonal cross-section because of large surface reconstruction.

The band gap increases with decrease in the diameter of the NW because of the quantum confinement. In the case of the unpassivated smallest diameter hexagonal WZ-H(6,6) NW, the surface states fill the energy gap and makes the NW metallic. On the contrary, the unpassivated smallest diameter triangular cross-section WZ-T(14,12) NW remains semiconducting even after the presence of the surface states in the band gap. The thick triangular cross-section WZ-T (50, 42) NW, on the other hand, shows a semi-metallic behaviour possessing a direct band gap of 70 meV. A small DOS appears at both the edges of the fundamental electron energy gap in all the H-passivated hexagonal and triangular cross-sectional NWs because of the large dispersion of the states. The no-phonon optical absorption in the unpassivated SiC NW's is seen to be quite strong in the UV region but there appears appreciable absorption in the infra-red and visible regions arising from the occurrence of the surface states.

ACKNOWLEDGEMENTS

The authors express thanks to the Defence Research Development Organization, New Delhi and University Grants Commission, New Delhi for financial assistance and to Dr. P. S. Yadav for providing us the computation facilities.

REFERENCES

1. Von Munch, in Landolt-Bornstein, edited by O. Madelung, M. Schulz, and H. Weiss, New Series, Groups IV and III-V, Vol. **17**, Pt. A (Springer, Berlin, 1982), and references therein.
2. R. W. G. Wyckoff, Crystal Structures, *Wiley New York*, p. 113 (1963).
3. C. Cheong, R. J. Needs and V. Heine, *J. Phys. C* **21**, 1049 (1988).
4. J. J. A. Shaw and V. Heine, *J. Phys. Condens. Matter* **2**, 1049 (1988).
5. C. Cheong, V. Heine, and I. L. Jones, *J. Phys. Condens. Matter*. **2**, 5097 (1990).
6. C. Cheong, V. Heine, and R. J. Needs, *J. Phys. Condens. Matter* **2**, 5115 (1990).
7. P. A. Ivanov and V. E. Chelnokov, *Semicond. Sci. Technol.* **7**, 863, (1992).
8. Y. Zhang, M. Nishitani-Gamo, C. Xiao, T. Ando, *J. Appl. Phys.* **91**, 6066 (2002).
9. G. W. Ho, A. S. W. Wong, D. J. Kang, and M. E. Welland, *Nanotechnology* **15**, 996 (2004).
10. D. Zhang, A. Alkhateeb, H. Han, H. Mahmood, D. N. Mcillroy, and M. Grant Norton, *Nano Lett.* **3**, 983 (2003).
11. N. Troullier and J. L. Martins, *Phys. Rev. B* **43** 1993 (1991).
12. L. Kleinman and D. M. Bylander, *Phys. Rev. Lett.* **48** 1425 (1982).
13. J. P. Perdew, K. Burke and M. Ernzerhof, *Phys. Rev. Lett.* **77** 3865 (1996).
14. M. Fuchs and M. Scheffler, *Comput. Phys. Commun.* **119** 67 (1999).
15. C. D. Spataru, S. I. Beigi, L. X. Benedict and S. G. Louie, *Phys. Rev. Lett.* **92**, 77402 (2004).
16. M. Bruno, M. Palummo, A. Marini, R. D. Sole, V. Olevano, A. N. Kholod and S. Ossicini, *Phys. Rev. B*, **72**, 153310 (2005).

Solid State Gas Sensor for Volatile Organic Vapours

Sk. Khadeer Pasha[a], K Chidambaram[a]*

School of Science and Humanities, Physics Division, VIT University, Vellore - 632014
Tamilnadu, India

Abstract. Nano Crystalline ZnO powders were synthesized by a Solvothermal method using propan-1ol as the reaction medium. Wurtzite – type ZnO powder has been synthesized by a Solvothermal method at low temperature with an aim to obtain nano crystalline ZnO and to study its ethanol and methanol gas sensing properties have been examined by nano crystalline Zinc oxide sensor in this study. The resulting powder was characterized by FT-IR and Powder X-ray diffraction (XRD) techniques which reveals the structure without any significant change in the cell parameters and the synthesized ZnO having average crystallite size of 10-25nm, respectively. The gas sensor was made on alumina substrate. Various preparation parameters are optimized for maximum sensitivity. The sensitivity for different volatile organic gases such as ethanol and methanol was investigated at different temperatures and concentrations. The sensors fabricated from the nano sized powder exhibited higher ethanol sensing properties at a working temperature of 300°C and similar characteristics were observed for methanol also. ZnO based gas sensor exhibits high sensitivity for ethanol and methanol. Results demonstrated that the ZnO were promising for gas sensor with best sensing characteristics.
Keywords: ZnO; Solvothermal; Gas Sensing; Nano particles; VOCs
PACS: 07.07.D

INTRODUCTION

With the emerging importance in the detection of toxic and flammable gases, the significance of gas sensing in military and industrial applications has been emphasized. In this area, Zinc oxide (ZnO) sensor material has been widely used as a gas sensor material due to its high sensitivity and low cost [1]. Alcohol vapor has been one of the most extensively studied for these sensors [2-4], particularly due to the need for portable practical devices to detect alcohol on the human breath or to detect leaks in distribution lines of industry.

Semiconductor metal oxides as gas sensing materials have been extensively studied for a long time due to their advantageous features, such as good sensitivity to the ambient conditions and simplicity in fabrication [5-7]. Nanophase materials have been prepared using physical vapour deposition methods. Chemical methods have shown several distinct advantages for the synthesis of nano phase particles. Several chemical methods for the synthesis of zinc oxide and mixed metal oxides have been reported [8-12]. The synthesis of

CP1147, *Transport and Optical Properties of Nanomaterials—ICTOPON - 2009,* edited by M. R. Singh and R. H. Lipson
© 2009 American Institute of Physics 978-0-7354-0684-1/09/$25.00

zinc oxide from organic solutions at low temperature has also been reported [13]. This method also being effectively used to synthesize variety of other compounds. The products were usually crystalline, and they did not need post heat treatment at high temperatures. In this paper, we report on the solvothermal synthesis of nano crystalline ZnO-based powders for gas sensors applications. Our purposes were to prepare nanocrystalline ZnO by a solvothermal process, under solvothermal conditions, and investigate the gas response for volatile organic compounds (VOCs). The role of construction of the sensor assembly along with the measuring set-up on the sensing properties has also been discussed.

.

EXPERIMENTAL

2A. Preparation of ZnO powders

All of the chemical reagents were of analytical grade and used without further purification in this experiment. Zinc Acetate dihydrate was dissolved in propan-1-ol and pH is maintained at 9.0. The solution was continuously stirred for 1h at room temperature and followed by heat treatment and the precursor was transferred to an oil bath at 80^0C for 6h. The obtained product was washed with double distilled water for several times and dried for 60^0C in oven over the night.

2B. Fabrication of the gas sensors

The structure of the gas sensor fabricated is illustrated in Fig. 1(A). The sensors were setup in a glass chamber and kept under continuous flow of fresh air before measurement. When sensors were tested, a measured quantity of gases was injected into the test chamber. The heating voltage (V_h) was supplied to the coils for heating the sensor and the circuit voltage (Vc) was supplied across the sensor and load resistor (R_L) connected in series. The signal voltage (Vout) across the load, which changed with the species and concentration of gas, was measured. The sensitivity $[(R_{air}-R_{gas})/R_{air}]$ defined as the change of the film resistance in presence of the test gas was calculated for each temperature and concentration of the test gas. All control data acquisition functions were handled by a PC with suitable interfaces.

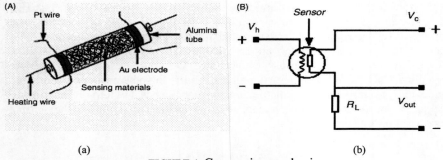

(a)   (b)

FIGURE 1. Gas sensing mechanism

RESULTS AND DISCUSSIONS

3A. FT-IR Spectroscopy

As can be seen, there is a strong absorption peak at 430-450 cm^{-1} attributed to Zn-O Vibrations. No absorption peaks of hydroxy group and Zinc hydrate of peroxide appear in the spectrum Fig. 2. It is apparent that the precipitate had fully decomposed and the absorption peaks matches with the reported value of ZnO.

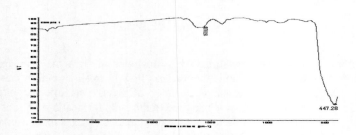

FIGURE 2. FT-IR Spectrum of the as-obtained ZnO

3B. X-ray Diffraction

X-ray diffraction studies confirmed that the synthesized materials were ZnO of the wurtzite phase and all the crystal structures agreed with the reported JCPDS data (Card No. 5 - 664) Fig. 3 shows the powder XRD pattern of the sample synthesized at 80°C for 6h as a representation of all sample. A definite line broadening of the diffraction peaks is

378

an indication that the synthesized materials are in nanometer range. The average crystallite size was calculated from Scherrer formula applied to the major peaks and was found to be around 25 nm. The obtained zinc oxide synthesized powders were characterized by using a X- ray diffractometer with monochromatic CuKα radiation (λ =1.5418Å). An estimate crystallite sizes was calculated using the Scherrer equation [14] applied to the major peaks and was found to be around 25nm.

D $_{crystallite}$ = 0.9 λ / βcos($θ_B$), where λ represents the x-ray wavelength, β is observed fullwidth half maximum (FWHM), and $θ_B$ is the Bragg angle.

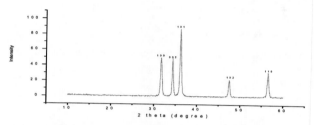

FIGURE 3. XRD pattern of the as obtained ZnO nano powder

3C. Gas sensing properties of the prepared samples

In general, the sensitivity of gas sensors is affected by the working temperature. Fig. 4 and Fig. 5 show the sensor response (S) uses operating temperature for the as synthesized Zinc oxide nano particles towards volatile organic gases like methanol and ethanol. The experiments were carried out to study the various concentrations and response of the sensor to other interfering gases like acetone and propanol, but it is observed that the sensitivity to these gases was very low compared to ethanol and methanol. The various concentrations of the volatile organic gases (VOCs) like ethanol and methanol studied. From the Fig. 6 it is observed that the synthesized nano powder has good concentration than commercial.

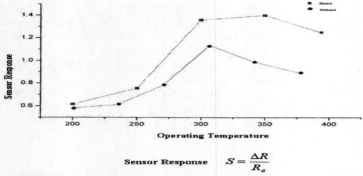

Sensor Response $\quad S = \dfrac{\Delta R}{R_a}$

FIGURE 4. Response of the ZnO (synthesized nano powder) sensors to 100 ppm ethanol and methanol at different working temperatures

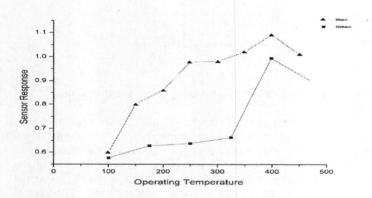

FIGURE 5. Response of the ZnO (Commercial nano powder) sensors to 100 ppm ethanol and methanol at different working temperatures

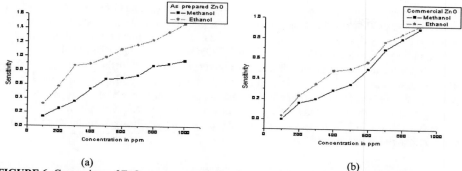

(a)   (b)

FIGURE 6. Comparison of ZnO sensor response (Synthesized & commercial nano powders) towards ethanol and methanol

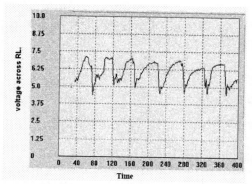

FIGURE 7. Response and Recovery time for ZnO ethanol concentration at 300°C

The response time for the sensor was also calculated from the Fig. 7. The rise time defined as the time required for the conductance to reach 90% of the equilibrium value after the gas is injected. The fall time is the time needed by the sensor to acquire 10% of the above the original value in air after the gas is removed. In Fig. 7 we display the variation in voltage of the sensor at the different concentration ranging from 5 to 65 ppm at 300°C. The average response time of the ethanol gas sensor was measured to be about 30 sec while the fall or recovery time of the sensor was observed to be 10 sec.

CONCLUSIONS

A simple solvothermal route was developed to synthesize nanocrystalline ZnO. Powders are obtained at 80°C with the average particle size of 25 nm by refluxing hydrated zinc hydroxide precipitate for 8h. The gas sensing measurements showed that the ZnO had excellent potential applications as a gas sensor. With its sensitivity being superior to other oxides reported so far, ZnO seems to be a very progressing material as a semiconducting oxide for sensor especially for EtOH detection. The rise and the fall time for sensing were measured to be 30 and 10 sec respectively. Further study about this work is in progress in our lab. It is highly expected that such a simple, cheap and mild solvothermal way could be extended to prepare many other important semiconducting oxide crystallites with novel and useful properties for technologies. Our ultimate aim is to develop the sensors at low temperature and also prove the synthesized material has good sensitivity as well as response.

ACKNOWLEDGEMENTS

The authors are thankful to TBI and the administration of VIT University, Vellore for providing instrumental facilities and financial support to complete this work.

REFERENCES

1. J. F. McAleer, P. T. Moseley, J.O. W. Norris and D. E. Williams, *J. Chem. Soc. Faraday Trans. I* **83**, 1323-1327 (1987).
2. P. Lesark, and M. Sriyudthsak, *Sens. Actuators B* **25**, 504-506 (1995).
3. G. Williams, and G. S. V. Coles, *J. Mater. Chem.* **8**, 1657-1666 (1998).
4. P. K. Heesook, *Sens. Actuators B* **14**, 511-512 (1993).
5. T. Siyama and A. Kato, *Anal. Chem.* **34**, 1502-1503 (1962).
6. J. Q. Xu, Q. Y. Pan, Y. A. Shun and Z. Li, *J. Inorg. Chem.* **14**, 355-359 (1998).
7. A. A. Tomchenko, G. P. Harmer, B. T. Marquis and J. W. Allen, *Sens. Actuators B* **93**, 126-134 (2003).
8. L. Vayssieres, K. Keis, S. E. Lindquist and A. Hag Feldt, *J. Phys. Chem. B* **105** 3350-3352 (2001).
9. S. Maensiri, P. Laokul and V. Promarak, *J. Crys.Growth* **289**, 102-106 (2006).
10. M. S. Arnold, P. Avouris, Z. W. Pan, and Z. H. Wang, *J. Phys. Chem. B.* **107**, 659-663 (2003).
11. Z. Pan, Z. H. Wang, E. Comini, G. Fuglia and G. Sberveglieri, *Appl. Phys. Lett.* **81**, 1869-1871 (2002).
12. Sk. KhadeerPasha., and K. Chidambaram, *International Conference on Sensor and Related Networks (Allied Publisher Proceedings)* (2007) pp.239-241).
13. Baomei Wen, Yizhong Huang and John J. Boland, *J. Phys. Chem. C* **112**, 106-111 (2008).
14. B. D. Cullity, *Elements of X-ray Diffraction (Addison Wesley*, London, (1978) pp.102)

Electron transmission in the Aharonov-Bohm Interferometer with (and without) a spin impurity: Effect of finite width of quantum wires

Chandan Setty*, Satyaprasad P. Senanayak*, K. Vijay Sai*,
K. Venkataramaniah* and Debendranath Sahoo†

*Sri Sathya Sai University, Prashanthi Nilayam, A.P. 515134, India
†Bangalore Academy of Specialized Education,
27, Bull Temple Road, Basavanagudi, Bangalore 560 004, India

Abstract. The Aharonov-Bohm interferometer (ABI) in a two-terminal configuration is a reliable tool to study electron transport in nanoscopic systems. Characterization of the ABI using the Griffith boundary conditions at the junctions of the leads and the ring reported earlier by Kumar and Sahoo (Inter. Jour. Mod. Phys. B 19, 3483 (2005)) are reanalyzed using improved boundary conditions recently proposed by Voo and Chen (Phys Rev B 73, 035307 (2006)). Transmittance (T) of electron as the electron momentum k is varied shows sharp localized peaks. The minimum of T as k is varied over a large range shows progressive increase with k and eventually saturates. Modification of T near the values of k corresponding to the resonances of the ring (without the leads) is not significant. However, noticeable changes do appear off the resonance positions. For the ABI with a spin half impurity–the problem investigated earlier by Joshi Sahoo and Jayannavar (Phys. Rev. B 64, 075320-1 (2001))–we find qualitative differences between large and small k behavior similar to the perfect ABI case. However, in the presence of magnetic flux, the interferometer becomes opaque to electrons for small k values. Our results suggest experiment tests not carried out so far to our knowledge.

Keywords: Aharonov-Bohm, electron wave-guide, ballistic transport, spin-flip mechanism, Heisenberg exchange, width effect
PACS: PACS numbers: 73.23.-b, 03.67.Mn, 72.10.-d, 85.35.Ds

INTRODUCTION

Spin-dependent transport properties of an Aharonov-Bohm ring connected with an incoming and an outgoing lead maintained at different chemical potentials (also known as the Aharonov-Bohm interferometer (ABI)) has potential applications to spintronics. Earlier, some interesting features of ABI were reported[1] and ref.[2] reported the effect of a spin half magnetic impurity (localized on one arm of the ABI) on the transmittance of the electron. The schematic diagram of a perfect ABI is given in fig.1. Investigation of the ABI shown in the left side of fig.1 was carried out in ref.[1] and that of the ABI shown in the right side was the subject of ref.[2]. Both these investigations assumed strictly one dimensional wires. Recently Voo et. al.[3] showed that the width of the quantum wires of the ABI has drastic influence on the transmittance. Moreover, they suggested a simple

CP1147, *Transport and Optical Properties of Nanomaterials—ICTOPON - 2009*, edited by M. R. Singh and R. H. Lipson
© 2009 American Institute of Physics 978-0-7354-0684-1/09/$25.00

way to incorporate the width effect by modifying the Griffith boundary conditions (see for example,[4]). This method agrees with the results of more

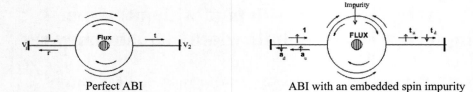

Perfect ABI   ABI with an embedded spin impurity

FIGURE 1. Schematic diagrams of the ABI. The incident amplitude is taken as unity. In the perfect ABI, r denotes the reflection amplitude and t, the transmission amplitude. In the right diagram, the incident electron with its spin 'up' has unit amplitude, a_u and a_d denote respectively the reflection amplitudes with spin 'up' and spin 'down' in the input arm. t_u and t_d denote the transmission amplitudes with spin 'up' and spin 'down' respectively, in the output arm.

detailed quasi-one-dimensional calculations[5]. Voo et. al.[3] have confined their study to variation with electron momentum only. Thus it is natural to extend the scope of their work to magnetic flux effect. Our purpose here is to carry out this remaining task. We report here modification of earlier results of ref.[2] arising due to the width effect. Experimental realization of the system we study has already been achieved by Bagraev et. al.[6]. They have assembled a Silicon ring with an embedded quantum point contact containing exactly one electron. However, they have not focused attention on the aspects of interest to us. In a related context, Shelykh et. al.[7] have proposed a somewhat similar model. Our investigation concerns a different aspect. As is well known[8], the spin-flip cross-section of an electron interacting by exchange and passing through a magnetic impurity is greatly enhanced in a situation of resonant transmission. Whereas Selykh et al[7] use the (delta) attractive (repulsive) potential of the triplet (singlet) coupled spin state and use the statistical weight factors in calculating the conductance, we deal with the amplitude of flipping of the incident electron by explicitly solving the Schrödinger equation with improved Griffith boundary conditions[3] by formulating the problem in the form of scattering matrices and using the composition of scattering matrices, we compute the transmission amplitude and the transmittance. In this manner, the effect of spin flipping on the transmission of electron is investigated. We also compute spin polarization. We now report our results.

RESULTS

Extensive theoretical analysis of the ABI already exists[9, 10]. The ring of circumference 1 acts like a periodic lattice of period 1 and at the values of the momenta

$2\pi/l$ corresponding to the bound states of discrete energies of the electron in the ring, there appear resonances which manifest as transmission maxima or transmission ones' in the ABI. The ring without the leads serves as the "unperturbed" reference state for the ABI. In dealing with the Schrödinger equation for the electron propagation in the ABI Amaresh Kumar and Sahoo[1] and Joshi, Sahoo and Jayannavar[2] used the Griffith boundary conditions at the ring-lead junctions assuming ballistic propagation. Continuity of wave functions at the node imply $\psi_1(0) = \psi_2(0) = \psi_3(0)$. Here $\psi_i(x)$ denotes the wave function on the i-th branch. Voo et. al.[3] wrote the time independent Schrödinger equation in the tight-binding model and showed that it assumes the discretized form

$$(\psi_1 - \psi_0) + (\psi_2 - \psi_0) + (\psi_3 - \psi_0) + \frac{E - V_0 + \varepsilon}{t} \psi_0 = 0,$$

where the wave function ψi is the tight-binding wave function at the site i, the label 0 denotes the Y-junction, and 1, 2, 3 label the sites closest to 0. Here t is the hopping parameter $2\hbar^2/2w$, w being the channel width. E is the energy at the band bottom and can be assumed to be zero. Lead by this expression, Voo et. al.[3] proposed the condition that ψ should satisfy at the Y-junction:

$$\psi_1'(0) + \psi_2'(0) + \psi_3'(0) + c\psi_1(0) = 0, \tag{1}$$

where the derivatives along the branch axes are denoted by primes and are evaluated at the node 0; $c = 2v/w$, w = width of ring, v = phenomenological "tuning" parameter. In this manner, the effect of the width of the quantum wire is incorporated through the parameter c in the above boundary condition. In the present calculation, the results using eq.(1) will be referred to as the "finite width" case and the results using the same equation, but with c = 0 will be referred to as the "zero width" case. We formulate this problem in terms of the Scattering Matrix formalism. The S-matrix at the ring-lead junctions turns out to be unitary and k−dependent. We first consider the width effect on the transmittance of electron for the ABI alone. Next we consider the ABI with an embedded spin impurity (fig.1) with its spin pointing in a direction opposite to that of the incident electron is of particular interest as there is a possibility of the incident electron flipping its spin after transmission through the ring. We use the condition of continuity of the wave functions at the impurity site and the relation $\psi'(0+) - \psi'(0-) = G\,(\vec{\sigma}.\vec{S})\psi(0)$, where we have assumed the Heisenberg exchange interaction $V = J\,\vec{\sigma}.\vec{S}$ [2] $(G = 2mJ/\hbar^2)$ between the electron and the impurity. The S-matrix of the impurity is then worked out. We assume ballistic propagation of electron all throughout the system and follow the rule of composition of S-matrices for each individual component of the ABI. The transmission amplitude is explicitly obtained using MATHEMATICA. We omit details of these cumbersome expressions and discuss only the salient features of our findings.

For the ABI without the spin impurity, Voo et al[3] computed the transmittance for values of k till k = 5 (in unit of 2p=l, where l is the ring circumference) and shown that it agrees remarkably with the results obtained by a laborious quasi-continuum method which incorporates discrete lattice features. We obtain yet another striking feature: for large k values, the minima of transmittance T increase progressively and ultimately attain a constant value which is characteristic of the zero width case. This behavior is seen in fig.2. We feel this observation is of significance and should be experimentally tested. We first reproduced the $k_i T$ plots of Xia[4] for the zero field case for different arm lengths of the ring. Then we computed the corresponding plots with the width effect taken into account. These two cases are shown in fig.3. The effect of flux (however small it may be) is to produce dips at the maxima of $k_i T$ plots. This is shown in the fig.4. Note that we have used the value of c which corresponds to the value of n that was used by Voo et. al.[3] in their work. The behavior of T as a function of flux (in the unit

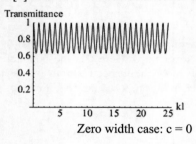
Zero width case: c = 0

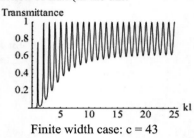
Finite width case: c = 43

FIGURE 2. Transmittance T versus k for the perfect ABI.

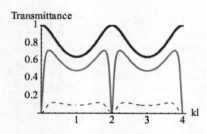

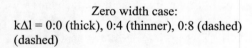
Zero width case:
kΔl = 0:0 (thick), 0:4 (thinner), 0:8 (dashed) (dashed)

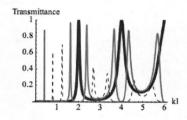
Finite width case:
kΔl = 0:0 (thick), 0:4 (thinner),0:8 (dashed)

FIGURE 3. Transmittance T as a function of k for the perfect ABI

of $2\pi/l$) is shown in fig.5. Here k is held constant. Note that the width effect near the resonance values of k is noticed on the shape of the maxima and there is a small increase in width of the resonance. However, if we consider the above behavior at a value of k away from its resonance value, then there is a noticeable change.

We now consider the case of the ABI with a spin half impurity placed on its upper arm. The incident electron is assumed to have its spin pointing 'up' and the impurity has a spin pointing 'down'. Then there is a probability of the electron flipping its spin after transmission through the interferometer. This problem was considered in ref.[2]. Here we include the effect of the width also and investigate if there are major changes.

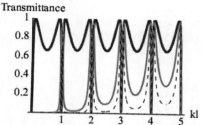

FIGURE 4. Effect of flux on k-T for the perfect ABI: thick-c = 0, thin-c = 43, dashed-c = 85.

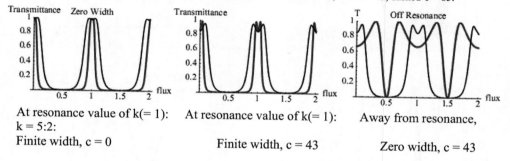

At resonance value of k(= 1):   At resonance value of k(= 1):   Away from resonance,
k = 5:2:
Finite width, c = 0   Finite width, c = 43   Zero width, c = 43

FIGURE 5. Flux effect on transmittance for k =constant for the perfect ABI

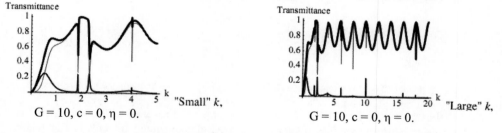

"Small" k,   "Large" k,
G = 10, c = 0, η = 0.   G = 10, c = 0, η = 0.

FIGURE 6. "Small" and "large" k behavior of T for zero width (c = 0) ABI with the spin impurity.

Assuming the incident electron to be pointing with its spin in the 'up' direction, the thinner (thinnest) line shows the dependence of T_u (T_d) and also that of the total transmittance $T = T_u + T_d$. Here Td corresponds to the "spin flip" component arising due to the Heisenberg exchange interaction and T_u denotes the "no spin-flip" component.

Figure 6 displays the variation of T for "small" and "large" k values for the zero width case in the absence of flux. For small k (with no flux) it is irregular and for large k (with no flux) it is rather regular.

The trend of attaining a constant minimum is exhibited as k is increased. However, when flux is present, the large k behavior is strikingly different–the interferometer becomes almost opaque! This behavior is displayed in fig.7 This observation is also worth verifying experimentally. Figure 8 shows the effect of impurity (characterized by G) on the flux variation of T.

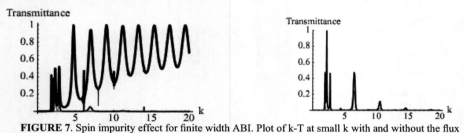

FIGURE 7. Spin impurity effect for finite width ABI. Plot of k-T at small k with and without the flux

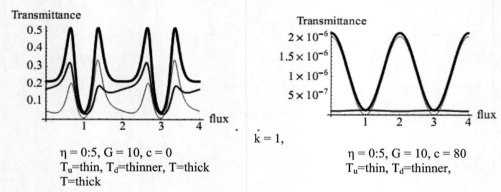

FIGURE 8. Flux variation of Tu, Td, and T for k = 1 for the ABI with the spin impurity.

Note that both T_u and T_d are asymmetric under the reversal of flux whereas T is perfectly symmetric under flux reversal. This was already noted in ref.[2]. The width effect suppresses the contribution of T_d significantly and the total T has a simple periodic variation of a fixed peak value. However, the magnitude of the peak is too small to be of any practical significance. We have results for spin polarization and also some interesting results on the complex plane plots of the transmission amplitude. However, these require more careful presentation and are not presented here for reasons of space restriction.

ACKNOWLEDGMENTS

One of the authors (DS) thanks H.S. Nagaraja for his support at BASE, Bangalore where this work was carried out.

REFERENCES

1. M. V. Amaresh Kumar and D. Sahoo, *Int. J. Mod. Phys. B* **19**, 3483 (2005).
2. S. K. Joshi, D. Sahoo and M. Jayannavar, *Phys. Rev. B* **64**, 075320-1 (2001).
3. K. K. Voo, S.-C. Chen, C.-S. Tang and C.-S. Chu, *Phys. Rev. B* **73**, 035307 (2006).
4. J.-B. Xia, *Phys. Rev. B* **45**, 3593 (1992).
5. J.-B. Xia and S.-S. Li, *Phys. Rev. B* **66**, 035311 (2002).
6. N. T. Bagraev, N. G. Galkin, W. Gehlhoff, L. E. Klyachkin, A. M. Malyarenko and I. A. Shelykh, *J. Phys. Condens. Mat.* **18**, L567 (2006).
7. I. A. Shelykh, M. A. Kulov, N. G. Galkin and N. T. Bagraev, *J. Phys. Condens. Mat.* **19**, 246207 (2007).
8. O. L. T. de Menezes and J. S. Helman, *Am. J. Phys.* **53**, 1100 (1985).
9. M. Bütiker, Y. Imry and M. Ya. Azbel, *Phys. Rev. A* **30**, 1982 (1984).
10. Y. Gefen, Y, Imry and M.Ya. Azbel, *Phys. Rev. Lett.* **52**,129 (1984).

Quantum Size Effects In Chemically Synthesized CdS Nanoclusters

Rhituraj Saikia[a] and P. K. Kalita[b]

[a]Department of Physics, Kaliabor College, Kuwaritol-782137, India
[b]Department of Physics, Guwahati College, Guwahati-781021,India
E-mail: send2rhituraj@gmail.com

Abstract. Nanoclustered CdS have been prepared through chemical bath deposition method in by preparing a matrix solution from an equivolume and equimolar solution of $CdSO_4$ and thiourea at room temperature. Polyvinyl alcohol (PVA) was taken as capping agent to see the quantum size effect. The samples were characterized by XRD, UV, PL, FTIR. XRD pattern shows zinc blende type structure of CdS nanocluster with average particle size 5.8nm for water matrix and 4.26nm for PVA matrix. Optical absorbance study of thin film of CdS deposited on glass substrate shows the blue shift of the band gap energy of 3.33eV for water matrix and 3.78eV for PVA matrix which is attributed to quantum size effect. Analysis of PL spectra gives the luminescence peak at 550nm for water matrix and 540.5nm for PVA matrix. IR-Spectra used to explain the various streaching of bonding in both the samples.

Keywords: Chemical synthesis, nanoclusters, optical absorbance, photo-luminescence.
PACS: 62.23.Eg, 63.22.Kn, 42.50.Ex, 42.70.Qs, 61.46.Bc, 61.46.Hk

INTRODUCTION

The optical and electrical properties have played a key role in the study of semiconductor nanoparticles. Nanocluster of CdS has been attracting great interest among II-VI semiconductor due to excellent optical conductivity, luminescence properties and quantum size effect. These are used to fabricate Field effect transistor, Light emitting diodes, Solar cell, Biological sensors, Photo detectors etc. Nanoclustered CdS has also shows the sensitive dependence of their optical and electrical properties for size, shape, size distribution and morphologies[1]. The CBD offers an inexpensive and simple means to synthesize such particles with good control over size and size distribution by optimizing various parameters. The dependence on molarity is an important aspect on quantum size effect. Most of the workers have shown an reduction of particle size on decreasing molarity and growth temperature[3]. Keeping above aspect an experimental work has been undertaken to synthesized CdS nanoparticles through chemical route at room temperature.

CP1147, *Transport and Optical Properties of Nanomaterials—ICTOPON - 2009*, edited by M. R. Singh and R. H. Lipson
© 2009 American Institute of Physics 978-0-7354-0684-1/09/$25.00

Experimental

CdS nanocrystalline thin films were deposited on the cleaned glass substrates by Chemical Bath Deposition (CBD) method. The deposition was carried out in a matrix solution of $CdSO_4$ and thiourea in an alkaline medium. The PVA was used as capping agent for this synthesis process in an aqueous solution (3%) with constant stirring at room temperature. Films are deposited with or without PVA to observe the size effect. The molarity of solution was kept at 1.0 M. The pH of the water matrix solution was maintained at around 8.5 while for PVA matrix solution was maintained at 9.5 by slowly adding ammonium hydroxide solution[5]. Prior to the synthesis the equimolecular solution of Cadmium Sulphate mixed to prepare the samples and glass substrates were introduced vertically into the solution. The substrates were kept in the solution for 24 hour at room temperature for deposition of films. After deposition, the substrates were taken out thoroughly washed and rinsed with doubly distilled water and dried. The characterization of films was done by employing the XRD, UV absorption spectroscopy, PL and IR spectra respectively.

Results and Discussions

XRD Analysis

The XRD patterns of CdS nanoclusters are shown for water matrix and PVA matrix in Fig 1(a) and Fig 2(b) respectively.The XRD shows the deposited films are polycrystalline having cubic zinc blende type fcc structure. A detail of structural parameters are given in Table1. The d values are well agreement with standard JCPDS datas. The lattice constant 'a' is calculated from the usual relation.

$$d = a / (h^2 + k^2 + l^2)^{1/2} \qquad (1)$$

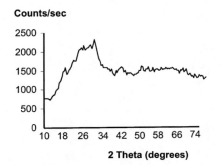

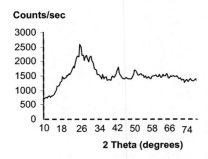

FIGURE 1. (a) XRD pattern water matrix   **FIGURE 1.** (b) XRD pattern for PVA matrix

The prominent peak along (111) direction have been utilized to estimate the grain size of the samples with the help of Scherer formula,

$$D = K\lambda / \beta Cos\theta \qquad (2)$$

where D is the average crystalline size , K is a constant ($\sim$ 1) , β is the full width at half maximum and θ is the Bragg's angle. The average grain sizes estimated for water matrix found to be 5.68nm and for PVA matrix it is 4.26nm.

TABLE 1. XRD ANALYSIS EXPERIMENTAL DATA

Samples	Observed Values of 2θ degrees	Plane (h k l)	Spacing (d) A⁰	Lattice Parameter (a) A⁰	Grain Sizes (D) nm
Water Matrix	27^0,	(111),	3.2984	5.7120	5.68 nm
	31.5^0,	(220),	2.8367		
	39^0	(311).	2.3067		
PVA Matrix	26.5^0,	(111),	3.3595	5.8188	4.26 nm
	29.5^0,	(220),	3.0243		
	35.5^0	(311)	2.5257		

Optical Studies

The formation of the CdS nanoclusters can also be identified from UV-VIS spectroscopy of the samples. The UV-Vis absorption spectrum of the both samples are shown in respectively

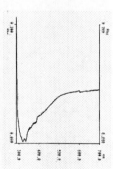

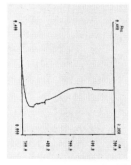

FIGURE 2. (a) UV-VIS for water matrix   **FIGURE 2.** (b) UV-VIS for PVA matrix

It can be seen from the figures that the strongest absorption edge of CdS prepared in water matrix is around 375 nm and that of PVA matrix, it is nearly 330 nm. The band gap corresponding to this absorption edge can be calculated by using the formula

$$Eg = hc / \lambda \qquad (3)$$

The band gap calculated for water matrix is 3.33 eV and for PVA matrix is 3.78 eV respectively. This indicates that a large blue shift in the both of the samples. Semiconductor crystallites in the diameter range of a few nanometers show a three dimensional quantum size effect in their electronic structure[2]. Therefore a decrease in crystalline size can enhance the band gap. The absorption spectroscopy result support our XRD observed data.

The polymer plays an important rule in size reduction mechanism through chemical route. PVA is extensively used to produce semiconductor nanostructures because of its potential to act as a strong capping agent and thereby reduces the aglomarism which is common in chemically deposited film[4]. In the present case a large blue shift using PVA in preparation of precursor solution clearly indicates a strong quantum confinement.

However some additional absorption in the form of blug peak in the next subsequent wavelength region around 370 nm onwards in both the films are found. This absorption may be due to higher concentration of the solution containing impurities

PL Studies

The PL spectra indicate the luminescence peak around 550 nm for water and 540.5 nm for PVA. This shows an usual behavior of CdS nanoclusters. The observed PL spectra for both the samples are shown below

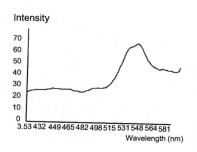

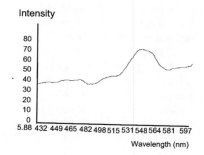

FIGURE 3. (a) PL Analysis of CdS in Water Matrix   FIGURE 3. (b) PL Analysis of CdS in PVA Matrix

393

Generally PL peak in CdS is found to be between 500nm -700 nm. In the present experiment the PL peak corresponding to CdS embebbed in PVA slightly shifted to lower wavelength over water matrix. The PL intensity also slightly increases for CdS/PVA film. This also supported the formation CdS nanocluster. This type of PL spectra in CdS nanostructures is also reported by various workers[6]. No additional peak is found apart from maximum intense peak.

FTIR-Studies

The IR peak recorded at room temperature for both the samples are as shown below.

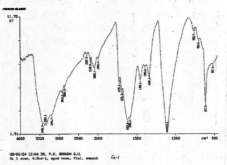
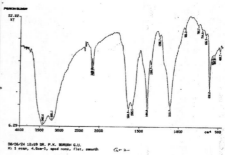

FIGURE 4. (a) FT-IR of CdS   **FIGURE 4.** (b) FT-IR of CdS/PVA

The IR spectra of CdS nanoparticle for water matrix and PVA matrix are shown in figure. The band at 3432.3 cm^{-1} for water matrix sample and the band at 3422.6 cm^{-1} in PVA matrix sample are due to O-H stretching. The band at 2912.6 cm^{-1} in sample I is due to C-H stretching. The bending vibration is appeared at band 1623.6 cm^{-1} for water matrix while for PVA matrix, it is found at 1634.9 cm^{-1}. Asymmetric stretching are found at band 1111.1 cm^{-1} and 1110.7 cm^{-1} for water matrix and PVA matrix respectively. Cd-S stretching found at band 617.3 cm^{-1} for water matrix and it is 618.2 cm^{-1} for PVA matrix. The band 483.7 cm^{-1} of PVA matrix indicates the asymmetric bending[8]. The vibration absorption peak at the CdS band which should be at 405 cm^{-1} and 265 cm^{-1} as reported by other workers could not be observed since it was beyond the extent of our measurement.

CONCLUSIONS

CdS nanoclusters have been prepared by using Chemical Bath Deposition (CBD) method in water matrix as well as in PVA matrix at room temperature. The XRD analysis shows the prepared samples are in cubic zinc blende type structure. The particle sizes obtained in a few nanometer range which clearly indicate that both the sample will show strong quantum size effect. The grain size found for water matrix is 5.68 nm and for PVA matrix is 4.26 nm. The value calculated the lattice parameter and interplaner spacing for both of the samples are agree with the theoretical value. The reduction of particle size is also supported by UV-VIS analysis. The absorption spectra show a clear blue shift. The absorption peak obtained from UV-Vis spectra is 375 nm for water matrix and 330 nm for PVA matrix respectively. The calculated band gap corresponding to these peaks is found to be 3.335 eV for water matrix and 3.78 eV for PVA matrix. The luminescence peak found in PL spectra indicates slight shift towards lower wavelength side as well as an increase of intensity for PVA capped CdS film. The formation of various O-H, C-H, Cd-S bond stretching for both of the samples were explained by FT-IR studies.

ACKNOWLEDGEMENTS

The authors would like to acknowledge a sincere thanks and gratitude to Dr. K. Boruah and Mr. U. Das of Tezpur University as well as Dr. P. K. Boruah of Gauhati University for their great help during the work.

REFERENCES

1. Yadong Yin and A. Paul Alivisatos, *Nature* **437**, 664-670 (2005).
2. L. Banyai and S. W. Koch, *Semiconductor Quantum Dots*, Singapore, World Scientific, 2005, pp.13-36.
3. Zheng Bin Sun, Wei Qiang Chen, Xian Zi Dong, and Xuam Ming Duan, *Chem. Lett.* **36**,156-158 (2007).
4. Ching-Fuh Lin, Eih – Zhe Liang, Sheng-Ming Shih and Wei-Fang Su, *Jpn. J. Appl. Phys.* **42**, L610-612 (2003).
5. R. Devi, P. K. Kalita, P. Purkayastha and B. K. Sharma, *Bull. Mater. Sci.* **30**, 123-128 (2007).
6. Rong He, Xue Feng Qian and Jie Yin, *Mat. Lett.* **57**, 1351-1352 (2003).
7. Huixian Tang and M Abdul Khadar, *Bull. Mater. Sci.* **31**, 511-515 (2008).
8. Eucai Hao, Haiping Sun, Zheng Zhou, Jonqiu Liu, Bai Yang and Jiacong Shen, *Chem. Mater.* **11**, 3096-3102 (1999).

Ellipsometric study of Atomic Layer Deposited TiO$_2$ thin films

Piyush Jaiswal, G. V. Kunte, A. M. Umarji and S. A. Shivashankar*

Materials Research Centre, Indian Institute of Science, Bangalore-560012
*Author for correspondence

Abstract. Atomic layer deposition was used to obtain TiO$_2$ thin films on Si (100) and fused quartz, using a novel metal organic precursor. The films were grown at 400^{0}C, varying the amount of oxygen used as the reactive gas. X-ray diffraction showed the films to be crystalline, with a mixture of anatase and rutile phases. To investigate their optical properties, ellipsometric measurements were made in the UV-Vis-NIR range (300-1700 nm). Spectral distribution of various optical constants like refractive index (n), absorption index (k), transmittance (T), reflectance (R), absorption (A) were calculated by employing Bruggemann's effective medium approximation (BEMA) and Maxwell-Garnet effective medium approximation, in conjunction with the Cauchy and Forouhi-Bloomer (FB) dispersion relations. A layered optical model has been proposed which gives the thickness, elemental and molecular composition, amorphicity and roughness (morphology) of the TiO$_2$ film surface and and the film/substrate interface, as a function of oxygen flow rate The spectral distribution of the optical band gap (E$_g$^{opt}), complex dielectric constants (ε' and ε"), and optical conductivity (σ_{opt}), has also been determined.

Keywords: Atomic Layer Deposition, TiO$_2$, Ellipsometry
PACS: 78.Ci

INTRODUCTION

Titanium dioxide has many attractive physicochemical properties and thus offers itself to numerous applications such as optical coatings for antireflection[1], high dielectric layer for electronic devices[2], biocompatible coatings for biomaterials[3], and photosensitive layer for photocatalysts, and solar cells[4, 5]. Due to its high refractive index and chemical durability in the visible and near IR region[6], it is used as an optical coating material. TiO$_2$ occurs in nature in three polymorphic forms: anatase (tetragonal), rutile (tetragonal), and brookite (orthorhombic).

Because of their many applications thin films of TiO$_2$ have been deposited using various physical and chemical deposition techniques. Among these techniques, atomic layer deposition (ALD) has recently received great attention for the preparation of TiO$_2$ thin films[6, 7]. ALD is a modified form of chemical vapor deposition (CVD), with the main difference between them being the method of introducing reactants into the chemical reactor. CVD uses simultaneous introduction of reactants and their continuous flow, and can utilize both gas and surface reactions, while ALD uses discrete introduction of reactant pulses with an intermediate purging step, restricting

CP1147, *Transport and Optical Properties of Nanomaterials—ICTOPON - 2009*, edited by M. R. Singh and R. H. Lipson
© 2009 American Institute of Physics 978-0-7354-0684-1/09/$25.00

the reaction to absorbed species on the substrate surface, which leads to self-limited layer-by-layer growth. This makes it possible to control film thickness and composition precisely, and also to achieve excellent film uniformity and conformality over large areas with low defect concentrations.

For accurate measurement of optical constants and thickness of thin films, many attempts have been made and theories established. Chiu et al.[8], Zheng and Kikuchi[9], Ramadan et al.[10], Hamza et al.[11], have tried to improve upon established optical techniques to obtain various optical properties of thin films, but spectroscopic ellipsometry[12], has found favor because of its non-destructive nature, high sensitivity and accuracy. In this study, TiO_2 thin films were prepared on p-Si(100) substrates through the ALD process. Spectroscopic ellipsometry was used to determine the optical constants, composition, and the thickness of TiO_2 thin films. The refractive index, the extinction coefficient, and the thickness of TiO_2 thin films were fitted according to an appropriate optical model. The influence of the flow rate of oxygen on the optical constants and thickness of TiO_2 thin films has been evaluated.

EXPERIMENTAL

Thin films of TiO_2 were deposited using a home-made five-channel atomic layer chemical deposition system with a vertical, cold=wall reactor. Bis-isopropoxide-bis tert-butyl 2-oxo butanoato titanium, $[Ti(O^iPr)_2(tbob)_2]$ was used as the precursor[13], Which was characterized by FT-IR spectroscopy, NMR spectroscopy and thermal analysis before use[14]. The substrates were cleaned in conc. HCl, water, acetone and finally in boiling trichloroethylene before use. UHP argon (99.99% purity) was used as the purge gas and the carrier gas. UHP oxygen (99.99% purity) was used as the reactant gas. The ALD process parameters are summarized in table 1.

The resulting films were characterized for their crystallinity by X-ray diffraction. Their microstructure was studied by scanning electron microscopy (SEM) in a Cambridge Stereoscan S-360 electron microscope, and an FEI Instruments ESEM Quanta microscope. The thickness of the films was determined by cross-sectional SEM.

Ellipsometry measurements were carried out using a Sentech Instruments spectroscopic ellipsometer SE 850. The spectral range was 200-1700 nm (5-1.5 eV), and all measurements were made at an angle of incidence of 70° from the sample normal, with a spot size of 5 mm.

RESULTS AND DISCUSSION

The thin films were deposited using different flow rates of the reactant gas, *i.e.,* oxygen gas flow was varied from 0 to 150 sccm, keeping all other parameters constant. It was observed that the XRD patterns of the resulting films looked similar, except that the amorphous hump became less intense as the oxygen flow rate was increased. This can be interpreted as the result of a reduction in the amorphous carbon content of the film when oxygen flow is increased.

TABLE 1. Summary of deposition parameters

Deposition paremeters	Range
Substrate temperature	400^0C
Precursor vaporizer temperature	$100\text{-}150^0C$
Purge gas (Ar) flow rate	20 sccm
Carrier gas (Ar) flow rate	20 sccm
Reactant gas (O_2) flow rate	0-150 sccm
Reactor pressure	3 torr
Pulse time (in sec) and sequence (Precursor-purge-rective gas-purge)	2-1-2-1
No. of deposition cycles	200

Correspondingly, the as-deposited films were observed to be lighter in colour, when oxygen flow was higher. The XRD patterns of these films are shown in figure 1. It is to be noted that the crystallinity of the film, as indicated by the peak intensities, increases as the oxygen flow rate is increased. This would also cause a change in the other properties of the film, such as morphology, and especially optical and electrical properties, which may be expected to depend strongly on the crystalline / non-crystalline nature of the film.

A variation in film microstructure was also observed with a change in the flow rate of oxygem, the reactant gas. It is observed that at zero oxygen flow, the film is much rougher, than when oxygen is used. Increasing the oxygen flow favoured the formation of smoother films. The micrographs of these films are shown in the figures 2.a – d.

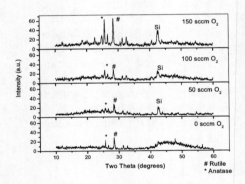

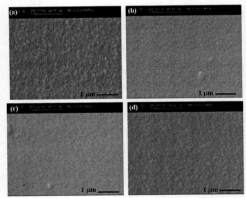

FIGURE 1. XRD patterns of films grown at different oxygen flow rates on Si(100)

FIGURE 2. Electron micrographs of films grown on Si(100) with (a) 0 sccm (b) 50 sccm (c) 100 sccm (d) 150 sccm

To interpret the ellipsometry data on these films, a three-layered optical model has been assumed, as shown in figure 3, considering the surface and cross-sectional SEM micrographs of the film. In this optical model, the film is assumed to have surface roughness and a film-substrate interface layer.

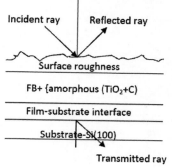

FIGURE 3. Assumed optical model

For the main film body, Bruggemann's effective medium approximation (EMA) has been used with Forouhi-Bloomer inclusions in the "host" made of amorphous (TiO_2+Carbon). Surface roughness is calculated by Maxwell-Garnet EMA, with voids as inclusions in a host of the main film body. Bruggemann's EMA has been used for the compositional analysis of film-substrate interface layer, with Cauchy SiO_2 as inclusion in a host of main film body. The variation (Table 2) in roughness, thickness and composition of the films using the proposed optical model indicates clearly an increase in the crystallinity and of voids (percentage porosity), and a reduction in both the carbon content amorphous TiO_2 content, as well as a reduction in surface roughness layer thickness, and the total film thickness. All of this is to be expected with the increasing flow rate of oxygen (for example, carbon deposition reduced due to increased oxygen flow). An anomaly is observed at 150 sccm of O_2, as seen in Table 2, wherein the fit to the elliposometric data leads to a high degree of amorphicity, contrasting with the improved crystallinity revealed by the XRD data (Fig.1). This may be due to the formation of a silicate-like layer at the film-substrate interface.

TABLE 2. Roughness, thickness and composition of the films using the proposed optical model

Flow rate of O_2 (sccm)	% of FB (nanocrystalline TiO_2) inclusions	% of amorphous (TiO_2+C)	% of voids in surface roughness	Surface roughness layer thickness (nm)	Film thickness (nm)	Film surface interfacial layer thickness (nm)	Total thickness (nm)
0	23.6	83.1	11.6	8.1	504.0	26.0	590.0
50	45.4	37.2	35.8	73.6	567.0	8.5	578.0
100	68.3	0.0	79.4	97.7	532.0	9.0	546.5
150	41.4	56.4	97.7	99.9	411.0	34.0	504.0

Figure 4-7. shows the spectral distribution of refractive index (n), attenuation constant (k), reflectance (R), transmittance (T), absorbance (A), real (ε_1) and imaginary (ε_2) of the complex dielectric funtion (ε), and optical conductivity (σ).

Table 3. shows the variation of direct $[(\alpha h v)^2]$ and indirect $[(\alpha h v)^{1/2}]$ energy band gaps of the oxide films with photon energy, at different flow rates of O_2. The absorption coefficient α has been calculated from the relation $\alpha = 4\pi k/\lambda$. The optical absorption edge was analyzed by the relation $\alpha h v = A(h v - E_g^{opt})^m$, where A is the

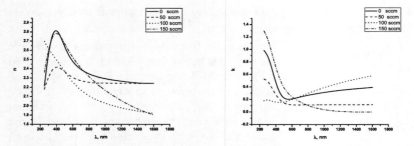

Figure 4. Variation of refractive index (n) and attenuation factor (k) with wavelength (λ)

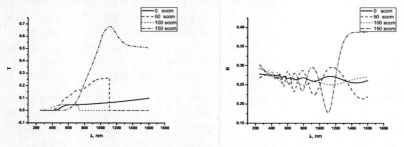

Figure 5. Variation of transmittance (T) and reflectivity (R) with wavelength (λ)

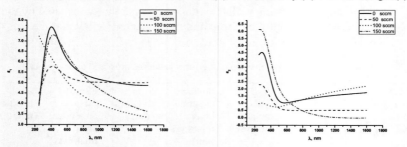

Figure 6. Variation of real (ε_1) and imaginary (ε_2) part of the complex dielectric function (ε) with wavelength (λ)

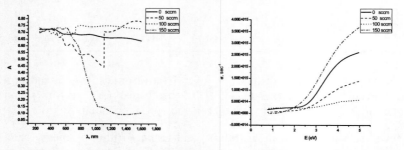

Figure 7. Variation of absorbance (A) and optical conductivity (σ) with wavelength (λ)

edge width parameter representing the film quality, E_g^{opt} is the optical energy gap of the material, and m determines the type of transition, with m = ½ for direct allowed transitions and m = 2 for indirect allowed transitions. E_g^{opt} is deterrmined by plotting $(\alpha h v)^m$ against hv and then extrapolating the intercept of the linear part to zero absorption.

Table 3. Variation of direct $(\alpha h v)^2$ and indirect $(\alpha h v)^{1/2}$ energy band gaps with photon energy

O_2 flow rate → E_g ▼	0 sccm	50 sccm	100 sccm	150 sccm
Direct$(\alpha h v)^2$	3.6 eV	2.9 eV	1 eV	3.8 eV
Indirect$(\alpha h v)^{1/2}$	1.59 eV	1.42 eV	0.39 eV	1.31 eV

The films show different behavior in UV-Vis and near IR range as their tranmittance and absorption characteristics differ with changing wavelength of incident electromagnetic radiation.

CONCLUSIONS

Carbonaceous and poorly crystallline films of TiO_2 were deposited using a novel metal-organic β-diketonate precursor, by the atomic layer deposition process. Anatase, rutile ,and amorphous TiO_2 were found to be present in the films which were grown at different O_2 flow rates. The optical and electronic properties of the films were studied using spectroscopic ellipsometry, using a new optical model. Qualitatively correct deduction about the composition, thickness and surface roughness of the films could be made using this model. A study of the optical constants reveals that the graphitic carbon present in the films greatly affects the optical properties of the films.

REFERENCES

1. H.A. Durand et al., *Appl. Surf. Sci.* **86**, 122, (1995).
2. C.W. Wang et al., *J. Appl. Phys.* **91**, 9198, (2002).
3. M. Keshmiri et al., *J. Non-Crystal. Solids* **3243**, 289, (2003).
4. T. Inoue et al., *Nature*, 277, 637, (1979).
5. H. Kim et al., *Appl. Phys. Lett.* **85**, 64, (2004).
6. H. Shin et al., *Adv. Mater.* **16**, 1197, (2004).
7. J.W. Lim et al., *Electrochem. And Solid-State Lett.* **7**, F73, (2004).
8. M.H. Chiu et al., *Appl. Opt.* **36**, 2936–2939, (1997).
9. Y. Zheng, K. Kikuchi, *Appl. Opt.* **36**, 6325–6328, (1997).
10. W.A. Ramadan, E. Fazio, M. Bertolotti, *Appl. Opt.* **35**, 6173–6178, (1996).
11. A.A. Hamza et al., *Opt. Commun.* **225**, 341–348, (2003).
12. T.E. Jenkins, *J. Phys. D: Appl. Phys.* **32**, R45–R56, (1999).
13. R. Bhakta et al., *Chem. Vapor Depos.* **9**, 295-298, (2003).
14. G.V. Kunte, S.A. Shivashankar and A.M. Umarji, *Meas. Sci. Technol.* **19**, 25704, (2008).

Electrical Percolation studies of Polystyrene-multi wall carbon nanotubes Composites

Ravi Bhatia, V. Prasad and Reghu M.

Department of Physics, Indian Institute of Science, Bangalore-560012, India

Abstract. Composites of Polystyrene-multi wall carbon nanotubes (PS-MWNTs) were prepared with loading up to 7 wt% of MWNTs by simple solvent mixing and drying technique. MWNTs with high aspect ratio ~ 4000 were used to make the polymer composites. A very high degree of dispersion of MWNTs was achieved by ultrasonication technique. As a result of high dispersion and high aspect ratio of the MWNTs electrical percolation was observed at rather low weight fraction ~ 0.0021. Characterization of the as prepared PS-MWNTs composites was done by Electron microscopy (EM), X-ray diffraction technique (XRD) and Thermogravimetery analysis (TGA).

Keywords: Polymer composite, carbon nanotube, aspect ratio.
PACS: 61.48.De, 81.05.Dk, 81.07.-b

INTRODUCTION

Polymer composites have attracted a great attention in past few decades [1-3]. The materials prepared by adding filler of nanometer dimensions in the polymer matrices have attracted many researchers across the world. Incorporating the fillers in small content to the polymer matrices results in composite materials having superior electrical and mechanical properties [4, 5]. Carbon black, metal particles, graphitic nanoparticles and carbon fibers were traditionally used as filler materials for polymer composites [1, 2, 6]. In recent years carbon nanotubes (CNTs) are thought to be most promising materials to be used as reinforcing material in the preparation of polymer composites. It is because of the unique structural, electrical, mechanical properties and extremely large length to diameter ratio (aspect ratio) as well as low density that make them excellent candidates for the production of conductive polymer composites. Both single- and multi-walled carbon nanotubes (SWNTs and MWNTs) are regarded as promising fillers, since they can provide electrical percolation at very low concentrations [10, 11]. To date, MWNTs rather than SWNTs have been predominantly used as conductive fillers due to their lower cost, better availability and easier dispersibility. Also MWNTs based polymer composites possess better mechanical properties than SWNTs based polymer composites [12]. Different processes such as solvent method, melt mixing, in situ chemical oxidative polymerization, latex technology and twin screw extrusion have been adopted for the preparation of nanocomposites [7-9, 13-15]. The host materials include a wide

CP1147, *Transport and Optical Properties of Nanomaterials—ICTOPON - 2009,* edited by M. R. Singh and R. H. Lipson
© 2009 American Institute of Physics 978-0-7354-0684-1/09/$25.00

spectrum of polymers ranging from conducting polymers such as polyaniline (PANI), polypyrrole (PPy), poly(p-phenylenevinylene) to insulating polymers such as polypropylene, polystyrene, polymethylmetha-acrylate (PMMA) and polyvinylalcohol (PVA), epoxy, polycarbonate, polyurethane, polyethylene, nylon 6 etc. [2-15]. The polymer composites can be used in many potential applications such as gas sensing, electromagnetic shielding and field emission displays etc.

The electrical conductivity of polymer composites is given by classical percolation theory [16]. According to the theory, the electrical conductivity of polymer composite depends on content of the conducting filler material. Electrical conductivity of the polymer composite increases abruptly as the filler content just crosses a particular value p_c, called percolation threshold when the conductive path is established with in the polymer matrix. It is called electrical percolation. The electrical percolation in the polymer/CNT composite depends on degree of dispersion, aspect ratio, waviness of CNTs and on thickness of composite film as well [17-19].

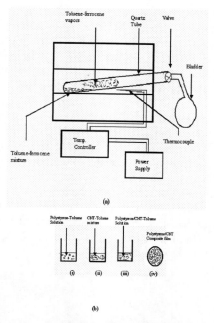

FIGURE 1. The two step preparation of PS-MWNTs composites. First step involves the synthesis of MWNTs and the set up employed for the synthesis is described in Fig. 1(a). Second step involves the dispersion of MWNTs into the polystyrene matrix by sonication method and casting into a PS-MWNTs composite film by solvent mixing and casting method which is systematically shown in Fig. 1(b).

EXPERIMENTAL DETAILS

Fig. 1(a) shows a simple set up for growing high quality MWNTs by thermally assisted chemical vapor deposition (CVD) method using ferrocene-toluene mixture. To prepare polymer composite, Polystyrene (PS) (supplied by Sigma Aldrich) was dissolved in toluene in one beaker and MWNTs were unbundled separately in another beaker by ultrasonication method and then mixed with solution of polystyrene (PS). Now this mixture of PS and MWNTs was again sonicated until the MWNTs were well dispersed into PS matrix. Fig. 1 (b) shows the step by step process employed for preparation of PS-MWNTs composites. Composites with different weight percent of MWNTs ranging from 0.1 to 7 wt% were prepared. A very high degree of dispersion of MWNTs into the PS matrix has been noticed by Scanning Electron Microscopy (SEM). XRD technique was used to confirm the graphitic nature of the MWNTs present in the PS matrix. The Keithley Model 2000 Multimeter, a high performance digital multimeter and a Keithley model 220 programmable current source were used for the electrical conductivity measurements of composites. A linear four probe method was used to measure electrical conductivity.

RESULTS AND DISCUSSION

Among different factors that affect the electrical and mechanical properties of polymer composites, aspect ratio of the filler material and degree of dispersion are most important. Therefore we produced MWNTs of high aspect ratio and better structural properties in order to prepare PS-MWNTs composites with enhanced properties. The approximate diameter and length of MWNTs is 50-80 nm and 100-300 μm respectively. Densely packed MWNTs with less amount of a-C can be seen in SEM micrograph as shown in fig 2. High degree of dispersion of MWNTs was achieved by ultrasonication method. An extreme level of care was taken while the sonication. Time to time unbundling of MWNTs was noticed. The adhesion and interaction between MWNTs and PS interface may be attributed to the toluene present in the solution. While sonication some of the toluene molecules get adsorbed onto the surface of the MWNTs, which facilitates the adhesion and interaction between MWNTs and long polymer molecules and hence MWNTs get embedded into the polymer matrix. Fig. 3 shows the high magnification SEM micrograph of one of the composite sample. The SEM studies confirm the high degree of homogeneous dispersion of fillers into the PS matrix. The homogeneous scattering of filler material into the polymer matrix is very much

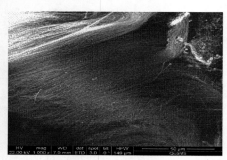

FIGURE 2. SEM micrograph of the MWNTs used in the composite preparation. The diameter and length of as prepared MWNTs lie in the range of 50-80nm and 200-300 μm respectively.

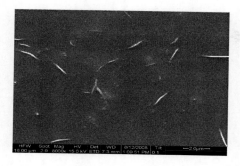

FIGURE 3. SEM micrograph of PS-MWNT composite of 7 wt%. It shows that the CNTs are very well dispersed. The white thread like structures represents MWNTs.

required due to a simple reason that all the physical properties which are to be enhanced depend solely on the reinforcing material. If there is an inhomogeneous dispersion or agglomeration of the fillers at some selected portions in the composite sample then the properties of that small portion only will be changed and not of the whole sample. For example the electrical conductivity of a polymer/MWNTs composite sample will be adversely affected by the agglomeration of MWNTs. Because the sample will be more conducting in a region where more number of MWNTs are present and making the conducting network easily than that region where fewer number of MWNTs are present. Thus a homogeneous distribution of the MWNTs is an important requirement to make conducting network all over the sample so that the property of whole sample is same every where. For further characterization transmission electron microscopy (TEM) was used. TEM samples were made by dissolving a small part of composite into tetra hydro furan (THF). Fig. 4 shows the TEM micrograph of one of the composite sample. PS molecules are seen grafted onto the surface of the MWNTs. TEM micrographs reveal the presence of Fe nanorods inside the MWNTs. The length of Fe nanorods vary from ~ 50-200 nm. X-ray diffraction patterns of the PS-MWNTs composites of 1 and 7 wt % are shown in fig. 5. XRD pattern shows a peak nearly at 26.56° for both the PS-MWNTs composite samples which correspond to MWNTs dispersed in the PS matrix. Also the Intensity of the MWNT peak is dominant for 7 wt% composite over 1 wt% composite which is quite obvious. A very weak peak is observed at 44°. It indicates the presence of Fe particles which are embedded in the MWNTs and it supports TEM analysis. Thermal stability of the composites was studied using the Thermogravimetery analysis (TGA). Fig. 6 shows the thermograph of composites of 3, 5 and 7 wt% of MWNTs. The composite samples were found to be thermally stable from room temperature to ~ 365 °C which lies in between the thermal stabilities of the host PS and MWNTs. Thus the inclusion of MWNTs enhances the thermal stability of the composites.

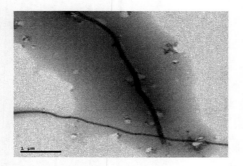

FIGURE 4. TEM micrograph of composite of 7 weight percent. The polymer grafting is clearly seen on to the surface of MWNT. The black spots inside the MWNTs are the nanosized iron particles.

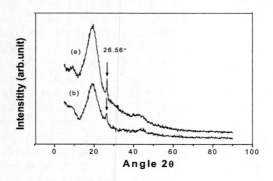

FIGURE 5. XRD patterns of PS-MWNTs composites of (a) 7 wt % and (b) 1 wt % of MWNTs. In both the peak at 26.56° is observed which corresponds to carbon nanotubes.

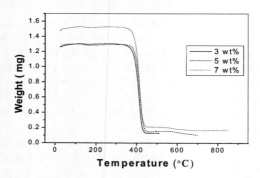

FIGURE 6. TGA curves for composites of 3, 5 and 7 wt %. The composites are thermally stable up to 365 ºC.

ELECTRICAL CONDUCTIVITY

The standard linear four probe method was used to measure the conductivity of the composites. The sample dimensions were 5mm×2mm. The Thickness of all the samples lies in the range 50-70 μm as measured by the digital micrometer. The electrodes were made of copper wire of thickness 50 μm and the contacts were made using silver paste. To study the percolative behavior the classical percolation equation is used which is given by

$$\sigma = \sigma_0 \, (p-p_c)^{\,t} \tag{1}$$

where σ is the actual conductivity of the composite sample, σ_0 is the intrinsic conductivity of the composite sample
p is the wt% of MWNTs in the composite, p_c is percolation threshold, t is the critical exponent. Fig. 7 shows the plot

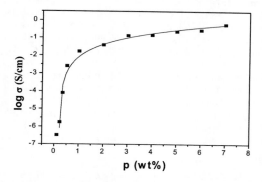

FIGURE 7. The plot between log σ and p (weight fraction). The dots represent the experimental points and the solid line is a fitted one. The percolation threshold (p_c) estimated from this plot is 0.21 wt %.

of log σ vs. p. The value of the percolation threshold p_c is estimated as 0.21±0.02) wt% by fitting method, which is quite low for a composite with MWNTs as fillers. The reason for the small value of percolation threshold is the high aspect ratio and homogeneous scattering of MWNTs.

CONCLUSIONS

In summary, we have prepared good quality polystyrene-multiwall carbon nanotubes (PS-MWNTs) composites by solvent mixing and casting method. Homogeneous dispersion of MWNTs in the PS matrix is observed with the help of SEM. XRD pattern shows presence of Fe particles inside the MWNTs which is been noticed by TEM as well. The composites are thermally stable up to an elevated temperature of ~ 365 °C. Percolation threshold is achieved for a low weight fraction of 0.0021 which is because of the high aspect ratio and homogeneous distribution of the MWNTs.

ACKNOWLEDGMENTS

We gratefully acknowledge the Institute of Nanoscience Initiative, IISc Bangalore for the SEM and TEM characterization.

REFERENCES

1. L. Nicodemo, L. Nicolais,G. Romeo and E. Scarfora, *Polym. Eng. Sci.* **18**, 293-298 (1978).
2. K. Miyasaka, K. Watanabe, E. Jojima, H. H. Aida, M. Sumita and K. Ishikawa, *J. Mater. Sci.* **17**, 1610-1616 (1982).
3. M. Reghu, C.O. Yoon, C.Y. Yang, D. Moses, P. Smith and A. J. Heeger, *Phys. Rev. B* **50**, 13931-13941 (1994).
4. D. Shi, J. Lian, P. He, F. Xiao, L. Yang, M. J. Schulz *et al.*, *Appl. Phys. Lett.* **83**, 5301-5303 (2003).
5. R. Ramasubermaniam, J. Chen and H Liu, *Appl. Phys. Lett.* **83**, 2928-2930 (2003).
6. F. Carmona. *Physica A* **157**, 461-469 (1989)
7. M. A. L. Manchado, L. Valentini, J. Biagotii and J. M. Kenny, *Carbon* **43**, 1499-1505 (2005).
8. Q. Zhang, S. Rastogi, D. Chen, D. Lippits and P. J. Lemskra, *Carbon* **44**, 778-85 (2006).
9. J. Xiong, Z. Zheng, X. Qin, M. Li, H. Li and X. Wang. *Carbon* **44**, 701-707 (2006).
10. T. E. Chang, A. Kisliuk, S. M. Rhodes, W. J. Brittain and A. P. Sokolov. *Polymer* **47**, 7740-746 (2006).
11. H. M. Kim, M. S. Choi and J. Joo, *Phys. Rev. B* **74**, 054202(7pp) (2006).
12. T. D. Fornes, J. W. Baur, Y. Sabba and E. L. Thomas, *Polymer* **47**, 1704-1714 (2006).
13. Y. Long and Z. Chen, *Appl. Phys. Lett.* **85**, 1796-1798 (2004).
14. Yu. K. Lu, E. Sourty, N. Grossiord, C. E. Koning and J. Loos. *Carbon* **45**, 2897-2903 (2007).
15. A. I. Isayev, R. Kumar and T. M. Lewis, *Polymer* **50**, 250-260 (2009).
16. A. Aharony and D. Stauffer, *Introduction to Percolation Theory* 2nd ed. (Taylor and Francis, London, 1993).
17. F. Du, J. E. Fischer and K. I. Winey, *Phys. Rev. B* **72**,121404(4pp)
18. Q. Wang, J. Dai, W. Li, Z. Wei and J. Jiang, *Compos. Sci. Technol.* **68**, 1644-1648 (2008).
19. C. Li, E. T. Thostenson and T. W. Chou, *Compos. Sci. Technol.* **68**, 1445-1452 (2008).

Phase dependence of secondary electron emission at the Cs-Sb-Si (111) interface

Govind[a], Praveen Kumar[a] and S.M. Shivaprasad[a,b]

[a]Surface Physics and Nanostructures Group, National Physical Laboratory, New Delhi -110012
[b]Jawaharlal Nehru Centre for Advanced Scientific Research, Bangalore- 560064

Abstract. The multi-alkali antimonides adsorption on Si (111) surface has drawn much attention of several surface science studies due to its importance in both, fundamental and technological aspects of night vision devices & photocathodes. We report the formation of alkali metal antimonide ternary interface on Si(111)-7x7 surface and in-situ characterization by X-ray Photoelectron Spectroscopy (XPS). The results show that Cs adsorption on clean Si(111) surface follows the layer-by-layer (Frank van der Merwe) growth mode at low flux rate, while Sb grows as islands (Volmer-Weber) on Cs/Si surface. The changes in the Si (2p) and Cs (3d) core level spectra show the formation of a ternary interface (Sb/Cs/Si) at room temperature, which is further confirmed by changes in the density of states in the valence band spectra. The temperature controlled desorption of ternary interface, by monitoring the chemical species remnant on the surface after annealing at different temperatures, reveal that the Sb islands desorb at < 550°C while Cs monolayer desorbs at temperatures >750°C, which implies a stronger Cs-Si bond to Cs-Sb bond. The work function changes from 3.9 eV to 0.8 eV for Cs adsorption on Si, which further reduces to 0.65eV after Sb adsorption on the Cs/Si interface. The changes in work function corresponds to the compositional and chemical nature of the interface and thus indicate that the secondary electron emission is an extremely phase dependent phenomena.

Keywords: X-ray Photoelectron Spectroscopy (XPS), Cesium antimonide, Si(111)
PACS: 68.49.Uv, 68.47.Fg

INTRODUCTION

Alkali metal (AM) antimonide compounds are an interesting material with different stoichiometry (MSb and M_3Sb, with M an alkali metal), which have highest known photo sensitivity in visible light; only complex devices, such as negative electron affinity III-V photo cathodes and low work function night vision devices can improve on their performance [1-3]. The good optical absorption in the visible region of the light spectrum and the low electron escape energy make the alkali antimonide compounds very suitable as photoemitters, [4, 5] therefore, this class of compounds has been studied extensively for many years. To understand the formation of alkali metal antimonide on Si surface the role of AM/Si interface plays an important role. In the recent past alkali metals adsorption on Si surfaces has received considerable attention, since they possess simple hydrogen-like electronic structures and are regarded as prototype systems [6]. A number of studies have been performed to understand AM/semiconductor interfaces, electronic property and the nature of the chemical bonds. The effects of charge transfer from AM to the Si substrate and the surface metallization have been well understood based on photoelectron emission

CP1147, *Transport and Optical Properties of Nanomaterials—ICTOPON - 2009*, edited by M. R. Singh and R. H. Lipson
© 2009 American Institute of Physics 978-0-7354-0684-1/09/$25.00

spectroscopy (PES) and work function (WF) studies [7]. However, surprisingly, the basic adsorption geometry of AM on Si surfaces, particularly at the sub-monolayer (ML) coverage, has not been well understood. In view of this, we have revisited adsorption and desorption kinetics of Cs and Sb on Si (111)-7x 7 surfaces in the sub monolayer regime and studied the formation of Sb/Cs/Si ternary interface *in-situ* in a UHV chamber which consist of X-ray photoelectron spectroscopy (XPS). XPS scans were used to monitor the elemental composition and chemical shifts due to cesium antimonide formation and the spectra of the valence band region was examined to elucidate the changes in electronic structure. The results pertaining to the core-level spectra with secondary electron emission and work function changes are presented and discussed in the following sections.

EXPERIMENTAL

The experiments were performed in a Perkin Elmer ultra high vacuum XPS chamber (Model 1257) at a base pressure of 5×10^{-10} torr. The chamber is equipped with a dual anode Mg-K_α and Al-K_α X-ray source and a high-resolution hemispherical energy analyzer for energy resolved electron detection. The chamber also includes an electron beam heating facility to anneal the sample upto 1200°C. The sample is mounted on a high precision sample manipulator that enables its positioning for growth and analysis. A 20x5 mm^2 piece is cut from a p-type boron doped Si (111) wafer having a resistivity of 10-15Ω-cm. The sample was cleaned by modified Shiraki process [8] before inserting in the vacuum chamber. Sample was mounted on a 4-axis manipulator which clamped home made Ta sample holder. After inserting the sample in the vacuum chamber it is degassed to 600°C for 12 hours followed by repeated flashing up to 1100°C for 5 sec by electron beam heating followed by cooling to RT at a very slow rate of 2°C/sec [9]. The sample temperature was monitored with the help of a pyrometer. The atomic cleanliness of a sample was ascertained by the absence of carbon and other contamination on the surface by XPS survey scan. Cs deposition was made by the home made tantalum assembly which includes cesium cromate dispenser (Seas Getters), while Sb was deposited from a home made tantalum-Knudsen cell. The flux rate of Cs and Sb adsorption was controlled by regulating the current to the cell and measured in terms of adsorbed monolayer. For optimal resolution survey scans and core level spectra are obtained at 100eV and 40eV pass energy respectively.

RESULTS

Cesium is adsorbed on clean Si (111)-7x7 surface at room temperature (RT) and the evolution of the Cs/Si interfaces was monitored by XPS. Fig. 1 shows the uptake curve for the adsorption of Cs on Si(111)-7x7 surface at RT, which plots the ratio of Cs(3d) to Si(2p) peak intensity with deposition time. The curve shows a linear increase in the ratio of Cs (3d) to Si (2p) peak intensity upto a deposition time of 9 min. On further deposition the slope of Cs/Si intensity ratio increases, which is due to attenuation of the Si (2p) XPS signal by the second layer Cs adatoms, suggests a layer-by-layer (Frank van-der Merve) growth mode. The break in the slope of Cs/Si ratio

($R_{Cs/Si}$) after 9 min of deposition time ($R_{Cs/Si}$ = 0.25), suggests the formation of 1 ML of Cs and also indicates that the adsorption rate is 0.11 ML/min [10]. The inset of figure 1 shows the change in the Cs (3d) core level peak position with Cs coverage on Si (111) surface. For the Cs/Si (111) system, initially a shift of 0.2eV is observed in Cs (3d) peak position upto coverage of 1ML, which suddenly increases to 0.4eV for next Cs deposition. On further deposition upto 2ML it increases monotonically upto 0.96eV. The ML shift in the core level peak positions during the formation of metal-semiconductor (Cs/ Si) interfaces can be attributed either to band bending at the interface or chemical interaction between the adsorbate and the substrate.

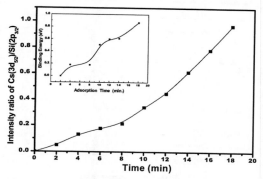

FIGURE 1. XPS uptake curve showing Frank-van der Merwe growth mode, with change in slope for the XPS intensity ratio of Cs ($3d_{5/2}$) to Si (2p) at a deposition time of about 9 min corresponding to 1ML Cs coverage on Si (111) surface.

The thermal stability of this layer-by-layer grown system is shown in Figure 2, which plots the XPS intensity ratio of Cs (3d) to Si (2p), as the RT adsorbed Cs/Si (111) system annealed to increasing temperatures. It is evident from the figure that as the substrate temperature increases from RT to 200°C the Cs/Si ratio falls sharply up to a value of 0.25, corresponding to coverage of about 1ML. The fall in the Cs/Si intensity ratio is attributed to the Cs multilayer desorption from Si surface and the formation of one monolayer (1ML). The Cs/Si ratio remains constant for the temperature range of 200-650°C, which reveals the stability of Cs monolayer after multilayer desorption from Si (111) surface. On further increase in the temperature from the 650°C, the $R_{Cs/Si}$ decreases, which shows the desorption of Cs monolayer and for temperature of

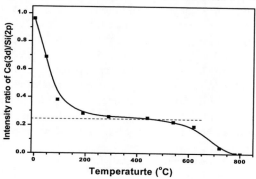

FIGURE 2. Desorption Curve of Cs adsorbed Si (111) surface plots the Cs (3d)/Si (2p) XPS intensity ratio as a function of temperature. The dashed straight line at the ratio of 0.25 corresponds to 1.0-ML Cs coverage.

~780°C the ratio becomes zero, which shows the complete Cs desorption from the Si (111) surface.

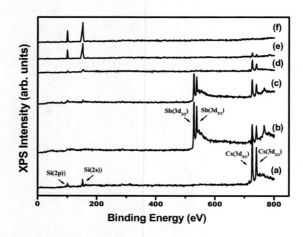

FIGURE 3. Part of the XPS survey scan (a) for the 2ML Cs adsorbed Si(111) surface, (b) after 4 min of Sb adsorption on Cs/Si interface and curve (c), (d), (e) and (f) incorporate the desorption of Sb and Cs from the Si(111) surface annealed to 350°C, 550°C, 750°C and 800°C respectively.

Figure 3 shows a part of the XPS survey scan (450eV-850eV), where peak corresponds to the Cs (3d), Sb (3d), Si (2p) and Si(2s) are incorporated. Curve (a) shows the spectra of Cs adsorbed (~2ML) Si (111) surface which comprise of Si (2p) at 101eV, Si (2s) at 151eV, Cs ($3d_{5/2}$) at 726eV and Cs ($3d_{3/2}$) at 740eV. Curve (b) shows the survey scan after 4 minute Sb adsorption on Cs/Si surface, which inclueds the Sb ($3d_{5/2}$) at 529eV and Sb ($3d_{3/2}$) at 538eV. The the deconvolution of core level of Cs (3d) and Si (2p) and the binding energy shift show the formation of ternary compound [10]. To determine the thermal stability of Sb/Cs/Si ternary interface system the sample was annealed from RT to 800K. Curve (c), (d), (e) and (f) incorporate the desorption of Sb and Cs from the Si(111) surface annealed to 350 °C, 550 °C, 750 °C and 800 °C respectively. From curve (c), it is evident that Sb reduces significantly due to annealing of the surface up to 350°C and it is observed that Sb completely desorb from the Cs/Si interface below 550°C (Curve d). As the temperature increses to 750°C, Cs reduces to aproximatly 0.2ML and for 800°C it desorb completely from the Si surface, which is also supported by figure 2. During the desorption process of Sb/Cs/Si ternary interface, it is observed that Sb desorbed at much loewer temperture than Cs which may be attributed to the weak bonding of Cs-Sb than the Cs-Si. The weak bonding of Cs-Sb lead to the Sb islands growth (Volmer

Waber) on Cs/Si interface. However our recent studies of Sb adsorption on various Si surfaces shows the layer-by-layer growth mode and a complete desorption of Sb from the Si surface above 850°C [9, 11].

In order to measure secondary electron emmision (SEE), take off edge E_0 is calculated which represent the direct measurement of work function. A negative bias of 15V was applied to separate the spectrometer work fuction to that of sample. For the clean Si(111) surface SEE onset comes at 1241.0 eV and the work function mesured is 3.9 eV. The change in SEE emmision is ploted in figure 4, by calculating area under the SEE curve, after appling the bias. The curve shows as Cs coverage on the Si surface increases, SEE increses linearly. After 2ML Cs deposition on Si(111) surface, Sb was adsorbed on Cs/ Si interface, which reprented in the curve by last two data points (20 mins and 24 mins), signify the SEE from the ternary interface. The change in SEE during the Cs desorption process from the Si surface is also presented as an inset in figure 4. From the SEE spectra the change in the work function has also been calculated during the Cs and Sb adsorption/desorption on Si(111) and Cs/Si interface respectively. As the Cs coverage increases on Si surface the work function changes from 3.9 eV to 0.8 eV, which further reduces to 0.65eV after Sb adsorption on the Cs/Si interface. The changes in work function corresponds to the compositional and chemical nature of the interface and thus indicate that the secondary electron emission is an extremely phase dependent phenomena.

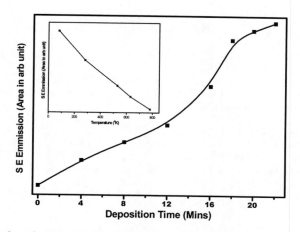

FIGURE 4. Secondary electron emission spectra during Cs and Sb adsorption on Si (111). Inset shows the SEE during Sb/Cs/Si system annealed from RT to higher temperature.

CONCLUSIONS

In summary, the adsorption and desorption of Cs and cesium antimonide formation on Si (111)-7x7 surface has been studied by X-ray photoelectron spectroscopy. The results show the layer-by-layer growth of Cs on Si (111) surface, while Sb grows as islands on Cs/Si interface. The XPS analysis shows the formation of ternary interface after adsorbing Sb on Cs/Si interface. The SEE and work function studies emphasis the conversion of photons into detectable electrons occurs by the excitation of a valence-band electron to a conduction-band state after the absorption of a photon and the subsequent emission of a photoelectron. Results show the increase in SEE and decrease in the work function with the increase in Cs coverage, which further continue for the ternary interface formation. The results also reveal that Sb desorbed at much lower temperature from the Cs/Si interface than from the Si surface which suggest a weak Sb-Cs bonding than that of Sb-Si. In conclusion, we mention that the formation of ternary interface and reduction in work function has promising use in photoemitter devices. The formation of other multialkali antimonide (M_1M_2Sb) interfaces is underway.

ACKNOWLEDGMENTS

One of the authors (Govind) is grateful to the Department of Science and technology Delhi for providing financial assistance under SERC-FTYS scheme and (P Kumar) for University Grant Commission, India for Fellowship. The authors are also thankful to the Director National Physical Laboratory, for his keen interest and constant encouragement.

REFERENCES

1. A.H. Sommer, *Photoemissive Materials* (Wiley, New York, 1968).
2. G. Gosh, *Phys. Thin Films* **12**, 53 (1982).
3. L. Galan, E. Elizalde and E. Martinez, *Phys. Rev. B* **37**, 4225 (1988).
4. R. Winter, *Thermodynamics of Alloy Formation*, edited by Y. Chang and F. Sommer (Minerals, Metals and Materials Socity London 1997).
5. A.H. Sommer, *J. Appl. Phys.* **29**, 1568 (1958).
6. K.Wu, *Sci. & Tech. of Adv. Mat.* **6**, 789 (2005).
7. K. D. Lee, J. R. Ahn and J. W. Chung, *Appl. Phys. A.* **68**, 115 (1999).
8. Y. Anta, S. Suzuki, S. Kono and T. Sakamoto, *Phys. Rev. B*. **39**, 56 (1989).
9. V. K. Paliwal, A. G. Vedeshwar and S. M. Shivaprasad, *Phys. Rev. B*. **66**, 245404 (2002).
10. Govind, P. Kumar and S. M. Shivaprasad, *Communicated*.
11. M. Kumar, Govind, V.K. Paliwal and S. M. Shivaprasad, *Surf. Sci.* **600**, 2745 (2006).

Exciton Plasmon Coupling in Hybrid Semiconductor - Metal Nanoparticle Monolayers

L. N. Tripathi, M. Haridas and J. K. Basu

Department of Physics, Indian Institute of Science, Bangalore, India
email: basu@physics.iisc.ernet.in

Abstract. Hybrid s e m i c o n d u c t o r - metal nanoparticles monolayer of Cadmium Selenide and gold nanoparticles has been prepared, using Langmuir - Blodgett technique. The near field photoluminescence spectra from such monolayer f i l m s , shows red shift ~75 meV with respect to CdSe QDs monolayer film and splitting ~57 meV. The composite spectra are much broader ~330 meV compared to the corresponding emission spectra of CdSe monolayer ~ 165 meV. The possible explanation for the observed features are provided in terms of exciton – Plasmon interaction .

Keywords: Excitons, Plasmons, Near-field s c a n n i n g m i c r o s c o p y
PACS: 71.35.Gg, 78.67.Hc, 68.37.Uv

INTRODUCTION

Hybrid semiconductor quantum dots (SQDs) and metal nanoparticles (NP) [1, 2, 3, 4] are interesting system for the study of optical properties. Such a superstructure takes advantage of the properties of its component material systems. These hybrid materials are useful for sensor and light harvesting properties. In such a system consisting of an emitter l i k e SQDs and metal NPs, excitons and plasmons interact via coulomb forces. Emission of semiconducting nanoparticles in the presence of metal NPs can be suppressed or enhanced depending upon the size and organization of the nano assembly [5, 6, 7, 8]. This originates due to several reasons: (a) Reduction of exciton life time due to enhanced radiation rates and energy transfer to the metal nanoparticles (b) The amplified absorption in the presence of plasmon resonance in the metal NPs and (c) Modified density of states of photons. Exciton Plasmon interaction between an emitting dipole and metal nanoparticles leads to energy transfer, leading to a shift of energy of quantum emitter [9, 10, 11] and increase in broadening of quantum well exciton resonances[12].

There are many methods to organize the quantum size particles onto the solid substrates, either by the direct synthesis of nanocrystals on the solid surface [13, 14, 15] or by arranging prepared nanoparticles through self assembly [16, 17, 18]. Although the self assembly technique have advantage being versatile and fast, Langmuir Blodgett (LB) deposition offers the possibility of making the film compact, ordered and controllable structures [19, 20]. This technique allows us to form the

CP1147, *Transport and Optical Properties of Nanomaterials—ICTOPON - 2009,* edited by M. R. Singh and R. H. Lipson
© 2009 American Institute of Physics 978-0-7354-0684-1/09/$25.00

hybrid nanostructure where we can control the inter particle spacing, and therefore, control the surface density of nanoparticles array. Here we present the near field photoluminescence spectra (PL) from hybrid (gold and Cadmium Selenide (CdSe)) nanoparticle film, prepared using LB technique. The transferred monolayer film was imaged using Atomic Force Microscope (AFM) and near field scanning optical microscope (SNOM). The near field PL spectra collected from hybrid monolayer film shows striking differences compared to the SQD (CdSe) monolayer film. The possible explanations for the observed phenomena are provided in result and discussion section.

SAMPLE PREPARATION AND CHARACTERIZATION

Synthesis of PMMA capped Gold nanoparticles

Gold nanoparticles were synthesized by reducing Gold(III)chloride trihydrate, in the presence of Poly methyl methacrylate (PMMA) (Molecular weight-15k). All the chemicals were purchased from Sigma-Aldrich and for gold nanoparticle synthesis, deionized water ((18.2MΩ/m) Barnsted) was used. Gold(III)chloride trihydrate was dissolved in a solution containing ethanol and water, 5 : 1 by volume which also contained the dissolved PMMA. An aqueous solution of Sodium borohydride (NaBH4) (0.5mM) was freshly prepared and rapidly added to the above solution. The color of the solution changes to reddish brown indicating the formation of gold nanoparticles The nanoparticles powder was extracted from the solution by controlled evaporation, as discussed earlier [21, 22, 23].

Synthesis of TOPO capped CdSe Quantum Dots

CdSe quantum dots (QDs) were synthesized using the method developed by Peng et al [24]. Cadmium oxide (CdO) were dissolved in Trioctylphosphine Oxide (TOPO) on being heated at 320 ^{0}C. After reducing the temperature to 270 ^{0}C, Selenium solution in Trioctylphosphine (TOP) was injected rapidly to this solution. Color of the solution changed to reddish indicating the quantum dots formation. The solution was kept at 270 ^{0}C for one hour. The nanoparticles powder was extracted from the reaction mixture by adding methanol after cooling down to room temperature. The reddish powder settled at the bottom of the vessel was collected.

Langmuir- Blodgett film preparation and atomic force microscopy

The hybrid NP film and CdSe QDs film were prepared using LB (KSV instruments) technique. 20 µl of 1mg/ml solution containing gold nanoparticles and CdSe QDs with number ratio 1:1 was spread on the LB trough. In order to make a homogeneous film, the monolayer was compressed and decompressed slowly (10mm/min), and isotherm was recorded, monitoring the variation of surface pressure with area. The monolayer was transferred on to a glass substrate by Langmuir Schaefer method at surface pressure 17 mN/m. Before transferring the film, the glass substrates were cleaned using standard RCA cleaning procedure, which makes the film hydrophilic. CdSe QDs films were also prepared using the same method transferred at a surface pressure 15 mN/m. The films were subsequently imaged using AFM (Veeco) in contact mode.

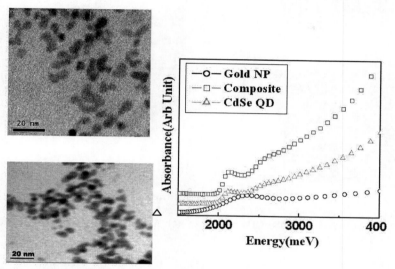

FIGURE 1. TEM micrograph of (a) CdSe QDs and (b) gold NPs, (c) UV- visible absorption spectra of gold NP (O), CdSe QD (Δ) and hybrid NP (□) solution dispersed in chloroform.

Scanning Near-field Optical Measurements.

The near field PL measurements on our samples were performed with (WITec alpha SNOM), a setup, using the blue line (488 nm) of an Argon ion laser, in transmission mode, at room temperature. The instrument uses an aperture (~ 100nm) cantilever, kept close to the sample and laser light passing through the aperture excites the sample. The emitted light is collected using high efficiency objective (Nikon, NA- 1.25, magnification 100X), and finally guided to Photo multiplier tube (PMT) or spectrograph.

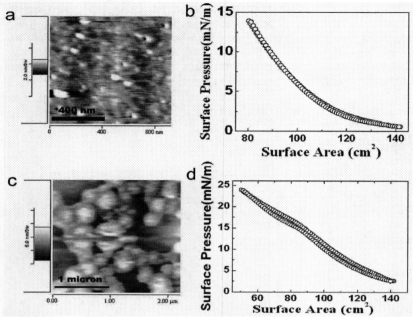

FIGURE 2. (a) AFM topography image of CdSe QDs monolayer transferred to glass substrate, (b) Pressure - area isotherm of the CdSe QDs monolayer, (C) AFM topography image of hybrid NP monolayer transferred to glass substrate and (d) Pressure - area isotherm of the hybrid NP monolayer.

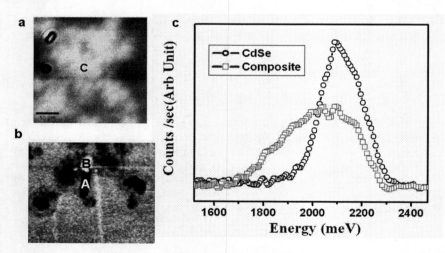

FIGURE 3. SNOM Light intensity image from (a) CdSe QDs monolayer transferred to glass substrate, (b) hybrid NP monolayer, (c) Near field PL spectra collected from of hybrid NP film (□) and CdSe film (○) from regions B and C as indicated in (a) and (b) above.

418

RESULTS AND DISCUSSION

Fairly mono dispersed gold NPs and CdSe QDs of diameters 6 nm and 5.5 nm respectively were prepared by above discussed methods. The UV visible absorption spectra from the gold NP, CdSe QD and gold - CdSe hybrid NP solutions in chloroform were measured and shown in FIGURE 1. The absorption spectra from hybrid NP solution do not show much difference compared to the spectra from CdSe QD solution.

To see the effect of metal nanoparticles on PL spectra, we transferred mono-layers of CdSe QD and hybrid NP. The isotherms recorded from CdSe QD and hybrid NP monolayer are shown in FIGURE 2b and FIGURE 2d. Almost no hysteresis in the isotherms (FIGURE 2b and FIGURE 2d), indicate that nano particles are well ordered on the water surface. Respective monolayers were transferred at suitable surface pressures (CdSe QD - 15mN/m and Hybrid NP - 17mN/m) on to glass substrates. The transferred films were imaged using AFM, as shown in FIGURE 2c.

AFM image gives the information about the overall arrangement of the nano particles, although distinction between gold NP and CdSe QD cannot be made there. SNOM light intensity image provides the positional information for the distribution of CdSe QDs and gold NP. Since the excitation wavelength of 488nm, which is nearer to surface plasmon resonance wavelength of gold NP (520nm), absorbs the radiation and leads to dark spots in the light intensity image. The CdSe QD can be identified with the bright centers corresponding to emission PL spectra. The region marked A in FIGURE 3b represents the gold nanoparticles and region B represents the CdSe QDs.

The near field PL spectra collected from region B shows splitting ~ 57 meV, with a large red shift (75 meV) compared to spectra from CdSe QD film, which is obvious in FIGURE 3, The splitting of the PL spectra from hybrid nanostructure is the signature of strong exciton plasmon coupling [25]. The overall spectra seems to be broadened (~ 330 meV) compared to the CdSe QD (~165 meV) film, indicating decrease in exciton's lifetime [12,1] as a result of energy transfer between gold NP and CdSe QD. The red shift in the PL arises because of the attractive interaction between gold NP and CdSe QD [25]. However some groups have seen a blue shift in the hybrid Cadmium Telluride nanowire and gold nano particle [11], which shows that the hybrid NP system is not a completely understood system and more work needs to be done to unravel the mystery.

CONCLUSIONS

We have prepared a hybrid nanoparticle monolayer using LB technique and studied the PL spectra from it using SNOM. Our result shows a significant red shift, splitting and broadening in emission spectra from the hybrid nanoparticle mono layer as compared to emission spectra from CdSe QDs mono layer, which reveals the interaction between exciton and plasmon in such hybrid system. Further work is being done to study the dependence on various factors like nanoparticles shape, size and inter particle distance on the extent of coupling between QDs and metal NPs, in such hybrid quantum systems.

ACKNOWLEDGMENTS

We would like to acknowledge DST India, for providing financial assistance and Institute Nanoscience Initiative IISc for the facilities for this work. M. Haridas acknowledges UGC for financial support.

REFERENCES

1. A. O. Govorov, J. Lee and N. A. Kotov, *Phys. Rev. B* **76**, 125308 (1-16) (2007).
2. G. Heliotis, G. Itskos, R. Murray, M. D. Dawson, I. M. Watson and D. D. C. Bradley, *Adv. Mater.* **18**, 334-338 (2006).
3. Y. Fedutik, V. Temnov, U. Woggon, E. Ustinovich, and, M. Artemyev, *J. Am. Chem. Soc.* **129**, 14939-14945 (2007).
4. J. Lee, A. O. Govorov, J. Dulka, and N. A. Kotov, *Nano Lett.* **4**, 2323-2330 (2004).
5. C. D. Geddes and J. R. Lakowich, *J. Fluorescence* **12** 212–213 (2002).
6. P. Anger, P. Bhardwaj and L. Novotny, *Phys. Rev. Lett* **96**, 113002 (1-6) (2006).
7. A.O. Govorov, G. W. Bryant, W. Zhang, T. Skeini, J. Lee, N. A. Kotov, J. M. Slocik and R. R. Naik, *Nano Lett.* **6**, 984-994 (2006).
8. T. Pons, I. L. Medintz, K. E. Sapsford, S. Higashiya, A. F. Grimes, Doug S. English, and H. Mattoussi, *Nano Lett.* **7**, 3157-3164 (2007).
9. B. N. J. Persson and N. D. Lang, *Phys. Rev. B* **26**, 5409 (1982).
10. H. T. Dung, L. Knoll and D. G. Welsch, P h y s. *Rev. A* **62**, 053804 (2000).
11. J. Lee, P. Hernandez, J. Lee, A. O. Govorov, and N. A. Kotov, *Nature Mater.* **6**, 291 – 295 (2007).
12. P. Vasa, R. Pomraenke, S. Schwieger, Y. I. Mazur, V. Kunets, P. Srinivasan, E. Johnson, J. E. Kihm, D. S. Kim, E. Runge, G. Salamo and C. Lienau, *Phys. Rev. Lett.* **101**, 116801 (2008).
13. Y. Mastai and G. Hodes, *J. Phys.Chem. B* **101**, 2685-2690 (1997).
14. M. A. Anderson, S. Gorner and R. M. Penner, *J. Phys. Chem. B* **101**, 5895-5899 (1997).
15. G. S. Hsiao, M. G. Anderson, S. Gorner, D. Harris and R. M. Penner, *J. Am. Chem. Soc.* **119**, 1439-1448 (1997).
16. H. Miyake, H. Matsumoto, M. Nishizava, T. Sakata, H. Mori, S. Kuwabata and H. Yoneyama, *Langmuir* **13**, 742-746 (1999).
17. T. Nakanishi, B. Ohtani and K. Uosaki, *J. Phys . Chem. B* **102**, 1571 (1998).
18. L. Sheeney -Haj-Ichia, J. Wasserman and I. Wilner, *Adv. Mater.* **14**,1323-1326 (2002).
19. M. Kerim, A. Gattás, C.A. Constantine, M. J. Lynn, D .A. Thimann, X. Ji and R. M. Leblanc, *J. Am. Chem. Soc.* **127**, 14640 – 14646 (2005).
20. Y. J. Shen, Y. L. Lee and Y. M. Yang. *J. Phys. Chem. B* **110**, 9556-9564 (2006).
21. S. Srivastava and J. K. Basu, *Phys. Rev. Lett.* **98**,165701 (2007).
22. S. Srivastava and J. K. Basu, *J. Nanoscience Nanotechnology*, **7**, 2101-2104 (2007).
23. M. Haridas, S. Srivastava and J. K. Basu, *Eur. Phys. J. D* **49**, 93-100 (2008).
24. Z. A. Peng and X. Peng, *J. Am. Chem. Soc.* **123**, 183-184 (2001).
25. A. Trügler and U. Hohenester, *Phys. Rev. B* **77**, 115403 (2008).

Hydrosilylation of 1-dodecene on Nanostructured Porous Silicon Surface: Role of Current Density and Stabilizing Agent

Shalini Singh[a,b] Shailesh N. Sharma[a], Govind[a], Mukhtar A. Khan[b] and P.K. Singh[a]

[a]Electronic Materials Division, National Physical Laboratory,
Dr. K.S. Krishnan Marg, New Delhi-110012, India.
[b]Faculty of Life Science, Aligarh Muslim University, Aligarh -202001, India

Abstract. We report the formation of nanostructured PS on boron doped p-type silicon wafer (100) by electrochemical anodization using aqueous hydrofluoric acid and isopropyl alcohol solution at different current densities 20 mAcm^{-2} and 50 mAcm^{-2} with pore size being 20-30 nm and 50-60 nm, respectively. For organic functionalization of PS surface 1-Dodecene treatment under UV light was done. PL, FTIR and XPS studies have been carried out to characterize the PS after surface modification using dodecene. Stability studies were performed under normal ambience and humid condition for as-anodized (Fresh PS) and dodecene-treated samples. It is observed that the dodecene functionalized samples were more stable than as-anodized porous silicon. The present study demonstrates that nanoporous silicon can provide chemically modified stable and high surface area for the sensing applications of PS.

Keywords: Dodecene, Hydrosilylation, FTIR, XPS
PACS: 73.63.Nm, 78.20.-e, 78.55.Mb, 78.66.-w

INTRODUCTION

The reliability of a biosensor strongly depends on the functionalization process depending on its fastness, simplicity, homogeneity and its repeatability [1-3]. Stabilization of porous silicon (PS) is a key requirement for many of its potential sensing applications. In order to prevent the deterioration of the structural and optical properties of PS, proper chemical and electronic termination of the PS surface is necessary, thereby limiting the presence of midgap states that act as non-radiative recombination centers [5]. Although methods like thermal nitridation, halogenation, wet and chemical oxidation and metal incorporation into the pores have been attempted in the past but even then the complexity of the PS surface and its interaction with various surface-adsorbates has not been fully understood [5,6]. Recently, it has been shown that organic-modified silicon surfaces of single crystal Si (111) through silicon-carbon (Si-C) and silicon-oxygen-carbon (Si-O-C) are remarkably stable in different organic and aqueous solutions [8,9]. The morphology of PS particularly the porosity and the pore size depends on the anodization parameters such as HF concentration, current density, wafer type and resistivity [1]. However, current density plays an important role in the morphology of PS which helps in the classification of bionert, bioactive and biodegradable materials attached to PS. Hydrosilylation of PS

CP1147, Transport and Optical Properties of Nanomaterials—ICTOPON - 2009, edited by M. R. Singh and R. H. Lipson
© 2009 American Institute of Physics 978-0-7354-0684-1/09/$25.00

retains the photoluminescence activity to an extent, enabling a wider range of applications.

EXPERIMENTAL

Boron doped p-type (100) monocrystalline silicon wafers were used for PS formation. A thin aluminum layer was evaporated on the back of the wafers and sintered to form an ohmic contact. PS was formed by the electrochemical anodization process using Si as the anode and Pt as the counter electrode in an acid-resistant Teflon container. The anodization was carried out at different current densities $I_d \sim 20$ and 50 mA cm^{-2} for 30 min, in a solution of H_2O: HF:IPA (1:1:2). After etching, the samples were rinsed with deionized (DI) water (18MΩ) and dried under a stream of dry high-purity nitrogen prior to use.

Surface modification studies were done on fresh PS, immediately after formation using 1-dodecene under UV light for about 30 min. at room temperature. Finally, the substrates were thoroughly rinsed with DI water and dried with nitrogen.

The PL was measured using a home assembled system consisting of a two stage monochromator, a photomultiplier tube (PMT) with a lock-in amplifier for PL detection, and an Ar+ ion laser operating at 488 nm and 5 mW (corresponding to 0.125 W cm-2) power for excitation. Fourier Transform Infrared (FTIR) spectra were recorded with a Perkin Elmer Model (Spectrum BX) spectrophotometer. X-ray photoelectron spectroscopy (XPS) measurements were performed in an ultra-high vacuum chamber (PHI 1257) with a base pressure of $\sim 4 \times 10^{-10}$ torr. The XPS spectrometer is equipped with a high resolution hemispherical electron analyzer (279.4 mm diameter with 25 meV resolution) and a Al (K_α) (hv = 1486.6 eV) X-ray excitation source.

RESULTS AND DISCUSSION

Fig. 1(a-b) shows Scanning Electron Micrographs (SEM) of PS corresponding to I_d ~20 and 50 mA cm^{-2} with pore sizes of ~20-30 (porosity ~68%) and 50-60 nm (porosity ~74 %) respectively. The porosity of PS films are estimated from the gravimetric measurements [1]. These results are in accordance with other studies where in general, an increase in the current density or the anodization potential leads to an increase in the pore-size and thus porosity [1,2].

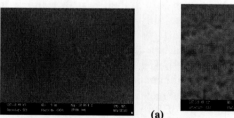

(a)   (b)

FIGURE 1. SEM of PS corresponding to different I_d (a) I_d= 20 mA cm^{-2}; and (b) I_d= 50 mA cm^{-2}

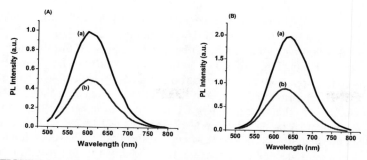

FIGURE 2. PL spectra of as-anodized (a) and dodecene-treated (b) PS at different I_d (A) I_d= 20 mA cm^{-2} and (B) I_d= 50 mA cm^{-2}.

Fig. 2 (A & B) shows PL spectra of fresh PS (curve a) and dodecene-treated PS (curve b) prepared at I_d ~20 and 50 mA cm^{-2}, respectively. The dodecene-treated PS film for both I_d ~20 & 50 mA cm^{-2} shows PL quenching with considerable decrease in PL intensity. This irreversible quenching can be attributed to a chemical modification of the PS sample that generates non-radiative surface traps resulting in decrease in PL intensity upon 1-dodecene attachment [7]. Another noticeable thing is that upon hydrosilylation of PS surface, only the PL intensity decreases and there is no shift in PL peak position which implies that the morphology (pore size, porosity etc.) of porous silicon remains unaffected by the hydrosilylation mechanism.

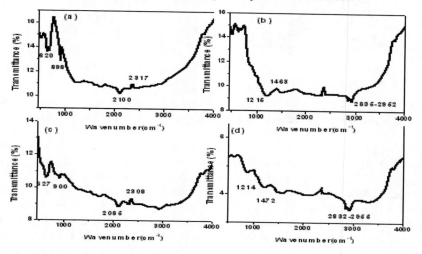

FIGURE 3. FTIR transmittance spectra of fresh PS and dodecene-treated PS at different I_d; (A) I_d~ 20 mA cm^{-2} and (B) I_d~ 50 mAcm^{-2}

Fig. 3 (a-d) shows FTIR spectra of Fresh PS and dodecene-treated PS corresponding to I_d~ 20 & 50 mA cm^{-2}. From Fig. 3 (a), for I_d~ 20 mA cm^{-2}, fresh PS surface shows the Si-H$_x$ stretching mode at ~2100 cm^{-1}, Si-H$_x$ bending modes at ~898 and 627 cm^{-1}

and O backbonded to Si in Si-H stretching mode ~ 2317 cm^{-1} [4, 8-9]. Similar features are exhibited by fresh PS at I_d~50 mA cm^{-2} Si-H$_x$ stretching mode at ~2085 cm^{-1}, Si-H$_x$ bending modes at ~ 900 and ~ 627 cm^{-1}, O backbonded to Si in Si-H stretching mode ~ 2308 cm^{-1} respectively (Fig. 3 (c)). The dodecene treated PS for I_d~ 20 mAcm^{-2} (Fig 3(b)), shows the appearance of new carbon-related aliphatic C-H$_x$ features ~2835-2952 cm^{-1}, C-H$_x$ deformation ~ 463 cm^{-1} and Si-O-Si stretching mode~1215 cm^{-1} respectively [4, 10-11]. The corresponding FTIR modes for dodecene-treated PS for I_d~50 mA cm^{-2} (Fig 3(d)) being aliphatic C-H$_x$ features ~2832-2955 cm^{-1}, C-H$_x$ deformation~ 1472 cm^{-1} and Si-O-Si mode~1214 cm^{-1}, respectively. The reduction in Si-H mode intensity upon dodecene treatment implies a hydrosilylation reaction that consumes Si-H bonds rather than cleaving Si-Si bonds.

Fig. 4(A & B) shows the Si (2p) core-level XPS spectra for fresh PS, dodecene-treated PS and their corresponding spectra under humid ambience corresponding to I_d ~ 20 & 50 mAcm^{-2}, respectively. From Fig. 4(A&B) for fresh PS for both I_d ~20 and

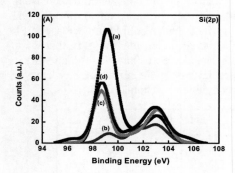
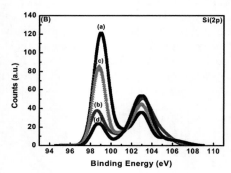

FIGURE 4. XPS Si (2p) core-level spectra (A) I_d~ 20 mA cm^{-2} and (B) I_d~ 50 mA cm^{-2} for fresh PS (a) dodecene-treated PS (c) and their corresponding spectra under humid ambience (b & d).

50 mAcm^{-2}, the Si(2p) core level spectra shows the doublet at 99.0 and ~103 eV which indicates that Si exists in two different environments pure Si and oxidized Si respectively [12]. The Si(2p) spectra for dodecene-treated PS corresponding to I_d~ 20 & 50 mA cm^{-2} shows a reduction in 99 eV peak and increase in 103 eV peak which indicates the hydrosilylation of PS surface with dodecene treatment. The presence of pure Si peak at ~99.0 eV after dodecene treatment indicates that mainly Si-H bonds break rather than cleavage of Si-Si bonds during hydrosilylation for both lower and higher porosity PS films. No shift of Si (2p) peaks towards high binding energy was observed upon dodecene treatment which shows that surface transformations occur without any further oxidation of the Si surface. Under humid ambience, the Si(2p) XPS peak intensity corresponding to pure Si (~ 99.1 eV) decreases drastically for fresh PS films and this effect is felt more for low porosity PS film (I_d ~20 mA cm^{-2}) as evident from Fig. 4(a). This result is in sharp contrast to that of dodecene-treated PS which infact shows no appreciable change in Si-core level XPS peak intensity corresponding to pure Si (~99.1 eV) under humid ambiance for both lower (I_d ~ 20

mA cm^{-2}) and higher porosity (I$_d$ ~50 mAcm^{-2}) PS films. Upon oxidation, the rate of decrease of XPS signal for Si(2p) core level is significant for fresh PS as compared to dodecene-treated PS which shows that dodecene-treated PS films are quite stable under oxidation conditions.

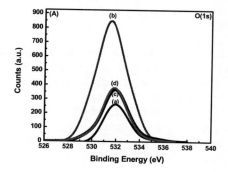 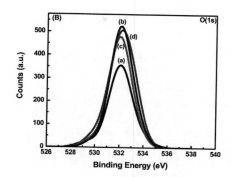

FIGURE 5. XPS O (1s) core-level spectra corresponding to (A) I$_d$~ 20 mA cm^{-2} and (B) I$_d$~ 50 mA cm^{-2} for fresh PS (a), dodecene-treated PS (c) and their corresponding spectra under humid ambience (b & d).

Fig. 5 (A & B) (a-d) shows the O (1s) core-level XPS spectra with peak at ~531-532 eV for fresh PS, dodecene-treated PS and their corresponding spectra under humid ambience corresponding to I$_d$ ~20 & 50 mA cm^{-2} respectively. The effect of oxidation for fresh PS is severe for lower porosity sample (I$_d$~20 mA cm^{-2}) as compared to higher porosity sample (I$_d$~50 mA cm^{-2}) under humid conditions (Fig. 5 (A&B) (curves a & b). A slight increment in O (1s) XPS signal upon dodecene treatment as compared to fresh PS indicates increment in Si-O-C like species [13] during hydrosilylation for both low and high porosity PS films (Fig. 5 (A&B) (curves a & c). From Fig. 5(A& B) (curves d), it is evident that under humid conditions, dodecene treated PS films are quite stable which commensurates well with other XPS core-level and FTIR studies.

The C (1s) core-level XPS spectra (figure not shown here) with peak at ~284.6 eV demonstrate that the fresh PS films exhibits lower C content as compared to dodecene-treated PS films which can be attributed to an increase in C-like species upon hydrosilylation for dodecene-treated PS film.

The drastic decrease in the PL intensity of PS with lower porosity (I$_d$ ~ 20 mA cm^{-2}) as compared to higher porosity (I$_d$~50 mA cm^{-2}) sample upon dodecene treatment can be attributed to the fact that the hydrosilylation reaction consumes a large fraction of the surface hydrides (SiH$_n$ species; n= 1-3) and breaking of weak Si-Si bonds leading to an overall reduction in the volume of the porous material and hence reduction in PL intensity [14]. This effect would be felt more for lower porosity sample since it is populated with more SiH$_x$ species, defects and microvoids as compared to higher porosity PS sample. FTIR and XPS results are also a testimonial to this fact. With increase in I$_d$, since surface to volume ratio increases and thus for I$_d$~50 mA cm^{-2}, the vast surface area of PS would facilitate higher degree of hydrosilylation mechanism and incorporation of Si-C species. This leads to increase in long-term stability of PS prepared at higher $I_d\sim50$ mA cm^{-2} as compared to that of lower I_d (~20 mA cm^{-2}) PS sample. PS films anodized at $I_d\sim50$ mA cm^{-2} with high porosity, high & stable PL, perfect morphology and chemically stable upon functionalization (dodecene treatment) would serve as an ideal platform for sensing applications.

CONCLUSIONS

Using a combination of SEM, PL, FTIR and XPS techniques, it was shown that high-quality monolayers can be formed in this manner at an optimized anodization current density ($I_d\sim50$ mA cm^{-2}). Long-term stability studies under humid conditions reveal that the PS surfaces particularly with high porosity ($I_d\sim50$ mA cm^{-2}) and hydrosilylated with 1-dodecene exhibits stability as compared to native hydride-terminated fresh PS samples. The hydrosilylation mechanism does not appear to apparently affect the nanostructured silicon size as its morphology essentially remains the same. This property is highly crucial for future technological applications of this functionalized material.

ACKNOWLEDGEMENTS

We thank Director NPL for providing the facilities for the successful completion of this research work. Shalini Singh gratefully acknowledges CSIR (New Delhi) for SRF fellowship.

REFERENCES

1. O. Bisi, S. Ossicini and L. Pavesi, *Surf. Sci. Rep.* **38**, 1-126 (2000).
2. J. Salonen and V.P. Lehto *Chem. Engg. J.* **13**, 162-172 (2008).
3. H. Ouyang, C.C. Striemer and P.M Fauchet, *Appl. Phys. Lett.* **88**, 163108 (2006).
4. O. Meskini, A. Abdelghani, A. Tlili, R. Mgaieth, N. Jaffrezic-Renault and C. Martelet, *Talanta* **71**, 1430-1433(2007).
5. L. De Stefano, L. Rotiroti, I. Rea, L Moretti, G. Di Francia, E. Massera, A. Lamberti, P. Arcari, C. Sanges and I. Rendina, *J. Opt. A: Pure Appl. Opt.* **8**, S540-S544 (2006).
6. J. N. Chazalviel and F. Ozanam, *M.R.S. Conf. Proc.* **536**, 155 (1999)
7. M. P. Stewart and J. M. Buriak, *Adv. Mater.* **12**, 859-869 (2000).
8. S. N. Sharma, R. K. Sharma and S. T. Lakshmikumar, *Physica E* **28**, 264-272 (2005).
9. S.N.Sharma, R. Banerjee, D. Das, S. Chattopadhyay and A.K. Barua, *Appl. Surf. Sci.* **182**, 333-337(2001).
10. B. Xia, S. J. Xiao, D. J.Guo, J. Wang, J. Chao, H. B. Liu, J. Pei, Y.Q. Chen, Y.C. Tang and J. N. Liu, *J. Mater. Chem.* **16**, 570-578 (2006).
11. C. A. Canaria, I. N. Lees, A. W. Wun, G.M. Miskelly and M. J. Sailor, *Inorg. Chem. Commun.* **5**, 560-564 (2002).
12. M. A. Hernandez, R.J.M. Palma, J. P. Rigueiro, J. P. Garcia-Ruiz, J. L. Garcia-Fierro and J. M. M. Duart, *Mat. Sci. Eng. C* **23**, 697-701 (2003).
13. S. Sharma, R.W. Johnson and T. A. Desai, *Biosens. Bioelectron.* **20**, 227-239 (2004).
14. N.Y. Kim and P. E. Laibinis, *J. Am. Chem. Soc.* **119**, 2297-2298 (1997).

Color Tunability and Raman Investigation in CdS Quantum Dots

Ashish K. Keshari* and Avinash C. Pandey

*Nanophosphor Application Centre, Department of Physics,
University of Allahabad, Allahabad-211 002, India.
Tel. /Fax: +91-532-2460675*
* Corresponding author e-mail: jeevaneshk26@yahoo.co.in

Abstract: CdS quantum dots (QDs) with improved luminescence properties places it among active area of research for their exploitation in appealing application in next generation opto-electronic devices and in photonics. We present here the tunability of emission spectra in CdS QDs in whole visible spectrum i.e. from violet to red region by proper choice of the synthesis temperature and photoluminescence excitation wavelength. Different luminescence behavior is observed at low temperature and at higher excitation energy. Raman spectra show no blue shift in longitudinal optical (LO) phonon bands due to phonon confinement with synthesis temperature. There is noticeable asymmetry in the line shape indicating the effect of phonon confinement which confirms the small crystallites of good quality.

Keywords: CdS; Quantum dots; Semiconductors; Photoluminescence; Raman spectra
PACS : 61.46.Hk, 78.67.Bf, 81.07.Bc

INTRODUCTION

The development of faster and smaller opto-electronic devices demands a continuous decrease of the element sizes and resulted in remarkable progress in electronics, data processing and communication techniques. The aim of this trend is not only to increase the integration level, but mainly, to increase the operation speed. Commercial requirements for miniaturized microelectronic devices provide strong motivation for exploring the synthesis of nanoscale systems using bottom-up techniques. II-VI nanocrystals (NCs) are receiving considerable attention for fundamental studies, as an example of zero-dimensional quantum confined material and for their exploitation in appealing applications in opto-electronics and photonics [1]. The NC based emitters can be used for many purposes such as optical switches, sensors, electro-luminescent devices [2], lasers [3] and biomedical tags [4]. Nanometer size quantum dots exhibit a wide range of electrical and optical properties that depends sensitively on the size of the nanocrystals and are of both fundamental and technological interest. It is, therefore, possible in principal to manipulate the properties of the nanomaterials for specific application of interests by designing and controlling the parameters that affect their properties. There is great departure in optical properties from their bulk counterparts [1, 5-6] due to quantum size effects in nanocrystals which leads to tunable blue shifts in both optical absorption

CP1147, *Transport and Optical Properties of Nanomaterials—ICTOPON - 2009*, edited by M. R. Singh and R. H. Lipson
© 2009 American Institute of Physics 978-0-7354-0684-1/09/$25.00

and emission spectra with decreasing nanocrystal size. Indeed it is possible to synthesize differently sized NCs emitting from blue to red and up to near IR with a very narrow band width [7-8]. The optical properties get modified dramatically due to the confinement of charge carriers within the nanoparticles. Similar to the effects of charge carriers on optical properties, confinement of optical and acoustic phonons leads to interesting changes in the phonon spectra. For this reason the control and improvement of the luminescence properties of quantum dots (QDs) has been a major goal in synthetic 'nanochemistry' and related preparative procedures [9]. The recombination processes of photogenerated carriers in semiconductor nanocrystals are important to their applications in opto-electronic devices. The PL technique is widely used to investigate both radiative and non-radiative transitions of carriers in nanocrystals. The band edge PL emissions were attributed to various recombination mechanisms such as the donar-acceptor pair emission [10] and the recombination luminescence of shallow traps [11-12] and excitons [12-13]. The CdS is an important semi-conducting material that has attracted much interest owing to their unique electronic and optical properties, and their potential applications in solar energy conversion, photoconducting cells, nonlinear optics and heterogeneous photocatalysis [14-16].

EXPERIMENTAL

In this work, we report a study of the temperature dependent PL spectra of CdS semiconductor NCs for different excitation wavelength. The color tunability of this semiconductor nanoparticle with proper choice of the synthesis temperature and excitation wavelength is one of the most attractive investigations. Details Raman investigation were also performed. The CdS nanocrystals at four different set temperatures viz. 4, 25, 35 and 45^0C were synthesized by chemical precipitation route. The synthesis procedure was reported elsewhere [17] by excluding the capping agent PVA from the reaction mechanism. It is observed that color of the precipitate obtained is changes from light green to orange-red as we change the Rx temperature from 4 to 45^0C.The materials were characterized by Photoluminescence and Raman spectroscopy immediate after the synthesis. PL spectra of all the prepared samples were recorded from Perkin-Elmer LS-55 spectrophotometer at 230, 325, 375 and 425 nm excitation wavelengths from Xenon lamp respectively. Raman spectra were recorded from Renishaw micro-Raman setup described elsewhere in more details [18]. For Raman excitation, radiation of wavelength 514 nm from Ar ion laser was used.

RESULTS AND DISCUSSION

The Raman spectra of as prepared CdS samples were recorded as shown in Fig. 3. The Raman signals are strong for CdS-like longitudinal optical (LO) phonon mode located at

297 and 594 cm^{-1} with overtone progression for all synthesis temperatures and in agreement with what expected by the two-mode behavior of the lattice vibrations in CdS alloy. There is no shift in the phonon band are observed with reaction (Rx) temperature however Raman intensity becomes stronger as we increase the Rx temperature. Both first and second harmonics of the LO CdS phonon can be observed by Raman spectroscopy. There is noticeable asymmetry in the line shape indicating the effect of phonon confinement. An asymmetric broadening of the Raman peak caused by the small crystal size can be observed as well. This shows that the semiconductor crystalline quality is fairly good in comparison with that reported [19]. The broadening of the half-widths compared with the bulk material could have two reasons: (1) the size distribution of the quantum dot (QD) and the presence of defects, (2) the contribution of phonon confinement. The phonon confinement only contributes to a slightly asymmetric line shape, which results from the relaxation of the k=0 conservation momentum rule. In view of these results the Raman spectra shown in Fig. 3 indicate a better crystal quality for QDs considering the half-widths and overall intensity. Shiang et al. [20] pointed out a connection between the width of the overtone lines and the vibrational relaxation. In accordance with this model a relaxation mechanism dominated by the decay of LO phonons into acoustic phonons can be concluded for our samples. A dephasing mechanism would result in a square dependence between half-widths and the order of the phonon bands.

Similar to the effects of charge confinement on the optical spectra, confinement of optical and acoustic phonons leads to the interesting changes in the phonon spectra. Confinement of the acoustic phonons may lead to the appearance of new modes in the low-frequency Raman spectra, whereas optical phonon line shapes develop marked asymmetry [21-22]. The spectra in Fig. 3 exhibit strong but broad peak at ~ 297 cm^{-1} corresponding to the LO phonon mode. This peak has also slightly shifted to lower frequency compared to the LO mode of bulk CdS (305 cm^{-1}). In bulk crystals, the eigenstate is a plane wave and the wave vector selection rule for first-order Raman scattering requires q ≈ 0. However, the confinement of the phonon to the volume of the nanocrystals results in the relaxation of the conservation of crystal momentum in the process of creation and decay of phonons. This relaxation of the q ≈ 0 selection rule results in the additional contribution of the phonon with q ≠ 0 which causes the asymmetric broadening and low frequency shift of first-order LO-Raman peak.

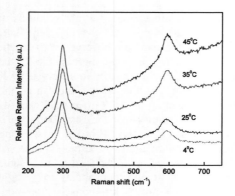

FIGURE 1. Raman spectra of CdS nanoparticles for different synthesis temperature. The wavelength of excitation laser line is 514 nm.

The photoluminescence spectra (PL) of the CdS nanoparticles synthesized at different temperatures viz. 4, 25, 35 and 45°C were recorded at excitation wavelength of 230, 325, 375 and 425 nm. The selected characteristics spectra are shown in Fig 2. It is seen that there is a pronounced change in the characteristics and intensity of the spectra for a particular combination of reaction temperature and excitation wavelength. There is a degradation in intensity is observed in almost all emission either with increasing reaction temperature or with increasing excitation wavelength while the general features of the excitonic emission remains unaltered. The different emission features observed at different condition are listed in the table1. The characteristic band edge green emissions are observed to be varying from 512 nm to 532 nm explicitly depends upon reaction temperature and excitation wavelength. The high energy band in the green region also referred to as band edge luminescence is related to various radiative mechanisms [10-11]. Decreasing the excitation energy results in a blue shift and broadening the band edge luminescence perhaps due to band filling. The temperature dependence of the exciton emission (band edge) energy is expected to follow the band gap of bulk material. The similar dependence may also be observed in the recombination processes of the shallow traps. Thus it is difficult to distinguish whether the band edge emission originates from the recombination of excitons or shallow traps. The recombination of shallow traps is dominant at weak excitation intensities and saturates at higher excitation intensities [23], where the Luminescence of excitons becomes dominant. The PL intensities of the band edge luminescence at different excitation wavelength decreases rapidly with increasing reaction temperature. Although the PL spectra of grown CdS generally depends on the growth characteristics, the band edge luminescence is possibly composed of recombination luminescence of shallow traps or defect level of interstitial sulfur [24], localized excitons and intrinsic excitons. The PL bands placed at lower energies indicate another type of defect such as Cd or S vacancies, interstitial Cd etc. In general, the variation in the energy gap with temperature is believed to results from the following two mechanisms: (1) lattice dilation, which causes a linear effect with temperature, and (2) temperature dependant electron-phonon interactions.

The spectra generated by the higher excitation energies at $4^{\circ}C$ display two strong, broad peak centered at 420 nm and 645 nm plus weak green emission at 525 nm (Fig 2(a)). Realizing that our CdS nanoparticles sample may not be strictly mono-dispersed. The low intensity of the spectrum for 325, 375 and 425 nm excitations indicate that the CdS nanoparticles are not sufficiently excited at this excitation energy: A few reports [25] on the effect of change in the excitation wavelength on PL spectra of CdS and CdSe nanoparticles are reported. Rodrigues et al.[26] has reported the band edge emission to be strongly dependent on the excitation photon energy. This idea was first reported by Tews et al.[27] to explain the dependence of exciton spectra on excitation photon energy. The observation of these selectively excited PL depends very much on the size distribution of nanocrystals. If the distribution is very broad, a large number of particles of different sizes will always be excited. Hence a broad PL spectrum with no distinct features will be observed independent of photon energy. But, if the distribution is extremely narrow, the emission peak will always be occur at the same energy determined by unique crystallite size. In the intermediate case of distribution which is not too broad or too narrow, appropriate excitation energy can excite several nanocrystals simultaneously producing PL spectrum which contains more than one peak. So the present sample may not be strictly mono-dispersed.

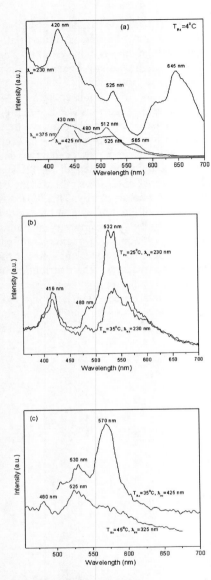

FIGURE 2. PL spectra of CdS nanoparticles recorded for different synthesis temperature at different excitation wavelengths of radiation from a Xenon lamp.

TABLE 1. The different emission features of CdS nanoparticles are observed for different synthesis temperature at different excitation wavelength. Corresponding visible color band are also shown.

Reaction temperature (^{O}C)	Excitation wavelength (nm)	Emission band (nm)	Visible region (Color)
4	230	420	Violet
		525	Green
		645	Orange-red
	375	430	Violet
		480	Blue
		512	Green
	425	525	Green
		565	Yellow
25	230	416	Violet
		480	Blue
		532	Green
35	230	416	Violet
		480	Blue
		532	Green
	425	530	Green
		570	Yellow
45	325	480	Blue
		525	Green

A lot of efforts have been spent to study the luminescence of CdS nanoparticles. Liu *et al.* reported that there were two emission bands. One is the green emission peak at 552 nm, the other is the broad red emission at 744 nm [28], Xu *et al.* found there were two luminescence peaks at 680 nm and 760 nm (IR), which were distributed to the formation of the sulfur vacancies and Cd-S composite vacancies respectively [29]. Moore *et al.* believed that Q-CdS showed the band edge PL peak centered at 450 nm [30].

CONCLUSIONS

Raman spectra show no blue shift in LO phonon bands due to phonon confinement with synthesis temperature. Line shape asymmetry due to phonon confinement confirms the prepared CdS nanoparticles are of small crystallites of good quality. Emission spectra of CdS QDs could be easily tuned in whole visible spectrum simply by proper choice of the synthesis temperature and PL excitation energy.

ACKNOWLEDGEMENTS

We thank Department of Science and Technology, India for funding the project under IRHPA in collaboration with Nanocrystals Technology, New York.

REFERENCES

1. A. P. Alivisatos, *Science* **271**, 933 (1996).
2. N. Tessler, V. Medvedev, M. Kazes, S. Kan and U. Banin, *Science*, **295**, 1506 (2002).
3. V.L. Klimov, A.A. Mikhailowsky, S. Xu, A. Malko, J. A. Hallingsworth, C.A. Leatherdale, H. J. Eisler and M.G. Bawendi, *Science* **290**, 314 (2000).
4. M. Han, X. Gao, J. Z. Su and S. Nie, *Nat. Biotechnol.* **19**, 631 (2001).
5. A.P. Alivisatos, *J. Phys. Chem. B* **100**, 13226 (1996).
6. Review articles on colloidal nanocrystals *Acc. Chem. Res.* **32**, 387 (1999).
7. X. Zhong, M. Han, Z. Dong, T.J. White, and W. Knoll *J. Am. Chem. Soc.* **125** 8589 (2003).
8. D. Battaglia, and X. Peng, *Nano Lett.* **2**, 1027 (2002).
9. L. Qu and X. Peng, *J. Am. Chem. Soc.* **124**, 2049 (2002).
10. N. Chestnoy, T.D. Haris, R. Hull and L.E. Brus, *J. Phys. Chem.* **90**, 3393 (1986).
11. A. Eychmuller, A. Hasselbarth, L. Katsikas and H. Weller, *J. Lumin.* **48&49**, 745 (1991).
12. M. O'Neil, J. Marohn and G. McLendon, *J. Phys. Chem.* **94**, 4356 (1990).
13. B.G. Potter Jr. and J.H. Simmons, *Phys. Rev.* **B37**, 10838 (1988).
14. K. Hu, M. Brust and A. J. Bard, *Chem. Mater.* **10**, 1160 (1998).
15. L. E. Brus, *J. Phys. Chem.* **90**, 2555 (1986).
16. A. Henglein, *Chem. Rev.* **89**, 1861 (1989).
17. A. K. Keshari, M. Kumar, P. K. Singh, and A. C. Pandey, *J. Nanosci. Nanotechnol.* **8**, 301-308 (2008).

18. B. Schreder, T. Schmidt, V. Ptatschek, U. Winkler, A. Materny, E. Umbach, M. Lerch, G. Muller and L. Spanhel, *J. Phys. Chem. B* **104**, 1677 (2000)
19. Y. A. Vlasov, V. N. Astratov, O. Z. Karimov, A .A. Kaplyanskii, V. N. Bogomolov, and A. V. Prokofiev, *Phys. Rev.* **B 55**, 13357 (1997)
20. J.J. Shiang, S.H. Risbud, A.P. Alivisatos, *J. Chem. Phys.* **98**, 8432 (1993)
21. I. H. XCampbell and P .M. Fauchet, *Solid State Commun.* **58**, 739(1986)
22. P. Nandakumar, C. Vijayan, M. Rajalakshmi, A. K. Arora and Y .V .G. S. Murti, *Physica* **E11**, 377 (2001)
23. J. Zhao, K. Dou, Y. Chen, C. Jin, L. Sun, S. Huang, J. Yu, W. Xiang and Z. Ding, *J. Lumin.* **66&67**, 332 (1996)
24. O. Vogil, I. Reich, M. Garcia-Rocha and O. Zelaya-Angel, *J. Vac. Sci. Technol.* A **15**, 282 (1997)
25. S. Okamotu, Y. Kanemitsu, H. Hosukawa, K. M. Koshi and S. Yanagida, *Solid State Commun.* **105**, 7 (1998)
26. P.A.M. Rodrigues, G. Tammulaitis, P. Y. Yu and S. H. Risbud, *Solid State Commun.* **94**, 583 (1995)
27. H. Tews, H. Venghaus and P.J.Dean, *Phys. Rev.* **B19**, 5178 (1979)
28. B. Liu, G.Q. Xu, L.M. Gan, C.H. Chew, W. S. Li, Z. X. Shen, *J. Appl. Phys.* **89**, 1059 (2001)
29. G.Q. Xu, B. Liu, S.J. Xu C. H. Chew, S.J.Chua, L.M. Gana, *J. Phys. Chem. Solids* **61**, 829 (2000)
30. D.E. Moore and K. Patel, *Langmuir* **17**, 2541 (2001)

Synthesis, Characterization and Application Of PbS Quantum Dots

Sweety Sarma[a], Pranayee Datta[a], Kishore Kr. Barua[b] and
Sanjib Karmakar[c]

[a] Department of Electronics Science, Gauhati University, Guwahait-781014, Assam, India

[b] Department of Physics, Central University, Tezpur-784028 , Assam, India,

[c] Department of Instrumentation and USIC, Gauhati University ,Guwahati-781014 , Assam, India.

Abstract. Lead Chalcogenides (PbS, PbSe, PbTe) quantum dots (QDs) are ideal for fundamental studies of strongly quantum confined systems with possible technological applications. Tunable electronic transitions at near – infrared wavelengths can be obtained with these QDs. Applications of lead chalcogenides encompass quite a good number of important field viz. the fields of telecommunications, medical electronics, optoelectronics etc. Very recently, it has been proposed that 'memristor' (Memory resistor) can be realized in nanoscale systems with coupled ionic and electronic transports. The hystersis characteristics of 'memristor' are observed in many nanoscale electronic devices including semiconductor quantum dot devices. This paper reports synthesis of PbS QDs by chemical route. The fabricated samples are characterized by UV-Vis, XRD, SEM, TEM, EDS, etc. Observed characteristics confirm nano formation. I-V characteristics of the sample are studied for investigating their applications as 'memristor'.

Keywords: Quantum dots, PbS
PACS: 81.07-b

INTRODUCTION

Quantum dots embedded in polymer have been extensively studied for visible spectral region luminescence and photo-voltaic properties. Such nano devices with suitable characteristics in the infrared region can be utilized as active optical elements at telecommunication wavelengths from 1.3 to 1.6μm. Lead Chalcogenides (PbS, PbSe, PbTe) quantum dots are ideal for fundamental studies of strongly confined systems with possible applications in the telecommunication band. Apart from the field of telecommunications, application of Lead Chalcogenides are possible in the fields of medical electronics, optoelectronics etc. Very recently , it has been proposed that 'memristor'(memory resistor) can be realized in nanoscale systems with coupled ionic and electronic transports.[5] The hysteresis characteristic of 'memristor'are observed in many nanoscale electronic devices including semiconductor quantum dot devices.[8] Important applications of memristor include ultradense, semi-non-volatile memories and learning networks that require synapse-like function.[5] In this paper an attempt has been made to study the properties of PbS QDs synthesized by chemical route and to explore its possible application as memristor.

CP1147, *Transport and Optical Properties of Nanomaterials—ICTOPON - 2009*, edited by M. R. Singh and R. H. Lipson
© 2009 American Institute of Physics 978-0-7354-0684-1/09/$25.00

EXPERIMENTAL

Synthesis

PbS quantum dots are synthesized by adopting chemical route with poly-vinyl alcohol (PVOH) as the matrix. Six samples of different pH are synthesized. The samples are characterized and results being shown in Table 1.

TABLE 1. Sample characteristics and results.

SL. NO	SAMPLE NAME	pH-VALUE	STIRRING TIME OF PVOH(hrs)	SIZE (nm)	UV-VIS BAND GAP (eV)
1.	S1	1.7	3	8.2	2.75
2.	S2	3.5	3	5.5	2.9
3.	S3	5.6	3	3.1	3.0
4.	S4	-	2	6.2	2.825
5.	S5	-	3	6.1	2.875
6.	S6	-	4	9.6	3.1

Experimental Setup for I-V Characteristics

Prepared PbS QD samples are deposited in the space between two copper wires that is kept only 0.1 cm apart with glass slides as the base. This device is inserted in a standard circuit for I-V characterization.

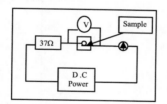

FIGURE 1. Setup for obtaining I-V characteristics.

Characterization

Optical absorption studies are made by HITACHI U-3210 Double beam spectrophotometer in the Department of Chemistry, Gauhati University, Assam. Photoluminescence recorded in FSP 920 spectrophotometer is performed in IIT, Guwahati, Assam. TEM is performed in RSIC, NEHU, Shillong. SEM and Energy dispersive X-ray spectroscopy are performed to study morphology and chemical composition in Central University, Tezpur Assam. XRD is performed in IIT, Guwahati as well as in Central University, Tezpur, Assam.

RESULTS AND DISCUSSION

UV-Vis Absorption Spectra

Optical absorption spectra of PbS quantum dots, taken at room   temperature, are shown in figure 2(a). The absorption curves are featureless i.e.   no   excitonic absorption structure is observed .This is due to inhomogeneous broadening caused by size distribution .Infact, particularly for PbS QDs, depending on the matrix and the particle shape, even if the size dispersion is less than 10 percent, the exciton peaks could not be resolved.[3,4] The bandgap of the samples are determined following the procedure of Tauc . [6] figure 2(b) .

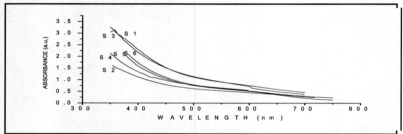

FIGURE 2(a). Optical absorption spectra of samples

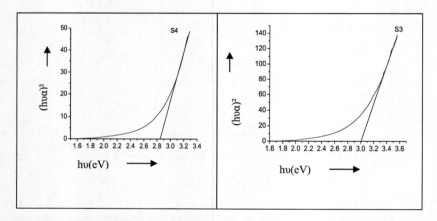

FIGURE 2(b). (hvα) ^{2} versus hv. Curve for band gap determination of S4 and S3

X-Ray Diffraction

Figure 3 shows the XRD spectrum of sample, S2. The first peak observed in all the XRD spectra is due to the PVOH matrix. The other peak positions show the sample to be PbS. The peak positions show the structure to be cubic fcc (JCPDS card no. 20-0596). However, the peaks are slightly deviated from the standard positions due to residual stress. Broadened peaks indicate lowering in size. Sizes calculated from the Debye-Scherrer formula are given in Table 1, wherein band gaps obtained from UV-Vis are also given. Significant variations of band gap with size, size with ph and size with stirring time of PVOH for samples are observed.

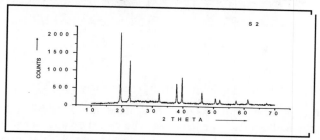

FIGURE 3. XRD spectrum of sample S2

Photoluminescence

Figure 4 shows the PL spectra of PbS quantum dots for all the samples of different sizes taken at room temperature. All the samples are photoexcited at 1.9125 eV (650 nm). Single PL emission peak in the near infrared region for each sample is found which is in good agreement with the works of other workers [1]. The Gaussian shape of the PL emission peaks is due to inhomogeneous broadening caused by size dispersion of the nanoparticles and recombination through surface states [2].

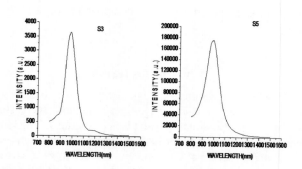

FIGURE 4. Photoluminescence spectra of samples S3 and S5

Transmission Electron Microscopy (TEM)

Figure 5 shows the TEM image of PbS QD sample S3. The well defined, nearly spherical particles were observed with an average size of 3 nm. The size obtained by TEM is almost same as that obtained by Debye Scherrer formula using

XRD spectra (table1).

FIGURE 5. TEM image of sample S3

Scanning Electron Microscopy (SEM)

Figure 6 shows the SEM image of PbS quantum dot sample S3. The picture shows the spherical shape of the particles. The size shown in this picture is larger than that obtained by other characterization techniques as SEM reveals only the topographical observations.

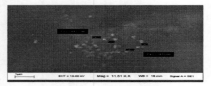

FIGURE 6. SEM image of sample S3

Energy-Dispersive-X-ray Spectroscopy (EDX)

EDX chemical characterization confirms composition of the PbS quantum dot sample.

I-V Characteristic

The I-V characteristics of the samples are investigated for possible applications as 'Memristor'. The characteristics are studied for increasing and decreasing voltage in the forward as well as in reverse direction shown as in Figure 7. The loop 2 is for the data recorded immediately after those for loop 1. Occurrence of such multiple loops indicates that the internal structure of the nanodevice rearranges as charge flows. This makes rather like a resistor with memory. Repeatability of the device characteristics can be investigated by switching of the power supply and allowing the device to get discharged for 20-30 minutes through a resistor [7]. The asymmetry between forward and reverse direction may be due to Schottky contact at the interface between the nanocrystallite samples and the metal electrode [1, 8]. Hysteresis behavior has been demonstrated by the PbS nanodevice fabricated in the present investigation. For loop L1, more hysteresis behavior is observed in the forward direction compared to that in the reverse direction. Whereas for loop L2, the observation is just reversed.

This behavior is similar as modeled by Strukov *et. al.* [5] for memristors. However, to arrive at more concrete conclusion regarding the memristive property, some more work on I-V characterization needs to be done.

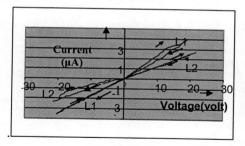

FIGURE 7. I-V characteristics of sample S3

CONCLUSIONS

1) UV-Vis observations confirm quantum confinement of the fabricated PbS samples.
2) PL emission obtained from 970 nm to 994 nm is attributed to surface states.
3) Size of the fabricated nanoparticles from XRD spectrum are found to be in the range from 3.1nm to 9.6nm
4) UV-Vis, XRD, PL, SEM, TEM & EDS observations confirm that we have successfully synthesized PbS QDs adopting chemical route.
5) The PbS nanodevice exhibits hysteresis characteristic of a memristor.

ACKNOWLEDGEMENTS

We acknowledge the help received from IITG, Assam; Central University, Tezpur, Assam; RSIC NEHU, Shillong, Meghalaya; Department of Chemistry, Gauhati University, Assam, for sample characterization.

REFERENCES

1. M. Hamaguchi, K. Aoyama, S. Asanuma, Y. Uesu and T. Katsufuji, *Appl. Phys. Lett.* **88**, 142508/1-142508/3 (2006).
2. A. Martucci, J. Fick, S. E. LeBlanc, M. LoCascio and A. Hache, *J. Non-Cryst. Solids* 639642 (2004).
3. L. Guo, K. Ibrahim, F. Q. Liu, X. C.Ai, Q. S. Li, H. S. Zhu and Y. H. Zou, *J. Lumin.* **83**, 111 (1999).
4. A. A. Patel, F. Wu, J. Z. Zhang, C. L.Torres-Martinez, R. K. Mehra, Y. Yang and S. H. Risbud, *J. Phys. Chem. B* **104**, **11598** (2000).
5. D. B. Strukov, G. S. Snider, D. R. Stewart, R. S. William, *Nature* **453**, 80-83 (2008).
6. R. K. Joshi, A. Kanjilal, H. K. Sehgal, *Appl. Surf. Sci.* **221**, 43-47(2004).
7. Syed Hassan Mujtaba Jafri, Sujira Promnimit, Chanchana Thanachayanoht and Joydeep Dutta, www.nano.ait.ac.th (2006).
8. Rinki Bhadra, P. Datta and K. C. Sarma, "Studies on some aspects of doped and undoped ZnS nanocrystals for applications in electronics", submitted to this conference.

Attachment Of Streptavidin-Biotin On 3-Aminopropyltriethoxysilane (APTES) Modified Porous Silicon Surfaces

Shalini Singh[1,3], Norman Lapin[2], P.K. Singh[1], Mukhtar A. Khan[3] and Yves J. Chabal[2,4]

[1]*Electronic Materials Division, National Physical Laboratory, Dr. K.S. Krishnan Road, New Delhi-110012, INDIA*
[2]*Department of Biomedical Engineering, Rutgers University, Piscataway, NJ 08854*
[3]*Faculty of Life Science, Aligarh Muslim University, Aligarh-202001, INDIA*
[4] *Department of Materials Science and Engineering, The University of Texas at Dallas, Richardson, TX 75080*

Abstract. Nanostructured porous silicon (PS) has a large surface area that can be well controlled and modified to have a high specificity for biomolecules. These properties make it a very promising biomaterial, in particular for sensing devices that need to be linked to the biological system and completely compatible with standard integrated circuit processes. We report the formation of nanostructured PS on boron doped, p-type silicon (100) wafers by electrochemical anodization, using aqueous hydrofluoric acid and isopropyl alcohol solutions and a constant current density 50 mA/cm². The pore diameter can be tuned by varying the etching conditions. The interaction of streptavidin with biotin was studied on 3-aminopropyltriethoxysilane (APTES) functionalized PS surfaces using infrared absorption spectroscopy to characterize the surface at each step, including subsequent reaction steps. These studies show that the streptavidin-biotin interaction on modified PS surfaces depends on the details of the APTES adsorption process.

INTRODUCTION

Porous silicon (PS) was first developed by Uhlir (1956) when performing electrochemical etching of silicon and it exhibits unique optical and electrical properties due to quantum confinement effects was reported by Canham (1990) and Cullis et al. (1997) [1-3].

Nanostructured porous silicon has been used for chemical and biological sensing due to its morphological and physical properties like tunable pore size. It is very sensitive to the presence of biochemical species that can penetrate inside the pores and it has a sponge-like structure with a specific area of the order of 200 – 500 m²/cm³ [4-8].

PS provides a large and often highly reactive surface area, which enables more effective capture and detection of biomolecules than bulk materials. Biomolecule detection using a porous silicon platform was pioneered by Sailor and coworkers in the

CP1147, *Transport and Optical Properties of Nanomaterials—ICTOPON - 2009,* edited by M. R. Singh and R. H. Lipson
© 2009 American Institute of Physics 978-0-7354-0684-1/09/$25.00

late 1990s [9]. DNA has been the most commonly detected target molecule, although there have been several demonstrations of enzyme, virus, and protein detection using various porous silicon structures and optical transduction methods [10-16].

The biotin/streptavidin system exhibits one of the strongest receptor–ligand interactions found in nature (17, 18). The high binding affinity and the symmetry of the biotin-binding pockets, positioned as pairs at opposite sides of the protein, [19] can be used to conjugate biotin with diverse biomolecules such as antibodies, enzymes, peptides, and nucleotides (20).

We report here streptavidin–biotin interaction on APTES modified PS. To create a biotin-functionalized surface for the capture of streptavidin, hydroxyl-terminated nanostructured PS was silanized with APTES to create amino groups on the surface. Next, the sulfo-NHS-biotin was immobilized and eventually used to bind streptavidin. The various stages of protein immobilization on modified silicon surfaces were observed by infrared spectroscopy. The attachment of streptavidin-biotin on modified silicon substrates and the ease of functionalization of substrates with biotin, make this system extremely useful in a wide range of biological sensing applications.

MATERIALS AND METHODS

Anhydrous 3-aminopropyltriethoxysilane (APTES), hydrofluoric acid (HF), isopropyl alcohol (IPA), anhydrous toluene, biotin 3-sulfo-N-hydroxy-succinimide ester sodium salt (NHS-biotin), N,N-dimethylformamide (DMF), streptavidin from Streptomyces avidinii, and Tween 20 were purchased from Sigma-Aldrich and 18.2 MΩ-cm^2 deionized water was used. All chemicals were used as received unless otherwise noted.

Boron doped, single crystalline, p-type silicon (100) wafers were used for preparing PS. They were treated with a standard SC1 (standard cleaning1)/SC2 (standard cleaning 2) cleaning method to remove organic and metallic contaminants on the sample surface [21,22]. A thin aluminum layer was evaporated on the back of the wafers and sintered to form an ohmic contact. PS samples were formed by electrochemical etching using Si as the anode and Pt as the counter electrode in a Teflon container. The anodization was carried out at different current density (I_d) 50 mA/cm^2 for 30 min, in a solution of H$_2$O:HF:IPA (1:1:2). After etching, the samples were rinsed with DI water and dried under a stream of dry high purity nitrogen prior to use. As-anodised hydride-terminated PS was then treated with SC2 to obtain a SiO$_2$ hydroxyl-terminated surface.

Anhydrous APTES (98%) and anhydrous toluene (99.8%, <0.001% water) were used for silanization. All APTES experiments were conducted inside a dry nitrogen-purged glovebox. For APTES treatment, PS samples were immersed in a 0.1% (v/v) solution of APTES in anhydrous toluene and incubated in the glovebox for a time period of 48 hours at 70 ° C. After the reaction, the functionalized Si sample was removed from the APTES solution and rinsed with fresh anhydrous toluene.

Next, NHS-biotin solvated in DMF was added to APTES-treated PS and incubated for about 1 hr. The sample was then rinsed twice with DMF to remove unbound biotin. Biotinylated substrates were treated with streptavidin in phosphate buffered saline (PBS) for 45 min. Subsequently, the substrates were rinsed with 0.05% Tween 20 in PBS twice, followed by thorough washing in PBS and DI water. Then samples were dried under a stream of purified air.

IR spectra were recorded with a Mercury-Cadmium-Telluride-A (MCT-A) detector over the 650 to 4000 cm^{-1} spectral range. All PS samples were colleted in transmission, with 4 cm^{-1} resolution. Five sets of one thousand scans each are typically recorded at a mirror velocity of 1.9 cm/second. Dry nitrogen gas is used to purge the spectrometer chamber during scans.

RESULTS AND DISCUSSION

Scheme 1 shows the immobilization steps of the streptavidin-biotin to APTES-modified porous silicon surfaces: First, freshly prepared PS with SiH$_x$ terminated surfaces is treated with SC2 to obtain hydroxyl-terminated oxide surfaces (A); Second, these hydroxyl-terminated oxidized PS surfaces are silanized with APTES in toluene resulting in amine-terminated surfaces (B); Third, these APTES-modified surfaces are reacted with NHS-biotin to produce biotinylated surfaces (C); Fourth, streptavidin is bound to biotinylated surfaces (D), respectively.

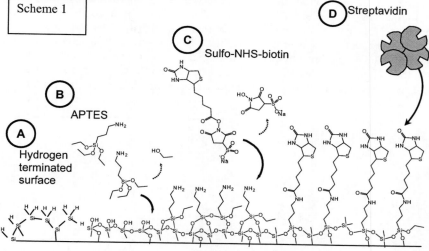

Scheme 1

D Streptavidin

C Sulfo-NHS-biotin

B APTES

A Hydrogen terminated surface

Porous silicon

445

The IR absorption spectra, collected after each step for streptavidin-biotin attachment on PS, are shown in Figures 1-4. The IR spectrum of a freshly hydride-terminated PS wafer is shown in Fig. 1. It exhibits the typical Si–H$_x$ stretching modes at 2073-2105 cm^{-1} (23). The hydrolysis of the H-terminated PS surfaces to form hydroxyl-terminated surfaces is accomplished for grafting biomolecules using a well-developed silanization process and subsequent chemical functionalization.

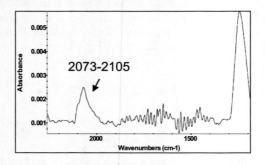

FIGURE 1. IR spectrum of freshly anodized PS referenced to SC1/SC2 cleaned wafer, showing hydrogen-terminated surface.

Fig. 2 (a) IR absorption bands are observed at (i) approximately 1580 cm^{-1}, assigned to the NH$_2$ scissoring vibration of APTES amine groups, (ii) 1647 cm^{-1}, corresponding to the asymmetric -NH^{3+} deformation mode and (iii) 1494 cm^{-1}, assigned to the symmetric NH^{3+} deformation mode [25]. Fig. 2 (b) shows the IR spectra of APTES functionalized PS. The presence of CH$_x$ stretching modes is observed between 2828 and 2961 cm^{-1}: the CH$_2$ asymmetric and symmetric stretching modes are observed at 2915 and 2828 cm^{-1}, respectively with the CH$_3$ asymmetric stretching mode at 2961 cm^{-1} [24].

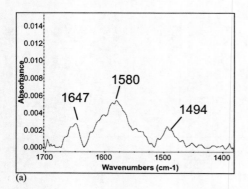

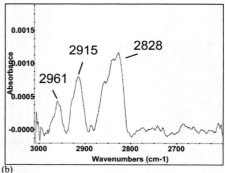

(a)   (b)

FIGURE 2. (a) (b) IR spectra of amine-terminated PS surface after APTES treatment. These spectra are referenced to the hydroxyl-terminated SiO₂ surface.

The cross-linking reactions of the amine-reactive surface with activated ester of NHS-biotin have been commonly applied for biomolecular immobilization on various substrates. The reaction takes place through a nucleophilic attack of the amine by ester C=O, eliminating NHS. The IR spectra of biotin-NHS (Fig. 3) show new peaks (i) approximately at 1714 cm⁻¹ attributable to the biotin ureido C=O and (ii) amide I at 1645 (C=O stretch) and amide II (N-H bend with C-N stretch) bands at 1552 cm⁻¹ [20,23] due to the covalent attachment of biotin to the amine-terminated surface.

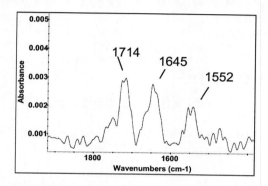

FIGURE 3. IR spectrum of biotinylated functionalized PS referenced to the APTES treated PS surface.

In Fig. 4, two bands at 1648 cm⁻¹ (amide I) and 1542 cm⁻¹ (amide II) are observed, which can be assigned to the amide functionalities of the peptide groups of streptavidin [20,26]. Each streptavidin tetramer has four equivalent sites for biotin (two on each side of the complex), which makes biotin a useful molecular linker. The specific binding of streptavidin to the biotinylated 3-APTES modified PS surface is evident in the IR spectrum in the appearance of the amide bands.

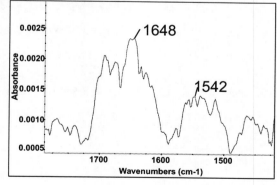

FIGURE 4. IR spectra of streptavidin attachment on PS referenced to the biotinylated surface.

CONCLUSIONS

Nanostructured PS surfaces with pore size of ~50-60 nm provide large and biocompatible surface areas for biospecific bonding of streptavidin. APTES functionalized PS reacts with biotin-NHS, at its amine group, and the biotinylated surface subsequently binds with streptavidin (shown using IR spectroscopy). The ability to monitor these important chemical and biochemical reactions and obtain a positive measure of streptavidin in porous silicon makes it possible to develop the use of PS for a broad range of applications in the field of biosensors. In addition it can be expected that a tailored porous structure could also act as a matrix for a large variety of biological and chemical molecules. Immobilization of proteins on functionalized PS surfaces constitutes a research area of considerable importance in emerging technologies employing biocatalytic and biorecognition events.

ACKNOWLEDGMENTS

This work was supported by the National Science Foundation (grant CHE-0415652) and Council for Scientific and Industrial Research, India

REFERENCES

1. A. Uhlir, *Bell Syst. Technol. J.* **35**, 333–347 (1956).
2. L. T. Canham, *Appl. Phys. Lett.* **57**, 1046–1048 (1990).
3. A. G. Cullis, L. T. Canham and P. D. J Calcott, *J. Appl. Phys.* **82**, 909–965 (1997).
4. H. Ouyang, M. Christophersen, R. Viard, B.L. Miller and P.M. Fauchet, *Adv. Funct. Mater.* **15**, 1851(2005).
5. S. M. Weiss, H. Ouyang, J. Zhang and P.M. Fauchet, *Opt. Express* **13** 1090 (2005).
6. F.P. Mathew and E.C. Alocilja, *Biosens. Bioelectron.* **20**, 1656 (2005).
7. F. Besseueille, V. Dugas, V. Vikulov, J. P. Cloarec and E. Souteyrand, *Biosensors Bioelectron.* **21**, 908 (2005).
8. C. Pacholski, M. Sartor, M. J. Sailor, F. Cunin and G.M. Miskelly, *J. Am. Chem. Soc.* **127**, 11636 (2005).
9. V. S. Y. Lin, K. Motesharei, K. P. S. Dancil, M. J. Sailor and M.R. Ghadiri, *Science* **278**, 840 (1997).
10. K. P. S. Dancil, D. P. Greiner and M.J. Sailor, *J. Amer. Chem. Soc.* **121**, 7925 (1999).
11. F. Cunin, T.A. Schmedake, J.R. Link, Y.Y. Li, J. Koh, S.N. Bhatia and M.J. Sailor, *Nature Mater.*, **1**, 39 (2002).
12. S. Chan, S. R. Horner, P. M. Fauchet and B.L. Miller, *J. Amer. Chem. Soc.* **123**, 11797 (2001).
13. H. Ouyang, M. Christophersen, R. Viard, B. L. Miller and P. M. Fauchet, *Adv. Funct. Mater.* **15**, 1851 (2005).

14. L. De Stefano, L. Rotiroti, I. Rendina, L. Moretti, V. Scognamiglio, M. Rossi and S. D'Auria, *Biosens. Bioelectron.* **21**, 664 (2006).
15. A. M. Rossi, L. Wang, V. Reipa and T. E. Murphy, *Biosens. Bioelectron.* **23**, 741 (2007).
16. I. Rendina, I. Rea, L. Rotiroti and L. De Stefano, *Physica E* **38**, 188–192 (2007).
17. L. Chaiet and F. J. Wolf, *Arch. Biochem. Biophys.* **106**, 1 (1964).
18. N. M. Green, *Adv. Protein Chem.* **29**, 85 (1975).
19. V. M. Mirsky, M. Riepl and O. S. Wolbeis, *Biosens. Bioelectron.* **12**, 977 (1997).
20. Z. Liu and M. D. Amiridis, *J. Phys. Chem. B* **109**, 16866-16872 (2005).
21. G. S. Higashi and Y. J. Chabal, *Handbook of Semiconductor Wafer Cleaning Technology*, W. Kern, Ed.; Noyes, Park Ridge, New Jersey, 433 (1993).
22. W. Kern, *J. Electrochem.* **137**, 1887-1892 (1990).
23. B. Xia, S. J. Xiao, D. J. Guo, J. Wang, J. Chao, H. B. Liu, J. Pei, Y. Q. Chen, Y. C. Tang and J. N. Liu, *J. Mater. Chem.* **16**, 570–578 (2006).
24. G. Socrates, *Infrared and Raman Characteristic Group Frequencies.* Wiley, New York (2001).
25. J. Blümel, *J. Am. Chem. Soc.* **117**, 2112-2113 (1995).
26. C. M. Pradier, M. Salmain, Z. Liu and G. Jaouen, *Surf. Sci.* **193**, 502-503 (2002).

Quantitative Analysis of Diameter Dependent Properties of Multi-walled Carbon Nanotubes

Dilip K. Singh[a], Parameswar K. Iyer[b] and P. K. Giri[a,c]

[a]Centre for Nanotechnology, Indian Institute of Technology Guwahati, Guwahati-781039
[b]Department of Chemistry, Indian Institute of Technology Guwahati, Guwahati-781039
[c]Department of Physics, Indian Institute of Technology Guwahati, Guwahati-781039

Abstract. Several exotic properties of Multiwalled carbon nanotubes (MWCNTs) can not be understood without the knowledge of average diameter distribution. Optical spectroscopy techniques have not been exploited much for studying MWCNTs. In this work, we use X-ray diffraction (XRD) and Raman spectroscopy as tools to study diameter distribution of MWNTs. Raman studies on the MWNTs show diameter dependence of G-band, with relative change in intensity of the metallic and semiconducting component of the G-band as a function of diameter. Diameter dependence of line shape in XRD pattern and Raman spectra is discussed in terms of lattice strain and atomic vibrations involving interaction among the walls in MWNTs. Electron paramagnetic resonance studies shows a Gaussian component whose g-value approaches free electron g-value with increasing diameter. Oxidation temperature is found to depend on diameter of MWCNTs by Thermogravemetic experiments.

Keywords: MWCNTs, XRD, Raman, EPR, TGA, Diameter dependence.
PACS: 61.48.De

INTRODUCTION

Diameter dependence of optical properties of single wall carbon nanotubes (SWCNTs) have been extensively studied.[1]With time Raman spectroscopy has evolved as a tool must to probe the properties of carbon nanotubes like diameter distribution, chirality, doping, nature like semiconducting or metallic etc.[2] Achiba et.al have done detailed analysis of mean diameter and diameter distribution of single-wall carbon nanotubes from their optical response.[3] Diameter grouping of bulk samples of single-walled carbon nanotubes from optical absorption spectroscopy has been illustrated.[4] Raman scattering intensity has been reported to depend on the diameter for single-wall carbon nanotubes.[5] But most of those techniques can't be applied to multiwalled carbon nanotubes (MWCNTs) due to multilayered structure or multiple absorptions and emission between concentric cylindrical structures. Only few of the studies are dedicated to Raman spectroscopic studies of MWCNTs. Lefrant et. al. have discussed the origin of low frequency (Radial breathing modes) modes in case of MWCNTs which are characteristic of diameter of SWCNTs.[6] C. Thomson believed that the inner (the more strongly curved) tubes of a MWCNT contribute to the SWCNT-like feature in the spectra, while the other tubes more closely resembled graphite.[7] Saito et. al. have studied G-band modes of MWCNTs and their splittings. They found that G-band consists of peak having linewidth 4 cm^{-1} from inner most tube

CP1147, *Transport and Optical Properties of Nanomaterials—ICTOPON - 2009,* edited by M. R. Singh and R. H. Lipson
© 2009 American Institute of Physics 978-0-7354-0684-1/09/$25.00

(when inner most diameter is less than 2 nm) and the Graphite like mode having linewidth ~ 20 cm^{-1} from outer cylinder wall.[8] Dresselhaus et.al have studied the variation in intershell spacing of MWCNTs using high resolution transmission electron microscopy.[9] Recently Hiroynki Nii have reported the influence of diameter on the Raman spectra of MWCNTs.[10] We have studied the effect of diameter on the properties of MWCNTs as bulk nanomaterial using XRD, Raman spectroscopy, ESR spectroscopy, Thermogravimetric analysis (TGA) and Transmission electron microscopy and for the first time have found a correlation between physical properties and diameter of the MWCNTS.

EXPERIMENTAL DETAILS

For the present study CVD grown commercially available carbon nanotubes (Shenzhen Nanotech, China) were used. Details of the samples used in study are give in Table-I. MWCNTs were 1-2µm long and of six diameter ranges less than 10nm, 10-20nm, 10-30nm, 20-40nm, 40-60nm and 60-100nm have been studied. XRD spectra were recorded using D8 Advanced tools Brucker in para focusing geometry of the Bragg-Brentano optics using lock coupled scan mode. Cu Kα_1 (λ=1.5452 A°) was used for recording X-ray diffraction spectra and the integration time was 1 sec/step and the step size was 0.02degree. Raman spectra was recorded in backscattering mode using 488 blue Ar$^+$ ion laser, a single grating monochromator, thermoelectrically cooled CCD detector, and Jovin-Yvon Triax550 monochromator for scattered light collection. TGA studies were done using Metler–Teledo TGA 1100. ESR measurements were done using JEOL, JES-FA 200, X-band spectrometer. Morphological characterization was done High resolution (HR) Transmission electron microscope JEOL 2010 TEM.

Sl. No.	Sample Name	Diameter range	Length range
\multicolumn{4}{c}{**TABLE: 1** Details of the samples of MWCNTs}			
1	CNT-A	Less than 10 nm	5-15 µm
2	CNT-B	10-20 nm	5-15 µm
3	CNT-C	10-30 nm	5-15 µm
4	CNT-D	20-40 nm	5-15 µm
5	CNT-E	40-60 nm	5-15 µm
6	CNT-F	60-100 nm	5-15 µm

RESULTS AND DISCUSSION

Transmission electron microscopy (TEM) of these sample (Fig: 1) gives the correct estimate of the diameter ranges of these nanotubes, but it provides information about only a very small volume of the sample. Transmission electron microscopy being costly, time consuming and the information obtained may not be correct representative of milli-grams of the sample volume renders its applicability for estimation of diameter distribution. High resolution TEM images (Fig: 2) shows that as the diameter of the MWCNTs increases the number of co-axial cylindrical layers increases. X-ray diffractometry (XRD) is a non-destructive method of characterization revealing information about the interlayer spacing, the structural strain and the impurities. However diameter and chirality distribution in carbon nanotubes samples makes multiple orientations with respect to the X-ray beam leading to a statistical result. Using the (002ℓ) peak position, the interlayer spacing is often found to be larger than that of HOPG. XRD peak profile of different diameter MWCNTS were fitted with voigt spectral function and their diameter dependency has been explored. Higher diameter MWCNTs shows higher Bragg angle (2θ) and low full-width at half maxima (FWHM) (Fig: 3). Dresselhaus et. al. have found that the inter-shell spacing of multi-walled carbon nanotube increases with decreasing tube diameter. This is attributed to

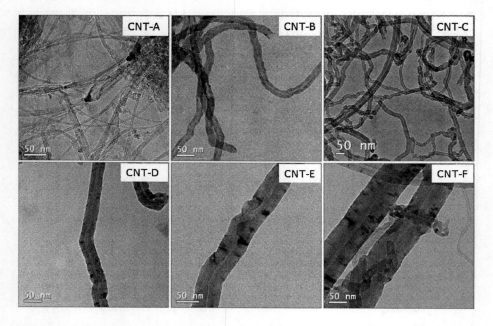

Figure 1. Transmission electron microscopy (TEM) images (a) CNT-A, d < 10nm   (b) CNT-B, 10-20 nm (c) CNT-C, 10-30 nm (d) CNT-D, 20-40 nm (e) CNT-E,   40-60nm (f) CNT-F 60-100 nm.

The curvature, resulting in an increased repulsive force, associated with the decreased diameter of the nanotube shells.[9] From their TEM images they found that the interlayer d_{002} spacing decreases from 0.39 nm with increasing diameter and becomes constant at 0.34 nm for MWCNTS of diameter 10 nm and higher. Whereas from Raman studies Kunishige et.al have shown that d- spacing of MWCNTs changes upto diameter of 20 nm and then it becomes constant.

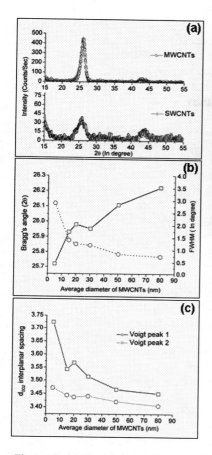

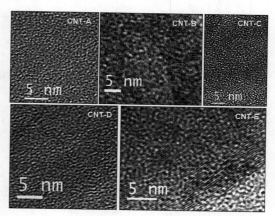

Figure 2. High Resolution Transmission electron microscopy (TEM) images (a) CNT-A, d < 10nm (b) CNT-B, 10-20 nm (c) CNT-C, 10-30 nm (d) CNT-D, 20-40 nm (e) CNT-E, 40-60nm

Figure 3. (a) Comparison of XRD spectra of MWCNTs. (b) voigt spectral profile of (002) peak (c) variation of d-spacing with increasing diameter.

But from X-ray diffraction studies they have shown that d_{002}-spacing monotonically decreases even for nanotubes having averaged diameter of 80 nm.[10] We have also found that the d-spacing of MWCNTs shows a very regular trend with diameter Fig:3 (c). But we noticed that the spectral profile requires two Voigt components to fit. Inter-planar spacing corresponding to both the peaks is plotted against the diameter. These two peaks shows d-spacing shifts to two different extents with diameter. We understand it in terms of inner walls of the MWCNTs is more strained than the outer walls. X-diffraction studies shows diameter dependence in diffraction peak profiles of MWCNTs in terms of up shift of Bragg's diffraction angle and decrease of peaks width with increase in MWCNTs diameter. Nanotubes with smaller innermost tube diameters often have fewer shells. Higher intershell spacings of the small diameter nanotubes are due to the higher lattice strain.

Raman spectroscopy is the best tool for characterization of carbon based materials. It has been intensively used for predicting properties of single walled carbon nanotubes. But it has not been explored much for characterizing MWCNTs. Only one recent study have shown that frequency increases with decreasing tube diameter for diameters less than 20nm.[10] Whereas we find that G- band frequency first increases upto 20 nm of diameter and then it starts to decrease as the MWCNTs diameter increases. The diameter dependent behavior of Raman shift of G-band is similar to that of changes in calculated d-spacing for MWCNTS of diameter larger than 20nm. G-band frequency first up shifts for MWCNTS of diameter up to 20 nm and then it decreases with further increase of diameter. The FWHM of this peak also shows similar variation with diameter. G-band frequency of MWCNTs is affected by the interlayer spacing as calculated by changes observed in X- ray diffraction profiles. The (002) d-spacing decreases with increasing diameter of the MWCNTs even for the largest diameter distribution nanotubes. The shift in the Bragg angle 2θ and the FWHM of the peak show clear diameter dependence. This is for the first time we have shown that X-ray diffractometry can give the estimate of diameter distribution of Multiwalled carbon nanotubes. This technique is not so useful in case of single walled carbon nanotubes because the number of atoms participating in the scattering event is comparatively too small as compared to MWCNTs and hence poor signal to noise ratio limits us. R. Saito *et. al.* have discussed about the origin of RBM features from innermost wall of MWCNTs and the low frequency Raman shift observed is found to be inversely proportional to the diameter of the innermost tube.[11] we did not observe any RBM features from our samples.

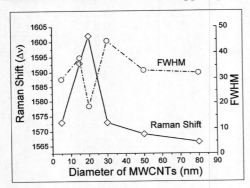

Figure 4. Dependence of G⁻ optical phonon mode Raman shift and FWHM on average diameter of multiwalled carbon nanotubes.

ESR spectra of carbon nanotubes were too broad. As observed broad spectra was deconvoluted into two Gaussian components. Gaussian peak profile gives better fit than the Lorentzian spectral profiles. These g-values were plotted against their average diameter and it shows diameter dependence. A huge shift in the g-value is observed for peak 1 from 4.48373 to 2.82805, whereas peak 2 show random behavior for lower diameter MWCNTs but for diameters above 20 nm it continuously increases from 2.06531 to 2.17057. The peak broadening of these two peaks shows complementary behavior. The high g-value (peak1) is attributed to the magnetic impurities present in the sample. The g-value of ~ 2.17057-2.06531 is attributed to the interaction between the conduction electrons in the nanotubes trapped at defects or magnetic ions site. The more the deviation from the free electron 'g' value, the more is the localization by defects. It indicates that the in case of low diameter nantoube the number of defects is random and more as compared to high diameter nanotubes.

Thermogravemetric analysis is a standard analytical technique for the determination of the amount of metal catalyst and relative amount of different forms of carbon. TGA data is analyzed in terms of combustion temperatures of SWCNTs, amorphous carbon and graphitic nanoparticles.[12] For the first time we have performed a systematic study of diameter dependence on oxidation temperature.[13] Lower diameter MWCNTs get oxidized at comparatively much lower temperature. This is due to larger strain between carbon bonds. XRD results also show similar results. Smaller diameter nanotubes are believed to oxidize at lower temperature due to higher curvature strain.

CONCLUSIONS

We have found systematic diameter dependence of physical properties of MWCNTs. X-ray diffraction profiles and Raman spectra shows diameter dependent spectral profiles in terms of peak position and line width. Broad ESR spectra of MWCNTs deconvoluted into two Gaussian peaks shows diameter dependence in terms of g-value shifts. One ESR around g-value of ~ 2.17057-2.06531 approaches to free electron g value as the diameter of MWCNTS increases. This shows that higher diameter MWCNTS have lesser defects or can be said that they are highly crystalline. Clear diameter dependence is shown in oxidation temperatures of these MWCNTS. Lower diameter MWCNTs shows lower oxidation temperature due to high lattice strain.

ACKNOWLEDGMENTS

We thank DST (Nos. SR/55/NM-01/2005) Govt. of India for supporting the TEM facility at CIF IITG. One of the authors (D. K. Singh) thanks CSIR for SRF Fellowship.

REFERENCES

1. (a) Michael J. O'connell, Sergei M. Bachilo, Chad B. Huffman, Valerie C. Moore, Michael S. Strano, Erik H. Haroz, Kristy L. Rialon, Peter J. Boul, William H. Noon, Carter Kittrell, Jianpeng Ma, Robert H. Hauge, R. Bruce Weisman and Richard E. Smalley, *Science*, **297**, 593-596 (2002); (b) Zhengtang Luo, Lisa D. Pfefferle, Gary L. Haller and Fotios Papadimitrakopoulos, *J. Am. Chem. Soc.* **128**, 15511-15516 (2006); (c) Sergei M. Bachilo, Michael S. Strano, Carter Kittrell, Robert H. Hauge, Richard E. Smalley and R. Bruce Weisman, *Science* **298**, 2361-2365 (2002).

2. (a) A. M. Rao, E. Richter, Shunji Bandow, Bruce Chase, P. C. Eklund, K. A. Williams, S. Fang, K. R. Subbaswamy, M. Menon, A. Thess, R. E. Smalley, G. Dresselhaus and M. S. Dresselhaus, *Science* **275**, 187-190 (1997); (b) M. S. Dresselhaus, G. Dresselhaus, R. Saito and A. Jorio, *Physics Reports* **409**, 47-99 (2005); (c) A. Jorio, R. Saito, J. H. Hafner, C. M. Lieber, M. Hunter, T. McClure, G. Dresselhaus and M. S. Dresselhaus, *Phys. Rev. Lett.* **86**, 1118-1121 (2001).

3. X. Liu, T. Pichler, M. Knupfer, M. S. Golden, J.Fink, H.Kataura and Y. Achiba, *Phys. Rev. B* **66**, 045411 (2002).

4. O. Jost, A. A. Gorbunov and W. Pompe, T. Pichler, R. Friedlein, M. Knupfer, M. Reibold, H.-D. Bauer, L. Dunsch, M. S. Golden and J. Fink, *Appl. Phys. Lett.* **75**, 2217-2219 (1999).
5. L. Alvarez, A. Righi, S. Rols, E. Anglaret and J. L. Sauvajol , *Phys. Rev. B* **63**, 153401 (2001).
6. J. M. Benoit, J. P. Buisson, O. Chauvet, C.Godon, and S. Lefrant, *Phys. Rev. B* **66**, 073417 (2002).
7. C. Thomsen, *Phys. Rev. B* **61**, 4542-4544 (2000).
8. Xinluo Zhao, Yoshinosi Ando, Lu- Chang Qin, Hiromichi Katura, Yutaka Maniwa and Riichirio Saito, *Appl. Phys. Lett.* **81**, 2550-2552 (2002).
9. C.-H. Kiang, M. Endo, P. M. Ajayan, G. Dresselhaus and M. S. Dresselhaus, *Phys. Rev. Lett.* **81**, 1869-1872.
10. Hiroyuki Nii, Yoshiyuki Sumiyama, Hamazo Nakagawa and Atsuhiro Kunishige, *Appl. Phys. Express* **1**, 064005 (2008).
11. X. Zhao,Y. Ando, L.-C. Qin, H. Katura, Y. Maniwa and R. Saito, *Physica B* **323**, 265-266 (2002).
12. M. Zhang, M. Yudasaka, A. Koshio and S. Iijima, *Chem. Phys. Lett.* **364**, 420-426 (2002).
13. Dilip K. Singh, Parameswar K. Iyer and P. K. Giri, Unpublished.

Room Temperature Growth Of Nano-crystalline InN Films On Flexible Substrates By Modified Activated Reactive Evaporation

S. R. Meher, Kuyyadi P. Biju and Mahaveer K. Jain

Department of Physics, Indian Institute of Technology, Madras, Chennai- 600036

Abstract. Nano-crystalline *c*-axis oriented indium nitride (InN) thin films were prepared on amorphous polycarbonate, polyimide and glass substrates by modified activated reactive evaporation (MARE) method without any intentional heating of the substrate. The films show strong visible photoluminescence (PL) peak at ~ 1.95 eV indicating the band edge transition which is in good agreement with the optical absorption. The shift in the band gap from reported value is mainly due to Burstein-Moss shift and presence of residual oxygen. InN film grown on inexpensive flexible substrates at room temperature opens opportunity for large scale device applications like solar cells and displays.

Keywords: InN, Modified activated reactive evaporation, flexible substrate
PACS: 01.30.Cc

INTRODUCTION

The group-III nitride semiconductors have evolved as one of the most promising candidates for optoelectronic device applications in the past decade. The indium nitride (InN) thin films have got potential applications in the fabrication of optoelectronic devices, low cost solar cells with high efficiency, optical coatings and various sensors [1,2,3]. Recently, the photoluminescence and optical absorption data for the InN films grown by molecular beam epitaxy (MBE) revealed the band gap energy of InN ~ 0.7 eV [4] in contrast to the band gap energy of ~ 1.9 eV [5,6] obtained for the films grown by reactive sputtering. This difference in the band gap is attributed to the Burstein-Moss shift, presence of oxide precipitates, the formation of indium clusters and other stoichiometric related defects. Flexible electronics has emerged as one of the challenging fields in the recent past. It has found enormous commercial applications in large area sensors, displays and wearable computers. In the present work, we have grown InN films by modified activated reactive evaporation (MARE) [7] method which is primarily a combination of conventional activated reactive evaporation (ARE) and biased sputtering. The low substrate temperature in MARE allowed us to grow these films on glass as well as on flexible substrates like polyimide and polycarbonate.

CP1147, *Transport and Optical Properties of Nanomaterials—ICTOPON - 2009*, edited by M. R. Singh and R. H. Lipson
© 2009 American Institute of Physics 978-0-7354-0684-1/09/$25.00

EXPERIMENTAL DETAILS

InN thin films on glass and flexible polymer substrates (polycarbonate and polyimide) were grown by MARE method. In conventional ARE [8,9], indium is thermally evaporated in the presence of rf plasma. But, in case of MARE, we have placed the substrate on the rf cathode itself (figure-1). Therefore, the substrate here is biased by the rf voltage of the cathode. MARE method provides independent control of indium species by controlling thermal evaporation rates and of nitrogen species by controlling the partial pressure and plasma power, as also in case of ARE. But, the key advantage in this method is that the energy of nitrogen species striking the substrate is much higher than ARE, as they are accelerated by the cathode voltage. The higher energy of nitrogen ions enables room temperature (no intentional substrate heating) growth of InN thin films. The higher mobility of the nitrogen ad-atoms due to its high kinetic energy, and the higher energy of deposited indium, as they are in excited states after passing through the plasma, ensures that nitrogen reaches all deposited indium metal and gets reacted and therefore no free metal clusters are left as in the case of conventional ARE and reactive sputtering.

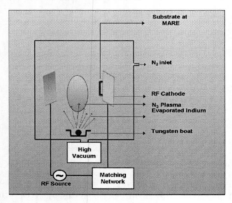

FIGURE 1. Schematic of InN thin film growth using MARE

Aluminium rf electrode was used as the cathode which has very low etch rate and was coated with a thick layer of indium in order to avoid impurity from sputtering of the cathode. Before the deposition, the vacuum chamber was evacuated to 1×10^{-6} Torr using a cryo pump, and repeatedly purged with ultra high pure (UHP) grade argon gas in order to reduce the residual gas impurity in the chamber. The UHP grade nitrogen gas was excited by an rf source (13.5 MHz) through a matching network coupled to parallel plates. After achieving stable plasma condition at 3 mTorr pressure, elemental indium was thermally evaporated at a constant rate of 2 Å/s. All the films were grown at an rf power of 100 W. The thickness of the InN films was found to be ~ 400 nm measured using the spectrophotometer (Filmetrics F20).

The structural properties of the films were studied using Phillips X'Pert Pro X-ray diffractometer with a Cu Kα radiation. The surface morphology was investigated using Digital Instruments Nanoscope IV atomic force microscope (AFM) in contact mode.

The transmission spectra in the wavelength range 300-2000 nm were recorded using a Jasco UV-Vis-NIR spectrophotometer. Raman spectra were obtained through Confocal Raman Instrument (CRM-200) with Ar-ion laser light source (514.5 nm). The photoluminescence (PL) measurements were carried out by an Ar-ion laser excitation source (514.5 nm).

RESULTS AND DISCUSSION

Figure-2 shows the X-ray diffraction pattern for the InN films grown on glass, polycarbonate and polyimide substrates. All the films exhibit hexagonal wurtzite structure of InN with highly preferential c-axis orientation. It can be easily inferred from the XRD pattern that no In traces are present in the film. The average crystallite size (D) was calculated using Scherrer's semi-empirical formula

$$D = \frac{0.9\lambda}{\beta \cos\theta} \tag{1}$$

for the (0002) orientation, where λ = 1.54 Å and $\beta = B - b$ (B being the observed FWHM and b is the instrument function determined from the monocrystalline silicon diffraction line). The average crystallite size for the films grown on different substrates was found to be ~ 20 nm. The lattice parameter c for the InN film grown on

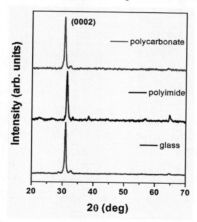

FIGURE 2. XRD of InN films grown on different substrates

glass and polycarbonate substrate was found to be 5.77 Å which is in accordance with the reported values of InN films grown on glass substrates [10]. However, for the film grown on polyimide substrate, the lattice parameter was found to be 5.69 Å which is mainly due to the in-built strain present in the polyimide substrate.

The AFM images of the InN films grown on different substrates are shown in figure-3. The 3-dimensional InN islands are clearly visible for all the films. The surface roughness was found to be more for the films grown on polymer substrates. This is mainly due to the fact that the surface of the polymer substrates becomes rougher due to the bombardment of the high energy nitrogen ions.

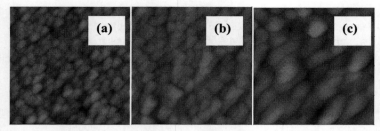

FIGURE 3. AFM images of InN films grown on different substrates (1 × 1 μm scan area). (a) Glass (b) Polycarbonate and (c) Polyimide.

The Raman spectra at room temperature in the spectrum range 300-800 cm^{-1} for the InN films grown on glass and polyimide substrate are shown in figure-4. The Raman

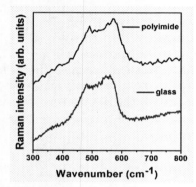

FIGURE 4. Raman spectra of InN films grown on different substrates

scattering was recorded in the backscattering geometry configuration using Ar-ion laser at 514.5 nm. The Raman spectra shows two phonon modes A_1 (LO) and E_2 (high) at 570 cm^{-1} and 490 cm^{-1} respectively. Gwo *et. al.* [11] have reported the A_1 (LO) and E_2 (high) peak to be at 580 cm^{-1} and 490 cm^{-1} respectively. The observed shift in the A_1 (LO) mode may be due to the crystallite size effect. The two phonon modes correspond to the hexagonal wurtzite structure of InN which is in consistent with the XRD results.

Figure-5 shows the UV-Vis-NIR transmission spectra for the InN films grown on glass and polycarbonate substrates. The transmission spectrum for the film grown on polyimide substrate is not shown here as polyimide starts to absorb at ~ 650 nm. Both the films are ~ 70% transparent in the visible region. Again, strong free carrier absorption is clearly seen for both the films. From the transmission spectra, the band gap values were calculated using the relations given below [12]:-

$$\alpha = \frac{2.303 \log(1/T)}{d} \tag{2}$$

$$\alpha h\nu = A(h\nu - E_g)^{1/2} \tag{3}$$

where, α is the absorption coefficient, T is the transmittance, d is the film thickness, E_g is the direct band gap energy and A is a constant. So, the energy band gap values can

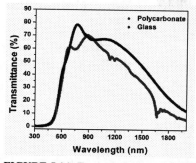 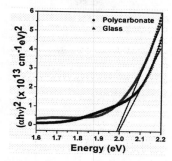

FIGURE 5 (a). Transmission spectra of InN films grown on different substrates (b) Corresponding squared absorption coefficient plots

be determined by plotting the squared absorption coefficient (α^2) as a function of photon energy ($h\nu$) and linearly extrapolating to $\alpha = 0$. The band gap for the films grown on glass and polycarbonate substrates was found to be 1.99 eV.

Figure-6 shows the PL spectra at room temperature under excitation by Ar-ion laser (514.5 nm) for the InN films grown on glass and polymer substrates. All the films

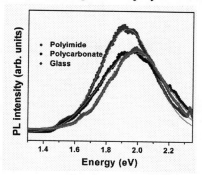

FIGURE 6. PL spectra of InN films grown on different substrates

exhibit strong band edge emission at ~ 1.95 eV which is also reported by several other authors [13,14,15]. The large departure of the observed band gap from the actual value of 0.7 eV is mainly attributed to the high carrier concentration (Burstein-Moss shift) [16], presence of residual oxygen and other stoichiometric related defects. According to Wu *et al.* [16], due to the small effective mass of InN, the Fermi surface in the conduction band shows a strong dependence on the free electron concentration. Thus, free electrons can shift the absorption edge and PL peak to higher energy due to the band filling effect and is known as the Burstein-Moss shift which as a function of carrier concentration is given by,

$$\Delta E = \frac{\hbar^2}{2m^*}\left(3\pi^2 n\right)^{2/3} \qquad (4)$$

Room temperature Hall effect measurement for our sample shows a free electron concentration of $\sim 7 \times 10^{20}$ cm^{-3}. For an electron carrier concentration (n) of 7×10^{20} cm^{-3}, the effective mass (m^*) = 0.26 [16]. So, the Burstein-Moss shift is calculated to be 1.1 eV. Hence, we can certainly attribute the observed band gap shift to Burstein-Moss shift. The obtained values of carrier concentration and band gap are in good agreement with the result proposed by Walukiewicz *et al.* [17]. However, little shift may also be due to the presence of residual oxygen in the InN films [18]. The oxygen content in InN films were found to be $\sim$ 7-8 % as measured from energy dispersive X-ray analysis (EDX).

CONCLUSIONS

In conclusion, we have successfully grown highly preferential *c*-axis orientated InN thin films on glass as well as flexible polymer substrates (polycarbonate and polyimide) by MARE technique. The low substrate temperature in MARE allowed us to grow these films on flexible substrates. The XRD and Raman spectra show the hexagonal wurtzite nature of the film. From the PL measurement and transmission spectra the band gap was found to be $\sim$ 1.95 eV. The large band gap observed is attributed to the Burstein-Moss shift and presence of residual oxygen. InN film grown by this method on inexpensive flexible substrates opens new opportunity for large scale device applications like solar cells and displays.

REFERENCES

1. S. Strite and H. Morkoc, *J. Vac. Sci. Technol. B* **10**, 1237 (1992).
2. K. L. Westra, R. P. W. Lawson and M. J. Brett, *J. Vac. Sci. Technol. A* **6**, 1373 (1988).
3. A.G. Bhuiyan, A. Hashimoto and A. Yamamoto, *J. Appl. Phys.* **94**, 2779 (2003).
4. V. PaCebutas, G. Aleksejenko, A. Krotkus, J. W. Ager III, W. Walukiewicz, H. Lu and W.J. Schaff, *Appl. Phys. Lett.* **88**, 191109 (2006).
5. H. J. Hovel and J. J. Cuomo, *Appl. Phys. Lett.* **20**, 71 (1972).
7. K. P. Biju, A. Subrahmanyam and M. K. Jain, *J. Phys. D: Appl. Phys.* **41**, 155409 (2008).
8. A. K. Mann, D. Varandani, B. R. Mehta and L. K. Malhotra, *J. Appl. Phys.* **101**, 084304 (2007).
9. S. J. Patil, D. S. Bodas, A. B. Mandale and S. A. Gaga, *Appl. Surf. Sci.* **245**, 73 (2005).
10. T. L. Tansley and C. P. Foley, *J. Appl. Phys.* **59**, 3241 (1986).
11. S. Gwo, C. L. Wu, Shen C.H. Shen, Chang W.H. Chang, T. M. Hsu, J.S. Wang, and J.T. Hsu, *Appl. Phys. Lett.* **84**, 3765 (2004).
12. J. Tauc, R. Grigorovici and A. Vancu, *Phys. Status Solidi (b)* **15**, 627 (1966).
13. V. M. Naik, R. Naik, D. B. Haddad, J. S. Thakur, G. W. Auner, H. Lu and W. Schaff, *Appl. Phys. Lett.* **86**, 201913 (2005).
14. Q. X. Guo, T. Tanaka, M. Nishio, H. Ogawa, X. D. Pu and W. Z. Shen, *Appl. Phys. Lett.* **86**, 231913 (2005).
15. T. Yodo, H. Yona, H. Ando, D. Nosei and Y. Harada, *Appl. Phys. Lett.* **80**, 968 (2002).
16. J. Wu, W. Walukiewicz, S. X. Li, R. Armitage, J. C. Ho, E. R. Weber, E. E. Haller, H. Lu, W. J. Schaff, A. Barcz and R. Jakiela, *Appl. Phys. Lett.* **84**, 2805 (2004).
17. W. Walukiewicz, J. W. III Ager, K. M. Yu, Z. L. Weber, J. Wu, S. X. Li, R. E. Jones and J. D. Denlinger, *J. Phys. D: Appl. Phys.* **39**, R83 (2006).
18. M. Yoshimoto, H. Yamamoto, W. Huang, H. Harima, J. Saraie, A. Chayahara and Y. Horino, *Appl. Phys. Lett.* **83**, 3480 (2003).

Relative Humidity Sensing Properties Of Cu₂O Doped Zno Nanocomposite

N. K. Pandey*, K. Tiwari, A. Tripathi, A. Roy, A. Rai, P. Awasthi

*Sensors and Materials Research Laboratory, Department of Physics,
University Of Lucknow, U.P., Pin-226007, India
E-Mail: nkp1371965@rediffmail.Com

Abstract. In this paper we report application of Cu_2O doped ZnO composite prepared by solid state reaction route as humidity sensor. Pellet samples of $ZnO\text{-}Cu_2O$ nanocrystalline powders with 2, 5 and 10 weight % of Cu_2O in ZnO have been prepared. Pellets have been annealed at temperatures of 200-500°C and exposed to humidity. It is observed that as relative humidity increases, resistance of the pellet decreases for the humidity from 10 % to 90 %. Sample with 5 % of Cu_2O doped in ZnO and annealed at 500°C shows best results with sensitivity of 1.50 MΩ/%RH. In this case the hysteresis is low and the reproducibility high, making it the suitable candidate for humidity sensing.

Keywords: ZnO, Cu_2O, Humidity Sensor, Electrical Properties, Annealing Temperature.
PACS: 81.05.Mh, 81.05.Rm, 81.07.Wx, 81.07.Bc, 81.16.Be, 81.20.Ev, 07.79.Cz,68.37.Yz

INTRODUCTION

Research has been going on to find suitable material that shows good sensitivity over large range of relative humidity (RH), low hysteresis and properties that are stable. Ceramic humidity sensors based on porous and sintered oxides have attracted much attention due to their chemical and physical stability [1-2]. Furthermore, the humidity sensors of thin film or pellet type having nanosize grains and nonporous structures have drawn much interest because of the high surface exposure for adsorption of water molecules. Yadav et. al [3] have reported a humidity sensor based on ZnO nanomaterials. Doping is an attractive and effective method for manipulating different applications of semiconductors [4-6]. Doped Zinc Oxide exhibits various properties, different types of morphologies and has many applications [7]. Copper oxide (Cu_2O) has been chosen for addition to ZnO because of the behavior of this as an acceptor impurity in n-type ZnO and has significant effect on the electrical and optical properties of ZnO [8-9]. Jayanthi et. al [10] have studied dopant induced morphology changes in ZnO nanocrystals. We report application of Cu2O doped ZnO prepared by solid-state reaction route for moisture sensing, which to the best of our knowledge, has not been reported so far in literature. The sensitivity is defined as the change in resistance per unit % RH.

CP1147, *Transport and Optical Properties of Nanomaterials—ICTOPON - 2009*, edited by M. R. Singh and R. H. Lipson
© 2009 American Institute of Physics 978-0-7354-0684-1/09/$25.00

EXPERIMENTAL

The nanocomposite ZnO-Cu$_2$O has been prepared by solid-state reaction method. The starting material is ZnO (Qualizen, 99.99% pure) and Cu$_2$O powder (Loba Chemie, 99.9%). 10 weight % glass powders have been used as binders. ZnO has been mixed with 2 weight % (Sample CZ-2), 5 weight % (Sample CZ-5) and 10 weight % (Sample CZ-10) of Cu$_2$O powder uniformly and made fine by grinding in mortar with pestle for two hours. The resultant powder has been pressed in a pellet shape by uniaxially applying pressure of 260 MPa in a hydraulic press machine (M.B. Instruments, Delhi, India) at 27°C (room temperature). The pellet sample prepared is in disc shape having a diameter of 8 mm and thickness of the sample 4 mm. The pressed powder pellet has been sintered in air at temperatures 200-500°C for 3 hours in an electric muffle furnace (Ambassador, India). After sintering, the sample has been exposed to humidity in a specially designed humidity chamber. Inside the humidity chamber, a thermometer (±1°C) and standard hygrometer (Huger, Germany, ±1% RH) are placed for the purpose of calibration. Variation in resistance has been recorded with change in relative humidity. Relative humidity has been measured using the standard hygrometer. Variation in resistance of the pellet has been recorded using a resistance meter (Sino meter, ±1 MΩ, model: VC-9808). Copper electrode has been used to measure the resistance of the pellet. The resistance of the pellet has been measured normal to the cross-section of the pellet. Standard solution of potassium sulphate has been used as humidifier and potassium hydroxide as de-humidifier.

RESULTS AND DISCUSSION

Variation in resistance with the change in relative humidity for the sensing elements of Cu$_2$O doped ZnO for temperatures 200-500°C have been plotted in figs.1 to 4. All these graphs show repeatable results. There is uniform decrease in the value of resistance with increase in the % RH for samples CZ-2 and CZ-5 for all the annealing temperatures. However, sample CZ-10 shows high value of sensitivity for the initial range of 10 to 40 % RH and a very low value of sensitivity for range 40-90 % RH for the annealing temperature of 200°C. But for annealing temperatures 300-500°C sample shows a uniform decrease in the value of resistance with increase in % RH.

To check the reproducibility and the effect of ageing, the samples have again been exposed to humidity after six months. Sensing element CZ-5, annealed at 400°C, shows highest value of sensitivity of 1.76 MΩ/%RH. However, in this case, the hysteresis is found to be high, limiting its usability. When CZ-5 is annealed at 500°C, the sensitivity is 1.50 MΩ/%RH, which is lower. But the results for this annealing temperature 500°C have been found to be reproducible. Moreover, the hysteresis in this case is very low. The repeatability graph for the sensing element CZ-5 annealed at 500°C is shown in fig.5. At 400°C when the % RH is decreased, the moisture absorbed in nano-pores due to capillary action is not fully removed. Hence high hysteresis is observed. At 500°C the moisture adsorbed is largely removed. Hence the hysteresis is low.

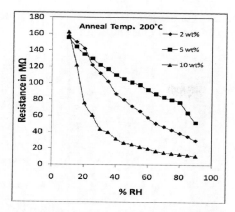

FIGURE 1. Variation in resistance of the samples CZ-2, CZ-5 and CZ-10 with relative humidity

FIGURE 2. Variation in resistance of the samples CZ-2, CZ-5 and CZ-10 with relative humidity

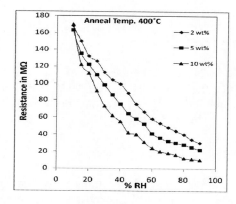

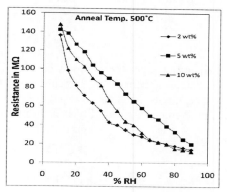

FIGURE 3. Variation in resistance of the samples CZ-2, CZ-5 and CZ-10 with relative humidity

FIGURE 4. Variation in Resistance of the samples CZ-2, CZ-5 and CZ-10 with relative humidity

SCANNING ELECTRON MICROSCOPE STUDY

The study of surface morphology of the samples CZ-2, CZ-5 and CZ-10 has been carried out using Scanning Electron Microscope (LEO-430, Cambridge, England). Micrographs show flakes of ZnO scattered throughout the whole substrate forming a network of pores and flakes. These pores are expected to provide sites for humidity adsorption. The SEM micrographs show that the porous structure is dependent on the composition. Each composition is characterized by a typical porous structure and small crystallites without inside pores but many inter grain pores. In addition, one can observe that the intergranular pores are linked through the large pores. The pore structures should be regarded as interconnected voids that form a kind of capillary tubes. This structure favors the adsorption and condensation of water vapors. The micrograph for the sensing element CZ-5 annealed at 500°C is shown in figure 6.

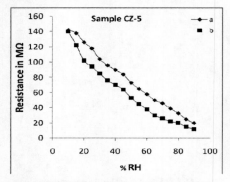

FIGURE 5. Variation in resistance of CZ-5 with % RH for Annealing Temperature 500°C after Six Months: a. Increasing Cycle; b. Another increasing Cycle.

FIGURE 6. SEM micrograph of sample CZ-5 annealed at 500°C

CONCLUSIONS

Sensing element CZ-5 shows best results for annealing temperature of 500°C. For the sensing element CZ-5, annealed at 500°C, the sensitivity is 1.50 MΩ/%RH. For this case, the hysteresis is low and the reproducibility high, making it the most suitable sensing element for humidity sensing in the 10 to 90 % R range.

ACKNOWLEDGMENTS

Authors would like to thank the University Grants Commission, India for providing the financial assistance and to the Geological Survey of India, Lucknow for extending XRD and SEM facilities.

REFERENCES

1. Y. Shimizu, M. H. Arai and T. Seiyama, *Chem. Lett.* **7**, 917 (1985).
2. J. G. Fagan and V. R. W. Amarakoon, *Am. Ceram. Soc. Bull.* **72**, 119 (1993).
3. B. C. Yadav, R. Srivastava, C. D. Dwivedi and P. Pramanik, *Sensors and Actuators B* **131**, 216–222 (2008).
4. M. Gosh, R. Seshadri and C.N.R. Rao, *J. Nanosci. Nanotechnol.* **4**, 136 (2004).
5. R.L. Edson, J. H. L. Eduardo and R. G. Tania, *J. Nanosci. Nanotechnol.* **4**, 774 (2004).
6. M. A. Barakat, G. Hayes and S. I. Shah, *J. Nanosci. Nanotechnol.* **5**, 759 (2005).
7. B. C.Yadav, R. Srivastava and C. D. Dwivedi, *Philosophical Magazine*, **88**, **7**,1113-1124 (2008).
8. R. E. Dietz, H. Kamimura, M. D. Sturge and A.Yariv, *Phys. Rev.,* **132**, 1559 (1963).
9. S. Kishimoto, T. Yamamoto and Y. Nakagawa, *Superlattices Microstruct.* **39**, 306 (2006).
10. K. Jayanti, S. Chawla , K. N. Sood, M. Chhibara and S. Singh, *Applied Surface Science* **255**, 5869-5875 (2009).

OTHER TOPICS IN NANOMATERIALS

Nanobarium Titanate As Supplement To Accelerate Plastic Waste Biodegradation By Indigenous Bacterial Consortia

Anil Kapri[a], M. G. H. Zaidi[b] and Reeta Goel[a]

[a]Department of Microbiology, G.B. Pant Univ. of Ag. & Tech., Pantnagar-263145, India.
[b]Department of Chemistry, G.B. Pant Univ. of Ag. & Tech., Pantnagar-263145, India.

Abstract. Plastic waste biodegradation studies have seen several developmental phases from the discovery of potential microbial cultures, inclusion of photo-oxidizable additives into the polymer chain, to the creation of starch-embedded biodegradable plastics. The present study deals with the supplementation of nanobarium titanate (NBT) in the minimal broth in order to alter the growth-profiles of the Low-density polyethylene (LDPE) degrading consortia. The pro-bacterial influence of the nanoparticles could be seen by substantial changes such as shortening of the lag phase and elongation of the exponential as well as stationary growth phases, respectively, which eventually increase the biodegradation efficiency. *In-vitro* biodegradation studies revealed better dissolution of LDPE in the presence of NBT as compared to control. Significant shifting in λ-max values was observed in the treated samples through UV-Vis spectroscopy, while Fourier transform infrared spectroscopy (FTIR) and simultaneous thermogravimetric-differential thermogravimetry-differential thermal analysis (TG-DTG-DTA) further confirmed the breakage and formation of bonds in the polymer backbone. Therefore, this study suggests the implementation of NBT as nutritional additive for plastic waste management through bacterial growth acceleration.

Keywords: NBT; LDPE; Biodegradation; Consortium; FTIR; TG-DTG-DTA.
PACS: 87.85.Rs

INTRODUCTION

Recently, the influence of different inorganic nanoparticles on bacterial growth has been documented by researchers worldwide. Silver nanoparticles have shown antibacterial and antiviral properties (Oka et al. 1994; Oloffs et al. 1994), while cobalt-ferrite nanoparticles have reported to increase the growth of *Escherichia coli* and *Corynebacterium xerosis* (Flores et al. 2004). Nanometric silicon particles have also shown to alter the growth-profiles of bacteria (Perez et al. 2003). Several inorganic nanoparticles, including silica, silica/iron oxide, and gold have been documented to exhibit no negative influence on the growth and activity of *E. coli* (Williams et al. 2006). In the present study, nanobarium titanate (NBT), an oxide of barium and titanium with the chemical formula, $BaTiO_3$, has been used to alter the growth profiling of LDPE-degrading consortium. This has been done with the view of accelerating the biodegradation potential of the consortium.

CP1147, *Transport and Optical Properties of Nanomaterials—ICTOPON - 2009*, edited by M. R. Singh and R. H. Lipson
© 2009 American Institute of Physics 978-0-7354-0684-1/09/$25.00

METHODOLOGY

NBT of size 38 nm was procured from polymer division, DRDO, Kanpur, India. Three potential bacterial strains *viz. Microbacterium* sp. strain MK3, *Pseudomonas putida* strain MK4 and *Bacterium Te68R* strain PN12 (accession numbers DQ318884, DQ318885 and DQ423487 respectively), were selected for consortium development and LDPE biodegradation, as standardized earlier (Satlewal et al. 2008). For the *in-vitro* biodegradation assay, 100 ml minimal broth (pH 7.0±0.2) was taken in 250 ml Erlenmeyer flasks containing powdered LDPE at a concentration of 5 mg/ml (Satlewal et al. 2008). The flasks were inoculated with 300 μl of active consortium. The assay was performed with respective positive (minimal broth + consortia) and negative (minimal broth + LDPE) controls with and without selected 10.6 nm sonicated NBT, respectively. Sonication was done at 50-60 Hz with 0.3 sec repeating duty cycles, for 2.5 min (LabsonicU B.Brown, USA) to remove the cluster formation in NBT. The flasks were incubated at 37^0C with continuous shaking (150 rpm). Bacterial growth was determined by measuring optical density (OD) at 600 nm using spectrophotometer (Perkin Elmer). Degraded samples were recovered from the broth after the consortium had attained stationary growth phase. Degraded compound was recovered from the broth after filtration and subsequent evaporation of the filtrate. The residue left after filtration was collected and centrifugation of the filtrate was done at 5000 rpm for 15 min to remove bacterial biomass. Further, supernatant was kept in oven at 60^0C for overnight to evaporate water and the residual sample was recovered and analyzed by FTIR and TG-DTG-DTA taking pure LDPE as control. FT-IR spectra were recorded on Perkin Elmer FT-IR Spectrophotometer in KBr, while the latter was performed over Perkin Elmer (Pyris Diamond) thermal analyzer under N_2 atmosphere (200 ml/min) from 28^0C to 500^0C at 5^0C/min.

RESULTS AND DISCUSSION

In-vitro Biodegradation Assay

The presence of LDPE in minimal broth brings about a significant increase in the bacterial growth which suggests that the consortium is using LDPE as the sole C-source for its growth. However, the growth of the consortium in LDPE is further enhanced by the supplementation of NBT (Fig. 1). This is verified by the shortening of the lag-phase (12 h) and elongation of the exponential phase (12 to 120 h). The alterations in growth profile suggest that the metabolically active state of the consortium was not only attained much earlier but was also prolonged by the action of NBT. This accelerated initial growth has also been observed in case of Nanometric silicon nanoparticles towards *C. xerosis* bacterial cultures (Perez et al. 2003). Further, shifting in the λ-max of LDPE was observed from 209 nm (pure LDPE) to 225.3 nm after 4 d in the absence of NBT. Similar shifts in the λ-max have been reported in case of LDPE and HDPE biodegradation (Satlewal et al. 2008; Soni et al. 2008). However, in the presence of NBT, the λ-max shifted to 224.11 nm within 2 d. This suggests changes in the polymer structure occurring faster in the presence of NBT as compared to its absence by the degradative action of consortium.

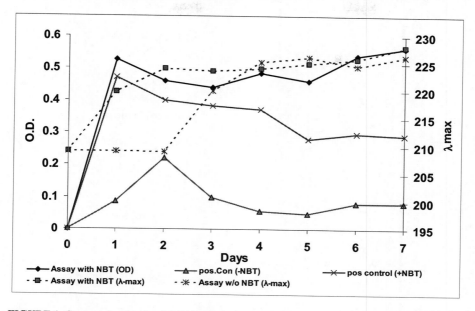

FIGURE 1. Comparative *in-vitro* LDPE biodegradation assay in the absence and presence of NBT.

FTIR Analysis

Pure LDPE has shown FT-IR absorptions (KBr, cm^{-1}) corresponding to ρ CH$_2$ (720.2), δ CH$_2$ (1465.2), δ CH$_3$ (symmetrical, 1352.6), >CH$_2$ deformation (1595.1), ν_s CH$_2$ (2850.6), ν_{as} CH$_2$ (2919.9), ν CH (3426.0), ν_{as} CH$_3$ (3030.8) along with a pair of combination bands due to δ CH$_2$ and ρ CH$_2$ at 2151.8 and 2368.0, respectively (Fig. 2). Biodegradation in the presence of consortium without NBT has rendered LDPE with absorptions corresponding to δ CH$_2$ (1443.5), >CH$_2$ deformation (1658.9), along with a combination band due to δ CH$_2$ and ρ CH$_2$ at 2370.3, respectively. This sample has shown absence of various absorptions corresponding to methyl and methylene groups as observed in pure LDPE. Further, introduction of ν C-O frequencies (1095.8) was observed in the biodegraded samples due to inclusion of O atoms into the hydrocarbon polymer backbone. However, pure LDPE did not show ν C-O frequencies. Similar changes in the FTIR spectra have been reported in case of LDPE and HDPE biodegradation (Satlewal et al. 2008; Soni et al. 2008). Further, shifts in the CH and C=O stretching frequencies in polycarbonate spectra have been reported due to biodegradation inflicted by *Arthrobacter* and *Enterobacter* species (Goel et al. 2008). Deviations in CH$_2$ frequencies have also been reported in the FTIR spectra of biodegraded LDPE by thermophilic bacterium *Brevibacillus borstelensis* (Hadad et al. 2005). The presence of NBT in consortia has induced biodegradation of LDPE with identical peaks of reduced transmittance. Further, inclusion of O-atoms into LDPE due to microbial action has introduced ν C-O frequencies corresponding to 1044.3 cm⁻¹.

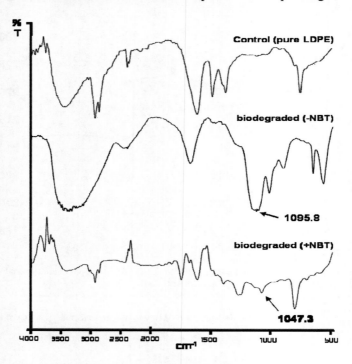

FIGURE 2. Comparative FTIR spectra of biodegraded LDPE in the absence and presence of NBT with reference to pure LDPE

Simultaneous TG-DTG-DTA

Decomposition of pure LDPE was observed in one-step with a steep weight loss in the temperature ranging 400 to 466°C (Fig. 3). The related thermal data has been reproduced in Table 1. However, prior to this temperature, LDPE has shown a DTA endotherm at 107°C with heat of decomposition (ΔH) = 129 mJ/mg. The steep weight loss range of LDPE was supported with a DTA endotherm at 457°C with ΔH = -14.6 mJ/mg and a DTG at 451°C with rate of decomposition 1.79 mg/min. LDPE degraded with consortium in the absence of NBT has shown two-step decomposition at 61°C and 200°C with weight losses of 5.33% and 9.74%, respectively. Formations of multiple DTG and DTA peaks have also been reported in case of consortial-degraded HDPE and LDPE (Satlewal et al. 2008). However, the presence of NBT has induced the thermal decomposition of biodegraded LDPE into four-steps, *viz.*60°C, 200°C, 274°C and 380°C, with weight losses amounting to 3.88%, 9.01%, 11.32% and 14.34%, respectively. The above given weight losses have been supported with DTA endotherms and DTG peaks at their respective temperature ranges (Table 1). These thermal data clearly indicate that a significant loss in the thermal stability of LDPE was observed due to consortia in the presence of NBT particles as compared to control (NBT absent).

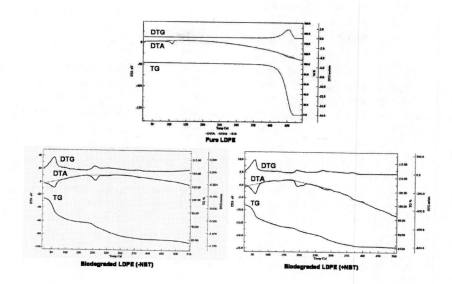

FIGURE 3. Comparative simultaneous TG-DTG-DTA of biodegraded LDPE in the absence and presence of NBT with reference to pure LDPE

TABLE 1. Comparative thermal analysis of biodegraded LDPE in the absence and presence of NBT with reference to their respective controls

Consortium	NBT	DTG peak temp		DTA Exotherm		DTA Endotherm	
		^0C	Rate (µg /min)	^0C	mJ/mg	^0C	mJ/mg
Control	-	451	1790.0	467	14.6	107 457	129 53
+	-	56 205	211.6 71.4	-	-	58 207, 248 362	143 120 5.03
+	+	56 190 273 368	142.3 36.7 38.4 18.5	397	2.31	58 192, 211 280	62.4 41.5 11.4

CONCLUSIONS

The study reveals that a stable suspension of NBT brings about an increase in the growth of LDPE-degrading microbial consortia which increases its biodegradation efficiency. The present investigation would therefore help to increase the efficacy of plastic biodegradation and help in better management of white pollution. The authors also propose the exploration of nanoparticles to influence various other microbial processes for commercial utilization.

ACKNOWLEDGMENTS

This work is supported by DBT grant to RG. Senior author (AK) also acknowledges JNMF, New Delhi, for providing financial assistance during the course of this study. We are also thankful to CDRI (SAIF), Lucknow; Instrumentation centre, IIT Roorkee and NCCS, Pune for FTIR, DTA-DTG-TG and 16S rDNA sequencing, respectively.

REFERENCES

1. M. Flores, N. Colón, O. Rivera, N. Villalba, Y. Baez, D. Quispitupa, J. Avalos, O. Perales, *Mat. Res. Soc. Symp. Proc.* **Vol 820**, Materials Research Society (2004).
2. R. Goel, M.G.H. Zaidi, R. Soni, K. Lata, Y.S. Shouche, *Int. Biodeter. Biodegrad.* **61**, 167-172 (2008).
3. D. Hadad, S. Geresh and A. Sivan, *J. Appl. Microbiol.* **98**, 1093-1100 (2005).
4. M. Oka, T. Tomioka, K. Tomita, A. Nishino and S. Ueda, *Metal-Based Drugs* **1**, 511 (1994).
5. A. Oloffs, C. Crosse-Siestrup, S. Bisson, M. Rinck, R. Rudolvh and U. Gross. *Biomaterials* **15**,753–758 (1994).
6. L. Perez, M. Flores, J. Avalos, L.S. Miguel, L. Fonseca and O. Resto. *Mat. Res. Soc. Symp. Proc.* **Vol 737**, Materials Research Society (2003).
7. A. Satlewal, R. Soni, M.G.H. Zaidi, Y. Shouche and R. Goel. *J. Microbiol. Biotechnol.* **18**, 477-482 (2008).
8. R. Soni, S. Kumari, M.G.H. Zaidi, Y. Shouche and R. Goel. Practical applications of rhizospheric bacteria in biodegradation of polymers from plastic wastes in *Plant Bacteria Interactions. Strategies and Techniques to Promote Plant Growth* edited by I. Ahmad, J. Pichtel and S. Hayat, Wiley-VCH, Weinheim, Germany, 2008, pp.235-243.
9. D.N. Williams, S.H. Ehrman and T.R.P. Holoman. *J. Nanobiotechnol.* **4** doi:10.1186/1477-3155-4-3. (2006).

Structural And Optical Study Of Er^{3+} In Sol-Gel Silicate Glass

S. Rai and P. Dutta

Department of Physics, Dibrugarh University, Dibrugarh -786 004, India

Email: srai.rai677@gmail.com

Abstract: The structural and optical properties of Er^{3+} doped with Al(NO$_3$)$_3$ sol-gel glass have been investigated. Thermo gravimetric analysis (TGA) shows an enormous mass loss between 40° to 190° C. Important structural changes on heating have been revealed by two endothermic peaks of Differential Thermal Analysis (DTA) which is further confirmed by Infra red (IR) spectroscopy and X-Ray Diffraction (XRD) analysis.. X-Ray Diffraction and Scanning Electron Micrograph of the sample heated to 800^{0}C suggests the particle size in nanometers. The oscillator strengths of the transitions in the absorption spectrum are parameterized in terms of three Judd–Ofelt (J.O) intensity parameters (Ω_2, Ω_4 and Ω_6).The radiative property for the potential lasing transition $^4S_{3/2} \rightarrow {}^4I_{15/2}$ of Er^{3+} is found to be very encouraging for the studied glass.

Keywords: Optical materials; rare-earth doped glasses; differential thermal analysis.
PACS: 81.05 Kf, 42.70 Ce, 78.55 Qr, 61.43 Fs

INTRODUCTION

Photonic devices are fabricated from a variety of materials, although the semiconductor and glasses are two constituents for making core components of fiber optics. Several papers have reported optical properties of luminescent species, such as semiconductor quantum size particles and rare earth ions in sol gel glasses which are characterized by quantum confinement effects[1,2]. Silica nanoparticles synthesized using sol gel techniques have generated considerable recent interest in applications. The nature and applications of optical materials have evolved rapidly in recent years. Their role as passive optical elements has been augmented by so called photonic systems. Nano particles prepared from lanthanides are attractive because they are photostable and usually present sharp emission spectra, large stokes shifts and long fluorescence lifetimes. In this work, Er^{3+} ions doped in Al(NO$_3$)$_3$-SiO$_2$ sol-gel glass has been characterized for the optical transitions, using the Judd-Ofelt theory. As these transition properties are sensitive to the surrounding environment of the rare earth ions, the structural composition relations in the Al(NO$_3$)$_3$-SiO$_2$ glass where emissions were detected is investigated using XRD, IR and Scanning Electron Microscopy (SEM) techniques.

CP1147, *Transport and Optical Properties of Nanomaterials—ICTOPON - 2009*, edited by M. R. Singh and R. H. Lipson
© 2009 American Institute of Physics 978-0-7354-0684-1/09/$25.00

EXPERIMENTAL

The glass samples for the study were prepared by using the sol-gel technique described in earlier paper [3].

The glass transition temperatures, crystallization temperatures and thermal stability were studied by subjecting the samples to DTA and TGA. IR spectra were recorded using KBr technique. To determine the crystalline phases formed during the heat treatment, XRD was performed. Information concerning the microstructure of the glass-ceramic material was obtained by studying the SEM micrograph.

The optical absorption of Er^{3+} doped in sol-gel glasses recorded in the wavelength range 400-1000nm. The photoluminescence spectrum was obtained by using excitation wavelength of 385 nm. All the measurements were done at room temperature.

RESULTS AND DISCUSSIONS

a) Structural Characterization:

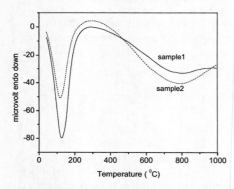

FIGURE 1(A). Comparative DTA plot of the samples at the heating rate of 20^0 C/min

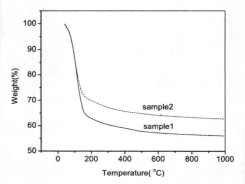

FIGURE 1(B). Comparative TGA plot of the samples at the heating rate of 20^0 C/min

Figures 1(A) and 1(B) shows the DTA and TGA plot for the Er^{3+} doped (sample 1) and Er^{3+} co-doped with $Al(NO_3)_3$ (sample 2) in the SiO_2 glass. The two endothermic peaks that appear in the DTA curves correspond to the glass transition (T_g) and crystallization peak (T_p) temperatures. The weight loss in the 40^0-190^0 C range, indicated by the TGA is due to loss of water and other hydroxyl groups. The characteristic glass transition temperature (T_g), onset of crystallization temperature (T_c) and the crystallization peak temperature (T_p) for the two glass samples are summarized in TABLE I. The value of T_c-T_g can be used to represent the stability against crystallization of the glasses. The larger the T_c-T_g is, the better is the stability against crystallization [4]. Moreover, the parameter $(T_c\text{-}T_g)/T_g$ can be used to represent the glass forming ability [5].The $T_c\text{-}T_g$ parameter for the present glass is 183.35, of the order of tellurite glass [6] which shows very good glass stability, indicating that the glass samples studied are very stable against devitrification, compared to fluoride and fluorophosphates glasses [7].

The IR spectra of Er^{3+}: $Al(NO_3)_3$- SiO_2 air dried glass and one heated to 500^0 C are seen in figure 2(A) and 2(B) respectively. The broad band at 3455 cm^{-1}, assigned to the fundamental stretching vibration of the –OH groups reveals the presence of the hydroxyl groups in the glass. The band around 1636 cm^{-1} corresponds to the bending mode of water molecules and indicates the presence of adsorbed water. The band around 1390 cm^{-1} is assigned to the vibrations of TEOS, ethoxy group. One prominent band in the spectra around 1080 cm^{-1} is due to Si-O-Si asymmetric

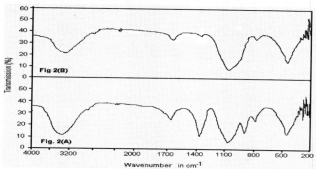

FIGURE 2. I-R spectra of the Er^{3+}: $Al(NO_3)_3$-SiO_2 sol-gel glass, (A) air dried and (B) heated to 500^0 C

stretching mode. Bands observed at 965 cm^{-1} and 800 cm^{-1} are assigned to the Si-O stretching and Si-O-Si symmetric stretching or vibrational modes of ring structures. Band at 475 cm^{-1} is observed due to Si-O-Si bending modes. Presence of absorption structures in the region between 400-200 cm^{-1} is indicative of the modifying effect of the Al on the glass network. In fig 2(B) the most remarkable change is the significant reduction of the TEOS ethoxy group and the merger of the 965 cm^{-1} Si-O stretching stretching bands with Si-O-Si asymmetric stretching mode. The result is consistent with the TGA result shown in fig 1(B) which shows an enormous mass loss within the 40^0 to 190^0C range.

The XRD spectrum is shown in figure 3. The harrow like pattern observed between 5^0 and 40^0 of 2θ in the air dried and sample heated to 500^0 C is attributive of the amorphous nature of silica gel. Appearance of strong peaks for the sample treated at 800^0 C indicates the transition to crystallization phase. The appropriate crystallite size at FWHM determined for the sharper peak at $2\theta \sim 26.67^0$ using the Scherrer formula $G = \lambda/D$ $Cos\theta$ was found to be 27 nm. In the Scherrer formula G is the grain size, λ is the wavelength of X-rays (1.5418 A) used, D is the width of the peak at half maximum and θ is the angle of incidence of the X-ray beam. SEM micrograph of the glass sample (heated at 800^0C) shown in figure 4 suggests the particle size to be in the range of 40 nm.

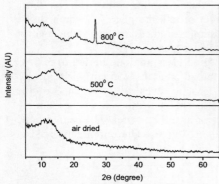

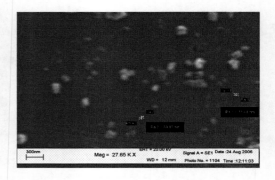

FIGURE 3. XRD Pattern of Er³⁺: Al(NO₃)₃-SiO₂ sol-gel glass for air dried and samples annealed at 500 and 800⁰C (The plots are vertically shifted to show the intensity peaks clearly)

FIGURE 4. SEM Micrograph of Er³⁺: Al(NO₃)₃-SiO₂ sol-gel glass annealed at 800⁰ C

3.2 Optical Characterization:

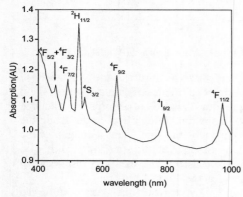

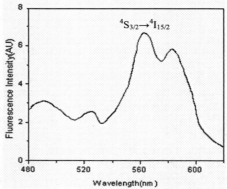

FIGURE 5. Absorption Spectra of Er³⁺: Al(NO₃)₃-SiO₂ glass in the VIS-NIR region. The different transitions from the Er³⁺ ground state ($^4I_{15/2}$) are labeled.

FIGURE 6. Fluorescence spectra of Er³⁺: Al(NO₃)₃-SiO₂ sol-gel glasses with 385 nm excitation

The Vis-NIR absorption spectrum of Er³⁺: Al(NO₃)₃- SiO₂ sol-gel glass is shown in figure 5. The seven bands observed in 400-1000 nm range are ascribed to electric dipole (e-d) $4f'$ to ground state ($^4I_{15/2}$) transitions of Er^{3+} terminating at $^4F_{5/2}+^4F_{3/2}$, $^4F_{7/2}$, $^2H_{11/2}$, $^4S_{3/2}$, $^4F_{9/2}$, $^4I_{9/2}$ and $^4F_{11/2}$ states.

The experimental oscillator strengths (f_{exp}) of the observed e-d transitions are determined from

$$f_{exp} = 4.318 \times 10^9 \int \varepsilon(\nu)d\nu \qquad (1)$$

which are co-related with its corresponding Judd's expression [8, 9] for oscillator strengths of e-d transitions between initial ΨJ and terminal $\Psi' J'$ states, in a $4f^n$ ground configuration.

The values of experimental and theoretical oscillator strengths are in good agreement with an r.m.s deviation of $\pm 0.256 \times 10^{-6}$ between them and the Judd Ofelt (J.O) intensity parameters are in the order of $\Omega_2 \rangle \Omega_4 \rangle \Omega_6$. Jorgensen and Reisfeld [10] noted that the Ω_2 is indicative of the amount of covalent bonding while the Ω_6 is related to rigidity of the host. The result indicates better covalency of Er–O bond in Er^{3+}: $Al(NO_3)_3$-SiO_2 glass. Apart from covalency, Ω_2 is also dependent on asymmetry in hyper-sensitive transitions i.e. transitions corresponding to $|\Delta J| = 2$. Ω_6, as well as Ω_4, rather insensitive to the surrounding environment of the ion are strongly dependent on the vibrational frequency of Ln^{3+} ions linked to ligand atoms. The high value of this vibronic dependent parameter in Er^{3+}: $Al(NO_3)_3$-SiO_2 glass indicates of its high rigidity. The ratio Ω_4 / Ω_6, defined as the spectroscopic quality factor of materials gain significance for systems like Nd^{3+} doped materials where the emission transitions are independent of Ω_2. For Er^{3+}: systems concerned with optical emission $^4S_{3/2} \rightarrow ^4I_{15/2}$, which are independent of Ω_2, the ratio may be used as a quality indicator.

J.O parameters are regarded as phenomenological parameters that characterize the radiative probabilities within ground state configuration. Using these, several radiative parameters including lifetime of states (τ) and branching ratios (β_r) have been calculated for the $^4S_{3/2} \rightarrow ^4I_{15/2}$ transition (shown in figure 6) in Er^{3+}: $Al(NO_3)_3$-SiO_2 glass and tabulated in TABLE III .

TABLE I. **Summarized glass transition temperatures, onset of crystallization temperatures and the crystallization peak temperatures**

	$T_g(^0C)$	$T_c(^0C)$	$T_p(^0C)$	T_c-T_g	$(Tc - Tg)/Tg$
Sample-1	125.69	291.50	801.63	165.81	1.319
Sample-2	118.16	301.41	798.03	183.25	1.550

TABLE II. Absorption bands, the oscillator strengths (f_{exp} and f_{cal}) of Er^{3+}: $Al(NO_3)_3$-SiO_2 sol gel glass.

Transitions	Wavelength (nm)	Energy(cm^{-1})	f_{exp} x 10^{-6}	f_{cal} x 10^{-6}	JO parameters (x10^{-20}cm$^{2)}$
$^4I_{15/2} \to {}^4F_{5/2}+{}^4F_{3/2}$	452	22123	1.171	1.322	
$^4I_{15/2} \to {}^4F_{7/2}$	488	20491	2.576	2.745	$\Omega_2 = 4.038$
$^4I_{15/2} \to {}^2H_{11/2}$	524	19084	6.974	6.888	
$^4I_{15/2} \to {}^4S_{3/2}$	542	18450	0.919	0.694	$\Omega_4 = 2.073$
$^4I_{15/2} \to {}^4F_{9/2}$	656	15243	3.316	2.748	
$^4I_{15/2} \to {}^4I_{9/2}$	786	12722	0.417	0.428	
$^4I_{15/2} \to {}^4F_{11/2}$	978	10226	0.743	0.796	$\Omega_6 = 1.855$

TABLE III. Laser characteristic parameters for $^4S_{3/2} \to {}^4I_{15/2}$ emission in the Er^{3+}: $Al(NO_3)_3$-SiO_2 sol gel glass.

Transition	$\Delta\lambda_{eff} (nm)$	$\lambda_p (nm)$	$A(s^{-1})$	β_r	$\sigma(\lambda_p).10^{20}$
$^4S_{3/2} \to {}^4I_{15/2}$	29	562	1477.692	0.672	0.24

CONCLUSIONS

The structural and optical properties of Er doped $Al(NO_3)_3$ silicate glasses are studied. The addition of aluminium increases the glass forming ability and shows good optical quality. This is attributed to the structural change in the xerogel because of the presence of Al^{3+} ions. The J.O intensity parameters of Er^{3+} ions that signifies the co-valent bonding and rigidity of the hosts is very satisfactory for $Al(NO_3)_3$- SiO_2 sol-gel glass compared to the other glass hosts. The radiative properties like peak emission cross section and branching ratio that characterize the lasing transition are found to be very encouraging for the $^4S_{3/2} \to {}^4I_{15/2}$ transition in our glass. The results of different studies indicate that the SiO_2-$Al(NO_3)_3$ doped with Er^{3+} ions are promising materials for sensors and amplifier in optical fibers.

REFERENCES

1. M. Yu, J. Lin and J. Fang, *Chem. Mater.*,**17** 1783 (2005).

2. P. S. J. Russell, *J. Lightwave Technol.*, **24** 4729 (2006).
3. S. Hazarika and S. Rai, Opt. Mater., **27**, 173 (2004).
4. M. Liao, S. Li, H. Sun, Y. Fang, L. Hu and J. Zhang, *Mater. Lett.*, **60**, 1783 (2006).
5. D. Dong, Z. Bo, J. Zhu and F. Ma, *J. Non Cryst. Solids*, **204**, 260 (1996).
6. G. Wang, S. Xu, S.Dai, J. Yang, L. Hu and Z. Jiang, *J. Non-Cryst. Solids*, **336**, 102 (2004).
7. N. Rigout, J.L. Adam and J. Lucas, *J. Non-Cryst. Solids*, **184** 319 (1995).
8. B. R. Judd, *Phys. Rev.*, **127**, 750 (1962).
9. G.S. Ofelt, *J. Chem. Phys.*, **37** 511 (1962).
10. C. K. Jorgensen, R. Reisfeld, *J. Less-Common Met.*, **93** 107 (1983).

Evidence of Phonon Condensers at Nanoscale

Abhay Abhimanyu Sagade, A. Ghosh, R. A. Joshi and Ramphal Sharma

Thin Film and Nanotechnology Laboratory, Department of Physics, Dr. Babasaheb Ambedkar Marathwada University, Aurangabad 431004, MS, India.

Abstract. The values of thermovoltage are observed to enhance in metal chalcogenide materials after heavy ion irradiation. The physical basis behind this process is unclear. Here we treated thermal and electrical currents equally and replaced two dimensional thermal system by electrical network incorporating resistors, connected in series and capacitors, connected in parallel. The physical essence, however, and units of the terms of the equations are different. This scheme is called as 'electro-thermal' analogy and it explains qualitatively the hampered motion of phonons from hot end to cold end. Restricting phonons in thermoelectric materials paves the way to enhancement of Seebeck coefficient.

Keywords: Thermoelectricity, Seebeck coefficient, Phonons, Metal chalcogenides.
PACS: 61.80.-x, 71.55.-I, 72.10.Bg, 73.50.Lw, 72.20.Dp, 72.20.Pa, 73.63.Bd, 73.90.+f

INTRODUCTION

Transfer of phonons from hot end to cold end with charge carriers during thermoelectric measurement is the major threat for reduction of Seebeck coefficient (S) in thermoelectric materials. The use of nanowires and quantum superlattices are the better solutions, since the diffuse interface scattering inside the nanostructure materials cannot only reduce the phonon mean free path but can also destroy the coherence of phonons. The possibilities and probabilities of electron-phonon decoupling have been discussed in more details by Ziman [1]. The enhancement in such metal chalcogenide structures (Bi_2Te_3/Sb_2Te_3, PbSeTe etc.) have been demonstrated successfully in the past years [1, 2, 3].

Phonons are quantized travelling elastic waves associated with the displacement of atoms from their equilibrium lattice positions. Zuckermann and Lukes [4] described the means of scattering of phonons from nanoparticles and methods to control the process. The obstacles for scattering includes other phonons, grain boundaries, impurity atoms, structural defects such as vacancies and dislocations, changes in atomic mass isotopes. In general, any features that change the bond stiffness, bond orientation, or mass of adjacent atoms from those of the host lattice are restricting phonon motion. This approach will reduce thermal conductivity by increasing scattering for a wide spectral range of phonons. Therefore we need a matrix with such a scattering centres (Fig. 1) which hamper phonons forward motion and propagate the electrons from hot end to cold end. The absorption of phonons through such a matrix has been investigated using molecular dynamics by Norris [5] on the basis of Mur's absorbing boundary conditions.

CP1147, *Transport and Optical Properties of Nanomaterials—ICTOPON - 2009*, edited by M. R. Singh and R. H. Lipson
© 2009 American Institute of Physics 978-0-7354-0684-1/09/$25.00

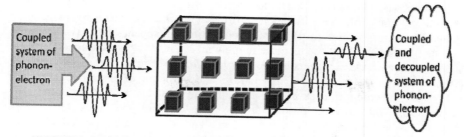

FIGURE 1. The process of breaking a coupling between electron (arrow) and phonon (wave packet). During the motion of their coupled system through the matrix containing embedded nanoparticles, phonons get scattered from the scattering centers (red cubes) and finally at the right end we get single electron, coupled and decoupled system of e-p with same or reduced coupling factor.

The best options available in creating such a matrix are the growth of material using molecular beam epitaxy of desired material with suitable scattering centers. Unfortunately it is a tiresome and tedious job to be performed. The other option is embedding single elemental atom/ion in the matrix using ion beam implantation. In this case the embedded atom forms its own deforming potential and bonding in the matrix causing unexpected changes in transport properties. Also low energy ions generate collision cascades, which disturbs periodicity of lattice. On the other hand swift heavy ion (SHI) bombardment method is useful in creating different zones (tracks) in the matrix with equal/different properties of same material. In the present study, phenomenological model of restricting phonon forward motion through such ion tracks in metal chalcogenide materials is discussed.

EXPERIMENTAL RESULTS

The irradiation experiments were carried out on cadmium sulfide (CdS) and bismuth sulfide (Bi_2S_3) thin films using 100 MeV gold SHI. The detailed experimental work and modifications in materials properties have been reported elsewhere [6].

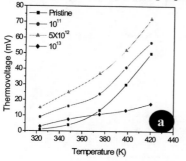
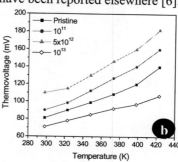

FIGURE 2. The variation of thermovoltage in CdS (a) and Bi_2S_3 (b) thin films with irradiation fluence.

It is observed that particle size, electrical resistivity and thermovoltage of these metal chalcogenide films are function of irradiation fluence. Generally irradiation causes diffusion of atoms across the grain boundaries and appends those two grains to form one bigger grain [6]. The increase in grain size plausibly decreases the electrical resistivity and increase carrier mobility. This task fulfills the requirements to enhance thermovoltage in the material. Fig. 2 shows the variation of thermovoltage as a function of irradiation fluence and temperature in CdS and Bi_2S_3 thin films. The irradiation of 100 MeV gold ions at 10^{11} ions/cm^2 fluence starts to increase thermovoltage in CdS and Bi_2S_3 from 2 to 6 mV and 80 to 90 mV, respectively. This behavior continues up to fluence of 5×10^{12} ions/cm^2 and after it there is sudden drop in thermovoltage at 10^{13} ions/cm^2 fluence in both the films.

DISCUSSION

Metal chalcogenides of bismuth (BiX) and cadmium (CdX, X = S, Se and Te) are very promising materials in solar cells, photodetectors, laser etc. applications. There thermoelectric materials have been extensively studied [7]. The main reason behind conductivity in these materials is the presence of X vacancies making them potential n-type semiconductors. The effects of SHI on their opto-electronic properties have been recently reported [6].

When SHI pass through a solid they loss their energy during their path into the matrix and create tracks of diameter ~ 10 nm. Figure 3 shows the schematic of heavy ion irradiation on solid matrix. The left part shows a passage of heavy ion and track created by it in solid (not to the scale).

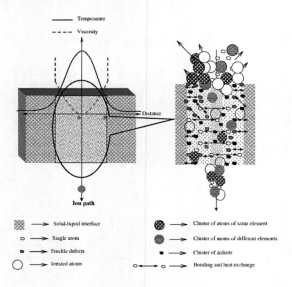

FIGURE 3. Schematic of thermal spike generation in solids.

The dark shaded area is the actual track with radius R from center O. Thermal spike generation in this nanodimensional region will reduce the viscosity of that region in comparison to the solid matrix around it. The temperature distribution in track is a Gaussian one; it has maximum at center O and decays out in the solid matrix near the track. If, for time being, the viscosity of matrix around the track is infinite, then the viscosity of this nanodimensional region is sufficiently less that it can be declared as a liquid. This nanodimensional region is explored in the right part of the figure. Because of high temperature, this region consists of single and ionized atoms, thermalized electrons, defects, etc., species. Hence there is a finite probability of heat exchange between all of these species and formation of new bonds. When cooling down occurs, due to frozen in, all the physical properties of this region get modified. The use of sufficient fluence will modify the whole solid matrix. If the fluences are increased excessively, an overlapping of tracks takes place and this phenomenon occurs for a number of times in particular nanodimensional regions [6].

Therefore, irradiated target consists of matrix/track/matrix (MTM) layered structure perpendicular to its thickness as shown in Fig. 4. These tracks may have smooth or rough interface with matrix in addition to same or different bonding between the atoms. Let us assume that these tracks may be formed at equal distance along X-axis so during thermoelectric measurement this layered structure may act as a quantum superlattice and hamper the motion of phonons from hot end to cold end. This can be explained by using a phenomenological model.

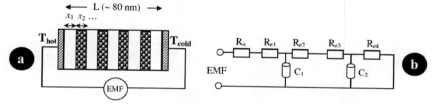

FIGURE 4. Schematic of (a) thermal system used in thermovoltage measurement and (b) equivalent electrical circuit of (a).

The thermal system is thus replaced by an electrical network incorporating resistors, connected in series and capacitors, connected in parallel (Fig. 2). A quantitative analysis at nanoscale is performed by treating both thermal and electric conduction processes equally. The physical essence, however, and units of the terms of the equations are different. This scheme is called as *electro-thermal analogy*. The phenomena of heat and electrical conduction obey the following equations

$$dQ = -\lambda(\partial T / \partial n_t)dA_t \quad \text{and} \quad dI = -\sigma(\partial E / \partial n_e)dA_e \qquad \qquad \dots (1)$$

where dQ and dI are elementary heat flow and electrical current per unit time per unit area dA_t, dA_e, in the direction of the perpendiculars n_t and n_e; T and E are temperature and electrical potential; λ and σ are the thermal and electrical conductivity. Here subscripts t and e indicate thermal and electrical phenomena, respectively.

An application of these equations to a two-dimensional problem of steady state conduction in time, in which the physical properties (λ and σ) are independent of temperature, leads us to the following Laplace differential equations:

$$(\partial^2 T / \partial x_t^2) + (\partial^2 T / \partial y_t^2) = 0$$
$$(\partial^2 E / \partial x_e^2) + (\partial^2 E / \partial y_e^2) = 0 \qquad \ldots (2)$$

These equations expressing temperature and electrical potentials are of identical structure. Analogous phenomena should develop in geometrically similar systems. The boundary conditions may be stated in following ways:
$-\lambda \nabla T = \alpha \Delta T$ or $-\nabla T = \Delta T / (\lambda / \alpha) = \Delta T / \ell_t$, and similarly, $-\nabla E = \Delta E / \ell_e$.

The quantitative relationship between analogous physical parameters may be established by reducing their mathematical descriptions to the dimensionless form. For this purpose a certain value Δt_0 can be taken as a scale for the temperature difference, Δu_0 for the electrical potential and for the linear dimensions ℓ_{t0} and ℓ_{e0}. Then the quantities expressed in a relative scale are

$$\frac{x_t}{\ell_{0t}} = X_t; \ \frac{y_t}{\ell_{0t}} = Y_t; \ \frac{l_t}{\ell_{0t}} = L; \ \frac{\Delta t}{\Delta t_0} = \Theta$$

Having substituted these relationships, eq. (2) acquire the dimensionless form

$$\frac{\Delta t_0}{l_{0t}^2}\left(\frac{\partial^2 \Theta}{\partial X_t^2} + \frac{\partial^2 \Theta}{\partial Y_t^2}\right) = 0 \quad \text{and} \quad \frac{\Delta e_0}{l_{0e}^2}\left(\frac{\partial^2 \Phi}{\partial X_e^2} + \frac{\partial^2 \Phi}{\partial Y_e^2}\right) = 0 \qquad \ldots (3)$$

which are identical at any choice of similar scales for the temperature and electrical potentials.

In investigating transient processes for 2D problems, the differential equations of thermal and electrical conduction have the form

$$\frac{\partial T}{\partial \tau_t} = a\left(\frac{\partial^2 T}{\partial x_t^2} + \frac{\partial^2 T}{\partial y_t^2}\right)$$
$$\frac{\partial E}{\partial \tau_e} = \frac{1}{R_e C_e}\left(\frac{\partial^2 E}{\partial x_t^2} + \frac{\partial^2 E}{\partial y_t^2}\right) \qquad \ldots (4)$$

Here R_e is the electrical resistance per unit length; C_e is the "electrical condenser". Comparison of these equations shows that analogy sets in only if the condition $a = 1/R_e C_e$ is observed.

Therefore, variation of electrical current is proportional to the capacity and variation in voltage

$$dI = C_e \frac{\partial T}{\partial \tau_e} d\tau_e, \qquad \qquad \dots (5)$$

variation in heat flow is proportional to the variation of "heat condenser" or "phonon condenser" of the system and the variation of temperature in time, i.e. by an analogy

$$dQ = C_t \frac{\partial T}{\partial \tau_t} d\tau_t \qquad \qquad \dots (6)$$

Considering graded thermoelectric materials with the thermal conductivity (λ), the Seebeck coefficient (S), and the electrical conductivity (σ) changing with position x, the heat equation at steady state is [8],

$$\frac{d}{dx}\left[k(x) \frac{dT(x)}{dx} \right] = -\frac{J^2}{\sigma(x)} + JT(x) \frac{dS(x)}{dx} \qquad \qquad \dots (7)$$

where J is the electrical current density. If we assume that $\lambda(x)$ and power factor $S(x)^2\sigma(x)$ are constant in a finite range of electrical conductivity values, then even though average S and σ do not change, the local thermoelectric efficiency (ZT) could become larger. The corresponding optimal Seebeck profile includes three sections,

$$S_{opt}(x) = \begin{cases} S_o, & 0 < x < 1/2 \\ (S_o/2)/(1-x), & 1/2 < x < 1 - S_o/2S_L \\ S_L, & 1 - S_o/2S_L < x < 1 \end{cases} \qquad \qquad \dots (8)$$

where position x is normalized with the material length (track separation) and S_0 and S_L represents Seebeck coefficients at the starting and ending positions, respectively.

In short, TMT system acts as "phonon filter" (in analogy with electron filter) in metal chalcogenides. The motion of phonon will get hampered at particular length of TMT layers. In the present investigation this threshold separation occurs at fluence of 5×10^{12} ions/cm^2. An increment in irradiation fluence may reduce the system length and hence the filtering action. At higher fluence, disorder in system will increase and there will be more scattering of electrons too. This will reduce the thermovoltage across the junction.

Hence it can be concluded that controlling local thermovoltage can increase average Seebeck coefficient and ZT. This task can be performed by creating tracks in the material using SHI irradiation at a particular fluence. This critical fluence will depend on the material.

ACKNOWLEDGMENTS

AAS is thankful to CSIR, India for awarding SRF.

REFERENCES

1. J. M. Ziman, *Electrons and Phonons*, Oxford University Press (2001); A. I. Boukai, Y. Bunimovich, J. Tahir-Kheli, J. Yu, W. A. Goddard III, *Nature* **451**, 168-171 (2008).
2. R. Venkatasubramanian, E. Silvola, T. Colpitts and B. O'Quinn, *Nature* **413**, 597-602 (2001).
3. I. A. Hochbaum, R. Chen, R. D. Delgado, W. Liang, E. C. Garnett, M. Najarian, A. Majumdar and P. Yang, *Nature* **451**, 163-167 (2008).
4. N. Zuckermann and R. Luckes, *Phys. Rev.* B **77**, 094302 (2008).
5. R. C. Norris, "Two-Dimensional Phononic Crystal Simulation and Analysis", Ph. D. Thesis, University of Waterloo, 2005.
6. R. R. Ahire, A. A. Sagade, N. G. Deshpande, S. D. Chavhan, F. Singh and R. Sharma, *J. Phys. D: Appl. Phys.* **40**, 4850-4854 (2007); R. R. Ahire, A. A. Sagade, S. D. Chavhan, V. B. Huse, Y. G. Gudage, F. Singh, D. K. Avasthi, D. M. Phase and R. Sharma, *Curr. Appl. Phys.* **9**, 374-379 (2009).
7. C. Jacome, M. Florez, Y. G. Gurevish, J. Giraldo and G. Gordillo, *J. Phys. D: Appl. Phys.* **34**, 1862-1867 (2001); L. D. Zhao, B. P. Zhang, W. S. Liu, H. L. Zhang, J. F. Li, *J. Solid State Chem.* **181**, 3278-3282 (2008).
8. Z. Bian, H. Wang, Q. Zhou, A. Shakouri, *Phys. Rev.* B **75**, 245208 (2007).

Spin relaxation of electrons in large quantum dots: a phenomenological model

R Mathew, K I von Königslöw and J A H Stotz

Department of Physics, Engineering Physics & Astronomy, Queen's University, Kingston, Ontario, Canada K7L 3N6

Abstract. We propose a phenomenological model designed to study electron spin relaxation in large quantum dots. The versatility of the model lies in its ability to incorporate both Rashba and Dresselhaus spin-orbit interactions with variable relative strengths and realistic confining potentials with only a marginal increase in computational complexity and time. Good agreement has been found with results from path-integral simulations of spin relaxation by Chang *et al.* (Phys. Rev. B, 70:245309, 2004).

Keywords: spin relaxation, quantum dot
PACS: 73.63.-b 72.25.Dc 72.25.Rb 72.50.+b

INTRODUCTION

The operation of spintronic-based devices typically relies on the long-range, coherent transport of spins or the strong confinement of spin-polarized particles. A novel mechanism that simultaneously enhances the electron spin lifetime through confinement while transporting spin-polarized electrons over long distances has been studied, which involves dynamic quantum dots (DQDs) produced by the piezoelectric field of a surface acoustic wave (SAW) [1]. The primary dephasing mechanism for electron spins confined by DQDs in clean semiconductor heterostructures at low temperature is the D'yakonov-Perel' (DP) mechanism [2]. During transport, the electron spins precess coherently about an internal magnetic field induced by the spin splitting of the conduction band. This spin orbit interaction (SOI) is a relativistic effect resulting from the motion of an electron in an electric field, part of which is transformed into an effective magnetic field in the rest frame of the electron. In this paper, we propose a phenomenological model for the numerical study of electronic spin relaxation in DQDs due to the Rashba SOI.

The action of time and spatial inversion symmetries in an elemental semiconductor, such as Si, can lead to electronic spin states that are doubly degenerate in energy such that $E_{\mathbf{k}\uparrow} = E_{\mathbf{k}\downarrow}$ and $E_{\mathbf{k}\uparrow} = E_{-\mathbf{k}\downarrow}$, where $\mathbf{k}$ is the wavevector of an electron with an up ($\uparrow$) or down ($\downarrow$) spin [3]. However, the symmetry, and hence the degeneracy, can be broken in a number of ways. For example, time reversal symmetry can be removed by the application of an external magnetic field. In III-V semiconductor heterostructures, such as GaAs, the spatial symmetry can be broken by either the structure inversion asymmetry of a confining potential, referred to as the Bychkov-Rashba SOI, or the bulk inversion asymmetry of the crystal structure, referred to as the Dresselhaus SOI. The spin-orbit coupling can be described through a $\mathbf{k}$-vector dependent magnetic field that acts on the electron's spin vector. The effective magnetic fields associated with the Rashba SOI and

CP1147, *Transport and Optical Properties of Nanomaterials—ICTOPON - 2009*, edited by M. R. Singh and R. H. Lipson
© 2009 American Institute of Physics 978-0-7354-0684-1/09/$25.00

Dresselhaus SOI for [001]-oriented quantum wells are given by (1) and (2), where $\hat{x}$ and $\hat{y}$ are the unit vectors in the [100] and [010] directions [4].

$$\mathbf{B}_R = \mathbf{B}_R(\hat{x}k_y - \hat{y}k_x) \tag{1}$$

$$\mathbf{B}_D = \alpha_{D1}(-\hat{x}k_x + \hat{y}k_y) + \alpha_{D3}(-\hat{x}k_xk_y^2 + \hat{y}k_yk_x^2) \tag{2}$$

In the absence of strain or external magnetic fields, the electron spin vector precesses around the resultant spin-orbit magnetic field $\mathbf{B}_{SO} = \mathbf{B}_R + \mathbf{B}_D$ at a constant frequency until it experiences a change in momentum. The precession of the spins can be characterized by the spin-orbit length L_{SO}, defined as

$$L_{SO} \equiv \frac{|\mathbf{v}|}{\hbar|\mathbf{B}_{SO}(\mathbf{k})|} \tag{3}$$

where $\mathbf{v}$ is the velocity of the electron [6]. The spin-orbit length is defined here as the distance travelled by the electron in the time that it takes for the spin to precess by one radian.

Previous studies of spin relaxation have primarily been restricted to systems with only Rashba SOI, partly due to its importance in spin control using electric fields. However, for transport of electrons by DQDs in symmetric heterostructures, the principal spin relaxation mechanism is a result of the Dresselhaus SOI. The time-step approach discussed in this article allows for the investigation of systems where the total magnetic field can have contributions from the Rashba, Dresselhaus SOIs as well as external magnetic fields. In addition, while previous studies have modelled the QD as an infinite well potential, the time step method allows for a more realistic parabolic potential that exerts a position dependent force on the electron. Other interactions such as electron-electron and electron-impurity scattering may also be considered. We demonstrate that our simulations show good agreement with path-integral calculations for spin relaxation in Rashba dominated systems [6] and that the model can be extended to more complicated geometries.

TIME-STEP ALGORITHM FOR SPIN PRECESSION

The semi-classical model employed here assumes that electrons are point particles travelling with velocity $\mathbf{v}$ in a 2-dimensional dot, while carrying a 3-dimensional spin vector $\mathbf{S}$. The electron behaviour is approximated by using a time-step method with segments Δt during which the electron travels a distance $\Delta l = |\mathbf{v}|\Delta t$. The resultant force acting on the electron is evaluated at each time step and is used to update the velocity vector for the successive interval; the resultant magnetic field $\mathbf{B}_R$ is then evaluated using Eq. 1 for Rashba SOI. The spin vector is updated by rotating it about $\mathbf{B}_R$ by an angle $\theta = \Delta l/L_{SO}$. For motion in an infinite well and in the absence of other forces, the electron continues on its ballistic trajectory, and the spin precesses about a constant magnetic field until it reaches the boundary of the dot, where it reflects specularly. The spin vector defined by the new $\mathbf{k}$-vector is then used as the input for the next iteration of the algorithm.

The behaviour of an ensemble of electrons was examined by modelling 1000 particles in a quantum dot of radius $1\,\mu m$. The electrons were initialized with identical thermal energies $E = 3k_B T/2$, where k_B is the Boltzmann constant and T is the temperature. At time $t = 0$, the particles were randomly distributed to produce a uniform area density, and their velocities were oriented in random directions. The spins were initialized in the $\hat{z}$ direction, thus mimicking experimental conditions for electron-hole pair generation using circularly polarized light incident perpendicular to the plane of the QD. The instantaneous spins of the ensemble were averaged to produce a time-evolving spin vector for the system, and the z-component of the average is shown in Fig. 1.

The time Δt, which determines the accuracy of the model, was chosen such that the electrons travel at most a distance $\Delta l = 1\,nm$, and in the limit that $\Delta t \to 0$, the fully semi-classical behaviour of the electron would be recovered. The discretization of the electron behaviour allows for a straightforward implementation of systems with time-varying and position dependent forces and magnetic fields, provided they vary slowly over the time or distance intervals used.

NUMERICAL RESULTS

Infinite Well Potential

Figure 1 shows the time evolution of the average z-component of the spin polarization $P_z(t)$ simulated for 1000 electrons in a Rashba-dominated ($\mathbf{B_D} = 0$), smooth, circular QD and varying spin-orbit lengths. During their random walk, the spins dephase through the DP mechanism due to the internal magnetic field. For confined electrons, the DP interaction is mitigated by the scattering of electrons on the walls of the infinite well potential. This, in turn, leads to a non-zero residual spin, the magnitude of which is determined by the spin-orbit length L_{SO}. The residual value ρ_z, is determined by a time average of the oscillations in $P_z(t)$ at long times. Simulations show that the magnitude of the oscillations and decay time decrease for increasing spin-orbit length.

The time-step implementation of the semiclassical model compares very well with the semi-classical path-integral results from Chang et al. [6]. The Chang model defines L_{SO} as the distance travelled by the electron in the time taken for the spin to precess by two radians rather than one. The smooth line in Fig. 2 shows the path-integral data from Chang et al., which corresponds to the z-component of the residual spin polarization ρ_z for a Rashba dominated, smooth, circular QD as a function of L_{SO}. Also shown as open and filled dots are the time-step data for the 2 radian and 1 radian definition of the spin orbit length, respectively. The time-step method is in good agreement with the path-integral method, with the 1 radian definition of L_{SO} yielding the expected increase in the residual spin values. All subsequent data presented here will assume the 1 radian definition of L_{SO}.

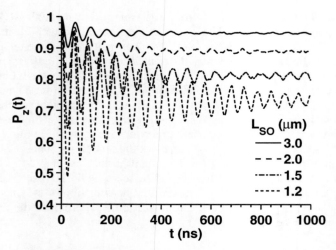

FIGURE 1. A plot of the average z-component of spin polarization $P_z(t)$ of 1000 electrons in a Rashba SOI dominated smooth, circular quantum dot for varying spin-orbit length.

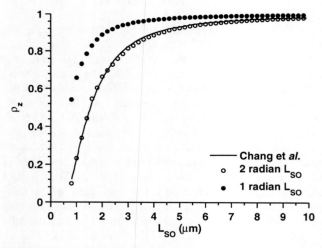

FIGURE 2. Residual spin as a function of spin-orbit length in a Rashba SOI dominated, smooth, circular quantum dot. The smooth curve represents the results of the Chang path-integral simulations and the dotted curve represents the time-step simulations.

Parabolic Potential

In order to expand the model to reflect experiments on DQDs, the infinite well potential was replaced by a parabolic potential of the form

$$V = \frac{1}{2}m^*\omega^2|\mathbf{r}|^2 \qquad (4)$$

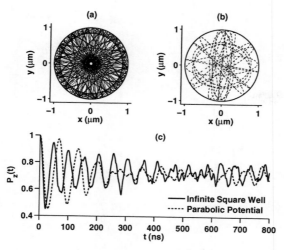

FIGURE 3. Paths of 5 and 15 electrons in (a) an infinite well and, (b) a parabolic potential with radius 1 μm respectively. (c) The time evolution of the z-component of spin polarization for the infinite well and the parabolic potential.

where $\mathbf{r}$ is the radial distance of the electron from the centre of the potential and ω is the strength of the potential. The potential exerts a position dependent force

$$\mathbf{F} = -\nabla V = -m^*\omega^2\mathbf{r} \tag{5}$$

that alters the acceleration of the electron at every time-step. Figure 3 (a) and (b) show the trajectories of 5 and 15 electrons in an infinite well with radius 1 μm and a parabolic potential respectively. The confinement in the parabolic potential was determined by setting ω to produce the same maximum displacement of the electrons from the center of the dot as in the infinite well. For the infinite well, the electrons follow linear trajectories within the QD whereas in the parabolic potential they follow harmonic trajectories determined by their initial position and velocity vectors. As seen in Fig. 3 (c), electrons confined in a parabolic potential also maintain a residual spin.

Figure 4 is a plot of the residual spin of 1000 electrons for a Rashba dominated, smooth, circular infinite well QD and a parabolic QD as a function of the spin orbit length L_{SO}. It was found that the residual spin of an ensemble of electrons confined in an infinite well is identical to that of an ensemble in a parabolic potential for $L_{SO} > 1$.

CONCLUSIONS

The phenomenological time-step model was developed to provide an accurate, yet flexible method to study the semi-classical behaviour of an ensemble electrons in a large, Rashba dominated, quantum dot. Model data shows good agreement with path-integral simulations by Chang *et al.* of electrons confined in an infinite well. The change in

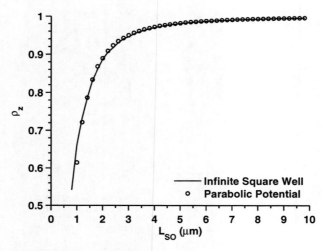

FIGURE 4. Residual spin as a function of spin-orbit length for 1000 electrons in a Rashba SOI dominated infinite well and parabolic potential.

confining potential from infinite well to an experimentally relevant parabolic potential produced an identical residual spin curve.

While the simulations presented here only show Rashba contribution for comparison with other studies, the model can be extended by including the Dresselhaus SOI, electron-electron interactions and non-elastic impurity scattering. Inclusion of the latter mechanisms may also permit the study of electron spin relaxation times T_1 of electrons as carried out by Koop *et al.* [4]. Future work should allow for the study of elliptical quantum dots as produced by the action of split-gates on a single SAW [7].

ACKNOWLEDGMENTS

The authors would like to thank the Natural Sciences and Engineering Research Council of Canada for their financial support.

REFERENCES

1. J. A. H. Stotz, R. Hey, P. V. Santos, and K. H. Ploog, *Nat. Mater.* **4**, 585–588 (2005), ISSN 1476-1122.
2. I. Žutić, *Reviews of Modern Physics* **76**, 323–410 (2004).
3. R. Winkler, *arXiv*:cond-mat/0605390v1 (2006).
4. E. J. Koop, B. J. van Wees, and C. H. van der Wal, *arXiv*:0804.2968v1 (2008)
5. R. Eppenga, and M. F. H. Schuurmans, *Phys. Rev. B* **37**, 10923–10926 (1988).
6. C.-H. Chang, A. G. Mal'shukov, and K. A. Chao, *Phys. Rev. B* **70**, 245309 (2004).
7. V. I. Talyanskii, J. M. Shilton, M. Pepper, C. G. Smith, C. J. B. Ford, E. H. Linfield, D. A. Ritchie, and G. A. C. Jones, *Phys. Rev. B* **56**, 15180–15184 (1997).

Rheological Properties of Iron Oxide Based Ferrofluids

M. Devi[a] and D. Mohanta[b]

Nanoscience Laboratory, Department of Physics, Tezpur University, Tezpur 784028, Assam, India

[a] manasidevi25@gmail.com; [b] best@tezu.ernet.in

Abstract. In the present work, we report synthesis and magneto-viscous properties of cationic and anionic surfactant coated, iron oxide nanoparticles based ferrofluids. Structural and morphological aspects are revealed by x-ray diffraction (XRD) and transmission electron microscopy (TEM) studies. We compare the rheological/magneto-viscous properties of different ferrofluids for various shear rates (2- 450 sec^{-1}) and applied magnetic fields (0-100 gauss). In the absence of a magnetic field, and under no shear case, the ferrofluid prepared with TMAH coated particle is found to be 12% more viscous compared to its counterpart. The rheological properties are governed by non-Newtonian features, and for a definite shear rate, viscosity of a given ferrofluid is found to be strongly dependent on the applied magnetic field as well as nature of the surfactant.

Keywords: Ferrofluids, surfactant; rheological properties.
PACS: 81.07.-b, 66.20.Ej, 75.75.+a

INTRODUCTION

In recent years, ferrofluid systems have emerged as technologically important candidates with prospective applications in magneto-sealing, magneto-shielding, biomedicine etc. [1-4]. These technological attractions are basically due to their unique magneto-viscous properties (magnetic control of their flow). Ferrofluid is a colloidal solution of magnetic nanoparticles (MNPs) and therefore, a ferrofluid can exhibit magneto-viscous effect [5-7]. In ferrofluid synthesis method, the most commonly used nanoparticles are magnetite (Fe_3O_4) and maghemite (γ- Fe_2O_3) and oil, water, kerosene etc. are used as dispersing media. In this kind of fluid, different long and short range forces result in sedimentation and agglomeration of the particles. The stability of this kind of fluid can be attained either by steric repulsion or by charging the particles electrically [8]. In case of steric repulsion, the MNPs are coated with a surfactant layer (anionic, cationic, zwitterionic etc). The surfactant molecule gets attached to the surface of the nanoparticles and helps in gaining the stability of the fluid by stopping agglomeration. Generally, particles of size 4 - 20 nm with a surfactant layer of thickness ~ 2 nm result in a good quality ferrofluid [9]. In a ferrofluid flow, the magnetic moment of the particles is aligned with the vorticity of the flow and in presence of a magnetic field the MNPs would try to reorient themselves. This is very important in deciding the rheological properties of the fluid and crucial for many devices.

Here, we report the synthesis of anionic and cationic surfactant coated two iron oxide (Fe_3O_4) based ferrofluids and compare their magneto-viscous characteristics for various shear rates, with and without application of magnetic fields.

CP1147, *Transport and Optical Properties of Nanomaterials—ICTOPON - 2009*, edited by M. R. Singh and R. H. Lipson
© 2009 American Institute of Physics 978-0-7354-0684-1/09/$25.00

EXPERIMENTAL DETAILS

As a prime step of ferrofluid preparation, first MNPs were synthesized. We have followed an inexpensive co-precipitation method for producing nano-sized Fe_3O_4. The chemical equation is as follows:

$$2FeCl_3 + FeCl_2 + 8NaOH \longrightarrow Fe_3O_4 + 8NaCl + 4H_2O$$

In the second step, the synthesized nanoparticles are coated with two kinds of surfactants namely, tetra methyl ammonium hydroxide (TMAH) and oleic acid. The former one is cationic and the latter one is anionic surfactant. In this work we take methanol as carrier fluid. The schematic synthesis procedure is presented below.

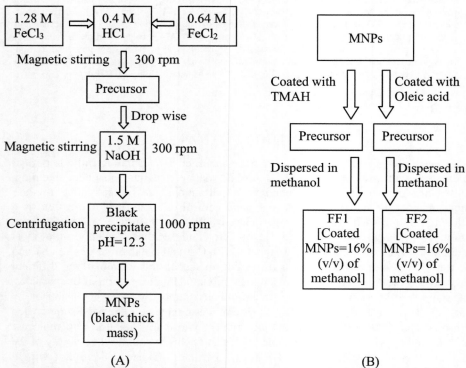

(A)   (B)

FIGURE 1. Block diagram of synthesis procedure of ferrofluid (A) 1st step (B) 2nd step

RESULTS AND DISCUSSION
Structural Studies

First, the synthesized magnetite particles are characterized by x-ray diffractometer (model: Rigaku Mini Flex 200). Fig.2 shows the typical XRD pattern of the synthesized uncoated magnetite particles. The peaks at 35.35°, 56.4°, 63.35° and 66.1° are designated as characteristics peaks of magnetite with preferred orientation along (311), (511), (531) and (442) planes, in consistency with the other works [10-11]. Peaks at 31° and 75.2° arise because of impurity. The size of the particles is calculated by the Scherrer's formula:

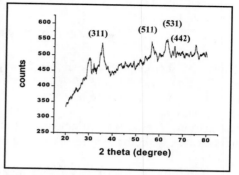

FIGURE 2. XRD pattern of uncoated magnetite sample

$$d = \frac{0.9\lambda}{\beta \cos\theta} \tag{1}$$

where d is the particle size, λ is the wavelength of the x-rays, β is the full width at half maxima in radian and θ being the diffraction angle. Considering the most prominent peak, the average size of the MNPs was calculated to be ~ 4.7 nm and with a strain value of -0.0012. The negative value of strain reflects contraction of atomic planes within a given nanoparticle. The formula exploited to calculate the strain of the particle is [12]

$$\frac{\beta \cos\theta}{\lambda} = \frac{1}{d} + \frac{s\sin\theta}{\lambda} \tag{2}$$

(s is the strain and other symbols are the same as in the Scherrer's formula)

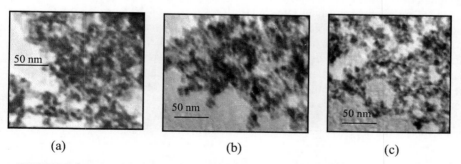

(a)	(b)	(c)

FIGURE 3. (a) Uncoated magnetite particles, particles coated with (b) TMAH and (c) oleic acid dispersed in methanol

497

Further, the samples are analyzed by transmission electron microscopy (TEM). Fig.3 depicts the TEM micrograph of uncoated, TMAH and oleic acid coated magnetite samples. It is found that the particles have an average size of 5±0.2 nm and are of nonspherical nature. Most of the structures include oblate, hexagonal and rhombohedral features.

Rheological properties

Rheological properties of the prepared samples are studied in a Brookfield dial reading viscometer (Model: M/00-151). Fig.4 (a) represents shear rate dependent variation of viscosities in the absence of a magnetic field. The non linear decay of viscosity with increasing shear rate confirm that the ferrofluid possess non Newtonian characteristics. The substantial amounts of shear thinning i.e. decrease of viscosity with shear rate, for both the ferrofluid can be expressed as:

$$y = y_1 e^{-x/t_1} + y_2 e^{-x/t_2} \tag{3}$$

Here y is the log(viscosity), y_1 and y_2 are viscosities at zero shear rate, x is the shear rate and t_1, t_2 are the decay parameter in sec^{-1}. These parameters correspond differently to FF1 and FF2 (Table 1). From the exponential equation it can be understood that the ferrofluids are undergoing two simultaneous decay equations- one of them is very fast with high decay parameter. The critical shear rates at which shear thinning slows down are 86.21 sec^{-1} and 118 sec^{-1} for FF1 and FF2 respectively. The overall viscous nature of the ferrofluids can be attributed to the arrangement of small dispersed chains of the MNPs [13]. With increasing shear rate some

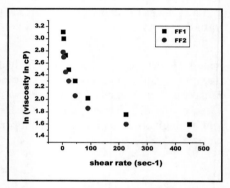

FIGURE 4(a). Variation of viscosity with shear rate For FF1 and FF2

kind of perturbation of these clusters occur leading to a decreasing trend of viscosity. It is evident that in the absence of any external force there are more number of clustering in FF1 than that of FF2 (fig.3(a)). In other words, oleic acid (anionic surfactant) coated particles are more dispersed than TMAH (anionic surfactant) coated particles in the ferrofluids. But with increasing shear rate oleic acid coated clusters respond to fragmentation more rapidly.

TABLE 1: Different parameters of FF1 and FF2

FF1		FF2	
y_1	0.67	y_1	0.64
y_2	0.99	y_2	0.86
t_1	10.99	t_1	15.36
t_2	117.74	t_2	168.73

Fig. 4 (b) and Fig.4 (c) demonstrate the magneto-viscous property of the as prepared ferrofluids. Pronounced non Newtonian behavior was observed even in the presence of magnetic fields (*H*). For a fixed shear rate, applied magnetic field could enhance the viscosity of the ferrofluids. But for a particular field, viscosity decreases with increasing shear rate. Other workers have argued that the formation of different field induced structures e.g. chain sequence, droplike etc. in real ferrofluids [14-15] might lead to such a variation. Particles larger than the critical size (~10 nm for magnetite particle) are more prone to this kind of structure formation [6]. In a ferrofluid the amount of such particles greatly influences the magneto-viscous property. With high shear rate these structures break down, which results in decreased viscosity. It is reported by Odenbach et. al. that the interaction between the magnetic moment and mechanical torque of the particles results in high magneto viscous effect. This is substantially owing to stronger orientation tendency of dipole moments from the direction of vorticity towards applied field [16]. In our case, the direction of the magnetic field is perpendicular to the vorticity of the fluid. As the field increases, the strength of the interaction between the field and the magnetic moments of the MNPs also increases, which results in shifting of the critical shear rate toward lower shear rate direction. The relative changes of viscosity (η_r) for both FF1 and FF2 at different

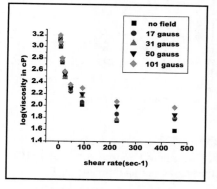

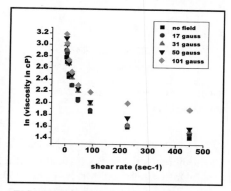

FIGURE 4(b). Variation of viscosity of FF1 with shear rate

FIGURE 4(c). Variation of viscosity of FF2 with shear rate

magnetic fields are shown in Fig. 4 (d) and Fig. 4(e). It is expressed by the equation

$$\eta_r = \frac{\eta(H) - \eta(0)}{\eta(0)} \tag{4}$$

It is seen that the FF2 responds more rapidly to the field than that of FF1 irrespective of shear rate. It implies that oleic acid coated particles easily interact with the field. TMAH coated particles are more stable in ferrofluid in comparison to oleic acid coated particles.

CONCLUSIONS

We prepared two chemically synthesized magnetite based ferrofluids. The ferrofluids differ from each other depending on the surfactant type used to stabilize the MNPs. The rheological study gives hints about the role played by the surfactants in deciding the viscous property of the ferrofluids. In consistent with other report, it is found that the magneto-viscous property of the ferrofluid exhibits shear thinning. Oleic acid coated particles in ferrofluid (FF2) are found to be more disperse in the absence of field and more rapidly responding to the applied magnetic field than that of TMAH coated particles (FF1). Understanding magneto-viscous properties in terms of role of surfactants, shear dependency, and size dispersity would find various application e.g., sealing, switching and lubricating agents etc.

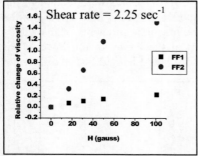

FIGURE 4(d). Relative change of viscosity with magnetic field

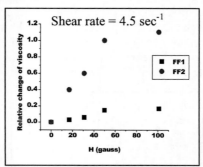

FIGURE 4(e). Relative change of viscosity with magnetic field

ACKNOWLEDGMENTS

The authors thank the Sophisticated Analytical Instrument Facility (SAIF), NEHU-Shillong, Dr. R. Dutta and Mr. H. Deka of Department of Chemical Sciences, TU.

REFERENCES

1. J. Popplewell, *Phys. Technol.* **15**, (1984).
2. Li -Ying Zhang, H. Chen Gu and Xu- Man Wang *J. Magn. Magn. Mater.* **311,** 228 (2007).
3. X. Mao, L. Yang, X.-Li Su and Y. Li *Biosens. Bioelectron.* **21**, 1178 (2006).
4. M. Strömberg, K. Gunnarsson, P. Svalizadeh, P. Svedlindh and M. Strømme, *J. Appl. Phys.* **101**, 023911 (2007).
5. S. Odenbach, *J. Phys.: Condens. Matter* **16**, R1135 - R1150 (2004).
6. S. Odenbach and K. Raj *Magnetohydrodynamics* **36**, 312 (2000).
7. S. Thurm and S. Odenbach, *Phys.Fluids* **15**, 1658 (2003).
8. C. Scherrer and A. M. Figueiredo Neto, *Braz. J. Phys.* **35**, 718 (2005).
9. S. Odenbach, *Magneto viscous effect in ferrofluid -Lecture Notes in Physics*, Springer, 2002.
10. R. Y. Hong, T. T. Pan, Y. P. Han, H. Z. Li, J. Ding and Sijin Han, *J. Magn. Magn. Mater.* **310**, 37 (2007).
11. D. K. Kim, Y. Zhang, W. Voit, K. V. Rao and M. Muhammed, *J. Magn. Magn. Mater.* **225**, 30 (2001).
12. S. B. Qadri, E. F. Skelton, D. Hsu, A. D. Dinsmore, J. Yang, H. F. Gray and B. R. Ratna, *Phys.*

Rev. B **60**, 9191(1999).

13. L. Mirela Pop, S. Odenbach, A. Wiedenmaan, N. Matoussevitch and H. Bönnemann, *J. Magn. Magn. Mater.* **289**, 303 (2005).
14. A. Yu. Zubarev and L. Yu. Iskakova, *Phys.Rev. E* **61**, 5415 (2000).
15. A. Yu. Zubarev, S. Odenbach and J. Fleischer, *J. Magn. Magn. Mater.* **252**, 241 (2002).
16. S. Odenbach and H. Störk, *J. Magn. Magn.Mater.* **183**, 188 (1998).

Chromium Doped ZnS Nanostructures: Structural and Optical Characteristics

D. P. Gogoi[a], U. Das[a], G. A. Ahmed[a], D. Mohanta[a], A. Choudhury[a], G. A. Stanciu[b]

[a]Nanoscience Research Laboratory , Department of Physics, Tezpur University, Naapam, Tezpur, Assam. [b]Centre for Microscopy Microanalysis & Image Processing, University "Politehnica" of Bucharest, Romania

Abstract. Chromium doped ZnS nanoparticles arranged in the form of fractals were fabricated by using inexpensive physico-chemical route. The Cr:ZnS samples were characterized by diffraction and spectroscopic techniques. Unexpected growth of fractals with several micrometer dimensions and of core size 1μm (tip to tip) was confirmed through TEM micrographs. At higher magnification, we found that individual fractals consist of spherical nanoparticles of average size < 30 nm. The mechanism leading to such organized structures describing fractal pattern is encountered in this work.

Keywords: Fractals, Self assembly, II-VI Semiconductor.
PACS: 61.47 Hv, 81.16.Dn, 81.05.Dz.

INTRODUCTION

Among II-VI semiconductor systems, ZnS is the most popular wide band gap system having a direct band gap of ~3.7eV at 300^0K. ZnS is a promising host material due to its thermal and environmental stability. Transition metal doped semiconductor nanostructures e.g. Mn:ZnS; Mn:ZnO; Cu:ZnS are believed to be potential candidates owing to strong emission due to impurity states [1-3]. Incorporation of both transition-metal ions and rare-earth ions ZnS nanostructures by adopting chemical and physical techniques have been reported in recent years [4-6]. It has been used as base material for cathode ray tube luminescent materials [7,8]. On the other hand, Cr doped ZnS has not received much attention owing to significant chemical incompatibility arising due to lattice mismatch at the host lattice. Over the years, chemically synthesized doped semiconductor nanostructures have attracted scientific community owing to inexpensive procedures for large scale production. Fabrication of micro structured fractals requires self assembly since direct manipulation of such self similar structures at the lower end of the scale is extremely difficult to achieve [9]. There is a growing interest and continuous demand in the fabrication of self similar materials, parts, features etc. for application in nanotechnology. One of the difficult tasks was to produce materials that incorporate multiple length scales simultaneously, from nano and micro, to micro scale. In fact, fractals could achieve this goal because of their self similarity and sustainability in the real world as well as in theory [10,11]. In this paper, we report physico-chemical synthesis of Cr:ZnS nanostructures and their

CP1147, *Transport and Optical Properties of Nanomaterials—ICTOPON - 2009*, edited by M. R. Singh and R. H. Lipson
© 2009 American Institute of Physics 978-0-7354-0684-1/09/$25.00

spectroscopic and structural characterizations. A theoritical approach on the formation of resulting fractals is also attempted.

EXPERIMENTAL

Cr-doped ZnS encapsulated in polyvinyl alcohol matrix (PVOH) was fabricated using a low cost colloidal solution casting route. For this 2% (w/v) PVOH in double distilled water was magnetically stirred at ~200 rpm at a constant temperature for six hours until a transparent solution was formed. Next, aqueous solution of $ZnCl_2$ was added to the PVOH matrix under stirring condition and then aqueous solution of CrO_3 solution was mixed at room temperature. To this precursor Na_2S solution was drop wise injected, which led to the growth of 'Cr:ZnS' nanoparticles. Optical properties of the samples were studied through PL-spectra. The structural aspects were revealed by XRD, TEM, AFM, MFM studies. The freshly prepared samples of Cr:ZnS, deposited on glass substrates are used for AFM, MFM, PL and XRD measurements, where as liquid samples were kept for TEM studies.

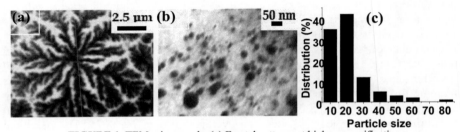

FIGURE 1. TEM micrographs (a) Fractal patterns at higher magnification, (b) spherical nanoparticles of Cr:ZnS in PVA matrix at closer inspection. (c) Distribution of particles with average grain size.

RESULTS AND DISCUSSION

At lower magnification unexpected growth of fractal-like patterns [10,11] were confirmed though TEM micrographs (Figure 1a), while at higher magnification these fractals were found to consist of individual nanoparticles of average size less than 30 nm (Figure 1b). In figure 1c, distribution (%) of nanoparticles with particle size in a 0.585 sq. μm area is shown. It is observed that the maximum number of particles lie within the range of 10 nm to 20 nm. The fractal like feature was further confirmed by systematically measuring its fractal dimension, typically defined by the divider formula [12]

$$D_f = \lim_{r \to 0} \frac{\log N(r)}{\log(r)}$$

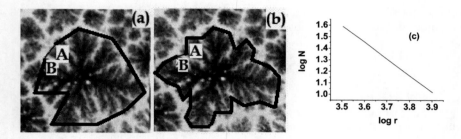

FIGURE 2. Analysis of the fractal patterns, (a) A straight rular of length r (thick line) , walks along the nanomaterial surface contour, starting from point A, and ends at point B, Left : r= 125 μm, and in figure (b) r = 50 μm, The distance between A and B (thin line) is measured to be d. The path length N is then defined as the sum of the number of steps and d/r . (c) A log-log graph of N versus r (μm) . The fractal dimension , D_f is the absolute value of the slope, in this case, D_f = 1.48.

where *r* is the length unit, *N(r)* is the size of the geomatric object measured with unit *r*. To measure D_f, We used a methode illustrated by Andrea Lomannder et. al. [9]. For a given value of r , the ruler usually cannot walk along the contour exactly with an integer number of steps, giving rise to the mismatch between the start and the endpoints   (points A and B in figure 3). In such cases, N was calculated as the sum of the number of full steps and the fractional length between the start and the end points with respect to r. In the example shown in figure 3(a), r=125 μm, it looks 10 steps (thick line) starting from A to proceed to the point B along the contour. The remaining distance between A and B (thin line) is 50 mμ, whose fractional length with respect to r is 50/125 = 0.4, thus the total contour length is 10.4. In figure 3(b) r = 50μm, the contour length yeilded an average value of N=38.75. The measured value of N depended on the location of the starting point of the contour. Inspired by the Nyquist sampling theorem [13], several values of *N* starting at different points on the contour were measured and averaged. The graph of log[r] vs. Log[N] gave the fractal dimension D_f = 1.48. This value corresponds to the fractal dimension D_f of a viscous fingering system [14], whose formation mechanism is known as diffusion-limited aggregation (DLA) [15,16]. In the plane, D_f = 1.71; however, in real systems, depletion effects may decrease D_f from this ideal value to 1.4 [13.14]. Thus, it can be interpreted that the fractal like feature was formed through DLA-like process. Change of pH and the presence of disulphide bond are important for the formation of fractals through DLA-like process [9]. Molecular self- assembly takes place via a subtle balance between non covalent bond interactions that result in the formation of well-defined structures [17-22]. The cysteine disulphide bond should remain stable at lower pH [23,24]. In a recent work [2], The authours have shown production of Cr:ZnS nanoparticles without formation of fractals. We have though followed the similar method replaced treatment of H_2S by acidic Na_2S treatment. We expect that, in our case the presence of excess disulfide bond could be responsible for the formation of fractals.

X-RAY DIFFRACTION STUDIES

The X-ray diffraction pattern of the Cr:ZnS nanoparticles embedded in PVOH matrix is shown in figure 3. It depicts cubic crystalline structure corresponding to three diffraction peaks (111), (220) and (311). The average particle size estimated is 12 nm as obtained by measuring full-width-at half maxima (FWHM) according to Scerrer's formula [25]

$$d = \frac{0.9\lambda}{\omega \cos \theta}$$

It has also been observed that maximum numbers of particles are oriented with crystalline state (111).

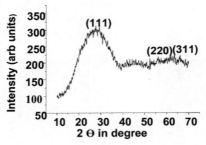

FIGURE 3. XRD pattern of Cr doped ZnS
nanostructures in PVA

Photoluminescence Study

Figure 4 represents photoluminescence response of Cr doped ZnS nanostructures in PVOH matrix at excitation wavelength 300 nm. The appearance of the PL peak with energy close to the band gap energy is recognized as band gap emission [26]. On the other hand emission peak corresponding to low energy values ascribed to the transition involving donors, acceptors, free electrons and holes. We ascribe the emission peak at ~380 nm as the band edge emission of ZnS and the peak at around 550 nm is expected due to d-electron transfer of Cr^{2+} into the ZnS host.

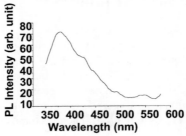

FIGURE 4. PL emission spectra of Cr doped ZnS nanostructures in PVOH matrix.

AFM AND MFM STUDIES

The surface morphology of the samples casted on glass substrates were studied by using scanning probe microscopy (SPM). A typical AFM and MFM micrographs for the sample is shown in Figure 5. AFM image of the sample exhibits uniform surface morphology as shown in figure 5(a). Magnetic force microscopy is a well established method to probe the micro-magnetic properties of samples with lateral resolution down to ~50 nm. The advantage of SPM is that less sample is needed, thinning or polishing of the sample is not necessary. Moreover, the technique yields information on both the structural (AFM mode) as well as the magnetic (MFM mode) aspects with regard to sample's surface. Therefore, the topology and magnetic domain structure of a sample can efficiently be correlated at the nanometer scale. Transition metals like Mn, Fe, Co, Ni doped ZnS, ZnTe systems are found to display spin glass state, where as V or Cr doped ZnS, ZnSe, ZnTe are ferromagnetic in nature [2].

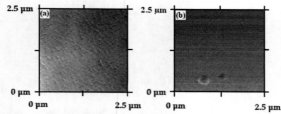

FIGURE 5. (a) AFM and (b) MFM micrographs of Cr doped ZnS embedded in PVA

To exploit magnetic properties, we have carried out MFM studies on the Cr doped ZnS samples. The figure 5(b) is basically a phase image of the sample; it shows clear response to the magnetic field. The average size of the magnetic cores (black spot) is measured as ~ 165 nm and white regions spreading over the cores are ascribed to region of influence by the respective particles. Thus, it is evident that ZnS:Cr nanoparticles can respond appreciably to magnetic force and fields and MFM in this regard, can be a very good tool to exploit magnetic domains and particle-particle interactions.

CONCLUSIONS

We have fabricated Cr-doped ZnS nanostructures by a simple and inexpensive chemical route. Photoluminescence shows a new emission peak ~ 550 nm due to incorporation of Cr^{+2} ions into ZnS host. Spherically isolated magnetic domains are found in MFM micrographs. A new direction for generation of fractals with II-VI based transition metal doped diluted magnetic semiconductor is reported. Here, the synthesized fractals are of dimension 1.48. This is evident that the growth of fractals were due to diffusion limited aggregation. The mechanism of fractal formation at nanoscale level is in progress.

ACKNOWLEDGMENTS

The authors acknowledge support from the DST, Govt. of India and Ministry of Education and Research, Govt. of Romania under the Indo-Romania Joint programme of co-operation. AFM & MFM was performed at the Centre for Microscopy, Microanalysis and Image Processing, University "Politehnica" of Bucharest. We also gratefully acknowledge to SAIF, NEHU, Shillong for TEM analysis.

REFERENCES

1. W. Lu and C.M. Lieber, *J. Phys. D* **39**, R387 (2006).
2. U. Das, D. Mohanta and A. Choudhury, *Ind. J. Phys.* **81**, 155-159 (2007).
3. D. Mohanta. S.S. Nath and A. Choudhury, *Bull. Mater. Sci.* **26**, 289 (2003).
4. Y. Wu, K. Arai, N. Kuroda, T. Yao, A. Yamammoto, M. Y. Shan and T.Goto, *Jpn. J. Appl. Phys.* **36**, LI 648 (1997).
5. K. Shibata, K. Takabayashi, I. Souma, J. Shen, K. Yanata and Y. Oka, *Physica E* **10**, 358 (2001).
6. C. S. Kim, S. Lee, J. Kossut, J. K. Furdyna and M. Dobrowolska, *J. Cryst. Growth* **395**, 214-215 (2000).
7. R. Vacassy, S. M. Scholz, J. Dutta, H. Hoffmann, C. J. C. Plummer, G. Carrot, J. Hilborn and M. Alkine, *Matter. Res. Soc. Symp. Proc.* **501**, 369 (1998).
8. P. Calandra, M. Goffred and V. T. Livery, *Colloid Surf. A* **160**, 9-13 (1999).
9. Andrea Lomander, Wonmuk Hwang, and Shuguang Zhang, *Nano Lett.* **5**, 1255-1260 (2005).
10. 10. B. Mandelbrot, *The Fractal Geometry of Nature*, 3rd ed.; W.H. Freeman and Company, New York, (1983).
11. B. Mandelbrot, *Fractal and Chaos*, The Mandelbrot Set and Beyond; Springer-Verlag, New York, (2004).
12. 12. C. Beck and F. Scho"gl, *Thermodynamics of Chaotic Systems,* Cambridge University Press: Cambridge, (1993).
13. D. Skoog and J. Leary, *Principles of Instrumental Analysis*, 4th Ed.; Saunders College Publishing, Philadelphia, (1992).
14. P. Meakin, *Phys. Rev. A* **27**, 1495-1507 (1983).
15. T. Witten and L. Sander, *Phys. Rev. Lett.* **47**,1400-1402 (1981).
16. T. Witten and L. Sander, *Phys. Rev. B* **27**, 5686-5697 (1983).
17. G. M. Whitesides, *Nat. Biotechnol.* **21**, 1161-1165 (2003).
18. S. Zhang, *Nat. Biotechnol.* **21**, 1171-1178 (2003).
19. M. Reches and E. Gazit, *Science* **300**, 625-627 (2003).
20. N. Kroger, R. Deutzmann and M. Sumper, *J. Biol. Chem.* **276**, 26066-26070 (2001).
21. L. L. Bratt and M. O. Stone, *Nature* **413**, 291-293 (2001).
22. S. Mann and H. Co"lfen, *Angew. Chem., Int. Ed.* **42**, 2350-2365 (2003).
23. P. Gupta, H. Rizwan, H. Khan and M. Saleemuddin, *Int. J. Biol.*
24. J. Horng, S. Demarest, and D. Rsleigh, *Proteins: Struct. Funct. Genet.* **52**, 193-202 (2003).
25. P. Scherre, *Gott. Nachr.* 2 98 (1990).
26. A. K. Pal, *Bull. Mater. Sci.* 22341 (1999).

Influence of Ultrasonic Irradiation on PbS Nanocrystals

Monika Mall[b]* and Lokendra Kumar[a,b]

[a]Nanophosphor Application Centre,
[b]Physics Department, University of Allahabad, Allahabad-211 002, India
E-mail: monikamall03@gmail.com

Abstract. Ultrasonic waves significantly influence the characteristic of the nanomaterials. In present paper, we synthesized the PbS nanocrystals by chemical co-precipitation method and studied the effect of ultrasonic irradiation on the properties of as prepared nanocrystals. X-Ray Diffraction results showed that the ultrasonically treated samples have better crystallinity as compared to pristine sample. Purity of samples and vibrational modes were demonstrated by Raman spectroscopy. UV-Vis spectroscopy revealed slight shift in excitonic transition of sonicated PbS nanocrystals as compared to untreated sample. Room temperature Photoluminescence (PL) spectra of both the samples showed the emission peaks at 437 nm, 472 nm and 510 nm, with the excitation wavelength of 350 nm. An enhancement in PL intensity was observed in the ultrasonically irradiated samples.

Keywords: Ultrasonic irradiation, semiconducting nanocrystals, Raman spectroscopy.
PACS: 78.67.-n, 78.30.-j, 78.40.-q, 43.35.Vz

INTRODUCTION

In recent years, semiconducting nanoclusters have drawn significant attention due to their unique optical, electrical and chemical properties originating from their small size [1]. When size of these particles is comparable to their corresponding Bohr exciton radius, they exhibit quantum confinement effect with significant blue shift in absorption spectra [2,3]. In addition to this, the nanoparticles are also having high thermal stability, increased quantum efficiency for luminescence and radiative lifetime shortening with decreasing particle size as compare to bulk [1].

Among the IV-VI compound semiconductors, PbS is an important direct band gap material. Due to large exciton Bohr radius (~18 nm) [4] and narrow band gap (E_g = 0.41eV) [5], PbS easily shows quantum confined effect and good absorption tunability throughout the near infrared region. These properties make it a suitable material for hybrid solar cells [6,7], infrared detectors, photo emitters and Pb^{2+} ion selective sensors [8]. Because of third order nonlinear optical properties [9], PbS nanoparticles can also be used in electroluminescent and optical devices [10]. Monodisperse PbS nanocrystals were firstly synthesized in PVA (Poly Vinyl Alchohol) matrix by Nenadovic et. al. in 1990 [11]. Subsequently several groups reported the synthesis methods of PbS nanomaterials with various size and shapes [12-15]. However, the fabrication of size controlled and monodispersed PbS nanoparticles is still a challenge. In this paper, we have synthesized PbS nanocrystals by chemical co-precipitation route and report the considerable improvement in characteristic properties after ultrasonic

CP1147, *Transport and Optical Properties of Nanomaterials—ICTOPON - 2009*, edited by M. R. Singh and R. H. Lipson
© 2009 American Institute of Physics 978-0-7354-0684-1/09/$25.00

treatment. The effects of sonication are studied by using XRD and spectroscopic techniques. Ultrasonic waves prevent the particle-particle contact or coagulation and help in formation of monodispersed particles with better crystallinity and enhancement in luminescence properties. Possible reasons of these improvements are also discussed.

2. EXPERIMENTAL SECTION

2.1 Chemicals

PbS nanoparticles were prepared by chemical method using Lead acetate dihydrate (Pb $(CH_3COO)_2$. $2H_2O$), and Thiourea (NH_2-CS-NH_2) as precursors. All the chemicals were of analytical grade, purchased from Merck India Limited and directly used without any further purification.

2.2 Sample Preparation

Synthesis method was simple and aimed at realizing the role played by ultrasonic waves on the properties of nanoparticles. A 50 ml solution of sulfur source with concentration of 1.5 M was prepared by dissolving thiourea in double distilled water at 80°C. Deep blackish solution of PbS colloids were formed immediately after the dropwise injection (at constant 80 °C temperature under continuous and vigorous stirring) of 0.1 M lead acetate into the above solution. To analyze the influence of ultrasonic irradiation, one part of as prepared sample was kept in sonochemical bath (33 kHz, 350 watt) for 3 hours. Finally, the resultants were centrifuged, washed in sequence with distilled water and ethanol and dried at 60 °C in vacuum oven to obtain the particles in powder form.

2.3 Instrumentation

The crystal structures and quality of synthesized PbS nanoparticles were examined by X-ray diffractometer (XRD) Rigaku D/MAX-2200 H/PC, Cu Kα radiation (λ= 0.1541 nm). In order to investigate the vibrational properties of PbS nanocrystallites, micro Raman measurements have been performed on Renishaw, RM 1000 using 514.8 nm line of Ar^+ ion laser as an excitation source. Absorption spectra were recorded on Perkin Elmer Lamda-35 UV-Vis spectrometer and Photoluminescence (PL) spectra have been achieved using Perkin Elmer Lambda-55 PL spectrophotometer. The PL studies were performed with the excitation wavelength of 350 nm.

3. RESULTS AND DISCUSSION

3.1 Structural and Vibrational Study

Crystallinity, size and phase of particles were characterized by XRD as shown in Fig.1. Peak positions indicate the formation of face centered cubic (fcc) rock salt structure with nine most preferred orientations [111], [200], [220], [311], [222], [400],

[331], [420] and [422]. No trace of PbO, S or any other phase indicating the purity of samples. Intense XRD peaks of sonicated PbS nanoparticles show that ultrasonic waves played a crucial role in an enhancement of particle crystallinity. The average size of irradiated and pristine PbS particles is about 14 nm, determined by Debye-Scherrer formula (Eq.1) [16]:

$$D = \frac{0.9\lambda}{B\cos\theta} \tag{1}$$

where λ is the wavelength of X-ray radiation, θ is the Bragg angle of the peak, B is the full width of the peak at half of its maximum intensity (FWHM).

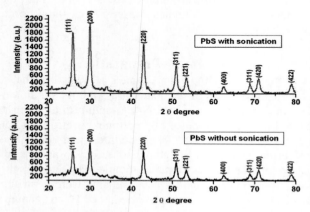

FIGURE 1. XRD pattern of sonicated and pristine PbS nanoparticles

Fig. 2 represents the room temperature Raman spectra of samples with 514 nm excitation wavelength. As can be seen, five bands at 137, 161, 270, 428 and 599 cm^{-1} are observed in both case.

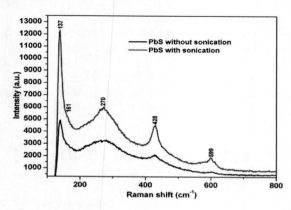

FIGURE 2. Raman spectra of PbS nanoparticles

A sharp and intense peak centered at about 137 cm^{-1} assigns to be a combination of longitudinal and transverse acoustic modes which is similar to 155 cm^{-1} band recorded with excitation at 632.8 nm [17]. A very weak band observed at about 161 cm^{-1} is attributed to the fundamental longitudinal optical phonon (LO) mode of PbS nanoparticles while two other bands at 428 and 599 cm^{-1} are first (2LO) and second (3LO) overtones respectively [18]. One small and broad band at 270 cm^{-1} is also observed which is in agreement with the 272 cm^{-1} peak recorded at 4.2 K by Krauss et. al.[19], using 581 nm wavelength laser as an excitation source. It is already known that the crystalline samples present sharp Raman peaks while amorphous or polycrystalline samples show line broadening [20]. Arora et. al. [21] reported that the line broadening also depends upon the grain size and defects present in material. In our case, it is clear that the size reduction can not be the reason of peak broadening because size calculated from XRD spectra does not show any significant difference in between pristine and irradiated sample. Small line broadening or sharp peaks in sonicated sample might be due to less defects and better crystallinity of particles which is also confirmed from PL and XRD results respectively.

3.2 Absorption and Emission Study

Fig.3 shows the UV-Vis absorption spectra of the PbS nanocrystals. A large blue shift is observed in the absorption spectra of PbS nanocrystals in comparison to its bulk (E$_g$ = 0.41 eV). This blue shift is due to the small effective mass as well as small size of the PbS nanocrystals. Since the shape of absorption spectra gives the idea about the particle size distribution. The sharp band edge of absorption spectra confirms the increased monodispersity in the sonicated samples.

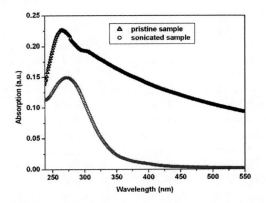

FIGURE 3. UV-Visible absorption spectrum of PbS nanoparticles

In Fig. 4(a), the comparison of emission spectra shows the considerable enhancement in the luminescence intensity of sonicated PbS nanocrystals and deconvoluted emission spectrum of untreated sample is shown in Fig. 4(b). A sharp peak centered at 437 nm is attributed to the transition of conduction band edge electrons to holes trapped at Pb^{2+} sites [17]. The other two small peaks, at 472nm and

510 nm may be associated with the defect levels which are more specified in unsonicated sample. Presence of more defects in non irradiated sample is also confirmed by line broadening in the Raman peaks.

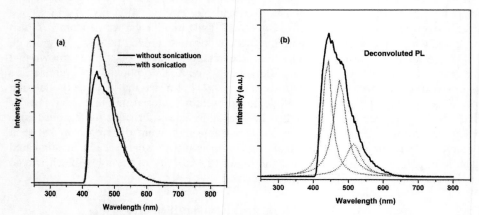

FIGURE 4. Room-temperature photoluminescence spectra of PbS nanoparticles (a) and Deconvoluted graph of unsonicated sample (b).

Better crystallinity and control over the non radiative transitions and size distribution of nanoparticles due to ultrasonic waves can be explained by using the property of acoustic cavitation: the formation, growth and implosive collapse of bubble in liquid. This process can affect the characteristics of particles as it provides a path to concentrate the ultrasonic energy into localized spot where high temperature and high pressure can be achieved. It has been found that three different regions are formed during the sonochemical process [15, 22].

a) The inner environment (gas phase) of the collapsing bubble, where elevated temperature (several thousands of degrees) and pressure (hundreds of atmospheres) are produced.
b) The interfacial region where the temperature is lower than that in the gas-phase region but still high enough to induce a sonochemical reaction.
c) The bulk solution region, which is at ambient temperature.

Among the above-mentioned three regions, it seems that the current improvement in particle properties occurs within the interfacial region. Because if it takes place inside the collapsing bubble, the product became amorphous in nature as a result of the cooling rates (> 10^{10} K/s). Whereas the bulk solution region, can not produce any change in the characteristics of the material due to such a low temperature. The case of interfacial region is quite different since in this region, cavitation process induces high velocity interparticle collisions which can result an effective melting at the point of collision. This environment provides good opportunity to rearrange atoms and one can expects to obtain crystalline product with less defects and monodisperse size distribution.

4. CONCLUSION

PbS nanoparticles have been synthesized successfully by the chemical co-precipitation method and studied the effect of ultrasonic irradiation. The XRD results showed the formation of fcc phase of PbS nanoparticles with 14 nm particle size. Raman bands located at 136, 161, 270, 428 and 599 cm^{-1} were attributed to the combination bands, fundamental LO phonon and their overtones (2LO, 3LO). Absorption and PL spectra confirm the improvement in monodispersity and enhancement in luminescence properties of ultrasonically irradiated samples. These good quality PbS nanoparticles can be directly used for organic solar cells and other optoelectronic devices.

ACKNOWLEDGEMENTS

Authors are very grateful and wish to express their gratitude to all the scientific members of Nanophosphor Application Centre, University of Allahabad, Allahabad, for the sample characterizations and kind cooperation. One of the authors (M. Mall) would like to thank Prof. Ram Gopal, Physics Department, University of Allahabad, Allahabad, for absorption measurements.

REFERENCES

1. R. N. Bhargava, D. Gallagher, X. Hong and A. Nurmikko, *Phys. Rev. Lett.* **72**, 416 (1994).
2. R. N. Bhargava, D. Gallagher, T. Welker, *J. Lumin.* **60**, 275 (1994).
3. L. Brus, *J. Phys. Chem.* **90**, 2555 (1998).
4. J. Machol, F. Wise, R. Patel and D. Tanner, *Phys. Rev. B* **48**, 2819 (1993).
5. M. Cardona and D. L. Greenaway, *Phys. Rev. A* **133**, 1685 (1964).
6. F. Zhang, W. Mammo, M.R. Andersson and O. Inganas, *Adv.Mater* **18**, 2169 (2006).
7. S. gunes, K. P. Fritz, H. Neugebauer, N. S. Sariciftci, S. Kumar and G. D. Scholes, *Solar Energy Materials & Solar Cells* **91**, 420 (2007).
8. H. Hirata, k. Higashyama, *Bull. Chem. Soc. Jpn* **44**, 2420 (1971).
9. Y. Wang, *Acc. Chem. Res.* **24**, 133 (1991).
10. R. S. Kane, R. E. Cohen and R. Silbey, *J. Phys. Chem.* **100**, 7928 (1996).
11. M. T. Nenadovic, M. I. Comor, and V. Vasic, *J. Phys. Chem.* **94**, 6390 (1990).
12. S. M. Zhou, Y. S. Feng and L. D. Zhang, *J. Mater. Res.* **18**, 1188 (2003).
13. J. H. Warner, A. R. Watt, M. J. Fernee and N. R. Heckenberg, *Nanotechnology* **16**, 479 (2005).
14. D. Liang, S. Tang, J. Liu, J. Liu and X. Lv, *Mater. Lett.* **62**, 2426 (2008).
15. D. G. Shchukin, H. Mohwald, *Phys. Chem. Chem. Phys.* **8**, 3496 (2006).
16. D. Song, P. Windenborg, W. Chin, A. Aberle, *Solar Energy Materials & Solar Cells* **73**, 269 (2002).
17. R. Sherwin, R. J. H. Clark, R. Lauck and M. Cardona *Solid State Commun.* **134**, 565 (2005).
18. H. Cao, G. Wang, S. Zhang and X. Zhang, *Nanotechnology* **17**, 3280 (2006).
19. T. D.Krauss and F. W. Wise, *Phys. Rev. B* **55**, 9860 (1997).
20. J. H. Chen, C. G. Chao, J. C. Ou and T. F. Liu, *Surface Science* **601**, 5142 (2007).
21. A. K. Arora, M. Rajalakshmi, T. R. Ravindran, *Encyclo. Nanosci.& Nanotech.* **10**, 1 (2003).
22. K. S. Suslick, D. A. Hammerton and R. E. Cline, *J. Am. Chem. Soc.* **108**, 5641 (1986).

Wide Temperature Range Thermopower in GaAs/AlGaAs Heterojunctions

M. D. Kamatagi[a,c], N. S. Sankeshwar[a] and B. G. Mulimani[b]

[a] *Department of Physics, Karnatak University, Dharwad – 580 003, Karnataka, India*
[b] *Gulbarga, University, Gulbarga – 585 106, Karnataka, India*
[c] *Govt. First Grade College, Sira -572 137, Karnataka, India*

Abstract. A systematic analysis of wide-temperature-range ($2 < T < 200$ K) thermopower (TP) data, on two-dimensional electron gas at GaAs/AlGaAs heterojunction (HJ) samples of Fletcher *et al.* [Phys. Rev.B, **50**, 14991 (1994)] and Cyca *et al.* [J. Phys. Condens. Mat, **4**, 4491 (1992)], is presented. Numerical results of TP are presented including the contributions from both diffusion TP, S_d and phonon drag TP, S_g. The phonons are considered to be scattered not only by sample boundaries, as assumed in literature, but also by impurities and other phonons. A good representation of the temperature dependence of TP, in all samples considered, is obtained. The parameters characterizing the intrinsic phonon scattering mechanisms are obtained from fitting thermal conductivity data, using modified Callaway model through linear regression method. The study provides a means of assessing the relative importance of carrier scattering mechanisms operative in GaAs HJs.

Keywords: Thermopower, thermal conductivity, GaAs, heterojunctions.
PACS: 74.25.Fy, 63.20.Kd, 74.78.Fk

INTRODUCTION

The search for more efficient thermoelectric energy conversion devices and the wide spread use of semiconductors in microelectronics has prompted increasing efforts in understanding charge and heat transport in semiconductor nanostructured material systems [1]. The suitability of a material for thermoelectric applications is quantified by the thermoelectric figure of merit, $Z = S^2 \sigma / \kappa$, where S, σ and κ are, respectively, the thermopower(TP), electrical conductivity and the thermal conductivity (TC). The physical limit of the ZT, which so far has been not much greater than unity, largely depends on TC and TP which are interrelated.

Thermopower, which is sensitive to the composition and structure of the system, is an interesting transport property for study in nanostructures [1]. In general, there are two contributions to the thermopower, namely, the diffusion thermopower, S_d and the phonon-drag thermopower, S_g. At the lowest temperatures ($T < 1$ K), TP is dominated by S_d often assumed to vary linearly with temperature [2,3]. S_g, which initially increases as T^3 and, overwhelms S_d in the liquid He temperature range is caused by the phonons being out of equilibrium. As the temperature increases, other phonon scattering mechanisms become increasingly effective in bringing the phonons back into equilibrium; this is expected to result in a peak in S_g and a subsequent rapid decrease in its magnitude. At higher temperatures, S_d will again become dominant, though a linear dependence on temperature may not be appropriate [4,5].

CP1147, *Transport and Optical Properties of Nanomaterials—ICTOPON - 2009*, edited by M. R. Singh and R. H. Lipson
© 2009 American Institute of Physics 978-0-7354-0684-1/09/$25.00

In literature, there exist extensive data on the TP of semiconductor heterostructures, the favourite system being GaAs/AlGaAs [1-6]. Except for some measurements on a series of GaAs/AlGaAs heterojunctions (HJs) up to 200 K [3,4], most of the experimental and theoretical work has been restricted mainly to low-temperature region (T<10 K) [2-5]. Further, in most of the experimental analyses of TP [2,3], the temperature variation of S_d is assumed to follow a simple linear variation given by Mott formula, and the contribution to S_g is considered in the phonon–boundary scattering limit, which is appropriate to a pure crystal at low temperatures. However, from wide-temperature thermal conductivity data [3,4], it is clear that the contributions from the scattering mechanisms other than phonon-boundary scattering also need to be considered.

In this article, considering the contributions from S_g and S_d we present a systematic analysis of the wide-temperature-range TP data available for GaAs/AlGaAs HJ samples of Fletcher et al [3] and Cyca et al [4]. To include realistic dependencies of the phonon mean free path on the phonon wave vector, we have, in our analysis, extracted the phonon scattering parameters from the available TC data. We obtain a good representation of temperature dependence of TP in all the samples considered.

The theory describing S_d and S_g is first given. The results of the detailed numerical calculations are then presented and discussed.

THEORY

We consider a 2DEG spatially confined along z-direction perpendicular to the interface at a HJ. The electron wavefunctions are given by [5]

$$\psi_n(r,z) = \xi_n(z)e^{ik.r} \tag{1}$$

where, $r=(x, y)$ and $k=(k_x, k_y)$ are, respectively, the 2D position and wave vectors of the electrons. Here, we assume that electrons occupy only the lowest subband $(n = 0)$ and $\xi_0(z)$, the confinement profile is described by Fang-Howard variational wave function [5].

Thermopower

Thermopower, S, is defined by the relation [1],

$$\mathbf{E} = S\ \nabla T \tag{2}$$

under open circuit conditions, where $\mathbf{E}$ is the effective electric field produced by the temperature gradient ∇T. In the presence of ∇T, the carriers diffuse through the specimen to produce the diffusion thermopower, S_d. In addition, the induced phonon momentum current drags electrons with it as a result of electron-phonon interaction to produce the phonon drag contribution, S_g.

Diffusion Thermopower

Employing the Boltzmann transport formalism, with ∇T in the plane of the 2DEG, the diffusion contribution to thermopower, S_d, in the relaxation time approximation, can be expressed as [1]

$$S_d = \left(\frac{1}{eT}\right)\left[-E_F + \left(\frac{\langle E \ \tau(E)\rangle}{\langle \tau(E)\rangle}\right)\right] \qquad (3)$$

with

$$\langle \ F(E) \ \rangle = \frac{\int F(E) \ E \ (-\partial f_o/\partial E) \ dE}{\int E \ (-\partial f_o/\partial E) \ dE}. \qquad (4)$$

Here $\tau(E)$ is the relaxation time of the electrons, $f_o(E)$ the Fermi Dirac distribution function and E_F the Fermi energy. Eq. (3) is used to evaluate S_d taking into account the relevant scattering mechanism(s). In the case of more than one scattering mechanism, the total S_d is calculated assuming the overall relaxation time $\tau(E)$ to be given by Matthiessen's rule: $\tau^{-1}(E) = \Sigma_i \tau_i^{-1}(E)$

In the temperature range of interest (T < 200 K), the processes contributing to electron scattering are the interactions with the acoustic phonons, background and remote impurities and surface roughness and charges. The expressions for the relaxation times for various electron scattering mechanisms are well documented [7].

Phonon Drag Thermopower

The phonon-drag component of thermopower, S_g, arising due to the quasi-elastic scattering of 2DEG by 3D acoustic phonons, is given by [3,5,6]

$$S_g = \frac{(2m^*)^{3/2}}{4(2\pi)^3 k_B T^2 n_s e\rho} \Sigma_s v_s^2 \int_0^\infty dq. \int_{-\infty}^\infty dq_z \frac{\tau_s^c(Q)\Xi_s^2(Q)q^2 Q^2 |I(q_z)|^2 \ G_s(Q)}{\varepsilon^2(q,T) \sinh^2(\hbar\omega_{Qs}/2k_B T)}. \qquad (5)$$

Here, $\hbar\omega_{Qs}$ is the energy of the phonons of wave vector $\mathbf{Q} \equiv (\mathbf{q}, q_z)$ and mode s(=L,T) and $\Xi_s(Q)$ denotes the effective acoustic phonon scattering potential which accounts for both deformation potential (DP) and piezoelectric (PZ) couplings [3,5,6]. v_s denotes velocity of phonons. $(\tau_s^c(Q))^{-1} = (\tau_s^R(Q))^{-1} + (\tau_s^N(Q))^{-1}$, where $\tau_s^N(Q)$, $\tau_s^R(Q)$ and $\tau_s^c(Q)$ are, respectively, relaxation times for normal phonon, resistive and combined processes.

The other symbols appearing in Eq. (5) have their usual meaning [5]. The explicit expressions for $G_s(Q)$, $I(q_z)$ and $\Xi_s(Q)$ for HJs are documented. In literature calculations of S_g are performed using approximate expressions for $G_s(Q)$, $I(q_z)$ and screening factor $\varepsilon(q,T)$ and compared with experiments. In our calculations presented below we use Eq.(5) without any approximation taking into account the relevant phonon scattering mechanisms. In case of more than one phonon scattering mechanism, the total S_g can be calculated assuming the overall relaxation time $\tau_s^c(Q)$ to be given by Matthiessen's rule. For temperatures 2<T<200 K, the scattering of phonons is assumed to be by sample boundaries, impurities, Umklapp and normal phonons.

Lattice Thermal Conductivity

The lattice TC component, κ_p, obtained by solving the Boltzmann transport equation for acoustic phonons in the relaxation time approximation, can be expressed as [8,9]

$$\kappa_p = T^3 \sum_s H_s \left(\int_0^{\theta_s} dx\, f(x)\, \tau_s^C(x) + \frac{\left(\int_0^{\theta_s} dx\, f(x)\, \tau_s^C(x)/\tau_s^N(x) \right)^2}{\int_0^{\theta_s} dx\, f(x) v_s\, \tau_s^C(x)/\tau_s^N(x)\tau_s^R(x)} \right) \tag{6}$$

where $f(x) = x^4 e^x / (e^x-1)^2$, $x = \hbar\omega/k_B T$, $H_s = k_B^4/6\pi^2\,\hbar^3 v_s$ and θ_s is mode Debye temperature.

The relaxation times for scattering of phonons by umklapp phonons, normal phonons, impurities, and boundaries are taken to be given, respectively, by [8,9]: $(\tau_U)^{-1} = B_{Us}\,\omega^2\,T\exp(-\theta_s/3T)$; $(\tau_N)^{-1} = B_{Ns}\omega^2 T^3$; $(\tau_I)^{-1} = A_s\,\omega^4$ and $(\tau_B)^{-1} = v_s/L_E$. Here, $A_s = (V\Gamma/4\pi\,v_s^3)$, with Γ denoting the strength of the phonon-impurity scattering and L_E the effective mean free path. The coefficients B_{Ns} and B_{Us} can be expressed as [8] $B_{Ns} = (k_B/\hbar)^3\,(\hbar\gamma_s V/M\,v_s^5)$ and $B_{Us} = \hbar\gamma_s^2/Mv_s^2\theta_s$, with V and M being the volume and average mass of an atom and γ_s is the mode Gruneisen parameter.

RESULTS AND DISCUSSIONS

We have performed numerical calculations of S_d and S_g using Eq. (3) – (5) for the parameters characteristic of GaAs/AlGaAs heterojunction sample of Fletcher $et\ al$ [3] and sample 2 of Cyca $et\ al$ [4]: $m^* = 0.067m_e$, $\rho = 5.31$ gmcm^{-3}, $v_L = 5.14 \times 10^5$ cms^{-1}, $v_T = 3.04 \times 10^5$ cms^{-1}, $E_d = 8.0$ eV, $\kappa_s = 12.91$, $h_{14} = 1.2 \times 10^7$ V/cm.

With a view to extract the sample dependent and the intrinsic phonon scattering parameters in determining the TP of the samples considered, we first analyze their TC data. For this, we employ the formalism used successfully in explaining the TC of GaN and GaAs HJ samples [8].

The TC of all the three samples considered is found to exhibit, broadly, the same temperature dependence typical of κ_p. Keeping in view the dominance of the various mechanisms, in the different temperature regions, we have performed calculations of κ_p using Eq. (6) by varying the parameters so as to obtain fits with data, for the sample

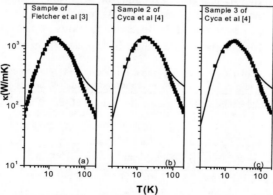

FIGURE 1. The thermal conductivity as a function of temperature. The curves represent fits to data of (a) Fletcher $et\ al$ [3] and (b and c) Cyca et al [4]. The filled squares are the experimental data.

of Fletcher *et al* [3], and the samples 2 and 3 of Cyca *et al* [4]. We have also kept in view the fact that the parameters L_E and Γ are specimen dependent whereas the parameters B_{sN} and B_{sU} are characteristic of the material. The results of our calculations along with the experimental data points are shown in Fig.1 (a), (b) and (c), respectively. Good fits are obtained over the entire temperature range for the values of the parameters given in Table 1.

Since the important sample-dependent phonon scattering mechanisms operative in the GaAs HJ samples, and which influence κ_p for $T < \sim T_{max}$ (~ 15 K), are the phonon-impurity and phonon-boundary scatterings, the fitting parameter L_E, which represents the phonon mean free path, was, determined by including phonon-impurity scattering and obtaining a good agreement with the measured TC data below about 10 K [8]. The impurity parameter Γ is then varied to obtain κ_p around the value of κ_{max} at T_{max}. Keeping the values of the relevant specimen-dependent resistive scattering parameters thus obtained fixed, the intrinsic scattering fitting parameters B_{sN} and B_{sU} are then determined following a non-linear regression procedure for each sample. Since B_{sN} and B_{sU} were adjustable to the extent of variation of the mode Gruneisen parameters γ_s, we used γ_s (instead of B_{sN} and B_{sU}) as adjustable parameters [8] to obtain fits with the high temperature ($T > T_{max}$) data. With a view to describe the TC of GaAs using a single set of the intrinsic phonon-scattering parameters, we have, by systematic variation of γ_L and γ_T, performed calculations of κ_p and obtained a good representation of the TC in the temperature range of 4 - 200 K, using $\gamma_L = 0.7$ and $\gamma_T = 0.2$, for all the samples studied. The deviation above 70 K may be due to additional scattering mechanisms operative in system and which we have not included in the present analysis [8,9].

Keeping in view the temperature dependence of TC of the samples, we next analyze their TP data. We have performed calculations of both S_d and S_g for the parameter characteristic of GaAs/AlGaAs samples and considering all the relevant scattering mechanisms. Figs.2 (a) and (b) show, respectively, the fits for sample of Fletcher *et* al [3] and sample 2 of Cyca *et al* [4]. In our analysis of TP data, we first obtain the fit for higher temperatures (T > 100 K) where S is assumed to be determined by only S_d and varying the electron scattering parameters. After obtaining a fairly good representation of data for T >100 K, we fit the low temperature data using S_g. A method similar as that for TC is adopted to fit the data. We use the same phonon scattering parameters obtained from TC data to obtain the fits for TP. We find that S_d shows an almost linear temperature dependence and is important at very low (T < 2K) and larger (T > 40 K) temperatures. S_g, which is dominant at low temperatures

TABLE 1. Sample dependent phonon scattering parameters determined from analysis of experimental TC data on GaAs/AlGaAs HJ samples and the electron scattering parameters used in the calculation of S_d

Sample	Phonon parameters		Parameters for S_d		
	L_E (mm)	Γ (10^{-4})	Background impurity concentration	Remote impurity concentration	Surface roughness parameters
of Fletcher *et al* [3]	2.7	0.40	10^{14} cm^{-3}	10^{17} cm^{-3}	$\Delta = 2$ A^0 $\Lambda = 16$ A^0
2 of Cyca *et al* [4]	2.5	0.28	10^{14} cm^{-3}	0.5×10^{17} cm^{-3}	$\Delta = 4$ A^0 $\Lambda = 20$ A^0
3 of Cyca *et al* [4]	1.8	0.30			

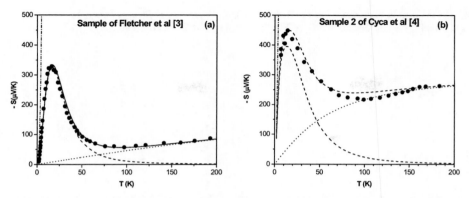

FIGURE 2. Temperature dependence of TP of (a) sample of [3] and (b) sample 2 of [4]. Dashed and dotted lines, respectively, represent S_g and S_d. The solid curve represents total TP. The circles represent the TP data. The dash-dotted curves depict the TP calculated using only boundary scattering.

(T<30 K) with a characteristic peak at T ~ 10 K, is determined by boundary scattering at very low (T <5 K) temperatures, phonon-phonon scattering at high (T > 40 K) temperatures and impurity scattering at intermediate temperatures.

CONCLUSIONS

We have, for the first time, made a detailed analysis of the wide-temperature range TP data on GaAs/AlGaAs HJ samples of Fletcher *et al* [3] and Cyca *et* al [4]. Contributions from S_d and S_g are considered. Good fits are obtained using the same set of intrinsic and sample dependent parameters as required for fits to TC data of the samples. The study provides better understanding of the behavior of TP and of carrier scattering mechanisms operative in GaAs HJs.

ACKNOWLEDGMENTS

The work is supported by UGC (India)

REFERENCES

1. L. Gallagher and P. N. Butcher, *"Handbook on semiconductors"*, vol.1. P. T. Landsberg ed.: Elsevier, Amsterdam, 1992.
2. R. Fletcher, V. M. Pudalov, Y. Feng, M. Tsaousidou and P. N. Butcher, *Phys. Rev. B* **56**, 12422-12428 (1997).
3. R. Fletcher, J. J. Harris, C. T. Foxon, M. Tsaousiduo and P.N. Butcher, *Phys. Rev.B* **50**, 14991 (1994).
4. B. R. Cyca, R. Fletcher and M. D'Iorio, *J. Phys. Condens. Mater.*, **4**, 4491 (1992).
5. M. D. Kamatagi, N. S. Sankeshwar and B. G. Mulimani, *Physica. Stat. Sol. (b)***242**, 2892 (2005)
6. M. D.Kamatagi, N. S. Sankeshwar and B. G. Mulimani, *IEEE Proc. of Int. Wksp. on 'Phys.of Semicond Devices'*, 446 (2007).
7. M. D. Kamatagi, N. S. Sankeshwar and B. G. Mulimani, *Phys. Rev. B* **71**, 1253341 (2005).

8. M. D. Kamatagi, N. S. Sankeshwar and B. G. Mulimani, *Diamond and Related Mater.* **16**, 98 (2007).

9. D. T. Morelli, J. P. Heremans, and G. A. Slack, *Phys. Rev. B.* **66**,1953041 (2002)

Formation of Oxygen Induced Nanopyramids on Rh(210) Surface

Govind[a]*, Wenhua Chen[b], Hao Wang[b] and T.E. Madey[b]

[a]Surface Physics and Nanostructures Group, National Physical Laboratory, Dr. K.S. Krishnan Road,
New Delhi -110012 (India))
[b]Department of Physics & Astronomy and Laboratory for Surface Modification, Rutgers, The State
University of New Jersey, 136 Frelinghuysen Road, Piscataway, NJ 08854, USA

Abstract. Formation of oxygen induced nano pyramids on atomically rough and morphological unstable Rh (210) surface has been studied using Auger electron spectroscopy (AES), low energy electron diffraction (LEED), and UHV-scanning tunneling microscopy (STM). Nanometer sized pyramids can be formed upon annealing the oxygen-covered Rh (2 1 0) surface at ≥550K in the presence of oxygen (10^{-8} Torr) and the facets of the pyramids are identified as two {731} face and a (1 1 0) face. The study suggests that the oxygen overlayer can be removed from the surface via catalytic reaction at low temperature using CO oxidation while preserving ("freezing") the pyramidal facet structure. The resulting clean faceted surface remains stable for $T\sim600$ K and for temperatures above this value, the surface irreversibly relaxes to the planar state. Atomically resolved STM measurements confirm the formation of nanopyramids with average pyramid size ~18nm. The nanopyramidal faceted Rh surface will be used as nanotemplates for growth of nanometer-scale clusters.

Keywords: Faceting, Rhodium, Low energy electron diffraction (LEED), Auger Electron Spectroscopy (AES), Scanning Tunneling Microscopy (STM)
PACS: 68.47, 68.43, 68.37

INTRODUCTION

Atomically rough clean metal surfaces generally have lower surface atom density and higher surface free energy than close-packed surfaces of the same metal. A variety of atomically-rough clean metal surfaces can be prepared as stable orientations, but the presence of (strongly interacting) adsorbates can cause changes in surface morphology through mechanisms such as reconstruction and facet formation [1–6]. These morphological changes are usually explained in terms of changes in surface free energy due to the presence of adsorbate [7,8]. Many studies have been performed on adsorbate induced faceting of atomically rough surfaces, e.g., bcc W(111), Mo(111), fcc Cu(210), Ir(210), Pt(210) and Ni(210) and hcp Re (12-31) and Re(11-21) surfaces [3-11].

In the present study, we focus on the fcc Rh (210) surface which is an important substrate for catalytic processes and has potential applications in surface chemistry. We study the oxygen induced morphological change of Rh(210) surface. Previous studies on other fcc (210) surfaces [5-9] indicate that the facets can be

CP1147, *Transport and Optical Properties of Nanomaterials—ICTOPON - 2009*, edited by M. R. Singh and R. H. Lipson
© 2009 American Institute of Physics 978-0-7354-0684-1/09/$25.00

induced due to the presence of various gaseous adsorbates for example oxygen and nitrogen can induce faceting of Cu (210) and Ni(210) while oxygen and CO provide favorable conditions to restructure the Pt (210) surface to form facets. Ir (210) [8] shows the formation of pyramidal type facets with {311} and (110) facets, when annealed in the presence of oxygen above 600K.

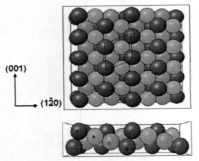

FIGURE 1. Hard sphere model of fcc (210) surface (top & side view) showing top four exposed layer

The Rh (210) surface is atomically rough; as shown in figure 1, the unreconstructed Rh (210) has four layers of atoms exposed and the top layer of the atoms shows mirror reflection symmetry along ($\bar{1}$20) plane. Here, we study the morphological changes induced by oxygen on Rh (210) and formation of nanopyramids. The nanofaceted surface is characterized by LEED, AES and STM.

EXPERIMENTAL DETAILS

The experiments were carried out in two different ultrahigh vacuum (UHV) chambers denoted as LEED and STM chambers respectively. All LEED images are obtained in the LEED chamber that also contains a quadrupole mass spectrometer (QMS) for residual gas analysis and a cylindrical mirror analyzer for AES. STM experiments described herein were performed in the STM chamber at room temperature using a hybrid variable temperature Omicron STM with tungsten tips. The Rh(210) crystal is cut from a single crystal Rh (99.99%) rod ~10 mm in diameter, ~1.5 mm thick, aligned within 0.5° of the (210) orientation and polished to a mirror finish. The sample is supported by two rhenium leads where a high current (up to 30A) can be passed through the leads to achieve temperatures up to 1600K. A C-type (W–5 at.% Re/W–26 at.% Re) thermocouple is spot welded directly to the rear of the sample for accurate temperature measurement. The sample support assembly also includes a tungsten filament for electron bombardment heating; a temperature up to 2000K can be achieved by flashing the sample in UHV. The sample was cleaned by repeated cycles of Argon ion bombardment (1keV, 3.5μA) at 300-650K (initially at 300K and then at increasingly higher temperature (<650K), annealing in UHV at 1200K, annealing in O_2 ($2x10^{-8}$ Torr) at 1000 -1200K followed by rapid flashes to ~1600K in UHV to desorbs the excess oxygen from the surface. The oxygen gas deposition was achieved by back-filling the chamber using research purity O_2 at background pressures 2×10^{-8} Torr for different time depending on the desired dose.

RESULTS

A typical LEED pattern obtained from the clean fcc Rh(2 1 0) surface is shown in Fig. 2a. The spots observed by LEED are consistent with the LEED pattern of an unreconstructed bulk fcc (2 1 0) plane. When electron beam energy (Ee) is increased the electron wavelength and, consequently, the diffraction angles decrease, and an apparent motion of the diffraction beams toward the specularly reflected (0,0) beam is observed. For the clean Rh(2 1 0) surface the (0,0) beam is perpendicular to the macroscopic surface plane and is in the center of the LEED pattern. This behavior of the LEED beams is an indication that the surface is both macroscopically and microscopically planar. No new beams appear in the LEED pattern as a result of oxygen adsorption at room temperature. However, the intensity of this background signal increases with exposure and is attributed to additional diffuse scattering from the oxygen overlayer and oxygen induced disorder in the topmost Rh layer. The behavior of the oxygen covered Rh(210) surface show significant changes on annealing the substrate to higher temperature. The LEED pattern gradually changes with temperature. As the sample temperature is slowly increased LEED beams of the initial (1 x 1) pattern diffuse and, seemingly, elongate in three different directions or three additional elongated beams appear around them. The transition between a clearly (1 x 1) pattern and the diffuse one is very gradual and it is difficult to pinpoint accurately the temperature at which it occurs; the lowest temperature at which the pattern is clearly different from the original (1 x1) is approximately 450 K. With further temperature increase the (1 x1) beams become progressively fainter, and the emerging new beams become gradually brighter and sharper. This process continues until the sample temperature has reached 550 K, when all trace of the (1 x 1) beams has disappeared, including the specularly reflected (0, 0) beam and a pattern appeared with a completely different geometry and roughly three times more spots than planar surface. The geometry of the LEED pattern does not change with further (above 550 K) temperature increase. Subsequent cooling of the sample to room temperature does not reverse the process and the new pattern remains stable and unchanged in the entire temperature range. For experiments performed without oxygen in the background gas surface cleanliness is critical: carbon contamination can obstruct or even completely prevent the occurrence of the described transformations. When Ee is varied the behavior of the new LEED pattern differs from that of the (1x1) (Fig. 2c) rather than converging on the center of the pattern, the diffracted beams move in directions that converge to three distinct position. Three LEED beams (labeled by small circles in the figure 2b & 2c) can be observed, each coinciding with one of the three convergence points. They can therefore be identified as specular reflection beams originating from surfaces that are tilted with respect to the (2 1 0) macroscopic surface plane. Moreover, the complete LEED pattern is the superposition of three distinct LEED patterns, whose respective specular directions are each inclined at unique angles (with respect to the macroscopic crystal plane), determined by the crystallographic orientation of the surfaces from which they originate. LEED patterns with similar revealing characteristics have been previously observed in similar experiments, involving oxygen and metallic overlayers on W(1 1 1) [3,4,12]. Subsequently, the emergence of these patterns has been interpreted to be a consequence of the formation of nanometer-size facets on the substrate surface.

The crystallographic orientation of these facets can be determined by calculating the corresponding facet tilt angles with respect to (210) plane, and the azimuthal orientation of facets with respect to each other. We calculate the tilt angle between a facet plane corresponding to the specular beam A & B and the normal direction of the (210) plane which is $8.1\pm0.1°$[13]. The azimuthal angle between these two faceted planes can be directly measured and is $\Phi =132\pm4°$. The tilt angle between the third facet, defined by the specular beam position C in figure 2(b) and the normal direction of the (210) plane is also calculated in similar manner. However, from figure 2(b), the exact position of the specular beam from third facet is difficult to observe as the specular beam corresponding to this facet is blocked by the leads of the sample support assembly. The estimated tilt angle between the facet and the normal direction of the (210) plane is found to be $18\pm2°$. Based on the tilt angle between the faceted plane and (210) plane as well as azimuthal angle between the two symmetric facets, the Miller indices of first two facets are identified as (731) and ($73\bar{1}$) planes and while the third facet is identified as (110) facet [13]. The observation of {731} and (110) facet is also supported by early study by Tucker [14] on Rh (210) surface. However, Tucker observed diffraction spots corresponding to {731} facet on Rh (210) when annealing the surface in oxygen background $1x10^{-7}$ Torr at 573K and these diffraction spot are completely replaced by (110) plane when annealing in oxygen pressure $1x10^{-6}$ Torr at the same temperature.

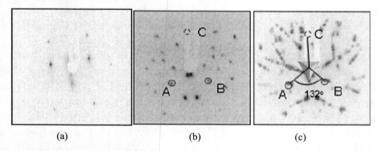

(a)　　　　　　　　　(b)　　　　　　　　　(c)

FIGURE 2. (a) LEED pattern from a clean surface of Rh(210) at incident electron beam 80eV (b) LEED pattern from faceted O/Rh(210) at electron beam energy 66eV(c) LEED pattern for electron beam energy 20-74ev. The pyramid in c illustrates the origin of the specular beams and azimuthal angle used to determine the facet.

The production of a clean or freezing of the facets is a new aspect to overlayer-induced faceting experiments of the faceted rhodium surface. From the literature the most straightforward method of removing oxygen from the rhodium surface is through thermal desorption, however, to produce a clean faceted rhodium surface is difficult because the faceted surface reverts to planar at 600 K (in the absence of oxygen background). This temperature is significantly lower than those required to completely remove the oxygen by desorption (~1400 K) making it impossible to remove the oxygen by desorption without destroying the faceted surface. However, our investigations have revealed that it is possible to chemically removing the oxygen overlayer at low temperature. For this we used the method to takes advantage of catalytic CO oxidation in the following manner: commencing with a faceted surface (Fig. 3a) completely covered with oxygen, the temperature of the sample is set to 390 K and then CO is leaked into the chamber at a pressure of 4×10^{-9} Torr. As expected (Eq. (1)), a surface reaction occurs,

$$CO_{(ad)} + O_{(ad)} \text{---------------------} CO_{2(g)} \qquad (1)$$

which results in an increase of the CO_2 signal in the residual gas mass spectrum. When the integrated CO dose has reached 0.48 L (after ~120 s) the CO is pumped from the chamber and the sample temperature is increased further (to 450 K) and held at that temperature (in UHV) for additional 2 min, in order to desorbs any excess CO. Upon cooling to room temperature the surface retains its faceted structure, as verified by LEED (Fig. 3b). AES studies confirm that an atomically clean faceted surface has been produced (Fig. 3c). The resulting clean faceted surface remains stable for $T\sim600$ K and for temperatures above this value, the surface irreversibly relaxes to the planar state.

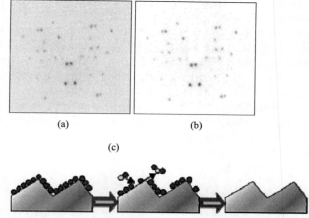

(a)　　　　　　　　　　(b)

(c)

FIGURE 3. LEED pattern form the faceted Rh(210) at electron energy 90eV. (a) oxygen–covered faceted surface and (b) clean faceted surface (c) illustrate the cleaning reaction

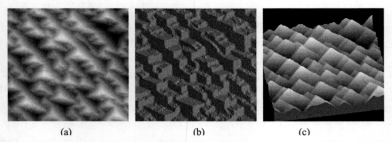

(a)   (b)   (c)

FIGURE 4. STM scan of the fully faceted Rh (210) surface by annealing to 850K and cooling to room temperature in oxygen background (a) The STM image of the nanopyramidal faceted surface san area 100nmx100nm (a) STM image of faceted Rh(210) surface with scan area 100nmx100nm (b) X-slope image with scan area 100nmx100nm, V=1.2mV, I=1nA (c)3D image of nanopyramids with facets of (731), (73-1) and reconstructed (110) surface

Figure 4(a) shows a typical STM image of faceted Rh (210) surface prepared by pre dosing of 10L oxygen at 300K followed by annealing at 850K for 2 min followed by cooling to 300K in the presence of oxygen (4×10^{-8} Torr) [13]. The faceted Rh surface is fully covered with well formed nano pyramids (three sided facets) with similar shape, which implies that they expose faces of identical crystal orientation. To reveal the atomic details, we differentiate the height of STM images along x-direction (x-slope); with this procedure the details of STM images can be enhanced at the cost of losing height information. The three dimensional image of the faceted surface is presented in figure 4(c).The orientation of the facets identified by LEED can be further confirmed by measuring azimuthal angles between the edge lines of each of the facet. The average pyramid size for the above surface is found to be ~18nm. Further, it is also observed that the size of the nanopyramids can vary from 12nm to 21 nm, depending upon the annealing temperature. The biggest average nano pyramid size can be obtained by flashing the Rh (210) surface in the presence of oxygen to 1600K. These nanopyramids can be used as a template to grow metallic nanoclusters and biological molecules.

CONCLUSIONS

In the present work, we focus on the adsorbate induced modification of the surface morphology of the Rh(210) substrate. We observed oxygen induced pyramidal faceting of Rh(210) surface with {731} and (110) facets when annealing Rh(210) in O_2 The average pyramid size ranges from 12nm to 21nm, which can be controlled by changing the annealing temperature from. The faceted Rh surfaces provide possible model systems to study structure sensitivity in Rh based catalytic reactions as well as potential nanotemplate to grow nanoclusters.

ACKNOWLEDGMENTS

This work has been supported by the U.S. Department of Energy (DOE), Office of Basic Energy Sciences (Grant DE-FG-02-93ER14331). Dr. Govind thanks Department of Science & Technology, Govt. India, New Delhi, India for BOYSCAST fellowship.

REFERENCES

1. G. Ertl, H. Knozinger and J. Weitkamp, Handbook of Heterogeneous Catalysis (Wiley, New York) 1997.
2. Q. Chen and N. V. Richardson, Surface, *Prog. Surf. Sci.* **73**, 59–77 (2003).
3. T.E. Madey, J.Guan, C. H. Nien, C. Z. Dong, H. S. Tao and R.A. Campbell, *Surf. Rev. Lett.* **3**, 1315-1328 (1996).
4. T.E. Madey, C.H. Nien, K. Pelhos, J.J. Kolodziej, I.M. Abdelrehim and H. S. Tao. *Surf. Sci.* **438**, 191-206 (1999).
5. R. E. Kirby, C. S. McKee and M. W. Roberts, *Surf. Sci.* **55**, 725–728. (1976).
6. R. E. Kirby, C. S. McKee and L. V. Renny, *Surf. Sci.* **97**, 457–477 (1980).
7. A. T. S. Wee, J. S. Foord, R. G. Egdell and J. B. Pethica, *Phys. Rev. B* **58**, R7548-R7551 (1998).
8. I. Ermanoski, K. Pelhos, W. Chen, J. S. Quinton and T. E. Madey, *Surf. Sci.* **549**, 1–23 (2004).
9. M. Sander, R. Imbihl, R. Schuster, J. V. Barth and G. Ertl, *Surf. Sci.* **271**, 159–169 (1992).
10. H. Wang, W. Chen and T. E. Madey, *Phys. Rev. B* **74**, 205426 (2006).
11. H. Wang, A.S.Y Chan, W. Chen, P. Kaghazchi, T. Jacob and T. E. Madey *ACS Nano* **1**, 449-455 (2007).
12. K.J. Song, J. C. Lin, M. Y. Lai and Y. L. Wang, *Surf. Sci.* **327**, 17 (1994).
13. Govind, Wenhua Chen, Hao Wang and T.E. Madey (communicated).
14. C.W. Tucker Jr., *Acta Metallurgica* **15**, 1465 (1967).

Fine Encapsulated ZnO Nanophosphors And Their Potential Antibacterial Evaluation On The Gram Negative Bacillus *Escherichia coli*

Ranu K. Dutta, Prashant K. Sharma and Avinash C. Pandey

Nanophosphor Application Centre, University of Allahabad, Allahabad-211002, India.
Email: ranu.dutta16@gmail.com
Tele/Fax: +91-532-2460675

Abstract. The spherical nanophosphors of ZnO of mean size 10-- 20 nm were synthesized at room temperature by simple co-precipitation method. Their size and shape were governed by several encapsulations using compounds like biotin, citric acid and Poly vinyl alcohol (PVA). Besides playing a critical role in determining their size and shape, these biomolecules served as effective capping agents. The antibacterial behavior of the suspension of fine ZnO nanophosphors was investigated against the gram negative bacillus *Escherichia coli*. The growth curve showed bacteriostatic activity against *Escherichia coli*. The antibacterial activities of the ZnO nanophosphors can be attributed to their total surface area, as increasing surface to volume ratio of nanophosphors provides more efficient means for enhanced antibacterial activity.

Keywords: ZnO Quantum dots, Toxicity, bactericidal effect, Colony forming unit.
PACS: 73.63 Bd, 78.67 Bf, 87.85 jf

INTRODUCTION

Nanometer scale composites having multifunctional properties, display unique, superior and indispensable properties and have attracted much attention for their distinct characteristics that are unavailable in conventional macroscopic materials. In the recent years the scientific communities have paid much attention to the synthesis, characterization and applications of these nanoparticles. Because of their excellent unique size dependent properties[1-3], quantum dots[4] such as ZnO, TiO_2, CdS, CdSe, CdTe, CdSe@ZnS and carbon nanotubes are widely used and studied for numerous applications in various aspects of life. Due to their excellent physical and chemical properties, these nanoparticles have wide range of applications in light-emitting devices, display devices, photo detectors, photodiodes, transparent UV protection films, biological systems[5] (drug delivery system, bio imaging, *in vivo* cell imaging etc.), cosmetic products and chemical sensors.

Physical parameters such as surface area, particle size, surface charge, and zeta potential are very important for providing mechanistic details in the uptake, persistence, and biological

CP1147, *Transport and Optical Properties of Nanomaterials—ICTOPON - 2009*, edited by M. R. Singh and R. H. Lipson
© 2009 American Institute of Physics 978-0-7354-0684-1/09/$25.00

toxicity of nanoparticles inside living cells. Some nanoparticles exhibit selective uptake through cell membranes. Some engineered nanoscale particles are in the size range of biological molecules, such as proteins and intracellular constituents that are critical to cellular functions, and thus have the potential to disrupt critical biological processes. Although many studies on the biological activity of ZnO have been carried out, most of these pertain to the antimicrobial effect of bulk ZnO with a large particle size. Yamamoto studied the bacterial activity of ZnO with various particle sizes in the range of 0.1–1μm. The surface roughness contributes to the mechanical damage of the cell membrane of *E coli*. Wang *et al* proposed that the orientation of ZnO can also affect bioactivity.

In the present work, we have conducted its bactericidal studies of ZnO. For the purpose we have synthesized the ZnO nanophosphors of mean size 10-20 nm. Smaller sized ZnO were synthesized using various biomolecules, which served as effective capping agents besides controlling the size of nanoparticles. We carried out experiments to investigate the bactericidal activity of ZnO nanophosphors on the gram negative bacterium *Escherichia coli* (*E coli* DH 5α). In the present study, *E coli* were used as a model microorganism for bactericidal assay of nanophosphors. Bactericidal activity of these ZnO nanophosphors as a function of nanophosphor concentration was carried out in Luria-Bertani medium as well as on solid agar medium. The effect of ZnO on *E coli* has been investigated. The discussion in this article is directed toward providing an insight into the potential of ZnO and encapsulated ZnO for any antibacterial activity.

MATERIALS AND METHOD

Synthesis

For synthesis of ZnO nanophosphors, the zinc acetate di-hydrate (99.2 %) $Zn(CH_3COO)_2.2H_2O$, potassium hydroxide KOH, methanol, biotin, citric acid, poly vinyl alcohol (PVA) and ethanol were procured from E. Merck Limited, Mumbai-400018, India. Bacterial culture *Escherichia coli* (DH 5α) were obtained from Centre for Cellular and Molecular Biology (CCMB), Hyderabad, India. Pyrex petri dishes were purchased from Tarson. Agar, Yeast extract, Peptone and NaCl were purchased from Alfa Aesar, Germany and were of bacteriological grade. All chemicals were of AR grade and were directly used without special treatment. Synthesis of ZnO nanophosphors was carried out using the same technique followed by OD Jayakumar et al. with little modification after optimization of reaction parameters and conditions. Appropriate amount of zinc acetate di-hydrate was taken in 100 ml of methanol and dissolved while continuous stirring for two hours at room temperature (Solution A). Simultaneously, 140 mmol KOH solution was prepared in 100 ml of methanol with refluxing through water condenser with constant stirring for two hours at 50 °C (Solution B). Now, we mixed both the solution A and solution B with constant stirring for two hours. This mixing was done while refluxing through water condenser at 50 °C. Final solution was allowed to cool at room temperature and aged overnight. This solution was centrifuged and washed several times with absolute ethanol and water in order to remove unnecessary impurities. The obtained white product was placed in a vacuum oven for 24 hours at 50 °C to get white powders of ZnO. For encapsulating ZnO nanoparticle with biotin, citric acid and poly vinyl alcohol (PVA), we added biotin, citric acid and poly vinyl alcohol (PVA) in separate reactions before adding KOH solution. Rest of the method is same as above reaction.

Characterizations Used

The prepared samples were characterized by X-Ray Diffraction (XRD) and Transmission Electron Microscopy (TEM). XRD was performed on Rigaku D/max-2200 PC diffractometer operated at 40 kV/20 mA using CuK$_{\alpha 1}$ radiation with wavelength of 1.54 Å in wide angle region from 25° to 70° on 2θ scale. The size and morphology of prepared nanophosphors were recorded by transmission electron microscope model Tecnai T20 G ^{2} S-Twin electron microscope operated at 200 KV accelerating voltage. The density of bacterial cells in the liquid cultures was estimated by optical density (O.D) measurements at 600 nm and was maintained at 0.8-1.0, which is the ideal optical density of the cells. The *E coli* treated with ZnO were further subjected for antibacterial test by recording optical density (OD) at 600 nm using Perkin Elmer Lambda 35 UV-Visible spectrometer. The effect of interaction of ZnO nanomaterial with *E coli* cells were observed with a scanning electron microscope model Tecnai Quanta 200MK2 operated at 25 KV in Environmental SEM (ESEM) mode on the ZnO treated *E coli* films.

RESULTS AND DISCUSSION

Characterization of synthesized ZnO nanoparticle

Figure 1 shows the XRD pattern of the synthesized ZnO nanoparticle. XRD spectra show broad peaks at the positions of 31.63°, 34.50°, 36.25°, 47.50°, 56.60°, 62.80°, 66.36°, 67.92° and 68.91°, which are in good agreement with the standard JCPDS file for ZnO (JCPDS 36-1451, a = b = 3.249 Å, c = 5.206 Å) and can be indexed as the hexagonal wurtzite structure of ZnO having space group P6$_{3mc}$. The particle size estimation was well supported by the TEM results. TEM images were recorded by dissolving the as synthesized powder sample in ethanol and then placing a drop of this dilute ethanolic solution on the surface of copper grid. The morphology of the products is shown in figure 2. Figure 2 clearly shows spherical particles with little agglomeration having mean size 10- 20 nm.

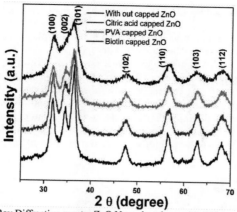

FIGURE 1. X-Ray Diffraction spectra ZnO Nanophosphors with different capping agents.

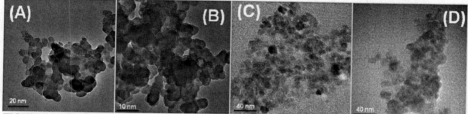

FIGURE 2. TEM images of ZnO Nanophosphors (A) Uncapped (B) Citric Acid capped (C) PVA capped and (D) Biotin capped.

Antibacterial Characterizations

Growth Kinetics

Bactericidal effect of ZnO nanoparticles was studied against Gram-negative bacteria *E coli* (DH 5α). These nanoparticles were dispersed in autoclaved deionized water by ultrasonication. Aqueous dispersions of ZnO nanoparticles of desired concentrations were made. An axenic culture of *E coli* (DH5α) was grown in liquid nutrient broth medium containing NaCl 5 g, peptone 5 g and yeast extract 2.5 g. For this experimental investigation, freshly grown bacterial inoculums of *E coli* was incubated in the presence of a range of ZnO nanoparticle concentrations of 20 µg/ml, 40 µg/ml, 60 µg/ml, 80 µg/ml and 100 µg/ml added in each flask to observe the bacterial cell growth at 37°C with constant shaking. One flask was taken as control, with no nanoparticle load. Shaking provided bacteria aeration and homogeneity. Total solution volume in each flask was kept 50 ml. In liquid medium, the growth of *E coli* was indexed by measuring optical density (OD) at 600 nm against abiotic control after every 2 h up to 24 h. Samples were collected for colony-forming units (CFU) measurements. Samples treated with nanoparticles were spread on nutrient agar plates and after incubation at 37 °C for 24 h, the number of CFU were counted, which is illustrated graphically (Fig:4). Bacterial cells were coated onto glass slides by using standard spin coating unit to form thin and uniform films for ESEM analysis.

Control flask containing all the initial reaction components except the ZnO nanoparticles showed no antibacterial activity. The Growth curve (Figure 3) shows mitigation of growth gradually from 20 µg/ml to 80 µg/ml and almost complete inhibition of bacterial growth at concentration of 100 µg/ml. Decrease in CFU counts (Figure 3) was observed on increasing nanophosphor concentration. Optical densities as a function of time measured periodically up to 24 h of the control and solutions containing different concentrations of ZnO nanoparticles treated *E coli* are shown in Fig.3. Bacterial cells grow by a process called binary fission in which one cell doubles in size and splits into halves to produce two identical daughter cells. Bacterial cell growth enhances the turbidity of the liquid nutrient medium and as a result the absorption increases.

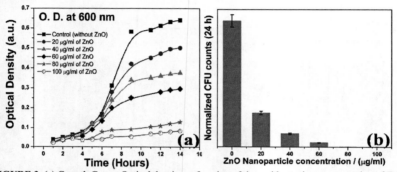

FIGURE 3. (a) Growth Curve: Optical density as function of time with varying concentration of ZnO nanophosphors treated *E coli* in solution state. (b) CFU after 24 hour incubation time with varying concentration of ZnO nanophosphors on solid agar plates (error bar 5 %).

It has been observed that optical density of the growth medium decreased in comparison to the control with increasing concentration of ZnO nanoparticles. This has been attributed to the reduced growth of bacterial cells. ZnO nanoparticles at concentrations 80 µg/ml and higher were found effective bactericides, and there was virtually no bacterial growth as optical absorption was insignificant. Figure 4 shows the normalized number of bacterial colonies grown on nutrient agar plates as a function of concentration of ZnO nanoparticles. The bacterial cell colonies on agar-plates were detected by viable cell counts. Viable cell counts are the counted number of colonies that are developed after a sample has been diluted and spread over the surface of a nutrient medium solidified with agar and contained in a petri dish. The number of CFU reduced significantly with increasing the concentration of ZnO nanophosphors. There was virtually no CFU observed in the samples containing 80 µg/ml and higher ZnO nanoparticles. The bacterial growth inhibition trend observed from CFU data are in accordance with the results of optical density. Similar decrease in growth curve was observed with ZnO capped with citric acid and PVA. Higher antibacterial effect was seen with biotin capped.

Surface morphology of *E coli* cells treated with ZnO nanoparticles were observed by ESEM measurements. ZnO nanoparticles were observed on the cell wall surface as well as in its interior. Figure 6 shows ESEM micrograph of *E coli* treated with (a) Control i.e. without ZnO: showing intact *E coli* cell, (b) 20 µg/ml ZnO: showing changes in the surface feature of the *E coli* cell, (c) 40 µg/ml ZnO: starting damage and disorganization in the cell wall was observed, (d) 60 µg/ml ZnO: considerable damage with much disorganized cell wall with affected morphology was observed, (e) 80 µg/ml ZnO: cellular internalization of few ZnO nanoparticles and observed cell wall disorganization, (f) 100 µg/ml ZnO: showing cellular internalization of large number of ZnO nanoparticles with extensively damaged shrinked *E coli* cell and loss of membrane integrity. For 80 µg/ml and 100 µg/ml concentrations of ZnO, the *E coli* cell membrane is extensively damaged (Figure 6 (e) and figure 6 (f)) and it is likely that the intracellular content has leaked out hence shrinkage of cell is observed (figure 6 (f)) and probably cell death. Preliminary results of cellular internalization of ZnO nanoparticles and cell wall disorganization have already been shown in available literatures[12].

This paper shows *E coli* cells being damaged, showing a Gram negative triple membrane disorganization and, consequently, ZnO internalization after contact with ZnO nanoparticles. Several other studies[13-16] have reported the cytotoxicity of ZnO, DNA damage and oxidative lesions determined using the comet assay. ZnO showed effects on cell viability as well as DNA damage[17], and intracellular production of reactive oxygen species (ROS). Smaller size ZnO has higher antibacterial activity as determined by N Padmavathy and R Vijayaraghavan[18]. Several explanations have been given for the damage caused by ZnO to bacterial cells. One explanation is based on the reactive oxygen species released on the surface of ZnO, which cause fatal damage to the cells[19]. The highly reactive oxygen species generation of such as OH^-, H_2O_2 and O_2^{2-} is explained as follows. Since ZnO with defects can be activated by both UV and visible light, electron-hole pairs (e⁻h+) can be created. The holes split H_2O molecules (from the suspension of ZnO) into OH^- and H^+. Dissolved oxygen molecules are transformed to superoxide radical anions (˙O^{-2}), which in turn react with H^+ to generate (HO_2˙) radicals, which upon subsequent collision with electrons produce hydrogen peroxide anions (HO_2^-). They then react with hydrogen ions to produce molecules of H_2O_2. The generated H_2O_2 can penetrate the cell membrane and kill the bacteria[20].

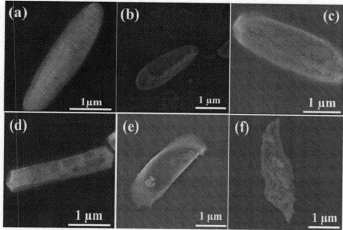

FIGURE 4. ESEM micrograph of *E coli* treated with (a) Control i.e. without ZnO: showing an intact *E coli* cell, (b) 20 µg/ml ZnO: showing changes in the surface feature of the *E coli* cell, (c) 40 µg/ml ZnO: starting damage and disorganization in the cell wall was observed, (d) 60 µg/ml ZnO: considerable damage with much disorganized cell

wall and affected morphology was observed, (e) 80 µg/ml ZnO: cellular internalization of few ZnO nanoparticles and observed cell wall disorganization, (f) 100 µg/ml ZnO: showing cellular internalization of large number of ZnO nanoparticles with extensively damaged shrinked *E coli* cell membrane.

ZnO and TiO$_2$, both ingredients of sunscreens and cosmetics showed cytotoxic or DNA damaging effects. It is established that there is a direct correlation between the reactive oxygen species (ROS) generating capability, surface area of nanoparticles. Jeng et al. showed cytotoxic effects and mitochondrial dysfunction after exposure to several metal oxide nanoparticles, out of which ZnO particles were most potent[13]. Nanoparticles internalization depends on the particle size, surface properties, and functionalization. It was seen that biotin encapsulated nanoparticles were smallest in size, about 5-8 nm and showed most effective antibacterial activity against bacterial cells. After internalization, the nanoparticle distribution in the body is a strong function of the nanoparticle's surface characteristics.

CONCLUSIONS

Thus we see that inside cells, nanoparticles might directly provoke alterations of membranes and other cell structures and molecules, as well as protective mechanisms. Indirect effects of nanoparticles depend on their chemical and physical properties and may include clogging effects, solubilization of toxic nanoparticle compounds or production of reactive oxygen species. Many questions regarding the bioavailability of environmental nanoparticles, their uptake by algae, plants, and fungi along with their toxicity mechanisms remain to be elucidated.

ACKNOWLEDGMENTS

The authors are grateful to CSIR, UGC and DST, India for financial Support in terms of facilities and fellowships.

REFERENCES

1. S. M. Prokes and K. L. Wang, *Mater. Res. Sci. Bull.* **24**, 13 (1999).
2. J. Hu, T. W. Odom and C. M. Lieber, *Acc. Chem. Res.* **32**, 435 (1999).
3. S. Nakamura, *Science* **281**, 956 (1998).
4. M. R. Wilson, J. H. Lightbody, K. Donaldson, J. Sales and V. Stone, *Toxicol. Appl. Pharmacol.* **184**, 172 (2002).
5. A. Nel, T. Xia, L. Madler and N. Li, *Science* **311**, 622–627 (2006).
6. W. H. Strehlow and E. L. Coo, *J. Phys. Chem. Ref. Data* **2**, 163 (1973).
7. K. A. D. Guzman et al., *Environl Sci. Technol.* **40**, 1401–1407 (2006).
8. J. R. Gurr et al., *Toxicology* **213**, 66–73 (2005).
9. K. Donaldson, V. Stone, A. Clouter, L. Renwick and W. Macnee, *Occup. Environ. Med.* **58**, 211 (2001).
10. K. Donaldson, V. Stone, P. S. Gilmour, D. M. Brown, W. Macnee, *Philos. T. Roy. Soc. A* **358**, 2741 (2000).
11. G. Oberdorster, *Philos. T. Roy. Soc. A* **358**, 2719–2739 (2000).
12. Roberta Brayner, Roselyne Ferrari-Iliou, Nicolas Brivois, Shakib Djediat, Marc F. Benedetti, and Fernand Fievet, *Nano Lett.* **6**, 866-870 (2006).
13. H. A. Jeng, and J. Swanson, *J. Environ. Sci. Heal. A* **41**, 2699–2711 2006.
14. S. Park, Y. K. Lee, M. Jung, K. H. Kim, N. Chung, E. K. Ahn, Y. Lim, and K. H. Lee, *Inhal. Toxicol.* **19**, 59–65, Suppl. 1 (2007).

15. T. J. Brunner, P. Wick, P. Manser, P. Spohn, R. N. Grass, L. K. Limbach, A. Bruinink, and W. J. Stark, *Environ. Sci. Technol.* **40**, 4374–4381 (2006).
16. A. Gojova, B. Guo, R. S. Kota, J. C. Rutledge, I. M. Kennedy, and A. I. Barakat, *Environ. Health Persp.* **115**, 403–409 (2007).
17. Hanna L. Karlsson, Pontus Cronholm, Johanna Gustafsson, and Lennart Mo¨ller *Chem. Res. Toxicol. article in press.*
18. Nagarajan Padmavathy and Rajagopalan Vijayaraghavan, *Sci. Technol. Adv. Mater.* **9**, (2008).
19. K. Sunada, Y. Kikuchi, K. Hashimoto and A. Fujshima, *Environ. Sci. Technol.* **32**, 726 (1998).
20. Fang M, Chen J H, Xu X L, Yang P H and Hildebrand H F *Int. J. Antimicrob. Ag.* **27**, 513 (2006).
21. K. Donaldson et al., "Nanotoxicology," *Occupational and Environmental Medicine,* **61** 9, pp. 727– 728 2004.
22. A. Nel et al., *Science,* **311**, 622–627 (2006).
23. G. Oberdorster et al., *Environ. Health Persp.* **113**, 823–839 (2005).

Magnetostatic Modes in Coupled Magnetic Bilayer Cylindrical Systems

T. K. Das and M. G. Cottam

Department of Physics & Astronomy, University of Western Ontario,
London, Ontario N6A 3K7, Canada

Abstract. Magnetostatic mode calculations are presented for bulk and surface spin waves in bilayer cylindrical systems where typically one material is ferromagnetic and the other is anti-ferromagnetic. The dispersion relations and the localization properties of the modes are shown to be modified compared to previous results for planar bilayers. Numerical examples for the dispersion relations are presented taking the ferromagnet Ni and the uniaxial antiferromagnet $GdAlO_3$ as the two constituents of the cylindrical bilayer. The case where Ni is the core material is shown to give a different behavior from the inverse structure with $GdAlO_3$ as the core.

Keywords: Magnetostatic spin waves; Bilayer magnetic systems; Magnetic interfaces; Cylind-rical geometries; Ferromagnets; Antiferromagnets.
PACS: 75.30.Ds; 75.70.Cn; 75.50.Cc; 75.50.Ee.

INTRODUCTION

Recently the properties of the magnetic excitations in magnetic nanostructures with cylindrical geometries have attracted considerable theoretical and experimental attention, due to their potential device applications (see, e.g., [1,2]). These nanostruct-ures have included arrays of ferromagnetic wires, disks and rings, as well as high-density arrays of magnetic nanotubes, which are essentially hollow cylinders composed of materials such as Ni or Permalloy [3]. The spin dynamics in many of these systems have been probed by Brillouin light scattering (BLS) and magnetic resonance techniques (see, e.g., [2,4]). There has also been interest in the dynamical properties of bilayer films of antiferromagnetic (AFM) and ferromagnetic (FM) mater-ials grown in direct contact [5,6], particularly with regard to the coupled magnetostatic modes, exchange bias effects, and the role of interface magnetic anisotropies. Mag-netostatic modes represent the spin-wave excitations of the system in a long-wavelength (small wave-vector) regime where the dynamic effects of exchange inter-actions become negligible compared the magnetic dipole-dipole interactions. The latter can then be treated using Maxwell's equations without retardation (see [7]).

Here a theory is developed for the surface and bulk magnetostatic modes in AFM / FM double structures with a cylindrical geometry. We expect, by comparison with the planar case [5,6] studied previously, that the magnetostatic modes near the curved interface may

CP1147, *Transport and Optical Properties of Nanomaterials—ICTOPON - 2009*, edited by M. R. Singh and R. H. Lipson
© 2009 American Institute of Physics 978-0-7354-0684-1/09/$25.00

have a modified frequency spectrum and localization properties in this geometry. This consideration has provided the motivation for our present work in which we generalize our previous calculations [8,9] where the magnetostatic mode properties were evaluated for cylindrical geometries (e.g., magnetic wires, antiwires, and tubes) that involved a single magnetic medium.

THEORY

We model the bilayer magnetic system as a long cylindrical tube of one magnetic material (see Fig. 1) interfaced with an inner core of another magnetic material. In most cases one of the material is chosen to be an AFM and the other is a FM, which leads to two distinct cases depending on which is chosen for the core. An external magnetic field is taken parallel to the cylindrical axis (the z axis) and a large length-to-diameter aspect ratio is assumed. The radii R_1 and R_2 are allowed to take general values, but are typically in the sub-micron range. The magnetostatic modes are characterized in terms of a wave number q along the cylinder axis of symmetry. Here we assume $q \sim 10^6$ or 10^7 m^{-1}, which falls within the magneto-static regime and is also typical of Brillouin or Raman light scattering if a 90° scattering geometry is employed.

Following previous calculations for a single magnetic material in a cylindrical geometry (see [8]), we now solve for the dynamic response within each magnetic material using Maxwell's equations (without retardation) and the non-diagonal magnetic susceptibility tensors. The latter quantities have a gyromagnetic form with the nonzero frequency-dependent components $\chi_{xx} = \chi_{yy} \equiv \chi_a$ and $\chi_{xy} = -\chi_{yx} \equiv i\chi_b$. This is applicable when the z axis (the cylindrical axis) coincides with the direction of the applied field H_0, the magnetization of the FM, and the sublattice magnetization of the AFM, as in Fig. 1. In the FM case the expressions (ignoring damping) are [10,11]

$$\chi_a = (\omega_0/\omega)\chi_b = \omega_0\omega_m/(\omega_0^2 - \omega^2). \tag{1}$$

Here ω denotes the angular frequency of the wave, while $\omega_0 = \gamma\mu_0 H_0$ and $\omega_m = \gamma\mu_0 M_{FM}$ with γ and M_{FM} denoting the gyromagnetic ratio and the saturation magnetization, respectively. For a uniaxial AFM χ_a and χ_b have analogous expressions [10,11] but the definition of ω_m involves the sublattice magnetization M_{AFM} and there are additional parameters ω_A and ω_{Ex} that relate to the bulk anisotropy and static exchange, respectively. The poles for ω in the AFM case occur at $\Omega_0 \pm \omega_0$, where

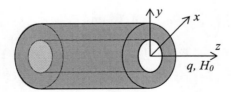

FIGURE 1. A cylindrical nanotube with inner and outer radii R_1 and R_2 of one magnetic material (e.g., a FM) with the core filled by a different magnetic material (e.g., an AFM), or vice versa, is surrounded by a nonmagnetic material. The external magnetic field H_0 and propagation wave number q are along the z axis, parallel to the magnetization (of the FM) and the sublattice magnetization (of the AFM).

$$\Omega_0 = \sqrt{\omega_A(\omega_A + 2\omega_{Ex})} \tag{2}$$

is the antiferromagnetic resonance (AFMR) angular frequency [11].

Briefly, the magnetostatic calculations proceed by generalizing our previous work [8]. The appropriate Maxwell's equations are re-expressed in terms of the magneto-static scalar potential ψ, which satisfies the Walker equation inside each of the magnetic materials and Laplace's equation in the nonmagnetic material outside. The solutions have the general form $f_{m,q}(r) \exp(im\theta + iqz)$ in cylindrical polar coordinates, where m is an integer and q is the wave number. The solutions for the radial function f involve the standard Bessel functions $I_m(\alpha_1 qr)$ for $r < R_1$, a combination of $I_m(\alpha_2 qr)$ and $K_m(\alpha_2 qr)$ for $R_1 < r < R_2$, and $K_m(qr)$ for $r > R_2$. The α parameters are ω-dependent quantities defined for each magnetic material by $\alpha^2 = 1/(1 + \chi_a)$. They are either real or imaginary for surface-like modes (localized near the interfaces) or bulk-like modes (with a wave-like behavior in the radial direction), respectively. At $r = R_1$ (the FM / AFM interface) it is important to take account of the anisotropy due to the two materials in contact. Following [6], for a FM material in contact with an AFM material, the effective external magnetic field in the FM material becomes

$$H_0^{FM} = H_0 + (M_{AFM}/M_{AV})H_I, \tag{3}$$

where H_I is the interface anisotropy field and M_{AV} is a volume-weighted average of M_{FM} and M_{AFM} for the bilayer. There is an analogous expression for the modified field in the AFM. Finally, on applying the magnetostatic boundary conditions at $r = R_1$ and R_2, we find the dispersion relation for the bulk and surface modes given by

$$\begin{vmatrix} I_m(\alpha_1 qR_1) & -c_1 & -d_1 & 0 \\ I_m(\alpha_1 qR_1)\Phi_{I11} & -c_1\Phi_{K21} & -d_1\Phi_{I21} & 0 \\ 0 & c_2 & d_2 & K_m(qR_2) \\ 0 & c_2\Phi_{K22} & d_2\Phi_{I22} & K'_m(qR_2) \end{vmatrix} = 0, \tag{4}$$

where $c_i = K_m(\alpha_2 qR_i)$, $d_i = I_m(\alpha_2 qR_i)$, with i and j denoting 1 or 2, and

$$\Phi_{Kji} = \left[K'_m(\alpha_j qR_i)/\alpha_j K_m(\alpha_j qR_i)\right] - \left[m\chi_b^{(j)}/qR_i\right]. \tag{5}$$

There is a similar definition for Φ_{Iji} in terms of the Bessel function $I_m(\alpha_j qR_i)$.

NUMERICAL RESULTS

We now apply the above theory to bilayer structures in which the FM is Ni (for which $\omega_m/2\pi = 18.7$ GHz) and the AFM is GdAlO$_3$ (for which $\omega_m/2\pi = 22.0$ GHz, $\omega_A/2\pi = 10.2$ GHz, and $\omega_{Ex}/2\pi = 52.6$ GHz), implying a relatively low AFMR freq-uency corresponding to $\Omega_0/2\pi = 34.4$ GHz). These parameter values are taken from [8,9]. There are two possible bilayer structures to consider, depending on which material forms the core, and we show below that they give rise to contrasting proper-ties for the magnetostatic modes.

In Fig. 2 we show calculations for a structure where the AFM forms the core and it is surrounded by the FM. The frequencies of the coupled surface magnetostatic modes for $|m| = 1$ and 2 are plotted versus wave number in terms of the dimensionless qR_2 for the above parameters and assuming R_1 / R_2 as a structure factor. Qualitatively the modes have some features that are similar to those for cylindrical structures with one magnetic material [8-10]. For example, the frequency of each branch decreases mono-tonically as q increases until there is a cut-off value above which no localized modes occur. Also the modes exist only within specific ranges of frequency, as indicated by the horizontal lines. However, quantitatively there are important differences including the existence of two bands of frequencies for each $|m|$. Broadly, these have the character of an upper perturbed AFM band and a lower perturbed FM band. Results for the frequencies, q dependence, and localization are strongly affected by the R_1/R_2 ratio and the interface coupling. For comparison we show in Fig. 3 the corresponding dispersion relations for the AFM core on its own and the FM tube on its own, each surrounded by a nonmagnetic (vacuum) region. We note that one of the branches to the dispersion relation for the FM tube is suppressed in the coupled structure (Fig. 2) because the localization condition can no longer be satisfied. Furthermore there is a shift to each band.

Next we show in Fig. 4 some calculations for the inverse bilayer structure to that of Fig. 2. Hence in this case the FM forms the core and it is surrounded by an AFM tube. The modification due to the coupling in this geometry is found to be more drastic than previously, essentially because the lower-frequency material now fills the core. In fact, for the example shown the lower range of frequencies is absent since the localization conditions are not satisfied. Again, for comparison, Fig. 5 shows the dispersion relations that would apply for the core and the tube separately.

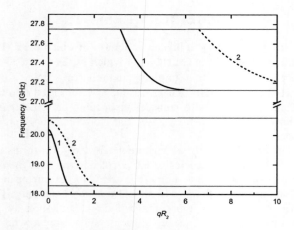

FIGURE 2. Frequencies of the surface modes in a Ni nanotube with a GdAlO$_3$ core plotted versus the dimensionless qR_2. The parameters are $\mu_0 H_0 = 0.3$ T, $\mu_0 H_1 = 0.05$ T, $R_1 / R_2 = 0.4$, and the labels 1 and 2 refer to the $|m|$ values.

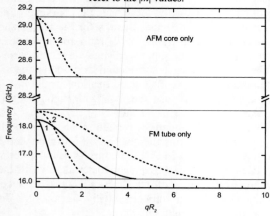

FIGURE 3. For comparison, the same as in Fig. 2 but separately for a Ni nanotube (with nonmagnetic core) and for a GdAlO$_3$ nanowire (with nonmagnetic outer layer).

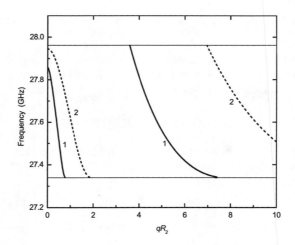

FIGURE 4. As in Fig. 2 but for the inverse structure consisting of a GdAlO$_3$ nanotube with a Ni core, using the same field values and the same radii. Note that, by contrast with Fig. 2, there is only one region of surface modes in this case.

CONCLUSIONS

We have presented calculations for AFM / FM double structures with a cylindrical geometry, generalizing previous studies for planar systems. Taking Ni and GdAlO$_3$ as examples we showed how the dispersion relations for surface magnetostatic modes are modified and how the inverse structure has different properties. Analogous conclus-ions follow regarding the coupled bulk modes of the bilayer. Calculations using other

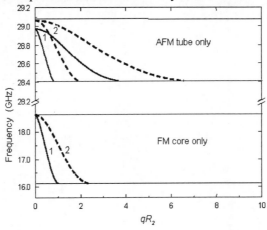

FIGURE 5. As in Fig. 3 but separately for a GdAlO$_3$ nanotube (with nonmagnetic core) and for a Ni nanowire (with nonmagnetic outer layer).

FM and AFM materials (such as Permalloy and MnF$_2$) have been carried out, leading to qualitatively similar results. As mentioned, inelastic light scattering provides a convenient experimental technique to study the modes described here.

ACKNOWLEDGMENTS

Partial support for this project has come from NSERC of Canada. We acknowledge helpful discussions with R. N. Costa Filho.

REFERENCES

1. M. L. Plumer, J. van Ek and D. Weller, *The Physics of Ultra-High-Density Magnetic Recording*, Berlin: Springer, 2001.
2. B. Hillebrands and K. Ounadjela, *Spin Dynamics in Confined Magnetic Structures I*, Berlin: Springer, 2001.
3. K. L. Hobbs, P. R. Larson, G. D. Lian, J. C. Keay and M. B. Johnson, *Nano Lett.* **4**, 167-171 (2004).
4. Z. K. Wang, H. S. Lim, H. Y. Liu, S. C. Ng, M. H. Kuok, L. L. Tay, D. J. Lockwood, M. G. Cottam, K. L. Hobbs, P. R. Larson, J. C. Keay , G. D. Lian and M. B. Johnson, *Phys. Rev. Lett.* **94**, 137208 /1-4 (2005).
5. R. L. Stamps and K. D. Usadel, *Europhysics Lett.* **74**, 512-518 (2006).
6. M. A. A. Monteiro, G. A. Farias, R. N. Costa Filho and N. S. Almeida, *Eur. Phys. J.* *B***61**,121-126 (2008).
7. M. G. Cottam and D. R. Tilley, *Introduction to Surface and Superlattice Excitations*, Bristol: Institute of Physics, 2005.
8. T. K. Das and M. G. Cottam, *J. Mag. Magn. Mat.* **310**, 2183-2185 (2007).
9. T. K. Das and M. G. Cottam, *Surf. Rev. Lett.* **14**, 471-480 (2007).
10. T. M. Sharon and A. A. Maradudin, *J. Phys. Chem. Solids* **38**, 977-981 (1977).
11. T. Wolfram and R. De Wames, *Prog. Surf. Sci.* **2**, 233-330 (1972).

AUTHOR INDEX

543

544

545

Y

Yadav, R. S., 282
Yakhmi, J. V., 347, 360
Yu, Y.-T., 152
Yundt, N., 108

Z

Zaidi, M. G. H., 469

1-MONTH